T0140479

Principal Component Analysis Networks and Algorithms

Xiangyu Kong · Changhua Hu
Zhansheng Duan

Principal Component Analysis Networks and Algorithms

Science Press
Beijing

Springer

Xiangyu Kong
Department of Control Engineering
Xi'an Institute of Hi-Tech
Xi'an
China

Changhua Hu
Department of Control Engineering
Xi'an Institute of Hi-Tech
Xi'an
China

Zhansheng Duan
Center for Information Engineering Science Research
Xi'an Jiaotong University
Xi'an, Shaanxi
China

ISBN 978-981-10-9738-6 ISBN 978-981-10-2915-8 (eBook)
DOI 10.1007/978-981-10-2915-8

Jointly published with Science Press, Beijing, China

© Science Press, Beijing and Springer Nature Singapore Pte Ltd. 2017
Softcover reprint of the hardcover 1st edition 2017
This work is subject to copyright. All rights are reserved by the Publishers, whether the whole or part of the material is concerned, specifically the rights of translation, reprinting, reuse of illustrations, recitation, broadcasting, reproduction on microfilms or in any other physical way, and transmission or information storage and retrieval, electronic adaptation, computer software, or by similar or dissimilar methodology now known or hereafter developed.
The use of general descriptive names, registered names, trademarks, service marks, etc. in this publication does not imply, even in the absence of a specific statement, that such names are exempt from the relevant protective laws and regulations and therefore free for general use.
The publishers, the authors and the editors are safe to assume that the advice and information in this book are believed to be true and accurate at the date of publication. Neither the publishers nor the authors or the editors give a warranty, express or implied, with respect to the material contained herein or for any errors or omissions that may have been made.

Printed on acid-free paper

This Springer imprint is published by Springer Nature
The registered company is Springer Nature Singapore Pte Ltd.
The registered company address is: 152 Beach Road, #22-06/08 Gateway East, Singapore 189721, Singapore

To all the researchers with original contributions to principal component analysis neural networks and algorithms

—Xiangyu Kong, Changhua Hu and Zhansheng Duan

Preface

Aim of This book

The aim of this book is to (1) to explore the relationship between principal component analysis (PCA), neural network, and learning algorithms and provide an introduction to adaptive PCA methods and (2) to present many novel PCA algorithms, their extension/generalizations, and their performance analysis.

In data analysis, one very important linear technique to extract information from data is principal component analysis (PCA). Here, the principal components (PCs) are the directions in which the data have the largest variances and capture most of the information content of data. They correspond to the eigenvectors associated with the largest eigenvalues of the autocorrelation matrix of the data vectors. On the contrary, the eigenvectors that correspond to the smallest eigenvalues of the autocorrelation matrix of the data vectors are defined as the minor components (MCs) and are the directions in which the data have the smallest variances (they represent the noise in the data). Expressing data vectors in terms of the minor components is called minor component analysis (MCA). Through PCA, many variables can be represented by few components, so PCA can be considered as either a feature extraction or a data compression technology. Now, PCA has been successfully applied to many data processing problems, such as high-resolution spectral estimation, system identification, image compression, and pattern recognition. MCA is mainly used to solve total least squares problem, which is a technology widely used to compensate for data errors in parameter estimation or system identification. However, how can we obtain the principal components or minor components from a stochastic data stream?

This book aims to provide a relatively complete view of neural network-based principal component analysis or principal subspace tracking algorithms and present many novel PCA algorithms, their performance analysis, and their extension/generalizations.

Novel Algorithms and Extensions

It is well known that many methods exist for the computation of principal components, such as the power method, eigenvalue decomposition (ED), singular value decomposition (SVD), and neural network algorithms. Neural network approaches on PCA pursue an effective "online" approach to update the eigen direction after each presentation of a data point, which possess many obvious advantages. Many neural network learning algorithms have been proposed to extract PC, and this has been an active field for around two decades up to now.

This book is not oriented toward all neural network algorithms for PCA, but to some novel neural algorithms and extensions of PCA, which can be summarized as follows.

(1) Compared with most neural principal component learning algorithms, the number of neural networks for minor component analysis is somewhat smaller. A norm divergence problem exists in some existing MCA algorithms. To guarantee the convergence, it is necessary to use self-stabilizing algorithms. In these self-stabilizing algorithms, the weight vector length converges to a fixed value independent of the presented input vector. In this book, the self-stabilizing algorithms are discussed in detail and some novel self-stabilizing MCA learning algorithms are introduced.

(2) Most neural PCA algorithms only focus on eigenvector extraction using uncoupled rules, and a serious speed-stability problem exists in most uncoupled rules. To overcome this problem, several coupled PCA algorithms are introduced and their performances are analyzed in this book.

(3) Most neural algorithms only deal with either principal component extraction or minor component extraction. Are there such algorithms as dual-purpose subspace tracking algorithm, which are capable of both PC and MC extractions by simply switching the sign in the same learning rule? This book will develop a few dual algorithms for such purposes.

(4) The convergence of PCA neural learning algorithms is a difficult topic for direct study and analysis. Traditionally, the convergence of these algorithms is indirectly analyzed via certain deterministic continuous-time (DCT) systems. The DCT method is based on a fundamental theorem of stochastic approximation theory, and some crucial conditions must be satisfied, which are not reasonable requirements to be imposed in many practical applications. Recently, deterministic discrete-time (DDT) systems have been proposed instead to indirectly interpret the dynamics of neural network learning algorithms described by stochastic discrete-time system. This book will discuss the DDT method in detail.

(5) It is well known that generalized eigen decomposition (GED) plays very important roles in various signal processing applications, and PCA can be seen as a special case of GED problem. The GED neural algorithms will be also discussed in detail.

(6) An important aspect of the generalization of classic PCA is the cross-correlation problem, which studies the maximization of the cross-correlation between two stochastic signals. The neural learning algorithm to extract cross-correlation feature between two high-dimensional data streams will be studied in this book as well.

Prerequisites

The mathematical background required for reader is that of college calculus and probability theory. Readers should be familiar with basic linear algebra and numerical analysis as well as the fundamentals of statistics, such as the basics of least squares, and preferably, but not necessarily, stochastic algorithms. Although the book focuses on neural networks, they are presented only by their learning law, which is simply an iterative algorithm. Therefore, no a priori knowledge of neural networks is required. Basic background in mathematics is provided in the review chapter for convenience.

Some of the materials presented in this book have been published in the archival literature over the last several years by the authors, and they are included in this book after necessary modifications or updates to ensure accuracy, relevance, completeness, and coherence. This book also puts effort into presenting as many contributions by other researchers in this field as possible. This is a fast-growing area, so it is impossible to make sure that all works published to date are included. However, we still have made special efforts to filter through major contributions and to provide an extensive bibliography for further reference. Nevertheless, we realize that there may be oversights on critical contributions on this subject. For these, we would like to offer our apology. More importantly, our sincere thanks go to the many researchers whose contributions have established a solid foundation for the topics treated in this book.

Outline of the Book

Chapter 2 reviews some important concepts and theorems of matrix analysis and optimization theory. We discuss some basic concepts, properties, and theorems related to matrix analysis, with the emphasis on singular value decomposition and eigenvalue decomposition. We also introduce some methods of gradient analysis and optimization theory, which are all important tools which will be instrumental for our theoretical analysis in the subsequent chapters.

In Chap. 3, we discuss the principal component analysis networks and algorithms. The first half of this chapter analyzes the problem, basic theorems, and SVD-based methods of principal component analysis. The second half of this chapter studies the principal component analysis networks in detail, which falls into the following classes, such as Hebbian rule-based, LMS error-based, optimization-based, anti-Hebbian rule-based, nonlinear, constrained, and localized PCA, providing a theoretical analysis of major networks.

Chapter 4 studies the minor component analysis networks and algorithms. First, we analyze the problem of the minor component analysis and its classical application in total least squares estimation. Second, we present some classical anti-Hebbian rule-based MCA algorithms, especially analyzing the divergence (sudden, dynamical, and numerical) property and self-stabilizing property of some MCA algorithms. This chapter concludes with a self-stabilizing MCA algorithm and a novel neural algorithm for total least squares filtering of ours, with the simulations and application presented as aid to the understanding of our algorithm.

Chapter 5 addresses the theoretical issue of the dual-purpose principal and minor component analyses. We analyze the merit of dual-purpose algorithms in application and theory analysis and introduce existing dual-purpose methods, such as Chen's, Hasan's, and Peng's algorithms. Two important dual-purpose algorithms of ours are presented. Also, the information criterion, its landscape and gradient flow, global convergence analysis, and numerical consideration are analyzed. This is one of the most important chapters in this book.

Chapter 6 deals with the stability and convergence analysis of PCA or MCA neural network algorithms. The performance analysis methods are classified into three classes, namely the deterministic continuous-time (DCT) system, the stochastic discrete-time (SDT) system, and the deterministic discrete-time (DDT) system, which are discussed in detail. We briefly review the DDT system of Oja's PCA algorithm and give a detailed analysis of the DDT systems of a new self-stabilizing MCA algorithm and Chen's unified PCA/MCA algorithm.

Chapter 7 studies the generalized feature extraction method. First, we review the generalized Hermitian eigenvalue problem. Second, a few existing adaptive algorithms to extract generalized eigen pairs are discussed. Third, a minor generalized eigenvector extraction algorithm and its convergence analysis via the DDT method are presented. Finally, we analyze a novel adaptive algorithm for generalized coupled eigen pairs of ours in detail, and a few simulation and application experiments are provided.

Chapter 8 analyzes the demerits of the existing uncoupled feature extraction algorithm, introduces Moller's coupled principal component analysis neural algorithm, and concludes with our two algorithms, the one of which is a unified and coupled self-stabilizing algorithm for minor and principal eigen pair extraction algorithms and the other an adaptive coupled generalized eigen pair extraction algorithms.

Chapter 9 presents the generalization of feature extraction from autocorrelation matrix to cross-association matrix. We briefly review the cross-correlation asymmetric network and analyze Feng's neural networks for extracting cross-correlation features. Then, an effective neural algorithm for extracting cross-correlation feature between two high-dimensional data streams is proposed and analyzed. Finally, a novel coupled neural network-based algorithm to extract the principal singular triplet of a cross-correlation matrix between two high-dimensional data streams is presented and analyzed in detail.

Suggested Sequence of Reading

This book aims to provide a relatively complete and coherent view of neural network-based principal component analysis or principal subspace tracking algorithms. This book can be divided into four parts, namely preliminary knowledge, neural network-based principal component learning algorithm, performance analysis of algorithms, and generalizations and extensions of PCA algorithms. For readers who are interested in general principal component analysis and future research directions, a complete reading of this book is recommended. For readers who are just interested in some specific subjects, selected chapters and reading sequences are recommended as follows.

(1) Numerical calculation of principal components

Chapter 2 → Chapter 3 → Chapter 4

(2) Performance analysis of neural network-based PCA algorithms

Chapter 3 → Chapter 4 → Chapter 6

(3) Neural network-based PCA algorithms and their extensions

Chapter 3 → Chapter 4 → Chapter 7 → Chapter 8 → Chapter 9

Acknowledgments

We are very grateful to Yingbin Gao and Xiaowei Feng, Ph.D., at Xi'an Institute of Hi-Tech, for their contributions in Chaps. 7 and 8, respectively. We also want to express our gratitude to Prof. Chongzhao Han, Jianfu Cao at Xi'an Jiaotong University, who have been hugely supportive throughout the pursuit of Ph.D. of the first author, and to Prof. Hongguang Ma at Xi'an Institute of Hi-Tech, for his help during the postdoctoral work of the first author, and also to Dean Junyong Shao, Director Xinjun Ding, Wang Lin, Vice-Director Zhicheng Yao, Prof. Huafeng He of Department of Control Engineering at Xi'an Institute of Hi-Tech, for their encouragements and supports of the work. We also express our gratitude to the Natural Science Foundation of China for Grants 61374120, 61673387, the National Outstanding Youth Science Foundation of China for Grant 61025014, and the Natural Science Foundation of Shaanxi Province, China for Grants 2016JM6015. Finally, we would like to thank the Department of Control Engineering at Xi'an Institute of Hi-Tech for offering a scholarly environment and convenient infrastructure for research.

Contents

About the Authors

Xiangyu Kong received the B.S. degree in optical engineering from Beijing Institute of Technology, P.R. China, in 1990, the M.S. degree in mechanical and electrical engineering from Xi'an Institute of Hi-Tech, in 2000, and the Ph.D. degree in electrical engineering from Xi'an Jiaotong University, P.R. China, in 2005. He is currently an associate professor in the Department of Control Engineering of Xi'an Institute of Hi-Tech. His research interests include adaptive signal processing, neural networks, and feature extraction. He has published three monographs (all first author) and more than 60 papers, in which nearly 20 articles were published in premier journal including *IEEE Transactions on Signal Processing, IEEE Transactions on Neural Networks and Learning Systems, and Neural Networks*. He has been PIs of two grants from the National Natural Science Foundation of China.

Changhua Hu received the Ph.D. degree in control science and engineering from Northwestern Polytechnical University, P.R. China, in 1996. He is currently a professor in the Department of Control Engineering of Xi'an Institute of Hi-Tech. His research interests include fault diagnosis in control systems, fault prognostics, and predictive maintenance. He has published three monographs and more than 200 papers, in which more than 50 articles were published in premier journal including *IEEE Transactions and EJOR*. In 2010, he obtained the National Science Fund for Distinguished Young Scholars. He was awarded national-class candidate of "New Century BaiQianWan Talents Program" and National Middle-aged and Young Experts with Outstanding Contributions in 2012. In 2013, he was awarded Cheung Kong Processor.

Zhansheng Duan received the B.S. and Ph.D. degrees from Xi'an Jiaotong University, P.R. China, in 1999 and 2005, respectively, both in electrical engineering. He also received the Ph.D. degree in electrical engineering from the University of New Orleans, LA, in 2010. From January 2010 to April 2010, he worked as an assistant professor of research in the Department of Computer Science, University of New Orleans. In July 2010, he joined the Center for

Information Engineering Science Research, Xi'an Jiaotong University, where he is currently working as an associate professor. His research interests include estimation and detection theory, target tracking, information fusion, nonlinear filtering, and performance evaluation. Dr. Duan has authored or coauthored one book, Multisource Information Fusion (Tsinghua University Publishing House, 2006), and 50 journal and conference proceedings papers. He is also a member of International Society of Information Fusion (ISIF) and the Honor Society of Eta Kappa Nu and is listed in *Who's Who in America 2015 and 2016*.

Chapter 1
Introduction

1.1 Feature Extraction

Pattern recognition and data compression are two applications that rely critically on efficient data representation [1]. The task of pattern recognition is to decide to which class of objects an observed pattern belonging to, and the compression of data is motivated by the need to save the number of bits to represent the data while incurring the smallest possible distortion [1]. In these applications, it is desirable to extract measurements that are invariant or insensitive to the variations within each class. The process of extracting such measurements is called *feature extraction*. It is also to say feature extraction is a data processing which maps a high-dimensional space to a low-dimensional space with minimum information loss.

Principal component analysis (PCA) is a well-known feature extraction method, while minor component analysis (MCA) and independent component analysis (ICA) can be regarded as variants or generalizations of the PCA. MCA is most useful for solving total least squares (TLS) problems, and ICA is usually used for blind signal separation (BSS).

In the following, we briefly review PCA, PCA neural networks, and extensions or generalizations of PCA.

1.1.1 PCA and Subspace Tracking

The principal components (PC) are the directions in which the data have the largest variances and capture most of the information contents of data. They correspond to the eigenvectors associated with the largest eigenvalues of the autocorrelation matrix of the data vectors. Expressing data vectors in terms of the PC is called PCA. On the contrary, the eigenvectors that correspond to the smallest eigenvalues of the autocorrelation matrix of the data vectors are defined as the minor components

© Science Press, Beijing and Springer Nature Singapore Pte Ltd. 2017

X. Kong et al., *Principal Component Analysis Networks and Algorithms*,
DOI 10.1007/978-981-10-2915-8_1

(MC), and MC are the directions in which the data have the smallest variances (they represent the noise in the data). Expressing data vectors in terms of the MC is called MCA. Now, PCA has been successfully applied in many data processing problems, such as high-resolution spectral estimation, system identification, image compression, and pattern recognition, and MCA is also applied in total least squares, moving target indication, clutter cancelation, curve and surface fitting, digital beamforming, and frequency estimation.

The PCA or MCA is usually one dimensional. However, in real applications, PCA or MCA is mainly multiple dimensional. The eigenvectors associated with the r largest (or smallest) eigenvalues of the autocorrelation matrix of the data vectors is called principal (or minor) components, and r is referred to as the number of the principal (or minor) components. The eigenvector associated with the largest (smallest) eigenvalue of the autocorrelation matrix of the data vectors is called largest (or smallest) component. The subspace spanned by the principal components is called principal subspace (PS), and the subspace spanned by the minor components is called minor subspace (MS). In some applications, we are only required to find the PS (or MS) spanned by r orthonormal eigenvectors. The PS is sometimes called signal subspace, and the MS is called noise subspace. Principal and minor component analyzers of a symmetric matrix are matrix differential equations that converge on the PCs and MCs, respectively. Similarly, the principal (PSA) and minor (MSA) subspace analyzers of a symmetric matrix are matrix differential equations that converge on a matrix whose columns' span is the PS and MS, respectively. PCA/PSA and MCA/MSA are powerful techniques in many information processing fields. For example, PCA/PSA is a useful tool in feature extraction, data compression, pattern recognition, and time series prediction [2, 3], and MCA/MSA has been widely applied in total least squares, moving target indication, clutter cancelation, curve and surface fitting, digital beamforming, and frequency estimation [4].

As discussed before, the PC is the direction which corresponds to the eigenvector associated with the largest eigenvalue of the autocorrelation matrix of the data vectors, and the MC is the direction which corresponds to the eigenvector associated with the smallest eigenvalue of the autocorrelation matrix of the data vectors. Thus, implementations of these techniques can be based on batch eigenvalue decomposition (ED) of the sample correlation matrix or on singular value decomposition (SVD) of the data matrix. This approach is unsuitable for adaptive processing because it requires repeated ED/SVD, which is a very time-consuming task [5]. Thus, the attempts to propose adaptive algorithms are still continuing even though the field has been active for three decades up to now.

1.1.2 PCA Neural Networks

In order to overcome the difficulty faced by ED or SVD, a number of adaptive algorithms for subspace tracking were developed in the past. Most of these techniques can be grouped into three classes [5]. In the first class, classical batch ED/SVD methods such as QR algorithm, Jacobi rotation, power iteration, and Lanczos method have been modified for the use in adaptive processing [6–10]. In the second class, variations in Bunch's rank-one updating algorithm [11], such as subspace averaging [12, 13], have been proposed. The third class of algorithms considers the ED/SVD as a constrained or unconstrained optimization problem. Gradient-based methods [14–19], Gauss–Newton iterations [20, 21], and conjugate gradient techniques [22] can then be applied to seek the largest or smallest eigenvalues and their corresponding eigenvectors adaptively. Rank revealing URV decomposition [23] and rank revealing QR factorization [24] have been proposed to track the signal or noise subspace.

Neural network approaches on PCA or MCA pursue an effective "online" approach to update the eigen direction after each presentation of a data point, which possess many obvious advantages, such as lower computational complexity, compared with the traditional algebraic approaches such as SVD. Neural network methods are especially suited for high-dimensional data, since the computation of the large covariance matrix can be avoided, and for the tracking of nonstationary data, where the covariance matrix changes slowly over time. The attempts to improve the methods and to suggest new approaches are continuing even though the field has been active for two decades up to now.

In the last decades, many neural network learning algorithms were proposed to extract PS [25–31] or MS [4, 32–40]. In the class of PS tracking, lots of learning algorithms such as Oja's subspace algorithm [41], the symmetric error correction algorithm [42], and the symmetric version of the back propagation algorithm [43] were proposed based on some heuristic reasoning [44]. Afterward, some information criterions were proposed and the corresponding algorithms such as LMSER algorithm [31], the projection approximation subspace tracking (PAST) algorithm [5], the conjugate gradient method [45], the Gauss–Newton method [46], and the novel information criterion (NIC) algorithm were developed [44]. These gradient-type algorithms could be claimed to be globally convergent.

In the class of MS tracking, many algorithms [32–40] have been proposed on the basis of the feedforward neural network models. Mathew and Reddy proposed the MS algorithm based on a feedback neural network structure with sigmoid activation function [46]. Using the inflation method, Luo and Unbehauen proposed an MSA algorithm that does not need any normalization operation [36]. Douglas et al. presented a self-stabilizing minor subspace rule that does not need periodically normalization and matrix inverses [40]. Chiang and Chen showed that a learning algorithm can extract multiple MCs in parallel with the appropriate initialization instead of inflation method [47]. On the basis of an information criterion, Ouyang et al. developed an adaptive MC tracker that automatically finds the MS without using the inflation method [37]. Recently, Feng et al. proposed the OJAm algorithm and extended it for tracking multiple MCs or the MS, which makes the corresponding state matrix tend to a column orthonormal basis of the MS [35].

1.1.3 Extension or Generalization of PCA

It can be found that the above-mentioned algorithms only focused on eigenvector extraction or eigen-subspace tracking with noncoupled rules. However, a serious speed stability problem exists in the most noncoupled rules [28]. This problem is that in noncoupled PCA rules the eigen motion in all directions mainly depends on the principal eigenvalue of the covariance matrix; thus, numerical stability and fast convergence can only be achieved by guessing this eigenvalue in advance [28]; in noncoupled MCA rules the speed of convergence does not only depend on the minor eigenvalue, but also depend on all other eigenvalues of the covariance matrix, and if these extend over a large interval, no suitable learning rate may be found for a numerical solution that can still guarantee stability and ensure a sufficient speed of convergence in all eigen directions. Therefore, the problem is even more severe for MCA rules. To solve this common problem, Moller proposed some coupled PCA algorithms and some coupled MCA algorithms based on a special information criteria [28]. In coupled rules, the eigen pair (eigenvector and eigenvalue) is simultaneously estimated in coupled equations, and the speed of convergence only depends on the eigenvalue of its Jacobian. Thus, the dependence of the eigenvalues on the covariance matrix can be eliminated [28]. Recently, some modified coupled rules have been proposed [48].

It is well known that the generalized eigen decomposition (GED) plays very important roles in various signal processing applications, e.g., data compression, feature extraction, denoising, antenna array processing, and classification. Though PCA, which is the special case of GED problem, has been widely studied, the adaptive algorithms for the GED problem are scarce. Fortunately, a few efficient online adaptive algorithms for the GED problem that can be applied in real-time applications have been proposed [49–54]. In [49], Chaterjee et al. present new adaptive algorithms to extract the generalized eigenvectors from two sequences of random vectors or matrices. Most algorithms in literatures including [49] are gradient-based algorithms [50, 51]. The main problem of this type of algorithms is slow convergence and the difficulty in selecting an appropriate step size which is essential: A too small value will lead to slow convergence and a too large value will lead to overshooting and instability. Rao et al. [51] have developed a fast recursive least squares (RLS)-like, not true RLS, sequential algorithm for GED. In [54], by reinterpreting the GED problem as an unconstrained minimization problem via constructing a novel cost function and applying projection approximation method and RLS technology to the cost function, RLS-based parallel adaptive algorithms for generalized eigen decomposition was proposed. In [55], a power method-based algorithm for tracking generalized eigenvectors was developed when stochastic signals having unknown correlation matrices are observed. Attallah proposed a new adaptive algorithm for the generalized symmetric eigenvalue problem, which can extract the principal and minor generalized eigenvectors, as well as their corresponding subspaces, at a low computational cost [56]. Recently, a fast and numerically stable adaptive algorithm for the generalized Hermitian eigenvalue problem (GHEP) was proposed and analyzed in [48].

Other extensions of PCA also include dual-purpose algorithm [57–64], the details of which can be found in Chap. 5, and adaptive or neural networks-based SVD singular vector tracking [6, 65–70], the details of which can be found in Chap. 9.

1.2 Basis for Subspace Tracking

In Sect. 1.1, we have reviewed the PCA algorithm and its extensions and generalizations from the viewpoint of the feature extraction. In this section, from another viewpoint of subspace, we will discuss the concept of subspace and subspace tracking method.

1.2.1 Concept of Subspace

Definition 1 If $\boldsymbol{S} = \{\boldsymbol{u}_1, \boldsymbol{u}_2, \ldots, \boldsymbol{u}_m\}$ is the vector subset of vector space $\boldsymbol{V}$, then the set $\boldsymbol{W}$ of all linear combinations of $\boldsymbol{u}_1, \boldsymbol{u}_2, \ldots, \boldsymbol{u}_m$ is called the *subspace* spanned by $\boldsymbol{u}_1, \boldsymbol{u}_2, \ldots, \boldsymbol{u}_m$, namely

$$\boldsymbol{W} = \mathrm{Span}\{\boldsymbol{u}_1, \boldsymbol{u}_2, \ldots, \boldsymbol{u}_m\} = \{\boldsymbol{u} : \boldsymbol{u} = \alpha_1 \boldsymbol{u}_1 + \alpha_2 \boldsymbol{u}_2 + \cdots + \alpha_m \boldsymbol{u}_m\}, \tag{1.1}$$

where each vector in $\boldsymbol{W}$ is called the generator of $\boldsymbol{W}$, and the set $\{\boldsymbol{u}_1, \boldsymbol{u}_2, \ldots, \boldsymbol{u}_m\}$ which is composed of all the generators is called the spanning set of the subspace. A vector subspace which only comprises zero vector is called a trivial subspace. If the vector set $\{\boldsymbol{u}_1, \boldsymbol{u}_2, \ldots, \boldsymbol{u}_m\}$ is linearly irrespective, then it is called a group basis of $\boldsymbol{W}$.

Definition 2 The number of vectors in any group basis of subspace $\boldsymbol{W}$ is called the *dimension* of $\boldsymbol{W}$, which is denoted by dim($\boldsymbol{W}$). If any group basis of $\boldsymbol{W}$ is not composed of finite linearly irrespective vectors, then $\boldsymbol{W}$ is called an infinite-dimensional vector subspace.

Definition 3 Assume that $\boldsymbol{A} = [\boldsymbol{a}_1, \boldsymbol{a}_2, \ldots, \boldsymbol{a}_n] \in \boldsymbol{C}^{m \times n}$ is a complex matrix and all the linear combinations of its column vectors constitute a subspace, which is called *column space* of matrix $\boldsymbol{A}$ and is denoted by Col($\boldsymbol{A}$), namely

$$\mathrm{Col}(\boldsymbol{A}) = \mathrm{Span}\{\boldsymbol{a}_1, \boldsymbol{a}_2, \ldots, \boldsymbol{a}_n\} = \left\{ \boldsymbol{y} \in \boldsymbol{C}^m : \boldsymbol{y} = \sum_{j=1}^{n} \alpha_j \boldsymbol{a}_j : \alpha_j \in \boldsymbol{C} \right\}. \tag{1.2}$$

Row space of matrix $\boldsymbol{A}$ can be defined similarly.

As stated in the above, the column space and row space of matrix $A_{m\times n}$ are spanned by n column vectors and m row vectors, respectively. If rank(A) is equal to r, then only r column or row vectors of matrix, which are linearly irrespective, can constitute column space Span(A) and row space Span(A^H), respectively. Obviously, it is an economical and better subspace expression method to use basis vector. The methods of constituting a subspace have primary transforms, and one can also use singular value decomposition to set up a normal orthogonal basis of base space.

Suppose that the data matrix A has measure error or noises, and define measure data matrix as

$$X = A + W = [x_1, x_2, \ldots, x_n] \in C^{m\times n}, \tag{1.3}$$

where $x_i \in C^{m\times 1}$. In the fields of signal processing and system science, the column space of measure data matrix $\mathrm{Span}(X) = \mathrm{Span}\{x_1, x_2, \ldots, x_n\}$ is called *measure data space*.

Define the correlation matrix as:

$$R_X = E\{X^H X\} = E\{(A+W)^H (A+W)\}. \tag{1.4}$$

Suppose that error matrix $W = [w_1, w_2, \ldots, w_n]$ is statistically irrespective of real data matrix A, then

$$R_X = E\{X^H X\} = E\{A^H A\} + E\{W^H W\}. \tag{1.5}$$

Define $R = E\{A^H A\}$ and $E\{W^H W\} = \sigma_w^2 I$, namely every measure noise is statistically irrespective and they have the same variance σ_w^2, it holds that

$$R_X = R + \sigma_w^2 I. \tag{1.6}$$

Define rank(A) = r, and the eigenvalue decomposition of matrix $R_X = E\{X^H X\}$ can be written as $R_X = U\Lambda U^H + \sigma_w^2 I = U(\Lambda + \sigma_w^2 I)U^H = U\Pi U^H$, where $\Pi = \Sigma + \sigma_w^2 I = \mathrm{diag}(\sigma_1^2 + \sigma_w^2, \ldots, \sigma_r^2 + \sigma_w^2, \sigma_w^2, \ldots, \sigma_w^2)$, $\Sigma = \mathrm{diag}(\sigma_1^2, \ldots, \sigma_r^2, 0, \ldots, 0)$, and $\sigma_1^2 \geq \sigma_2^2 \geq \cdots \geq \sigma_r^2$ are the nonzero eigenvalues of the real autocorrelation matrix $R = E\{A^H A\}$.

Obviously, if the signal-to–noise ratio is large enough, that is, σ_r^2 is obviously bigger than σ_w^2, then the first r largest eigenvalues of autocorrelation matrix R_X, namely $\lambda_1 = \sigma_1^2 + \sigma_w^2, \lambda_2 = \sigma_2^2 + \sigma_w^2, \ldots, \lambda_r = \sigma_r^2 + \sigma_w^2$ are called the principal eigenvalues, and the remaining $n - r$ small eigenvalues $\lambda_{r+1} = \sigma_w^2, \lambda_{r+2} = \sigma_w^2, \ldots, \lambda_n = \sigma_w^2$ are called the minor eigenvalues. Thus, the eigen decomposition of autocorrelation matrix R_X can be written as

$$R_X = [U_S \quad U_n]\begin{bmatrix} \Sigma_S & O \\ O & \Sigma_n \end{bmatrix}\begin{bmatrix} U_S^H \\ U_n^H \end{bmatrix} = S\Sigma_S S^H + G\Sigma_n G^H, \tag{1.7}$$

where $S \stackrel{\text{def}}{=} [s_1, s_2, \ldots, s_r] = [u_1, u_2, \ldots, u_r]$, $G \stackrel{\text{def}}{=} [g_1, g_2, \ldots, g_{n-r}] = [u_{r+1}, u_{r+2}, \ldots, u_n]$, $\Sigma_S = \text{diag}(\sigma_1^2 + \sigma_w^2, \sigma_2^2 + \sigma_w^2, \ldots, \sigma_r^2 + \sigma_w^2)$, $\Sigma_n = \text{diag}(\sigma_w^2, \sigma_w^2, \cdots, \sigma_w^2)$; $m \times r$ unitary matrix S is the matrix composed of the eigenvectors which correspond to the r principal eigenvalues, and $m \times (n - r)$ unitary matrix G is the matrix composed of the eigenvectors which correspond to the $n - r$ minor eigenvalues.

Definition 4 Define S as the eigenvector matrix which correspond to the first r largest eigenvalues $\lambda_1, \lambda_2, \ldots, \lambda_r$ of the autocorrelation matrix of the measurement data. Then its column space $\text{Span}(S) = \text{Span}\{u_1, u_2, \ldots, u_r\}$ is called the *signal subspace* of measurement data space $\text{Span}(X)$, and the column space $\text{Span}(G) = \text{Span}\{u_{r+1}, u_{r+2}, \ldots, u_n\}$ of the eigenvector matrix G which correspond to the $n - r$ minor eigenvalues is called the *noise subspace* of measurement data space.

In the following, we analyze the geometric meaning of the signal subspace and the noise subspace. From the constitution method of subspace and the feature of unitary matrix, we know that the signal subspace and noised subspace are orthogonal, that is,

$$\text{Span}\{s_1, s_2, \ldots, s_r\} \perp \text{Span}\{g_1, g_2, \ldots, g_{n-r}\}. \tag{1.8}$$

Since U is a unitary matrix, it holds that

$$UU^H = [S \quad G]\begin{bmatrix} S^H \\ G^H \end{bmatrix} = SS^H + GG^H = I,$$

that is,

$$GG^H = I - SS^H. \tag{1.9}$$

Define the projection matrix of signal subspace as

$$P_S \stackrel{\text{def}}{=} S\langle S, S\rangle^{-1} S^H = SS^H, \tag{1.10}$$

where the matrix inner product $\langle S, S\rangle = S^H S = I$.

Thus, $P_S x$ can be considered as the projection of vector x on the signal subspace, and $(I - P_S)x$ means the orthogonal projection of vector x on the signal subspace. From $\langle G, G\rangle = G^H G = I$, it holds that the projection matrix on the noise subspace is $P_n = G\langle G, G\rangle^{-1} G^H = GG^H$. Therefore, the following matrix

$$GG^H = I - SS^H = I - P_S \tag{1.11}$$

is usually called as the orthogonal projection matrix of signal subspace.

The subspace applications have the following characteristics [5, 71]:

(1) Only a few singular vectors or eigenvectors are needed. Since the number of larger singular values (or eigenvalues) of matrix $A_{m \times n}$ is smaller than the number of smaller singular values (or eigenvalues), it is more efficient to use the signal subspace with smaller dimension than the noise subspace.
(2) In many application occasions, one does not need to know the singular values or eigenvalues, and only needs to know the matrix rank and singular vectors or eigenvectors of matrix.
(3) In most instances, one does not need to know the singular vectors or eigenvectors of matrix well and truly, and only needs to know the basis vectors spanned by the signal subspace or noise subspace.

1.2.2 Subspace Tracking Method

The iterative computation of an extreme (maximal or minimum) eigen pair (eigenvalue and eigenvector) can date back to 1966 [72]. In 1980, Thompson proposed a LMS-type adaptive algorithm for estimating eigenvector, which correspond to the smallest eigenvalue of sample covariance matrix, and provided the adaptive tracking algorithm of the angle/frequency combing with Pisarenko's harmonic estimator [14]. Sarkar et al. [73] used the conjugate gradient algorithm to track the variation of the extreme eigenvector which corresponds to the smallest eigenvalue of the covariance matrix of the slowly changing signal and proved its much faster convergence than Thompson's LMS-type algorithm. These methods were only used to track single extreme value and eigenvector with limited application, but later they were extended for the eigen-subspace tracking and updating methods. In 1990, Comon and Golub [6] proposed the Lanczos method for tracking the extreme singular value and singular vector, which is a common method designed originally for determining some big and sparse symmetrical eigen problem $Ax = \lambda x$ [74].

The earliest eigenvalue and eigenvector updating method was proposed by Golub in 1973 [75]. Later, Golub's updating idea was extended by Bunch et al. [76, 77], the basic idea of which is to update the eigenvalue decomposition of the covariance matrix after every rank-one modification, and then go to the matrix's latent root using the interlacing theorem, and then update the place of the latent root using the iterative resolving root method. Thus, the eigenvector can be updated. Later, Schereiber [78] introduced a transform to change a majority of complex number arithmetic operation into real-number operation and made use of Karasalo's subspace mean method [79] to further reduce the operation quantity. DeGroat and

Roberts [80] developed a numerically stabilized rank-one eigen structure updating method based on mutual Gram–Schmidt orthogonalization. Yu [81] extended the rank-one eigen structure update to block update and proposed recursive update of the eigenvalue decomposition of a covariance matrix.

The earliest adaptive signal subspace tracking method was proposed by Owsley [7] in 1978. Using the stochastic gradient method, Yang and Kaveh [18] proposed a LMS-type subspace tracking algorithm and extended Owsley's method and Thompson's method. This LMS-type algorithm has a high parallel structure and low computational complexity. Karhumen [17] extended Owsley's idea by developing a stochastic approaching method based on computing subspace. Like Yang and Kaveh's extension of Thompson's idea to develop an LMS-type subspace tracking algorithm, Fu and Dowling [45] extended Sarkar's idea to develop a subspace tracking algorithm based on conjugate gradient. During the recent 20 years, eigen-subspace tracking and update has been an active research field. Since eigen-subspace tracking is mainly applied to real signal processing, these methods should be fast algorithms.

According to [71], the eigen-subspace tracking and updating methods can be classified into the following four classes:

(1) In some applications of eigen-subspace method such as MUSIC, one only needs to use the orthogonal basis of the noise subspace eigenvectors and does not need to use the eigenvector itself. This characteristic can predigest the adaptive tracking problem of a class of eigenvectors. The methods which only track the orthogonal basis of noise subspace are classified as the first class, and they are based on rank revealing URV [82] and rank revealing QR [83] decomposition of matrix, respectively.

(2) In the method conducting tracking and updating problem of the eigenvalues and eigen-subspace simultaneously, a common sight is to regard the covariance matrix of the nonstationary signal at the kth as the sum of the covariance matrix at the $k - 1$th and another rank-one matrix (the product of the conjugate transpose of measure vector and itself). Thus, tracking the eigenvalue decomposition of the covariance matrix has much to do with the so-called rank-one updating [81, 84].

(3) Regarding the determination of eigen-subspace as an optimization problem: The one is a constrained optimization problem, and the other is unconstrained optimization problem. The constrained optimization problem can be solved using the stochastic gradient [18] and conjugate gradient [45] methods. The unconstrained optimization problem presents a new explanation for the eigen-subspace, and its corresponding method was called projection approximation subspace tracking [5]. The other classical representative is that it is based on Lanczos algorithm, and to use the Lanczos iteration and stochastic approach concept to conduct on the computation of subspace of slowly changing data matrix [85]. Xu et al., proposed [86, 87] three Lanczos and dual Lanczos subspace tracking algorithms, and the former is suitable for the eigen decomposition of covariance matrix, and the latter is for the singular value

decomposition of data matrix, and at the processing of Lanczos iteration they can test and estimate the number of principal eigenvalues and principal singular values. In view of the close mathematics connections between the Lanczos algorithm and conjugate gradient, this algorithm, though it has not direct connections with the optimization problem, still falls into the third type of method.

(4) Modify and extend the classical eigen decomposition/singular value decomposition batch processing methods such as QR decomposition, Jacobi method, and power iteration to make them adaptive. For example, the singular value decomposition updating algorithm based on QR updating and Jacobi-type method [88] falls into this class.

1.3 Main Features of This Book

This book presents principal component analysis algorithms and its extensions using neural networks approach. Pertinent features include the following:

(1) A tutorial-style overview of neural networks-based principal component analysis algorithms, minor component analysis algorithms, principal subspace tracking, and minor subspace tracking.
(2) Analysis of self-stabilizing feature of neural-based PCA/MCA algorithms, and development of a self-stabilizing neural-based minor component analysis algorithm.
(3) Total least squares estimation application of MCA algorithms, and development of a novel neural-based algorithm for total least squares filtering.
(4) Development of a novel dual-purpose principal and minor subspace gradient flow and unified self-stabilizing algorithm for principal and minor components' extraction.
(5) Analysis of a discrete-time dynamics of a class of self-stabilizing MCA learning algorithms and a convergence analysis of deterministic discrete-time system of a unified self-stabilizing algorithm for PCA and MCA.
(6) Extension of PCA algorithm to generalized feature extraction and development of a novel adaptive algorithm for minor generalized eigenvector extraction and a novel multiple generalized minor component extraction algorithm.
(7) Development of a unified and coupled PCA and MCA rules and an adaptive coupled generalized eigen pairs extraction algorithm, based on Moller's coupled PCA neural algorithm.
(8) Generalization of feature extraction from autocorrelation matrix to cross-correlation matrix, and development of an effective neural algorithm for extracting cross-correlation feature between two high-dimensional data streams and a coupled principal singular triplet extraction algorithm of a cross-covariance matrix.

1.4 Organization of This Book

As reflected in the title, this book is concerned with three areas of principal component analysis method, namely neural-based algorithm, performance analysis method, and generalized/extension algorithm. Consequently, the book can be naturally divided into three parts with a common theme. In the three areas, many novel algorithms were proposed by us. To appreciate theses new algorithms, the conventional approaches and existing methods also need to be understood. Fundamental knowledge of conventional principal component analysis, neural-based feature extraction, subspace tracking, performance analysis methods, and even feature extraction based on matrix theory is essential for understanding the advanced material presented in this book. Thus, each part of this book starts with a tutorial type of introduction of the area.

Part I starts from Chap. 2, which provides an overview of some important concepts and theorems of decomposition and singular value decomposition related to principal component analysis. Chapter 3 serves as a starting point to introduce the neural-based principal component analysis. The key Hebbian network and Oja's network forming the core of neural network-based PCA algorithms can be founded in this chapter. Chapter 4 provides an introduction to neural network-based MCA algorithms and the self-stabilizing analysis of these algorithms, followed by a novel self-stabilizing MCA algorithm and a novel neural algorithm for total least squares filtering proposed by us. Part I ends on Chap. 5, which addresses the theoretical issue of the dual-purpose principal and minor component analysis. In this chapter, several important dual-purpose algorithms proposed by us are introduced, and their performance and numerical consideration are analyzed. Part II starts from a tutorial-style introduction to deterministic continuous-time (DCT) system, the stochastic discrete-time (SDT) system, the deterministic discrete-time (DDT) system, followed by a detailed analysis of DDT systems of a new self-stabilizing MCA algorithm and Chen's unified PCA/MCA algorithm in Chap. 6. Part III starts from Chap. 7. The generalized Hermitian eigenvalue problem and existing adaptive algorithms to extract generalized eigen pairs are reviewed, and then, a minor generalized eigenvector extraction algorithm and a novel adaptive algorithm for generalized coupled eigen pairs of ours are introduced and discussed. The other two chapters of Part III are devoted to coupled principal component analysis and cross-correlation feature extraction, respectively, in which our novel coupled or extension algorithms are introduced and analyzed.

Some of the materials presented in this book have been published in archival journals by the authors, and is included in this book after necessary modifications or updates (some modifications are major ones) to ensure accuracy, relevance, completeness and coherence. This portion of materials includes:

- Section 4.4 of Chapter 4, reprinted from Neural Networks, Xiangyu Kong, Changhua Hu, Chongzhao Han, "A self-stabilizing MSA algorithm in high-dimensional data stream", Vol. 23, 865–871, © 2010 Elsevier Ltd., with permission from Elsevier.

- Section 4.5 of Chapter 4, reprinted from Neural Processing Letter, Xiangyu Kong, Changhua Hu, Chongzhao Han, “A self-stabilizing neural algorithm for total least squares filtering”, Vol. 30, 257–271, © 2009 Springer Science+Business Media, LLC., reprinted with permission.
- Section 5.3 of Chapter 5, reprinted from IEEE Transactions on Signal Processing, Xiangyu Kong, Changhua Hu, Chongzhao Han, “A Dual purpose principal and minor subspace gradient flow”, Vol. 60, No. 1, 197–210, © 2012 IEEE., with permission from IEEE.
- Section 6.3 of Chapter 6, reprinted from IEEE Transactions on Neural Networks, Xiangyu Kong, Changhua Hu, Chongzhao Han, “On the discrete time dynamics of a class of self-stabilizing MCA learning algorithm”, Vol. 21, No. 1, 175–181, © 2010 IEEE., with permission from IEEE.
- Section 6.4 of Chapter 6, reprinted from Neural Networks, Xiangyu Kong, Qiusheng an, Hongguang Ma, Chongzhao Han, Qizhang, “Convergence analysis of deterministic discrete time system of a unified self-stabilizing algorithm for PCA and MCA”, Vol. 36, 64–72, © 2012 Elsevier Ltd., with permission from Elsevier.
- Section 7.3 and 7.4 of Chapter 7, reprinted from IEEE Transactions on Signal Processing, Gao Yingbin, Kong Xiangyu, Hu Changhua, Li Hongzeng, and Hou Li'an, “A Generalized Information Criterion for generalized Minor Component Extraction”, Vol. 65, No. 4, 947–959, © 2017 IEEE., with permission from IEEE.
- Section 8.3 of Chapter 8, reprinted from Neural Processing Letter, Xiaowei Feng, Xiangyu Kong, Hongguang Ma, and Haomiao Liu, “Unified and coupled self-stabilizing algorithm for minor and principal eigen-pair extraction”, doi: 10.1007/s11063-016-9520-3, © 2016 Springer Science+Business Media, LLC., reprinted with permission.
- Section 8.4 of Chapter 8, reprinted from IEEE Transactions on Signal Processing, Xiaowei Feng, Xiangyu Kong, Zhansheng Duan, and Hongguang Ma, “Adaptive generalized eigen-pairs extraction algorithm and their convergence analysis”, Vol. 64, No. 11, 2976–2989, © 2016 IEEE., with permission from IEEE.
- Section 9.3 of Chapter 9, reprinted from Neural Processing Letter, Xiang yu Kong, Hong guang Ma, Qiu sheng An, Qi Zhang, “An effective neural learning algorithm for extracting cross-correlation feature between two high-dimensional data streams”, Vol. 42, 459–477, © 2015 Springer Science+Business Media, LLC., reprinted with permission.

References

1. Diamantaras, K. I., & Kung, S. Y. (1996). *Principal component neural networks: Theory and application*. Wiley, INC.
2. Zhang, Q., & Leung, Y.-W. (2000). A class of learning algorithms for principal component analysis and minor component analysis. *IEEE Transactions on Neural Network, 11*(1), 200–204.

3. Washizawa, Y. (2010). Feature extraction using constrained approximation and suppression. *IEEE Transactions on Neural Network, 21*(2), 201–210.
4. Cirrincione, G., Cirrincione, M., Herault, J., & Huffel, S. V. (2002). The MCA EXIN neuron for the minor component analysis. *IEEE Transactions on Neural Network, 13*(1), 160–187.
5. Yang, B. (1995). Projection approximation subspace tracking. *IEEE Transactions on Signal Processing, 43*(1), 95–107.
6. Comon, P., & Golub, G. H. (1990). Tracking a few extreme singular values and vectors in signal processing. In *Processing of the IEEE* (pp. 1327–1343).
7. Owsley, N. L. (1978). Adaptive data orthogonalization. In *Proceedings of IEEE ICASSP* (pp. 100–112).
8. Tufts, D. W., & Melissinos, C. D. (1986). Simple, effective computation of principal eigenvectors and their eigenvalues and applications to high-resolution estimation of frequencies. *IEEE Transactions on Acoustic, Speech and Signal Processing, ASSP-34*, 1046–1053.
9. Shaman, K. C. (1986). Adaptive algorithms for estimating the complete covariance eigenstructure. In *Proceedings of IEEE ICASSP (Tokyo, Japan)* (pp. 1401–1404).
10. Moonen, M., Van Dooren, P., & Vandewalle, J. (1989). Updating singular value decompositions: A parallel implementation. In *PWC. SPlE 4dv. Algorithms Arthitectures Signal Processing* (San Diego, CA) (pp. 80–91).
11. Bunch, J. R., Nielsen, C. P., & Sorenson, D. (1978). Rank-one modification of the symmetric eigenproblem. *Numerical Mathematics, 31*, 31–48.
12. Karasalo, I. (1986). Estimating the covariance matrix by signal subspace averaging. *IEEE Transactions on Acoustic Speech and Signal Processing, ASSP-34*, 8–12.
13. DeGroat, R. D. (1992). Noniterative subspace tracking. *IEEE Transactions on Signal Processing, 40*(3), 571–577.
14. Thompson, P. A. (1980). An adaptive spectral analysis technique for unbiased frequency estimation in the presence of white noise. In *Proceeding 13th Asilnmur Conj:Circuit System. Computation* (pp. 529–533).
15. Oja, E. (1982). A simplified neuron model as a principal component analyzer. *Journal of Mathematical Biology, 15*(3), 267–273.
16. Karhunen, J., & Oja, E. (1982). New methods for stochastic approximation of truncated Karhunen-LoCve expansions. In *Prw. 6th International Conference: Putt. Recop.* (5Xk553).
17. Karhunen, J. (1984). Adaptive algorithms for estimating eigenvectors of correlation type matrices. In *Proceedings of IEEE ICASSP* (San Diego, CA) (pp. 14.6.1–14.6.3).
18. Yang, J., & Kaveh, M. (1988). Adaptive eigensubspace algorithms for direction or frequency estimation and tracking. *IEEE Transactions on Acoustic, Speech and Signal Processing, 36* (2), 241–251.
19. Kung, S. Y. (1993). *Digital neural processing*. Englewood Cliffs, NJ: Prentice Hall.
20. Reddy, V. U., Egardt, B., & Kailath, T. (1982). Least squares type algorithm for adaptive implementation of Pisarenko's harmonic retrieval method. *IEEE Transactions on Acoustic Speech and Signal processing, 30*(3), 399–405.
21. Bannour, S., & Azimi-Sadjadi, M. R. (1992). An adaptive approach for optimal data reduction using recursive least squares learning method. In *Pwc. IEEE ICASSP* (San Francisco, CA) (pp. 11297–11300).
22. Yang, X., Sarkar, T. K., & Arvas, E. (1989). A survey of conjugate gradient algorithms for solution of extreme eigen-problems of a symmetric matrix. *IEEE Transactions on Acoustic, Speech and Signal Processing, 37*(10), 1550–1556.
23. Stewart, G. W. (1992). An updating algorithm for subspace tracking. *IEEE Transactions on Signal Processing, 40*(6), 1535–1541.
24. Bischof, C. H., & Shroff, G. M. (1992). On updating signal subspaces. *IEEE Transactions on Signal Processing, 40*(1), 96–105.
25. Bannour, S., & Azimi-Sadjadi, R. (1995). Principal component extraction using recursive least squares learning. *IEEE Transactions on Neural Networks, 6*(2), 457–469.

26. Cichocki, A., Kasprzak, W., & Skarbek, W. (1996). Adaptive learning algorithm for principal component analysis with partial data. *Cybernetic Systems, 2*, 1014–1019.
27. Kung, S. Y., Diamantaras, K., & Taur, J. (1994). Adaptive principal component extraction (APEX) and applications. *IEEE Transactions on Signal Processing, 42*(5), 1202–1217.
28. Möller, R., & Könies, A. (2004). Coupled principal component analysis. *IEEE Transactions on Neural Networks, 15*(1), 214–222.
29. Ouyang, S., Bao, Z., & Liao, G. S. (2000). Robust recursive least squares learning algorithm for principal component analysis. *IEEE Transactions on Neural Networks, 11*(1), 215–221.
30. Sanger, T. D. (1989). Optimal unsupervised learning in a single-layer linear feedforward neural network. *Neural Networks, 2*(6), 459–473.
31. Xu, L. (1993). Least mean square error reconstruction principle for selforganizing neural-nets. *Neural Netwoks, 6*(5), 627–648.
32. Xu, L., Oja, E., & Suen, C. (1992). Modified Hebbian learning for curve and surface fitting. *Neural Networks, 5*, 441–457.
33. Oja, E. (1992). Principal component, minor component and linear neural networks. *Neural Networks, 5*(6), 927–935.
34. Feng, D. Z., Bao, Z., & Jiao, L. C. (1998). Total least mean squares algorithm. *IEEE Transactions on Signal Processing, 46*(6), 2122–2130.
35. Feng, D. Z., Zheng, W. X., & Jia, Y. (2005). Neural network learning algorithms for tracking minor subspace in high-dimensional data stream. *IEEE Transactions on Neural Networks, 16*(3), 513–521.
36. Luo, F. L., & Unbehauen, R. (1997). A minor subspace analysis algorithm. *IEEE Transactions on Neural Networks, 8*(5), 1149–1155.
37. Ouyang, S., Bao, Z., Liao, G. S., & Ching, P. C. (2001). Adaptive minor component extraction with modular structure. *IEEE Transactions on Signal Processing, 49*(9), 2127–2137.
38. Zhang, Q., & Leung, Y.-W. (2000). A class of learning algorithms for principal component analysis and minor component analysis. *IEEE Transactions on Neural Networks, 11*(2), 529–533.
39. Möller, R. (2004). A self-stabilizing learning rule for minor component analysis. *International Journal of Neural Systems, 14*(1), 1–8.
40. Douglas, S. C., Kung, S. Y., & Amari, S. (2002). A self-stabilized minor subspace rule. *IEEE Signal Processing Letter, 5*(12), 1342–1352.
41. Oja, E. (1989). Neural networks, principal components, and subspaces. *International Journal of Neural Systems, 1*(1), 61–68.
42. Williams R. J.(1985). *Feature discovery through error-correction learning*. Institute of Cognition Science, University of California, San Diego, Technical Report. 8501.
43. Baldi, P. (1989). Linear learning: Landscapes and algorithms. In D. S. Touretzky (Ed.), *Advances in neural information processing systems 1*. San Mateo, CA: Morgan Kaufmann.
44. Miao, Y. F., & Hua, Y. B. (1998). Fast subspace tracking and neural network learning by a novel information criterion. *IEEE Transactions on Signal Processing, 46*(7), 1967–1979.
45. Fu, Z., & Dowling, E. M. (1995). Conjugate gradient eigenstructure tracking for adaptive spectral estimation. *IEEE Transactions on Signal Processing, 43*(5), 1151–1160.
46. Mathew, G., Reddy, V. U., & Dasgupta, S. (1995). Adaptive estimation of eigensubspace. *IEEE Transactions on Signal Processing, 43*(2), 401–411.
47. Chiang, C. T., & Chen, Y. H. (1999). On the inflation method in adaptive noise subspace estimator. *IEEE Transactions on Signal Processing, 47*(4), 1125–1129.
48. Nguyen, T. D., & Yamada, I. (2013). Adaptive normalized quasi-newton algorithms for extraction of generalized eigen-pairs and their convergence analysis. *IEEE Transactions on Signal Processing, 61*(6), 1404–1418.
49. Chatterjee, C., Roychowdhury, V., Ramos, P. J., & Zoltowski, M. D. (1997). Self-organizing algorithms for generalized eigen-decomposition. *IEEE Transactions on Neural Networks, 8*(6), 1518–1530.

50. Xu, D. X., Principe, J. C., & Wu, H. C. (1998). Generalized eigendecomposition with an on-line local algorithm. *IEEE Signal Processing Letter, 5*(11), 298–301.
51. Rao, Y. N., Principe, J. C., & Wong, T. F. (2004). Fast RLS-like algorithm for generalized eigendecomposition and its applications. *Journal of VLSI Signal Processing, 37*(2–3), 333–344.
52. Wong, T. F., Lok, T. M., Lehnert, J. S., & Zoltowski, M. D. (1998). A linear receiver for direct-sequence spread-spectrum multiple-access systems with antenna arrays and blind adaptation. *IEEE Transactions on Information Theory, 44*(2), 659–676.
53. Strobach, P. (1998). Fast orthogonal iteration adaptive algorithms for the generalized symmetric eigenproblem. *IEEE Transactions on Signal Processing, 46*(12), 3345–3359.
54. Yang, J., Xi, H. S., Yang, F., & Zhao, Y. (2006). RLS-based adaptive algorithms for generalized eigen-decomposition. *IEEE Transactions on Signal Processing, 54*(4), 1177–1188.
55. Tanaka, T. (2009). Fast generalized eigenvector tracking based on the power method. *IEEE Signal Processing Letter, 16*(11), 969–972.
56. Attallah, S., & Abed-Meraim, K. (2008). A fast adaptive algorithm for the generalized symmetric eigenvalue problem. *IEEE Signal Processing Letter, 15*, 797–800.
57. Chen, T. P., & Amari, S. (2001). Unified stabilization approach to principal and minor components extraction algorithms. *Neural Networks, 14*(10), 1377–1387.
58. Hasan, M. A. (2007). Self-normalizing dual systems for minor and principal component extraction. In *Proceedings of ICASSP 2007 IEEE International Conference on Acoustic, Speech and Signal Processing* (15–20, Vol. 4, pp. IV-885–IV-888).
59. Peng, D. Z., Zhang, Y., & Xiang, Y. (2009). A unified learning algorithm to extract principal and minor components. *Digit Signal Processing, 19*(4), 640–649.
60. Manton, J. H., Helmke, U., & Mareels, I. M. Y. (2005). A dual purpose principal and minor component flow. *Systems Control Letter, 54*(8), 759–769.
61. Chen, T., Amari, S. I., & Lin, Q. (1998). A unified algorithm for principal and minor component extraction. *Neural Networks, 11*(3), 365–369.
62. Peng, D. Z., & Zhang, Y. (2007). Dynamics of generalized PCA and MCA learning algorithms. *IEEE Transactions on Neural Networks, 18*(6), 1777–1784.
63. Kong, X. Y., Hu, C. H., & Han, C. Z. (2012). A dual purpose principal and minor subspace gradient flow. *IEEE Transactions on Signal Processing, 60*(1), 197–210.
64. Kong, X. Y., Hu, C. H., Ma, H. G., & Han, C. Z. (2012). A unified self-stabilizing neural network algorithm for principal and minor components extraction. *IEEE Transactions on Neural Networks and Learning Systems, 23*(2), 185–198.
65. Ferali, W., & Proakis J. G. (1990). Adaptive SVD algorithm for covariance matrix eigenstructure computation. In *Proceedings of IEEE International Conference on Acoustic, Speech, Signal Processing* (pp. 176–179).
66. Cichocki, A. (1992). Neural network for singular value decomposition. *Electronic Letter, 28* (8), 784–786.
67. Diamantaras, K. I., & Kung, S. Y. (1994). Cross-correlation neural network models. *IEEE Transactions on Signal Processing, 42*(11), 3218–3223.
68. Feng, D. Z., Bao, Z., & Zhang, X. D. (2001). A cross-associative neural network for SVD of nonsquared data matrix in signal processing. *IEEE Transactions on Neural Networks, 12*(9), 1215–1221.
69. Feng, D. Z., Zhang, X. D., & Bao, Z. (2004). A neural network learning for adaptively extracting cross-correlation features between two high-dimensional data streams. *IEEE Transactions on Neural Networks, 15*(6), 1541–1554.
70. Kong, X. Y., Ma, H. G., An, Q. S., & Zhang, Q. (2015). An effective neural learning algorithm for extracting cross-correlation feature between two high-dimensional data streams. *Neural Processing Letter, 42*(2), 459–477.
71. Zhang, X. D. (2004) *Matrix analysis and applications*, Tsinghua University Press, Beijing.
72. Bradbury, W. W., & Fletcher, R. (1966). New iterative methods for solutions of the eigenproblem. *Numerical Mathematics, 9*(9), 259–266.

73. Sarkar, T. K., Dianat, S. A., Chen, H., & Brule, J. D. (1986). Adaptive spectral estimation by the conjugate gradient method. *IEEE Transactions on Acoustic, Speech, and Signal Processing, 34*(2), 272–284.
74. Golub, G. H., & Van Load, C. F. (1989). *Matrix computation* (2nd ed.). Baltimore: The John Hopkins University Press.
75. Golub, G. H. (1973). Some modified matrix eigenvalue problems. *SIAM Review, 15*(2), 318–334.
76. Bunch, J. R., Nielsen, C. P., & Sorensen, D. C. (1978). Rank-one modification of the symmetric eigenproblem. *Numerical Mathematics, 31*(1), 31–48.
77. Bunch, J. R., & Nielsen, C. P. (1978). Updating the singular value decomposition. *Numerical Mathematics, 31*(31), 111–129.
78. Schereiber, R. (1986). Implementation of adaptive array algorithms. *IEEE Transactions on Acoustic, Speech, and Signal Processing, 34*(5), 1038–1045.
79. Karasalo, I. (1986). Estimating the covariance matrix by signal subspace averaging. *IEEE Transactions on Acoustic, Speech, and Signal Processing, 34*(1), 8–12.
80. DeGroat, R. D., & Roberts, R. A. (1990). Efficient, numerically stabilized rank-one eigenstructure updating. *IEEE Transactions on Acoustic, Speech, and Signal Processing, 38* (2), 301–316.
81. Yu, K. B. (1991). Recursive updating the eigenvalue decomposition of a covariance matrix. *IEEE Transactions on Signal Processing, 39*(5), 1136–1145.
82. Stewart, G. W. (1992). An updating algorithm for subspace tracking. *IEEE Transactions on Signal Processing, 40*(6), 1135–1541.
83. Bischof, C. H., & Shroff, G. M. (1992). On updating signal subspace. *IEEE Transactions on Signal Processing, 40*(1), 96–105.
84. Champagne, B. (1994). Adaptive eigendecomposition of data covariance matrices based on first-order perturbations. *IEEE Transactions on Signal Processing, 42*(10), 2758–2770.
85. Fuhrmann, D. R. (1988). An algorithm for subspace computation with applications in signal processing. *SIAM Journal of Matrix Analysis and Applications, 9*(2), 213–220.
86. Xu, G., Cho, Y., & Kailath, T. (1994). Application of fast subspace decomposition to signal processing and communication problems. *IEEE Transactions on Signal Processing, 42*(6), 1453–1461.
87. Xu, G., & Kailath, T. (1994). Fast subspace decomposition. *IEEE Transactions on Signal Processing, 42*(3), 539–551.
88. Moonen, M., Dooren, P. V., & Vandewalle, J. (1992). A singular value decomposition updating algorithm for subspace tracking. *SIAM Journal of Matrix Analysis and Applications, 13*(4), 1015–1038.

Chapter 2
Matrix Analysis Basics

In this chapter, we review some basic concepts, properties, and theorems of singular value decomposition (SVD), eigenvalue decomposition (ED), and Rayleigh quotient of a matrix. Moreover, we also introduce some basics of matrix analysis. They are important and useful for our theoretical analysis in subsequent chapters.

2.1 Introduction

As discussed in Chap. 1, the PC or MC can be obtained by the ED of the sample correlation matrix or the SVD of the data matrix, and ED and SVD are also primal analysis tools. The history of SVD can date back to the 1870s, and Beltrami and Jordan are acknowledged as the founder of SVD. In 1873, Beltrami [1] published the first paper on SVD, and one year later Jordan [2] published his independent reasoning about SVD. Now, SVD has become one of the most useful and most efficient modern numerical analysis tools, and it has been widely used in statistical analysis, signal and image processing, system theory and control, etc. SVD is also a fundamental tool for eigenvector extraction, subspace tracking, and total least squares problem, etc.

On the other hand, ED is important in both mathematical analysis and engineering applications. For example, in matrix algebra, ED is usually related to the spectral analysis, and the spectral of a linear arithmetic operator is defined as the set of eigenvalues of the matrix. In engineering applications, spectral analysis is connected to the Fourier analysis, and the frequency spectral of signals is defined as the Fourier spectral, and then the power spectral of signals is defined as the square of frequency spectral norm or Fourier transform of the autocorrelation functions.

Besides SVD and ED, gradient and matrix differential are also the important concepts of matrix analysis. In view of the use of them in latter chapters, we will provide detailed analysis of SVD, ED, matrix analysis, etc. in the following.

© Science Press, Beijing and Springer Nature Singapore Pte Ltd. 2017

X. Kong et al., *Principal Component Analysis Networks and Algorithms*,
DOI 10.1007/978-981-10-2915-8_2

2.2 Singular Value Decomposition

As to the inventor history of SVD, see Stewart's dissertation. Later, Autonne [3] extended SVD to complex square matrix in 1902, and Eckart and Young [4] further extended it to general rectangle matrix in 1939. Now, the theorem of SVD for rectangle matrix is usually called Eckart–Young Theorem.

SVD can be viewed as the extension of ED to the case of nonsquare matrices. It says that any real matrix can be diagonalized by using two orthogonal matrices. ED works only for square matrices and uses only one matrix (and its inverse) to achieve diagonalization. If the matrix is square and symmetric, then the two orthogonal matrices of SVD will be the same, and ED and SVD will also be the same and closely related to the matrix rank and reduced-rank least squares approximations.

2.2.1 Theorem and Uniqueness of SVD

Theorem 2.1 *For any* $\mathbf{A} \in \Re^{m\times n}$ (or $\mathbb{C}^{m\times n}$), *there exist two orthonormal (or unitary) matrices* $\boldsymbol{U} \in \Re^{m\times n}$ (or $\mathbb{C}^{m\times m}$) *and* $\mathbf{V} \in \Re^{m\times n}$ (or $\mathbb{C}^{n\times n}$), *such that*

$$\boldsymbol{A} = \boldsymbol{U}\Sigma\boldsymbol{V}^{\mathrm{T}} \left(\text{or } A = \boldsymbol{U}\boldsymbol{\Sigma}\boldsymbol{V}^{H}\right), \tag{2.1}$$

where,

$$\boldsymbol{\Sigma} = \begin{bmatrix} \boldsymbol{\Sigma}_1 & 0 \\ 0 & 0 \end{bmatrix}$$

and $\boldsymbol{\Sigma} = \mathrm{diag}[\sigma_1, \sigma_2, \ldots \sigma_r]$, *its diagonal elements are arranged in the order:*

$$\sigma_1 \geq \sigma_2 \geq \cdots \geq \sigma_r \geq 0, \quad t = \mathrm{rank}(\boldsymbol{A})$$

The quantity $\sigma_1, \sigma_2, \ldots, \sigma_r$ *together with* $\sigma_{r+1} = \sigma_{r+2} = \cdots = \sigma_n = 0$ *are called the singular values of matrix* $\mathbf{A}$. *The column vector* $\boldsymbol{u}_i$ *of matrix* $\boldsymbol{U}$ *is called the left singular vector of* $\mathbf{A}$, *and the matrix* $\boldsymbol{U}$ *is called the left singular matrix. The column vector* $\mathbf{v}_i$ *of matrix* $\mathbf{V}$ *is called the right singular vector of* $\mathbf{A}$, *and the matrix* $\boldsymbol{V}$ *is called the right singular matrix. The proof of* Theorem 2.1 *can see* [4, 5]. The SVD of matrix $\boldsymbol{A}$ can also be written as:

$$\boldsymbol{A} = \sum_{i=1}^{r} \sigma_i \boldsymbol{u}_i \boldsymbol{v}_i^H. \tag{2.2}$$

It can be easily seen that

$$AA^H = U\Sigma^2 U^H \tag{2.3}$$

which shows that the singular value σ_i of the $m \times n$ *matrix* A *is the positive square root of the eigenvalue (these eigenvalues are nonpositive) of the matrix product* AA^{H}.

The following theorem strictly narrates the singular property of a matrix A.

Theorem 2.2 *Define the singular values of matrix* $\mathbf{A} \in \Re^{m\times n}\,(m > n)$ *as* $\sigma_1 \geq \sigma_2 \geq \cdots \geq \sigma_r \geq 0$.
Then

$$\sigma_k = \min_{E\in\mathbb{C}^{m\times n}} \left\{ \|E\|_{\text{spec}}\colon \operatorname{rank}(A + E) \leq (k-1) \right\}, \quad k = 1, 2, \ldots n \tag{2.4}$$

and there is an error matrix which meets $\|E_k\|_{\text{spec}} = \sigma_k$, *so that*

$$\operatorname{rank}(A + E_k) = r - 1, \quad k = 1, 2, \ldots, n.$$

Theorem 2.2 shows that the singular value of a matrix is equal to the spectral norm of the error matrix E_k which makes the rank of the original matrix reduce one. If the original $n \times n$ matrix A is square and it has a zero singular value, the spectral norm of error matrix whose rank reduces to one is equal to zero. That is to say, when the original $n \times n$ matrix A has a zero singular value, the rank of the matrix is $\operatorname{rank}(\mathbf{A}) \leq n - 1$ and the original matrix is not full-rank essentially. So, if a matrix has a zero singular value, the matrix must be singular matrix. Generally speaking, if a rectangle matrix has a zero singular value, then it must not be full column rank or full row rank. This case is called rank-deficient matrix, which is a singular phenomenon with regards to the full-rank matrix.

In the following, we discuss the uniqueness of SVD.

(1) The number r of nonzero singular values and their values $\sigma_1, \sigma_2, \ldots, \sigma_r$ is unique relative to matrix A.
(2) If rank(A) = r, the dimension of the sets of vector $x \in \mathbb{C}^n$ which meets $Ax = 0$, namely the zero space of matrix A, is equal to $n - r$. Thus, one can select orthogonal basis $\{v_{r+1}, v_{r+2}, \ldots, v_n\}$as the zero space of matrix A in $\mathbb{C}^n$. From this point, the subspace $\text{Null}(A)$of $\mathbb{C}^n$ spanned by column vectors of V is uniquely determined. However, as long as every vector can constitute the orthogonal basis of this subspace, they can be selected arbitrarily.
(3) The sets of $\mathbf{y}(\in \mathbb{C}^m)$ which can be denoted as $y = Ax$ constitute the image space ImA of matrix A, whose dimension is equal to r. The orthogonal supplement space $(\text{Im}A)^{\perp}$ of ImA is m-r dimensional. Thus, one can select $\{u_{r+1}, u_{r+2}, \ldots, u_m\}$as the orthogonal basis of $(\text{Im}A)^{\perp}$. The subspace $(\text{Im}A)^{\perp}$ of $\mathbb{C}^m$ spanned by the column vectors $u_{r+1}, u_{r+2}, \ldots, u_m$ of U is uniquely determined.

(4) If σ_i is single singular value $(\sigma_i \neq \sigma_j, \forall j \neq i)$, $\boldsymbol{v}_i$ and $\boldsymbol{u}_i$ is uniquely determined except discrepancy of an angle. That is to say, after $\boldsymbol{v}_i$ and $\boldsymbol{u}_i$ multiply $\mathrm{e}^{i\theta}(j = \sqrt{-1})$ and θ is real number) at the same time, they are still the right and left singular vectors, respectively.

2.2.2 Properties of SVD

Assume $\boldsymbol{A} \in \Re^{m\times n}$, $\boldsymbol{B} \in \Re^{m\times n}$, and $r_A = \mathrm{rank}(A)$, $p = \min\{m, n\}$. The singular values of matrix $\boldsymbol{A}$ can be arranged as follows: $\sigma_{\max} = \sigma_1 \geq \sigma_2 \geq \cdots \geq \sigma_{p-1} \geq \sigma_p = \sigma_{\min} \geq 0$, and denote by $\sigma_i(\boldsymbol{B})$the ith largest singular value of matrix $\boldsymbol{B}$. A few properties of SVD can summarized as follows [6]:

(1) The relationship between the singular values of a matrix and the ones of its submatrix.

Theorem 2.3 (interlacing theorem for singular values). *Assume* $\mathbf{A} \in \Re^{m\times n}$, *and its singular values satisfy* $\sigma_1 \geq \sigma_2 \geq \cdots \geq \sigma_r$, *where* $r = \min\{m, n\}$. *If* $\boldsymbol{B} \in \Re^{p\times q}$ *is a submatrix of* $\mathbf{A}$, *and its singular values satisfy* $\gamma_1 \geq \gamma_2 \geq \cdots \geq \gamma_{\min\{p,q\}}$, *then it holds that*

$$\sigma_i \geq \gamma_i, \quad i = 1, 2, \ldots, \min\{p, q\} \tag{2.5}$$

and

$$\gamma_i \geq \sigma_{i+(m-p)+(n-q)}, \quad i \leq \min\{p+q-m, p+q-n\}. \tag{2.6}$$

From Theorem 2.3, it holds that: If $\boldsymbol{B} \in \Re^{m\times(n-1)}$ is a submatrix of $\mathbf{A} \in \Re^{m\times n}$ by deleting any column of matrix $\mathbf{A}$, and their singular values are arranged in non-decreasing order, then it holds that

$$\sigma_1(\boldsymbol{A}) \geq \sigma_1(\boldsymbol{B}) \geq \sigma_2(\boldsymbol{A}) \geq \sigma_2(\boldsymbol{B}) \geq \cdots \geq \sigma_h(\boldsymbol{A}) \geq \sigma_h(\boldsymbol{B}) \geq 0, \tag{2.7}$$

where $h = \min\{m, n-1\}$.

If $\boldsymbol{B} \in \Re^{(m-1)\times n}$ is a submatrix of $\boldsymbol{A} \in \Re^{m\times n}$ by deleting any row of matrix $\boldsymbol{A}$, and their singular values are arranged as non-decreasing order, then it holds that

$$\sigma_1(\boldsymbol{A}) \geq \sigma_1(\boldsymbol{B}) \geq \sigma_2(\boldsymbol{A}) \geq \sigma_2(\boldsymbol{B}) \geq \cdots \sigma_h(\boldsymbol{A}) \geq \sigma_h(\boldsymbol{B}) \geq 0. \tag{2.8}$$

(2) The relationship between the singular values of a matrix and its norms.
The spectral norm of a matrix $\boldsymbol{A}$ is equal to its largest singular value, namely,

$$\|A\|_{\text{spec}} = \sigma_1. \tag{2.9}$$

According to the SVD theorem of matrix and the unitary invariability property of Frobenius norm $\|A\|_F$ of matrix A, namely $\|U^H A V\|_F = \|A\|_F$, it holds that

$$\|A\|_F = \left[\sum_{i=1}^{m}\sum_{j=1}^{n}|a_{ij}|^2\right]^{1/2} = \|U^H A V\|_F = \|\Sigma\|_F = \sqrt{\sigma_1^2 + \sigma_2^2 + \cdots + \sigma_r^2}. \tag{2.10}$$

That is to say, the Frobenius norm of any matrix is equal to the square root of the sum of the squares of all nonzero singular values of this matrix.
Consider the rank-k approximation of matrix A and denote it as A_k, in which $k < r = \text{rank}(A)$. The matrix A_k is defined as follows:

$$A_k = \sum_{i=1}^{k}\sigma_i u_i v_i^H, k < r,$$

Then the spectral norm of the difference between A and any rank(k) matrix B, and the Frobenius norm of the difference can be written, respectively, as follows:

$$\min_{\text{rank}(B)=r}\|A - B\|_{\text{spec}} = \|A - A_k\|_{\text{spec}} = \sigma_{k+1}, \tag{2.11}$$

$$\min_{\text{rank}(B)=r}\|A - B\|_F^2 = \|A - A_k\|_F^2 = \sigma_{k+1}^2 + \sigma_{k+2}^2 + \cdots \sigma_r^2. \tag{2.12}$$

The above properties are the basis of many concepts and applications. For example, the total least squares, data compression, image enhancement, the solution of linear equations, etc., all need to approximate A using a lower rank matrix.

(3) The relationship between the singular values of a matrix and its determinant. Define A as an $n \times n$ square matrix. Since the absolute value of the determinant of a unitary matrix is equal to one, from SVD theorem it holds that

$$|\det(A)| = |\det\Sigma| = \sigma_1\sigma_2\cdots\sigma_n. \tag{2.13}$$

If all σ_i are non-zero, then $|\det(A)| \neq 0$, which means that A is nonsingular. If at least one $\sigma_i (i > r)$ is equal to zero, then $|\det(A)| = 0$, namely A is singular.

(4) The relationship between the singular values of a matrix and its condition number.
For an $m \times n$ matrix A, its condition number can be defined using SVD as

$$\text{cond}(\boldsymbol{A}) = \sigma_1/\sigma_p, \quad p = \min\{m, n\}. \tag{2.14}$$

Since $\sigma_1 \geq \sigma_p$, the condition number is a positive number which is equal to or larger than one. Obviously, since there is at least one singular value which meets $\sigma_p = 0$, the condition number of a singular matrix is infinite. When the condition number, though not infinite, is very large, the matrix $\boldsymbol{A}$ is called to be close to singular. Since the condition number of unitary or orthogonal matrix is equal to one, the unitary or orthogonal matrix is of "ideal condition". Equation (2.14) can be used to evaluate the condition number.

(5) Maximal singular value and minimal singular value.
If $m \geq n$, for any matrix $\boldsymbol{A}_{m\times n}$, it holds that

$$\begin{aligned}\sigma_{\min}(\boldsymbol{A}) &= \min\left\{\left(\frac{\boldsymbol{x}^H\boldsymbol{A}^H\boldsymbol{A}\boldsymbol{x}}{\boldsymbol{x}^H\boldsymbol{x}}\right)^{1/2} : \boldsymbol{x} \neq 0\right\} \\ &= \min\left\{\left(\boldsymbol{x}^H\boldsymbol{A}^H\boldsymbol{A}\boldsymbol{x}\right)^{1/2} : \boldsymbol{x}^H\boldsymbol{x} = 1, \boldsymbol{x} \in \mathbb{C}^n\right\}\end{aligned} \tag{2.15}$$

and

$$\begin{aligned}\sigma_{\max}(\boldsymbol{A}) &= \max\left\{\left(\frac{\boldsymbol{x}^H\boldsymbol{A}^H\boldsymbol{A}\boldsymbol{x}}{\boldsymbol{x}^H\boldsymbol{x}}\right)^{1/2} : \boldsymbol{x} \neq 0\right\} \\ &= \max\left\{\left(\boldsymbol{x}^H\boldsymbol{A}^H\boldsymbol{A}\boldsymbol{x}\right)^{1/2} : \boldsymbol{x}^H\boldsymbol{x} = 1, \boldsymbol{x} \in \mathbb{C}^n\right\}.\end{aligned} \tag{2.16}$$

(6) The relationship between the singular values and eigenvalues.
Suppose that the eigenvalues of an $n \times n$ symmetrical square matrix $\boldsymbol{A}$ are $\lambda_1, \lambda_2, \ldots \lambda_n (|\lambda_1| \geq \|\lambda_2\| \geq \cdots \geq \|\lambda_n\|)$, and its singular values are $\sigma_1, \sigma_2, \ldots, \sigma_n$ $(\sigma_1 \geq \sigma_2 \geq \cdots \geq \sigma_n \geq 0)$. Then $\sigma_1 \geq |\lambda_i| \geq \sigma_n (i = 1, 2, \ldots, n)$ and cond($\boldsymbol{A}$) $\geq |\lambda_1|/|\lambda_n|$.

2.3 Eigenvalue Decomposition

2.3.1 Eigenvalue Problem and Eigen Equation

The basic problem of the eigenvalue can be stated as follows. Given an $n \times n$ matrix $\boldsymbol{A}$, determine a scalar λ such that the following algebra equation

$$\boldsymbol{A}\boldsymbol{u} = \lambda\boldsymbol{u}, \quad \boldsymbol{u} \neq 0 \tag{2.17}$$

has an $n \times 1$ nonzero solution. The scalar λ is called as an eigenvalue of matrix $\boldsymbol{A}$, and the vector $\boldsymbol{u}$ is called as the eigenvector associated with λ. Since the eigenvalue

λ and eigenvector $\boldsymbol{u}$ appear in couples, $(\lambda, \boldsymbol{u})$ is usually called as an eigen pair of matrix $\boldsymbol{A}$. Although the eigenvalues can be zeros, the eigenvectors cannot be zero.

In order to determine a nonzero vector $\boldsymbol{u}$, Eq. (2.17) can be modified as

$$(\boldsymbol{A} - \lambda\boldsymbol{I})\boldsymbol{u} = 0. \tag{2.18}$$

The above equation should come into existence for any vector $\boldsymbol{u}$, so the unique condition under which Eq. (2.18) has a nonzero solution $\boldsymbol{u} = 0$ is that the determinant of matrix $\boldsymbol{A} - \lambda\boldsymbol{I}$ is equal to zero, namely

$$\det(\boldsymbol{A} - \lambda\boldsymbol{I}) = 0. \tag{2.19}$$

Thus, the solution of the eigenvalue problem consists of the following two steps:

(1) Solve all scalar λ (eigenvalues) which make the matrix $\boldsymbol{A} - \lambda\boldsymbol{I}$ singular.
(2) Given an eigenvalue λ which makes $\boldsymbol{A} = \lambda\boldsymbol{I}$ singular, and to solve all nonzero vectors which meets $(\boldsymbol{A} - \lambda\boldsymbol{I})\boldsymbol{x} = \boldsymbol{0}$, i.e., the eigenvectors corresponding to λ.

According to the relationship between the singular values of a matrix and its determinant, a matrix is singular if and only if $\det(\boldsymbol{A} - \lambda\boldsymbol{I}) = \boldsymbol{0}$,, namely

$$(\boldsymbol{A} - \lambda\boldsymbol{I})x \text{ singular} \Leftrightarrow \det(\boldsymbol{A} - \lambda\boldsymbol{I}) = \boldsymbol{0}. \tag{2.20}$$

The matrix $(\boldsymbol{A} - \lambda\boldsymbol{I})$ is called as the eigen matrix of $\boldsymbol{A}$. When $\boldsymbol{A}$ is an $n \times n$ matrix, spreading the left side determinant of Eq. (2.20) can obtain a polynomial equation (power-n), namely

$$\alpha_0 + \alpha_1\lambda + \cdots + \alpha_{n-1}\lambda^{n-1} + (\ 1)^n\lambda^n - 0, \tag{2.21}$$

which is called as the eigen equation of matrix $\boldsymbol{A}$. The polynomial $\det(\boldsymbol{A} - \lambda\boldsymbol{I})$ is called as the eigen polynomial.

2.3.2 Eigenvalue and Eigenvector

In the following, we list some major properties about the eigenvalues and eigenvector of a matrix $\boldsymbol{A}$.

Several important terms about the eigenvalues and eigenvectors [6]:

(1) The eigenvalue λ of a matrix $\boldsymbol{A}$ is called as having algebraic multiplicity μ, if λ is a μ-repeated root of the eigen equation $\det(\boldsymbol{A} - \lambda\boldsymbol{I}) = \boldsymbol{0}$.
(2) If the algebraic multiplicity of eigenvalue λ is equal to one, the eigenvalue is called as single eigenvalue. Non-single eigenvalues are called as multiple eigenvalues.
(3) The eigenvalue λ of a matrix $\boldsymbol{A}$ is called as having geometric multiplicity γ, if the number of linear independent eigenvectors associated with λ is equal to γ.

(4) An eigenvalue is called half-single eigenvalue if its algebraic multiplicity is equal to geometric multiplicity. Not half-single eigenvalues are called as wane eigenvalues.
(5) If matrix $\boldsymbol{A}_{n\times n}$ is a general complex matrix and λ is its eigenvalue, the vector $\boldsymbol{v}$ which meets $\boldsymbol{Av} = \lambda\boldsymbol{v}$ is called as the right eigenvector associated with the eigenvalue λ, and the eigenvector $\boldsymbol{u}$ which meets $\boldsymbol{u}^H\boldsymbol{A} = \lambda\boldsymbol{u}^H$ is called as the left eigenvector associated with the eigenvalue λ. If $\boldsymbol{A}$ is Hermitian matrix and all its eigenvalues are real number, then it holds that $\boldsymbol{v} = \boldsymbol{u}$, that is to say, the left and right eigenvectors of a Hermitian matrix are the same.

Some important properties can be summarized as follows:

(1) Matrix $\boldsymbol{A}$ $(\in \Re^{n\times n})$ has n eigenvalues, of which the multiple eigenvalues are computed according to their multiplicity.
(2) If $\boldsymbol{A}$ is a real symmetrical matrix or Hermitian matrix, all its eigenvalues are real numbers.
(3) If $\boldsymbol{A} = \mathrm{diag}(a_{11}, a_{22},\ldots, a_{\mathrm{nn}})$, its eigenvalues are $a_{11}, a_{22},\ldots, a_{\mathrm{nn}}$; If $\boldsymbol{A}$ is a trigonal matrix, its diagonal elements are all its eigenvalues.
(4) For $\boldsymbol{A}$ $(\in \Re^{n\times n})$, if λ is the eigenvalue of matrix $\boldsymbol{A}$, λ is also the eigenvalue of matrix $\boldsymbol{A}^{\mathrm{T}}$. If λ is the eigenvalue of matrix $\boldsymbol{A}$, λ^* is the eigenvalue of matrix $\boldsymbol{A}^H$. If λ is the eigenvalue of matrix $\boldsymbol{A}$, $\lambda + \sigma^2$ is the eigenvalue of matrix $\boldsymbol{A} + \sigma^2\boldsymbol{I}$. If λ is the eigenvalue of matrix $\boldsymbol{A}$, $1/\lambda$ is the eigenvalue of matrix $\boldsymbol{A}^{-1}$.
(5) All eigenvalues of matrix $\boldsymbol{A}^2 = \boldsymbol{A}$ are either 0 or 1.
(6) If $\boldsymbol{A}$ is a real orthogonal matrix, all its eigenvalues are on the unit circle.
(7) If a matrix is singular, at least one of its eigenvalues is equal to zero.
(8) The sum of all the eigenvalues is equal to its trace, namely $\sum_{i=1}^{n} \lambda_i = \mathrm{tr}(\boldsymbol{A})$.
(9) The nonzero eigenvectors $\boldsymbol{u}_1, \boldsymbol{u}_2, \ldots, \boldsymbol{u}_n$ associated with different eigenvalues $\lambda_1, \lambda_2, \ldots \lambda_n$ are linearly independent.
(10) If matrix $\boldsymbol{A}$ $(\in \Re^{n\times n})$ has r nonzero eigenvalues, then it holds that $\mathrm{rank}(\boldsymbol{A}) \geq r$; If zero is a non-multiple eigenvalue, then $\mathrm{rank}(\boldsymbol{A}) \geq n - 1$; If $\mathrm{rank}(\boldsymbol{A} - \lambda\boldsymbol{I}) \geq n - 1$, then λ is an eigenvalue of matrix $\boldsymbol{A}$.
(11) The product of all eigenvalues of matrix $\boldsymbol{A}$ is equal to the determinant of matrix $\boldsymbol{A}$, namely $\prod_{i=1}^{n} \lambda_i = \det(\boldsymbol{A}) = |\boldsymbol{A}|$.
(12) A Hermitian matrix $\boldsymbol{A}$ is positive definite (or positive semi-definite), if and only if all its eigenvalues are positive (or non-negative).
(13) If the eigenvalues of matrix $\boldsymbol{A}$ are different, then one can find a similar matrix such that $\boldsymbol{S}^{-1}\boldsymbol{AS} = \boldsymbol{D}$(diagonal matrix) and the diagonal elements of $\boldsymbol{D}$ are the eigenvalues of matrix $\boldsymbol{A}$.
(14) (Cayley–Hamilton Theorem) : If $\lambda_1, \lambda_2, \ldots \lambda_n$ are the eigenvalues of an $n \times n$ matrix $\boldsymbol{A}$, then $\prod_{i=1}^{n} (\boldsymbol{A} - \lambda_i\boldsymbol{I}) = 0$.

(15) It is not possible that the geometric multiplicity of any eigenvalue λ of an $n \times n$ matrix $\boldsymbol{A}$ is larger than its algebraic multiplicity.

(16) If λ is an eigenvalue of an $n \times n$ matrix $\boldsymbol{A}$ and an $n \times n$ matrix $\boldsymbol{B}$ is not singular, then λ is also an eigenvalue of $\boldsymbol{B}^{-1}\boldsymbol{AB}$. However, the corresponding eigenvectors are usually different. If λ is an eigenvalue of an $n \times n$ matrix $\boldsymbol{A}$ and an $n \times n$ matrix $\boldsymbol{B}$ is a unitary matrix, then λ is also an eigenvalue of $\boldsymbol{B}^H\boldsymbol{AB}$. However, the corresponding eigenvectors are usually different. If λ is an eigenvalue of an $n \times n$ matrix $\boldsymbol{A}$ and an $n \times n$ matrix $\boldsymbol{B}$ is a orthogonal matrix, then λ is also an eigenvalue of $\boldsymbol{B}^T\boldsymbol{AB}$. However, the corresponding eigenvectors are usually different.

(17) The largest eigenvalue of an $n \times n$ matrix $\boldsymbol{A} = [a_{ij}]$ is less than or equal to the maximal of the sum of all the column elements of this matrix, namely $\lambda_{\max} \leq \max\limits_i \sum\limits_{j=1}^{n} a_{ij}$.

(18) The eigenvalues of autocorrelation matrix $\boldsymbol{R} = E\{\boldsymbol{x}(t)\boldsymbol{x}^H(t)\}$ of stochastic vector $\boldsymbol{x}(t) = [x_1(t), x_2(t), \ldots x_n(t)]^{\mathrm{T}}$ is within the maximal power of signal $P_{\max} = \max\limits_i E\left\{|x_i(t)|^2\right\}$ and its minimal power $P_{\min} = \min\limits_i E\left\{|x_i(t)|^2\right\}$, namely $P_{\min} \leq \lambda_i \leq P_{\max}$.

(19) The spread of eigenvalues in autocorrelation matrix $\boldsymbol{R}$ of a stochastic vector $\boldsymbol{x}(t)$ is $x(\boldsymbol{R}) = \lambda_{\max}/\lambda_{\min}$.

(20) If $|\lambda_i| < 1, i = 1, 2, \ldots, n$, the matrix $\boldsymbol{A} \pm \boldsymbol{I}_n$ is nonsingular. $|\lambda_i| < 1, i = 1, 2, \ldots, n$, is equivalent to the case in which the roots of $\det(\boldsymbol{A} - z\boldsymbol{I}_n) = 0$ is not on or at the interior of the unit circle.

(21) For $m \times n (n \geq m)$ matrix A and $n \times m$ matrix B, if λ is an eigenvalue of the product $\boldsymbol{AB}$, then λ is also an eigenvalue of the product $\boldsymbol{BA}$. If $\lambda \neq 0$ is an eigenvalue of the product $\boldsymbol{BA}$, then λ is also an eigenvalue of the product $\boldsymbol{AB}$. If $\lambda_1, \lambda_1, \ldots \lambda_m$ are eigenvalues of the product $\boldsymbol{AB}$, then the eigenvalues of matrix product $\boldsymbol{BA}$ are $\lambda_1, \lambda_2, \ldots \lambda_m, 0, \ldots, 0$.

(22) If the eigenvalue of matrix $\boldsymbol{A}$ is λ, then the eigenvalue of matrix polynomial $f(\boldsymbol{A}) = \boldsymbol{A}^n + c_1\boldsymbol{A}^{n-1} + \cdots + c_{n-1}\boldsymbol{A} + c_n\boldsymbol{I}$ is $f(\lambda) = \lambda^n + c_1\lambda^{n-1} + \cdots + c_{n-1}\lambda + c_n$.

(23) If λ is an eigenvalue of matrix $\boldsymbol{A}$, then the eigenvalue of matrix exponential function e^{A} is e^{λ}

Properties of an eigen pair which consists of an eigenvalue λ and its associated eigenvector $\boldsymbol{u}$ can be summarized as follows:

(1) If $(\lambda, \boldsymbol{u})$ is an eigen pair of matrix $\boldsymbol{A}$, then $(c\lambda, \boldsymbol{u})$ is an eigen pair of matrix $c\boldsymbol{A}$, where c is a nonzero constant.

(2) If $(\lambda, \boldsymbol{u})$ is an eigen pair of matrix $\boldsymbol{A}$, then $(\lambda, c\boldsymbol{u})$ is an eigen pair of matrix $\boldsymbol{A}$, where c is a nonzero constant.

(3) If $(\lambda_i, \boldsymbol{u}_i)$ and $(\lambda_j, \boldsymbol{u}_j)$ are eigen pairs of matrix $\boldsymbol{A}$ and $\lambda_i \neq \lambda_j$, then the eigenvector $\boldsymbol{u}_i$ and $\boldsymbol{u}_j$ are linearly independent.

(4) The eigenvectors of an Hermitian matrix associated with different eigenvalues are mutual orthogonal to each other, namely $\lambda_i \neq \lambda_j \Rightarrow \boldsymbol{u}_i^H \boldsymbol{u}_j = 0$.

(5) If λ is an eigenvalue of matrix $\boldsymbol{A}$ and the vectors $\boldsymbol{u}_1$ and $\boldsymbol{u}_2$ are the eigenvectors associated with λ, then $c_1\boldsymbol{u}_1 + c_2\boldsymbol{u}_2$ is also an eigenvector of matrix $\boldsymbol{A}$ associated with the eigenvalue λ, in which c_1 and $\boldsymbol{u}_2$ are constants and at least one of them is not zero.

(6) If $(\lambda, \boldsymbol{u})$ is an eigen pair of matrix $\boldsymbol{A}$ and $\alpha_1, \alpha_2, \ldots, \alpha_p$ are complex constants, then $f(\lambda) = \alpha_0 + \alpha_1\lambda + \cdots + \alpha_p\lambda^p$ is the eigenvalue of matrix polynomial $f(\boldsymbol{A}) = \alpha_0\boldsymbol{I} + \alpha_1\boldsymbol{A} + \cdots + \alpha_p\boldsymbol{A}^p$, and the associated eigenvector is still $\boldsymbol{u}$.

(7) If $(\lambda, \boldsymbol{u})$ is an eigen pair of matrix $\boldsymbol{A}$, then $(\lambda^k, \boldsymbol{u})$ is an eigen pair of matrix $\boldsymbol{A}^k$.

(8) If $(\lambda, \boldsymbol{u})$ is an eigen pair of matrix $\boldsymbol{A}$, then $(\mathrm{e}^{\lambda}, \boldsymbol{u})$ is an eigen pair of matrix exponential function $\mathrm{e}^{\boldsymbol{A}}$.

(9) If $\lambda(\boldsymbol{A})$ and $\lambda(\boldsymbol{B})$ are eigenvalues of matrices $\boldsymbol{A}$ and $\boldsymbol{B}$, respectively, and $\boldsymbol{u}(\boldsymbol{A})$ and $\boldsymbol{u}(\boldsymbol{B})$ are their associated eigenvectors, then $\lambda(\boldsymbol{A})\lambda(\boldsymbol{B})$ is an eigenvalue of matrix Kronecker product $\boldsymbol{A} \otimes \boldsymbol{B}$ with $\boldsymbol{u}(\boldsymbol{A}) \otimes \boldsymbol{u}(\boldsymbol{B})$ being the associated eigenvector, and $\lambda(\boldsymbol{A})$ and $\lambda(\boldsymbol{B})$ are the eigenvalues of matrix direct sum $\boldsymbol{A} \oplus \boldsymbol{B}$ with $\begin{bmatrix} \boldsymbol{u}(\boldsymbol{A}) \\ \boldsymbol{0} \end{bmatrix}$ and $\begin{bmatrix} \boldsymbol{0} \\ \boldsymbol{u}(\boldsymbol{B}) \end{bmatrix}$ being the associated eigenvectors, respectively.

(10) If an $n \times n$ matrix $\boldsymbol{A}$ has n linearly independent eigenvectors, then its ED is $\boldsymbol{A} = \boldsymbol{U}\boldsymbol{\Sigma}\boldsymbol{U}^{-1}$, where the $n \times n$ eigen matrix $\boldsymbol{U}$ consists of n eigenvectors of matrix $\boldsymbol{A}$, and the diagonal elements of the $n \times n$ diagonal matrix $\boldsymbol{\Sigma}$ are the eigenvalues of matrix $\boldsymbol{A}$.

The SVD problem of a matrix $\boldsymbol{A}$ can be transformed into its ED problem to solve, and there are two methods to realize this.

Method 2.1 The nonzero singular values of matrix $\boldsymbol{A}_{m\times n}$ are the positive square root of nonzero eigenvalue λ_i of $m \times m$ matrix $\boldsymbol{A}\boldsymbol{A}^{\mathrm{T}}$ or $n \times n$ matrix $\boldsymbol{A}^{\mathrm{T}}\boldsymbol{A}$, and the left singular vector $\boldsymbol{u}_j$ and right singular vector $\boldsymbol{v}_i$ of matrix $\boldsymbol{A}$ associated with σ_i are the eigenvectors of matrix $\boldsymbol{A}\boldsymbol{A}^{\mathrm{T}}$ and $\boldsymbol{A}^{\mathrm{T}}\boldsymbol{A}$ associated with nonzero eigenvalue λ_i, respectively.

Method 2.2 The SVD of matrix $\boldsymbol{A}_{m\times n}$ can be transformed into the ED of $(m+n) \times (m+n)$ augmented matrix $\begin{bmatrix} \boldsymbol{O} & \boldsymbol{A} \\ \boldsymbol{A}^{\mathrm{T}} & \boldsymbol{O} \end{bmatrix}$.

The following theorem holds for the eigenvalues of matrix sum $\boldsymbol{A} + \boldsymbol{B}$.

Theorem 2.4 (Wely theorem): *Suppose that* $\boldsymbol{A}, \boldsymbol{B} \in \mathbb{C}^{m\times n}$ *are Hermitian matrices, and their eigenvalues are arranged as an increasing order, namely,*

$$\lambda_1(\boldsymbol{A}) \le \lambda_2(\boldsymbol{A}) \cdots \le \lambda_n(\boldsymbol{A}),$$

$$\lambda_1(\boldsymbol{B}) \le \lambda_2(\boldsymbol{B}) \cdots \le \lambda_n(\boldsymbol{B}),$$

$$\lambda_1(\boldsymbol{A}+\boldsymbol{B}) \le \lambda_2(\boldsymbol{A}+\boldsymbol{B}) \le \cdots \le \lambda_n(\boldsymbol{A}+\boldsymbol{B}),$$

Then,

$$\lambda_i(\boldsymbol{A}+\boldsymbol{B}) \ge \begin{cases} \lambda_i(\boldsymbol{A}) + \lambda_1(\boldsymbol{B}) \\ \lambda_{i-1}(\boldsymbol{A}) + \lambda_2(\boldsymbol{B}) \\ \quad\vdots \\ \lambda_1(\boldsymbol{A}) + \lambda_i(\boldsymbol{B}) \end{cases} \tag{2.22}$$

and

$$\lambda_i(\boldsymbol{A}+\boldsymbol{B}) \le \begin{cases} \lambda_i(\boldsymbol{A}) + \lambda_n(\boldsymbol{B}) \\ \lambda_{i+1}(\boldsymbol{A}) + \lambda_{n-1}(\boldsymbol{B}) \\ \quad\vdots \\ \lambda_n(\boldsymbol{A}) + \lambda_i(\boldsymbol{B}). \end{cases} \tag{2.23}$$

where $i = 1, 2, \ldots u$.

Especially, when $\boldsymbol{A}$ *is a real symmetric matrix, and* $\boldsymbol{B} = azz^T$, *the interlace theorem in the following holds.*

Theorem 2.5 (Interlacing eigenvalue theorem): *Suppose that* $\boldsymbol{A} \in \Re^{n\times n}$ *is a symmetric matrix, and its eigenvalues* $\lambda_1, \lambda_2, \ldots, \lambda_n$, *meet* $\lambda_1 \ge \lambda_2 \ge \cdots \ge \lambda_n$, *and let* $z \in \Re^n$ *be a vector satisfying* $\|z\| = 1$. *Suppose that* a *is a real number and the eigenvalues of matrix* $\boldsymbol{A} + azz^{\mathrm{T}}$ meet $\zeta_1 \ge \zeta_2 \ge \cdots \ge \zeta_n$, *then it holds that*

$$\zeta_1 \ge \lambda_1 \ge \zeta_2 \ge \lambda_2 \ge \cdots \ge \zeta_n \ge \lambda_n, \quad a > 0 \tag{2.24}$$

or

$$\lambda_1 \ge \zeta_1 \ge \lambda_2 \ge \zeta_2 \ge \cdots \ge \lambda_n \ge \zeta_n, \quad a<0 \tag{2.25}$$

and whether $a > 0$ or $a<0$, *it holds that*

$$\sum_{i=1}^{n} (\zeta_i - \lambda_i) = a. \tag{2.26}$$

2.3.3 Eigenvalue Decomposition of Hermitian Matrix

All the discussions on eigenvalues and eigenvectors in the above hold for general matrices, and they do not require the matrices to be real symmetric or complex conjugate symmetric. However, in the statistical and information science, one usually encounter real symmetric or Hermitian (complex conjugate symmetric) matrices. For example, the autocorrelation matrix of a real measurement data vector $\boldsymbol{R} = E\{\boldsymbol{x}(t)\boldsymbol{x}^T(t)\}$ is real symmetric, while the autocorrelation matrix of a complex measurement data vector $\boldsymbol{R} = E\{\boldsymbol{x}(t)\boldsymbol{x}^H(t)\}$ is Hermitian. On the other hand, since a real symmetric matrix is a special case of Hermitian matrix and the eigenvalues and eigenvectors of a Hermitian matrix have a series of important properties, and it is necessary to discuss individually the eigen analysis of Hermitian matrix.

1. Eigenvalue and Eigenvector of Hermitian matrix.
   Some important properties of eigenvalues and eigenvectors of Hermitian matrices can be summarized as follows:

   (1) The eigenvalues of an Hermitian matrix $\boldsymbol{A}$ must be a real number.
   (2) Let $(\lambda, \boldsymbol{u})$ be an eigen pair of an Hermitian matrix $\boldsymbol{A}$. If $\boldsymbol{A}$ is invertible, then $(1/\lambda, \boldsymbol{u})$ is an eigen pair of matrix $\boldsymbol{A}^{-1}$.
   (3) If λ_k is a multiple eigenvalue of Hermitian matrix $\boldsymbol{A}^H = \boldsymbol{A}$, and its multiplicity is m_k, then $\operatorname{rank}(\boldsymbol{A} - \lambda_k\boldsymbol{I}) = n - m_k$.
   (4) Any Hermitian matrix $\boldsymbol{A}$ is diagonalizable, namely $\boldsymbol{U}^{-1}\boldsymbol{A}\boldsymbol{U} = \boldsymbol{\Sigma}$.
   (5) All the eigenvectors of an Hermitian matrix are linearly independent, and they are mutual orthogonal, namely the eigen matrix $\boldsymbol{U} = [\boldsymbol{u}_1, \boldsymbol{u}_2, \ldots, \boldsymbol{u}_n]$ is a unitary matrix and it meets $\boldsymbol{U}^{-1} = \boldsymbol{U}^H$.
   (6) From property (5), it holds that $\boldsymbol{U}^H\boldsymbol{A}\boldsymbol{U} = \boldsymbol{\Sigma} = \operatorname{diag}(\lambda_1, \lambda_2, \ldots, \lambda_n)$ or $\boldsymbol{A} = \boldsymbol{U}\boldsymbol{\Sigma}\boldsymbol{U}^H$, which can be rewritten as: $\boldsymbol{A} = \sum_{i=1}^{n} \lambda_i \boldsymbol{u}_i \boldsymbol{u}_i^H$. This is called the spectral decomposition of a Hermitian matrix.
   (7) The spread formula of the inverse of an Hermitian matrix $\boldsymbol{A}$ is

$$\boldsymbol{A}^{-1} = \sum_{i=1}^{n} \frac{\mathrm{I}}{\lambda_i} \boldsymbol{u}_i \boldsymbol{u}_i^H \tag{2.27}$$

   Thus, if one know the eigen decomposition of an Hermitian matrix $\boldsymbol{A}$, then one can directly obtain the inverse matrix $\boldsymbol{A}^{-1}$ using the above formula.
   (8) For two $n \times n$ Hermitian matrices $\boldsymbol{A}$ and $\boldsymbol{B}$, there exists a unitary matrix so that $\boldsymbol{P}^H\boldsymbol{A}\boldsymbol{P}$ and $\boldsymbol{P}^H\boldsymbol{B}\boldsymbol{P}$ are both diagonal if and only if $\boldsymbol{A}\boldsymbol{B} = \boldsymbol{B}\boldsymbol{A}$.
   (9) For two $n \times n$ non-negative definite Hermitian matrices $\boldsymbol{A}$ and $\boldsymbol{B}$, there exists a nonsingular matrix $\boldsymbol{P}$ so that $\boldsymbol{P}^H\boldsymbol{A}\boldsymbol{P}$ and $\boldsymbol{P}^H\boldsymbol{B}\boldsymbol{P}$ are both diagonal.

2. Some properties of Hermitian matrix.
   The ED of an Hermitian matrix $\boldsymbol{A}$ can be written as $\boldsymbol{A} = \boldsymbol{U\Sigma U}^H$, where $\boldsymbol{U}$ is a unitary matrix and it meets $\boldsymbol{U}^H\boldsymbol{U} = \boldsymbol{UU}^H = \boldsymbol{I}$.
   From the property of determinant and trace of a matrix, it holds that

$$\mathrm{tr}(\boldsymbol{A}) = \mathrm{tr}\left(\boldsymbol{U\Sigma U}^H\right) = \mathrm{tr}\left(\boldsymbol{U}^H\boldsymbol{U\Sigma}\right) = \mathrm{tr}(\boldsymbol{\Sigma}) = \sum_{i=1}^{n} \lambda_i, \tag{2.28}$$

$$\det(\boldsymbol{A}) = \det(\boldsymbol{U})\det(\boldsymbol{\Sigma})\det\left(\boldsymbol{U}^H\right) = \prod_{i=1}^{n} \lambda_i. \tag{2.29}$$

   For a positive definite Hermitian matrix $\boldsymbol{A}$, its inverse $\boldsymbol{A}^{-1}$ exists and can be written as

$$\boldsymbol{A}^{-1} = \boldsymbol{U}\mathrm{diag}\left(\lambda_1^{-1}, \lambda_2^{-1}, \ldots, \lambda_u^{-1}\right)\boldsymbol{U}^H. \tag{2.30}$$

   Let z_A be the number of zero eigenvalues of matrix $\boldsymbol{A} \in \mathbb{C}^{n\times n}$, then

$$\mathrm{rank}(\boldsymbol{A}) = n - z_n, \tag{2.31}$$

   That is to say, the rank of a Hermitian matrix is equal to the number of its nonzero eigenvalues.
3. Solving for maximal or minimal eigenvalue of Hermitian matrix.

In signal processing, one usually needs to compute the maximal or minimal eigenvalue of a Hermitian matrix $\boldsymbol{A}$. The power iteration method is a method for such purposes.

Select some initial vector $x(0)$, and iteratively repeat the following linear equation

$$\boldsymbol{y}(k+1) = \boldsymbol{Ax}(k) \tag{2.32}$$

to obtain $\boldsymbol{y}(k+1)$, then normalize it. It holds that

$$\boldsymbol{x}(k+1) = \frac{\boldsymbol{y}(k+1)}{\sigma_{k+1}}, \tag{2.33}$$

$$\sigma_{k+1} = \boldsymbol{y}^H(k+1)\boldsymbol{y}(k+1). \tag{2.34}$$

The iterative procedure continues until the vector $\boldsymbol{x}_k$ converges. The σ_k obtained at the last iteration is the maximal eigenvalue, and the $\boldsymbol{x}_k$ is its associated eigenvector. Only if the initial vector $\boldsymbol{x}(0)$ is not orthogonal to the eigenvector associated with the maximal eigenvalue, the convergence can be guaranteed.

If one needs to compute the minimal eigenvalue and its associated eigenvector, use $\boldsymbol{y}(k+1) = \boldsymbol{A}^{-1}\boldsymbol{x}(k)$, i.e., the iterative linear equation is $\boldsymbol{Ay}(k+1) = \boldsymbol{x}(k)$.

By combining the power iteration method and shrink mapping method, one can compute all eigenvalues and the associated eigenvectors of a Hermitian matrix $\boldsymbol{A}$. Suppose that one has obtained some eigenvalue σ using the power iteration method. The first step corresponds to the first maximal eigenvalue and uses the shrink mapping method to eliminate the eigenvalue. Then matrix $\boldsymbol{A}_k$ $(\text{rank}\boldsymbol{A}_k = k)$ is changed into matrix $\boldsymbol{A}_{k-1}(\text{rank}\boldsymbol{A}_{k-1} = k-1)$. Thus, the maximal eigenvalue of matrix $\boldsymbol{A}_{k-1}$ is the residual maximal eigenvalue of matrix $\boldsymbol{A}_k$, which is smaller than σ. It should be noted that the kth step corresponds to the kth maximal eigenvalue. New matrix can be obtained by using the above idea and the following spectral decomposition formula:

$$\left(\boldsymbol{A}_k - \sigma \boldsymbol{x}\boldsymbol{x}^H\right) = \boldsymbol{A}_{k-1}.$$

Repeat the above procedure, one can compute all eigenvalues of matrix $\boldsymbol{A}$ in turn.

2.3.4 Generalized Eigenvalue Decomposition

Let $\boldsymbol{A}$ and B both be $n \times n$ square matrices, and they constitute a matrix pencil or matrix pair, written as $(\boldsymbol{A}, \boldsymbol{B})$. Now we consider the following generalized eigenvalue problem. That is, to compute all scalar λ such that

$$\boldsymbol{A}\boldsymbol{u} = \lambda \boldsymbol{B}\boldsymbol{u} \tag{2.35}$$

has nonzero solution $\boldsymbol{u} \neq 0$, where the scalar λ and the nonzero vector $\boldsymbol{u}$ are called the generalized eigenvalue and the generalized eigenvector of matrix pencil $(\boldsymbol{A}, \boldsymbol{B})$, respectively. A generalized eigenvalue and its associated generalized eigenvector are called generalized eigen pair, written as $(\lambda, \boldsymbol{u})$. Equation (2.35) is also called the generalized eigen equation. It is obvious that the eigenvalue problem is a special case when the matrix pencil is chosen as $(\boldsymbol{A}, \boldsymbol{I})$.

Theorem 2.6 *$\lambda \in \mathbb{C}$ and $\mathbf{u} \in \mathbb{C}^n$ are respectively the generalized eigenvalue and the associated generalized eigenvector of matrix pencil $(\boldsymbol{A}, \boldsymbol{B})_{n\times n}$ if and only if:*

(1) $\det(\boldsymbol{A} - \lambda \boldsymbol{B}) = 0$.
(2) $\boldsymbol{u} \in \text{Null}(\boldsymbol{A} - \lambda \boldsymbol{B})$, and $\boldsymbol{u} \neq 0$.

In the natural science, sometimes it is necessary to discuss the eigenvalue problem of the generalized matrix pencil.

Suppose that $n \times n$ square matrices $\mathbf{A}$ and $\mathbf{B}$ are both Hermitian, and B is positive definite. Then $(\mathbf{A}, \mathbf{B})$ is called the regularized matrix pencil.

The eigenvalue problem of regularized matrix pencil is similar to the one of Hermitian matrix.

Theorem 2.7 *If $\lambda_1, \lambda_2, \ldots, \lambda_n$ are the generalized eigenvalues of a regularized matrix pencil ($\boldsymbol{A}$, $\boldsymbol{B}$), then*

(1) *there exists a matrix $\boldsymbol{X} \in \mathbb{C}^{n \times n}$, so that*

$$\boldsymbol{XBX}^H = \boldsymbol{I}_n, \quad \boldsymbol{XAX}^H = \mathrm{diag}(\lambda_1, \lambda_2, \ldots, \lambda_n),$$

or equivalently

$$\boldsymbol{X}^H\boldsymbol{BX} = \boldsymbol{I}_n, \quad \boldsymbol{AX} = \boldsymbol{BX\Lambda},$$

where $\boldsymbol{\Lambda} = \mathrm{diag}(\lambda_1, \lambda_2, \cdots, \lambda_n)$.

(2) *all generalized eigenvalues are real numbers, i.e., $\lambda_i \in \Re, i = 1, 2, \ldots, n$.*

(3) *Denote $\boldsymbol{X} = [\boldsymbol{x}_1, \boldsymbol{x}_2, \ldots, \boldsymbol{x}_n]$. Then it holds that*

$$\boldsymbol{Ax}_i = \lambda_i \boldsymbol{Bx}_i, \quad i = 1, 2, \ldots, n.$$

$$\boldsymbol{x}_i^H \boldsymbol{Bx}_j = \delta_{ij}, \quad i, j = 1, 2, \ldots, n.$$

where δ_{ij} is the Kronecker δ function.

Some properties of the generalized eigenvalue problem $\boldsymbol{Ax} = \lambda\boldsymbol{Bx}$ can be summarized as follows, see [7, pp. 176–177]:

(1) If we interchange matrices $\boldsymbol{A}$ and $\boldsymbol{B}$, then the generalized eigenvalue will be its reciprocal. However, the generalized eigenvector retain unaltered, i.e.,

$$\boldsymbol{Ax} = \lambda\boldsymbol{Bx} \quad \Rightarrow \quad \boldsymbol{Bx} = \frac{1}{\lambda}\boldsymbol{Ax}.$$

(2) If matrix $\boldsymbol{B}$ is nonsingular, then the generalized ED will be simplified to the standard ED

$$\boldsymbol{Ax} = \lambda\boldsymbol{Bx} \quad \Rightarrow \quad \left(\boldsymbol{B}^{-1}\boldsymbol{A}\right)\boldsymbol{x} = \lambda\boldsymbol{x}.$$

(3) If matrices $\boldsymbol{A}$ and $\boldsymbol{B}$ are both positive definite and Hermitian, then the generalized eigenvalues must be real numbers, and the generalized eigenvectors associated with different generalized values are orthogonal with respect to the positive definite matrices $\boldsymbol{A}$ and $\boldsymbol{B}$, i.e.,

$$\boldsymbol{x}_i^H = \boldsymbol{Ax}_j = \boldsymbol{x}_i^H\boldsymbol{Bx}_j = 0.$$

(4) If $\boldsymbol{A}$ and $\boldsymbol{B}$ are real symmetrical matrices, and $\boldsymbol{B}$ is positive definite, then the generalized eigenvalue problem $\boldsymbol{Ax} = \lambda\boldsymbol{Bx}$ can be changed into the standard eigenvalue problem,

$$(L^{-1}AL^{-T})(L^{T}x) = \lambda(L^{T}x),$$

where $\boldsymbol{L}$ is a lower triangular matrix, which is the factor of Cholesky Decomposition $\boldsymbol{B} = \boldsymbol{L}\boldsymbol{L}^{\mathrm{T}}$.

(5) If $\boldsymbol{A}$ and $\boldsymbol{B}$ are real symmetrical and positive definite matrices, then the generalized eigenvalues must be positive.

(6) If $\boldsymbol{A}$ is singular, then $\lambda = 0$ must be a generalized eigenvalue.

(7) If $\tilde{\boldsymbol{B}} = \boldsymbol{B} + (1/\alpha)\boldsymbol{A}$, where α is a nonzero scalar, then the following relationship holds between the generalized eigenvalue $\tilde{\lambda}$ of the modified generalized value problem $\boldsymbol{A}\boldsymbol{x} = \tilde{\lambda}\tilde{\boldsymbol{B}}\boldsymbol{x}$ and the original generalized eigenvalue λ, i.e.,

$$\frac{1}{\tilde{\lambda}} = \frac{1}{\lambda} + \frac{1}{\alpha}.$$

In the following, we introduce a few generalized ED algorithms for matrix pencil.

We know that if $n \times n$ square matrices $\boldsymbol{A}$ and $\boldsymbol{B}$ are both Hermitian, and $\boldsymbol{B}$ is positive definite, then the generalized ED Eq. (2.35) can be equivalently written as

$$\boldsymbol{B}^{-1}\boldsymbol{A}\boldsymbol{u} = \lambda\boldsymbol{u}, \tag{2.36}$$

That is to say, the generalized ED becomes the standard ED of a Hermitian matrix.

The following algorithm uses the shrink mapping to compute the generalized eigen pair $(\lambda, \boldsymbol{u})$ of an $n \times n$ real symmetrical matrix pencil ($\boldsymbol{A}$, $\boldsymbol{B}$).

Algorithm 2.1 Lanczos algorithm for generalized ED [8, p. 298].

Step 1 Initialization

Select vector $\boldsymbol{u}_1$ whose norm meets $\boldsymbol{u}_1^H\boldsymbol{B}\boldsymbol{u}_1 = 1$, and let $\alpha_1 = 0, z_0 = \boldsymbol{u}_0 = 0, z_1 = \boldsymbol{B}\boldsymbol{u}_1$.

Step 2 For $i = 1, 2, \ldots, n$, compute

$$\boldsymbol{u} = \boldsymbol{A}\boldsymbol{u}_i - \alpha_i z_{i-1}$$

$$\beta_i = \langle \boldsymbol{u}, \boldsymbol{u}_i \rangle$$

$$\boldsymbol{u} = \boldsymbol{u} - \beta_i z_i$$

$$\boldsymbol{w} = \boldsymbol{B}^{-1}\boldsymbol{u}$$

$$\alpha_{i+1} = \sqrt{\langle \boldsymbol{w}, \boldsymbol{u} \rangle}$$

$$\boldsymbol{u}_{i+1} = \boldsymbol{w}/\alpha_{i+1}$$

$$z_{i+1} = \boldsymbol{u}/\alpha_{i+1}$$

$$\lambda_i = \beta_{i+1}/\alpha_{i+1}.$$

The following is the tangent algorithm for generalized ED of a $n \times n$ symmetric positive definite matrix pencil (***A***, ***B***), which was proposed by Dramc in 1998 [9].

Algorithm 2.2 Generalized ED of symmetric positive definite matrix pencil.

Step 1 Compute $\varDelta_A = \mathrm{diag}(A_{11}, A_{22}, \ldots, A_{nn})^{-1/2}, A_5 = \varDelta_A A \varDelta_A$ and $\mathbf{B}_1 = \varDelta_A \mathbf{B} \varDelta_A$,
Step 2 Compute Cholesky Decomposition $\boldsymbol{R}_A^{\mathrm{T}} \boldsymbol{R}_A = \boldsymbol{A}_S$ and $\boldsymbol{R}_B^{\mathrm{T}} \boldsymbol{R}_B = \boldsymbol{\Pi}^{\mathrm{T}} \boldsymbol{B}_1 \boldsymbol{\Pi}$.
Step 3 By solving the matrix equation $\boldsymbol{F}\boldsymbol{R}_B = \boldsymbol{A}\boldsymbol{\Pi}$, compute $\boldsymbol{F} = \boldsymbol{A}\boldsymbol{\Pi}\boldsymbol{R}_B^{-1}$.
Step 4 Conduct the SVD $\boldsymbol{\Sigma} = \boldsymbol{V}\boldsymbol{F}\boldsymbol{U}^{\mathrm{T}}$.
Step 5 Compute $\boldsymbol{X} = \boldsymbol{\Delta}_A \boldsymbol{\Pi} \boldsymbol{R}_B^{-1} \boldsymbol{U}$.

Output: Matrix $\boldsymbol{X}$ and $\boldsymbol{\Sigma}$, which meets $\boldsymbol{A}\boldsymbol{X} = \boldsymbol{B}\boldsymbol{X}\boldsymbol{\Sigma}^2$.

When matrix B is singular, the above algorithms will be unstable. The generalized ED algorithm of matrix pencil (***A***, ***B***) under this condition was proposed by Nour-Omid et al. [10], whose main ideas is to make $(\boldsymbol{A} - \sigma\boldsymbol{B})$ nonsingular by introducing a shift factor.

Algorithm 2.3 Generalized ED when matrix ***B*** is singular [8, 10], p. 299].

Step 1 Initialization
Select the basis vector $\boldsymbol{w}$ of $\mathrm{Range}[(\boldsymbol{A} - \sigma\boldsymbol{B})^{-1}\boldsymbol{B}]$., compute $z_1 = \boldsymbol{B}\boldsymbol{w}, \alpha_1 = \sqrt{\langle \boldsymbol{w}, z_1 \rangle}$. Let $\boldsymbol{u}_0 = 0$.
Step 2 For $i = 1, 2, \ldots, n$, compute

$$\boldsymbol{u}_i = \boldsymbol{w}/\alpha_i$$

$$z_i = (\boldsymbol{A} - \sigma\boldsymbol{B})^{-1}\boldsymbol{w}$$

$$\boldsymbol{w} = \boldsymbol{w} - \alpha_i \boldsymbol{u}_{i-1}$$

$$\beta_i = \langle \boldsymbol{w}, z_i \rangle$$

$$z_{i+1} = \boldsymbol{B}\boldsymbol{w}$$

$$\alpha_{i+1} = \sqrt{\langle z_{i+1}, \boldsymbol{w} \rangle}$$

2.4 Rayleigh Quotient and Its Characteristics

The quotient of quadratic function of a Hermitian matrix is defined as Rayleigh quotient. As an important quantity in matrix algebra and physics, Rayleigh quotient is a ratio of quadratic functions expressed by eigenvalues and eigenvectors, which has been widely used in many areas such as optimization, signal processing, pattern recognition, and communication.

2.4.1 Rayleigh Quotient

Definition 2.1 The Rayleigh quotient (RQ) of an Hermitian matrix $\boldsymbol{C} \in \mathbb{C}^{n \times n}$ is a scalar, defined as

$$r(\boldsymbol{u}) = r(\boldsymbol{u}, \boldsymbol{C}) = \frac{\boldsymbol{u}^H \boldsymbol{C} \boldsymbol{u}}{\boldsymbol{u}^H \boldsymbol{u}},$$

where $\boldsymbol{u}$ is a quantity to be selected. The objective is to maximize or minimize the Rayleigh quotient.

The most relevant properties of the RQ are can be summarized as follows:

① Homogeneity: $r(\alpha \boldsymbol{u}, \beta \boldsymbol{u}) = \beta r(\boldsymbol{u}, \boldsymbol{C}) \quad \forall \alpha, \beta \neq 0$.

② Translation invariance: $r(\boldsymbol{u}, \boldsymbol{C} - \alpha \boldsymbol{I}) = r(\boldsymbol{u}, \boldsymbol{C}) - \alpha$.

③ Boundedness: Since $\boldsymbol{u}$ ranges over all nonzero vectors, $r(\boldsymbol{u})$ fills a region in the complex plane which is called the field of values of $\boldsymbol{C}$. This region is closed, bounded, and convex. If $\boldsymbol{C} = \boldsymbol{C}^*$ (selfadjoint matrix), the field of values is the real interval bounded by the extreme eigenvalues.

④ Orthogonality: $\boldsymbol{u} \perp (\boldsymbol{C} - r(\boldsymbol{u})\boldsymbol{I})\boldsymbol{u}$.

⑤ Minimal residual: $\forall \boldsymbol{u} \neq 0 \wedge \forall$ scalar μ, $\|(\boldsymbol{C} - r(\boldsymbol{u})\boldsymbol{I})\boldsymbol{u}\| \leq \|(\boldsymbol{C} - \mu \boldsymbol{I})\boldsymbol{u}\|$.

Proposition 2.1 (Stationarity) *Let* $\boldsymbol{C}$ *be a real symmetric n-dimensional matrix with eigenvalues* $\lambda_n \leq \lambda_{n-1} \leq \cdots \lambda_1$ *and associated unit eigenvectors* $z_1, z_2, \ldots, z_n$. *Then it holds that* $\lambda_1 = \max r(\boldsymbol{u}, \boldsymbol{C})$, $\lambda_n = \min r(\boldsymbol{u}, \boldsymbol{C})$. *More generally, the critical points and critical values of* $r(\boldsymbol{u}, \boldsymbol{C})$ *are the eigenvectors and eigenvalues of* $\boldsymbol{C}$.

Proposition 2.2 (Degeneracy): *The RQ critical points are degenerate because at these points the Hessian matrix is not invertible. Then the RQ is not a Morse function in every open subspace of the domain containing a critical point.*

Furthermore, the following important theorems also holds for RQ.

Courant–Fischer Theorem: Let $\boldsymbol{C} \in \mathbb{C}^{n \times n}$ be an Hermitian matrix, and its eigenvalues are $\lambda_1 \geq \lambda_2 \geq \cdots \leq \lambda_n$, then it holds that for $\lambda_k (1 \leq k \leq u)$:

$$\lambda_k = \min_{S, \dim(S)=n-k+1} \max_{\boldsymbol{u} \in S, \boldsymbol{u} \neq 0} \left(\frac{\boldsymbol{u}^H \boldsymbol{C} \boldsymbol{u}}{\boldsymbol{u}^H \boldsymbol{u}} \right).$$

The Courant–Fischer Theorem can also written as

$$\lambda_k = \min_{S, \dim(S)=k} \max_{\boldsymbol{u} \in S, \boldsymbol{u} \neq 0} \left(\frac{\boldsymbol{u}^H \boldsymbol{C} \boldsymbol{u}}{\boldsymbol{u}^H \boldsymbol{u}} \right).$$

2.4.2 Gradient and Conjugate Gradient Algorithm for RQ

If the negative direction of RQ gradient is regarded as the gradient flow of vector $\boldsymbol{x}$, e.g.,

$$\dot{\boldsymbol{x}} = -[\boldsymbol{C} - r(\boldsymbol{x})\boldsymbol{I}]\boldsymbol{x}$$

then vector $\boldsymbol{x}$ can be computed iteratively by the following gradient algorithm:

$$\boldsymbol{x}(k+1) = \boldsymbol{x}(k) + \mu \dot{\boldsymbol{x}} = \boldsymbol{x}(k) - \mu[\boldsymbol{C} - r(\boldsymbol{x})\boldsymbol{I}]\boldsymbol{x}.$$

It is worth noting that the gradient algorithm of RQ has faster convergence speed than the iterative algorithm of standard RQ.

In the following, the conjugate gradient algorithm for RQ will be introduced, where $\boldsymbol{A}$ in the RQ is a real symmetric matrix.

Starting from some initial vector, the conjugate gradient algorithm uses the iterative equation, e.g.,

$$\boldsymbol{x}_{k+1} = \boldsymbol{x}_k + \alpha_k \boldsymbol{P}_k \tag{2.37}$$

to update and approach the eigenvector, associated with the minimal or maximal eigenvalue of a symmetric matrix. The real coefficient α_k is

$$\alpha_k = \pm \frac{1}{2D} \left(-B + \sqrt{B^2 - 4CD} \right), \tag{2.38}$$

where "+" is used in the updating of the eigenvector associated with the minimal eigenvalue, and "−" is used in the updating of the eigenvector associated with the maximal eigenvalue. The formulae for parameters D, B, C in the above equations are

$$\begin{cases} D = P_b(k)P_c(k) - P_a(k)P_d(k) \\ B = P_b(k) - \lambda_k P_d(k) \\ C = P_a(k) - \lambda_k P_c(k) \\ P_a(k) = \boldsymbol{P}_k^{\mathrm{T}} \boldsymbol{A} \boldsymbol{x}_k / \left(\boldsymbol{x}_k^{\mathrm{T}} \boldsymbol{x}_k\right) \\ P_b(k) = \boldsymbol{p}_k^{\mathrm{T}} \boldsymbol{A} \boldsymbol{p}_k / \left(\boldsymbol{x}_k^{\mathrm{T}} \boldsymbol{x}_k\right) \\ P_c(k) = \boldsymbol{p}_k^{\mathrm{T}} \boldsymbol{x}_k / \left(\boldsymbol{x}_k^{\mathrm{T}} \boldsymbol{x}_k\right) \\ P_d(k) = \boldsymbol{p}_k^{\mathrm{T}} \boldsymbol{p}_k / \left(\boldsymbol{x}_k^{\mathrm{T}} \boldsymbol{x}_k\right) \\ \lambda_k = r(\boldsymbol{x}_k) = \boldsymbol{x}_k^{\mathrm{T}} \boldsymbol{A} \boldsymbol{x}_k / \left(\boldsymbol{x}_k^{\mathrm{T}} \boldsymbol{x}_k\right). \end{cases} \quad (2.39)$$

At the $k+1$th iteration, the search direction can be selected as

$$\boldsymbol{p}_{k+1} = \boldsymbol{r}_{k+1} + b(k)\boldsymbol{p}_k, \quad (2.40)$$

where $b(-1) = 0$and $\boldsymbol{r}_{k+1}$ is the residual vector at the $k+1$th iteration. $\boldsymbol{r}_{k+1}$ and $b(k)$ can be computed, respectively, as

$$\boldsymbol{r}_{k+1} = -\frac{1}{2}\nabla_x r(\boldsymbol{x}_{k+1}) = (\lambda_{k+1}\boldsymbol{x}_{k+1} - \boldsymbol{A}\boldsymbol{x}_{k+1}) / \left(\boldsymbol{x}_{k=1}^{\mathrm{T}} \boldsymbol{x}_{k+1}\right) \quad (2.41)$$

and

$$b(k) = -\frac{\boldsymbol{r}_{k+1}^{\mathrm{T}} \boldsymbol{A} \boldsymbol{p}_k + \left(\boldsymbol{r}_{k+1}^{\mathrm{T}} \boldsymbol{r}_{k+1}\right)\left(\boldsymbol{x}_{k+1}^{\mathrm{T}} \boldsymbol{p}_k\right)}{\boldsymbol{p}_k^{\mathrm{T}} (\boldsymbol{A}\boldsymbol{p}_k - \lambda_{k+1}\boldsymbol{I})\boldsymbol{p}_k}. \quad (2.42)$$

Equations (2.5)–(2.9) constitute the conjugate gradient algorithm for RQ, which was proposed in [11]. If the updated $\boldsymbol{x}_k$ is normalized to one and "+" (or "−") is selected in Eq. (2.6), the above algorithm will obtain the minimal (or maximal) eigenvalue of matrix $\boldsymbol{A}$ and its associated eigenvectors.

2.4.3 Generalized Rayleigh Quotient

Definition 2.3 Assume that $\boldsymbol{A} \in \mathbb{C}^{n\times n}, \mathbf{B} \in \mathbb{C}^{n\times n}$ are both Hermitian matrices, and $\boldsymbol{B}$ is positive definite. The generalized RQ or generalized Rayleigh–Ritz of the matrix pencil ($\boldsymbol{A}$, $\boldsymbol{B}$) is a scalar function, e.g.,

$$r(\boldsymbol{x}) = \frac{\boldsymbol{x}^H \boldsymbol{A}\boldsymbol{x}}{\boldsymbol{x}^H \boldsymbol{B}\boldsymbol{x}}, \quad (2.43)$$

where $\boldsymbol{x}$ is a quantity to be selected, and the objective is to maximize or minimize the generalized RQ.

In order to solve for the generalized RQ, define a new vector $\tilde{\boldsymbol{x}} = \boldsymbol{B}^{1/2}\boldsymbol{x}$, where $\boldsymbol{B}^{1/2}$ is the square root of the positive definite $\boldsymbol{B}$. Replace $\boldsymbol{x}$ by $\boldsymbol{B}^{-1/2}\tilde{\boldsymbol{x}}$ in (2.43). Then it holds that

$$r(\tilde{\boldsymbol{x}}) = \frac{\tilde{\boldsymbol{x}}^H \left(\boldsymbol{B}^{-1/2}\right)^H \boldsymbol{A} \left(\boldsymbol{B}^{-1/2}\right)^H \tilde{\boldsymbol{x}}}{\tilde{\boldsymbol{x}}^H \tilde{\boldsymbol{x}}}, \tag{2.44}$$

which shows that the generalized RQ of matrix pencil $(\boldsymbol{A}, \boldsymbol{B})$ is equivalent to the RQ of matrix product $\left(\boldsymbol{B}^{-1/2}\right)^H \boldsymbol{A} \left(\boldsymbol{B}^{-1/2}\right)^H$. From the Rayleigh–Ritz theorem, it is clear that when vector $\tilde{\boldsymbol{x}}$ is the eigenvector associated with the smallest eigenvalue $\lambda_{\min}$ of matrix product $\left(\boldsymbol{B}^{-1/2}\right)^H \boldsymbol{A} \left(\boldsymbol{B}^{-1/2}\right)^H$, the generalized RQ obtains $\lambda_{\min}$. And if vector $\tilde{\boldsymbol{x}}$ is the eigenvector associated with the largest eigenvalue $\lambda_{\max}$ of matrix product $\left(\boldsymbol{B}^{-1/2}\right)^H \boldsymbol{A} \left(\boldsymbol{B}^{-1/2}\right)^H$, the generalized RQ obtains $\lambda_{\max}$.

In the following, we review the eigen decomposition of matrix product $(\boldsymbol{B}^{-1/2})^H \boldsymbol{A} (\boldsymbol{B}^{-1/2})^H$, e.g.,

$$\left(\boldsymbol{B}^{-1/2}\right)^H \boldsymbol{A} \left(\boldsymbol{B}^{-1/2}\right)^H \tilde{\boldsymbol{x}} = \lambda \tilde{\boldsymbol{x}}. \tag{2.45}$$

If $\mathbf{B} = \sum_{i=1}^{n} \beta_i \boldsymbol{v}_i \boldsymbol{v}_i^H$ is an eigen decomposition of matrix $\boldsymbol{B}$, then

$$\mathbf{B}^{1/2} = \sum_{i=1}^{n} \sqrt{\beta_i} \boldsymbol{v}_i \boldsymbol{v}_i^H$$

and $\boldsymbol{B}^{1/2}\boldsymbol{B}^{1/2} = \boldsymbol{B}$. Since matrix $\boldsymbol{B}^{1/2}$and $\boldsymbol{B}^{-1/2}$have the same eigenvectors and their eigenvalues are reciprocals to each other, then it follows that

$$\mathbf{B}^{-1/2} = \sum_{i=1}^{n} \frac{1}{\sqrt{\beta_i}} \boldsymbol{v}_i \boldsymbol{v}_i^H,$$

which shows that $\boldsymbol{B}^{-1/2}$ is also an Hermitian matrix, e.g., $\left(\boldsymbol{B}^{-1/2}\right)^H = \boldsymbol{B}^{-1/2}$.

Premultiply both sides of (2.45) by $\boldsymbol{B}^{-1/2}$, and use $\left(\mathbf{B}^{-1/2}\right)^H = \boldsymbol{B}^{-1/2}$, then it holds that

$$\boldsymbol{B}^{-1}\boldsymbol{A}\boldsymbol{B}^{-1/2}\tilde{\boldsymbol{x}} = \lambda \boldsymbol{B}^{-1/2}\tilde{\boldsymbol{x}}$$

or

$$\boldsymbol{B}^{-1}\boldsymbol{A}\boldsymbol{x} = \lambda \boldsymbol{x}.$$

Since $\boldsymbol{x} = \boldsymbol{B}^{-1/2}\tilde{\boldsymbol{x}}$, thus the eigen decomposition of matrix product $\left(\boldsymbol{B}^{-1/2}\right)^H \boldsymbol{A}\left(\boldsymbol{B}^{-1/2}\right)^H$ is equivalent to the one of matrix $\boldsymbol{B}^{-1}\boldsymbol{A}$. The eigen decomposition of matrix $\boldsymbol{B}^{-1}\boldsymbol{A}$ is the generalized eigenvalue decompositions of matrix pencil $(\boldsymbol{A}, \boldsymbol{B})$. Thus, the conditions for the maximum and minimum of generalized RQ are

$$r(\boldsymbol{x}) = \frac{\boldsymbol{x}^H\boldsymbol{A}\boldsymbol{x}}{\boldsymbol{x}^H\boldsymbol{B}\boldsymbol{x}} = \lambda_{\max}, \quad \boldsymbol{A}\boldsymbol{x} = \lambda_{\max}\boldsymbol{B}\boldsymbol{x},$$

$$r(\boldsymbol{x}) = \frac{\boldsymbol{x}^H\boldsymbol{A}\boldsymbol{x}}{\boldsymbol{x}^H\boldsymbol{B}\boldsymbol{x}} = \lambda_{\min}, \quad \boldsymbol{A}\boldsymbol{x} = \lambda_{\min}\boldsymbol{B}\boldsymbol{x}.$$

That is to say, to maximize the generalized RQ, vector $\boldsymbol{x}$ must be the eigenvector associated with the largest generalized eigenvalue $\lambda_{\max}$ of matrix pencil $(\boldsymbol{A}, \boldsymbol{B})$. And to minimize the generalized RQ, vector $\boldsymbol{x}$ must be the eigenvector associated with the smallest generalized eigenvalue $\lambda_{\min}$ of matrix pencil $(\boldsymbol{A}, \boldsymbol{B})$.

2.5 Matrix Analysis

In the derivation and analysis of neural network-based PCA algorithm and its extensions, besides SVD, ED, etc., matrix gradient and matrix differential are also very necessary analysis tools. In this section, we will introduce some important results and properties of matrix gradient and matrix differential.

2.5.1 *Differential and Integral of Matrix with Respect to Scalar*

If $\boldsymbol{A}(t) = \{a_{ij}(t)\}_{m\times n}$ is a real matrix function of scalar t, then its differential and integral are, respectively, defined as

$$\begin{cases} \frac{\mathrm{d}}{\mathrm{d}t}\boldsymbol{A}(t) = \left\{\frac{\mathrm{d}}{\mathrm{d}t}a_{ij}(t)\right\}_{m\times n} \\ \int \boldsymbol{A}(t)\mathrm{d}t = \left\{\int a_{ij}(t)\mathrm{d}t\right\}_{m\times n} \end{cases}.$$

If $\boldsymbol{A}$(t) and $\boldsymbol{B}$(t) are, respectively, $m \times n$ and $n \times r$ matrices, then

$$\frac{\mathrm{d}}{\mathrm{d}t}[\boldsymbol{A}(t)\boldsymbol{B}(t)] = \left[\frac{\mathrm{d}\boldsymbol{A}(t)}{\mathrm{d}t}\right]\boldsymbol{B}(t) + \boldsymbol{A}(t)\left[\frac{\mathrm{d}\boldsymbol{B}(t)}{\mathrm{d}t}\right].$$

If $\boldsymbol{A}$(t) and $\boldsymbol{B}$(t) are both $m \times n$ matrices, then

$$\frac{\mathrm{d}}{\mathrm{d}t}[\boldsymbol{A}(t)+\boldsymbol{B}(t)] = \frac{\mathrm{d}\boldsymbol{A}(t)}{\mathrm{d}t} + \frac{\mathrm{d}\boldsymbol{B}(t)}{\mathrm{d}t}.$$

If $\boldsymbol{A}$(t) is a rank-n invertible square matrix, then

$$\frac{\mathrm{d}\boldsymbol{A}^{-1}(t)}{\mathrm{d}t} = -\boldsymbol{A}^{-1}(t)\frac{\mathrm{d}\boldsymbol{A}(t)}{\mathrm{d}t}\boldsymbol{A}^{-1}(t).$$

2.5.2 Gradient of Real Function with Respect to Real Vector

Define gradient operator $\nabla_{\boldsymbol{x}}$ of an $n \times 1$ vector $\boldsymbol{x}$ as

$$\nabla_{\boldsymbol{x}} = \left[\frac{\partial}{\partial x_1}, \frac{\partial}{\partial x_2}, \cdots, \frac{\partial}{\partial x_n}\right]^{\mathrm{T}} = \frac{\partial}{\partial \boldsymbol{x}},$$

Then the gradient of a real scalar quantity function $f(\boldsymbol{x})$ with respect to $\boldsymbol{x}$ is a $n \times 1$ column vector, which is defined as

$$\nabla_{\boldsymbol{x}} f(\boldsymbol{x}) = \left[\frac{\partial f(\boldsymbol{x})}{\partial x_1}, \frac{\partial f(\boldsymbol{x})}{\partial x_2}, \cdots, \frac{\partial f(\boldsymbol{x})}{\partial x_n}\right]^{\mathrm{T}} = \frac{\partial f(\boldsymbol{x})}{\partial \boldsymbol{x}}.$$

The negative direction of the gradient direction is called as the gradient flow of variable $\boldsymbol{x}$, written as

$$\dot{\boldsymbol{x}} = -\nabla_{\boldsymbol{x}} f(\boldsymbol{x}).$$

The gradient of m-dimensional row vector function $\boldsymbol{f}(\boldsymbol{x}) = [f_1(\boldsymbol{x}), f_2(\boldsymbol{x}), \ldots, f_m(\boldsymbol{x})]$ with respect to the $n \times 1$ real vector $\boldsymbol{x}$ is an $n \times m$ matrix, defined as

$$\frac{\partial \boldsymbol{f}(\boldsymbol{x})}{\partial \boldsymbol{x}} = \begin{bmatrix} \frac{\partial f_1(\boldsymbol{x})}{\partial x_1} & \frac{\partial f_2(\boldsymbol{x})}{\partial x_1} & \frac{\partial f_m(\boldsymbol{x})}{\partial x_1} \\ \frac{\partial f_1(\boldsymbol{x})}{\partial x_2} & \frac{\partial f_2(\boldsymbol{x})}{\partial x_2} & \frac{\partial f_m(\boldsymbol{x})}{\partial x_2} \\ \frac{\partial f_1(\boldsymbol{x})}{\partial x_n} & \frac{\partial f_2(\boldsymbol{x})}{\partial x_n} & \frac{\partial f_m(\boldsymbol{x})}{\partial x_n} \end{bmatrix} = \nabla_{\boldsymbol{x}} \boldsymbol{f}(\boldsymbol{x}).$$

Some properties of gradient operations can be summarized as follows:

① If $f(\boldsymbol{x}) = c$ is a constant, then gradient $\frac{\partial c}{\partial \boldsymbol{x}} = \boldsymbol{O}$.

② Linear principle: If $f(\boldsymbol{x})$ and $g(\boldsymbol{x})$ are real functions of vector $\boldsymbol{x}$, and c_1 and c_2 are real constants, then

$$\frac{\partial[c_1 f(\boldsymbol{x}) + c_2 g(\boldsymbol{x})]}{\partial \boldsymbol{x}} = c_1 \frac{\partial f(\boldsymbol{x})}{\partial \boldsymbol{x}} + c_2 \frac{\partial g(\boldsymbol{x})}{\partial \boldsymbol{x}}.$$

③ Product principle: If $f(\boldsymbol{x})$ and $g(\boldsymbol{x})$ are real functions of vector $\boldsymbol{x}$, then

$$\frac{\partial f(\boldsymbol{x}) g(\boldsymbol{x})}{\partial \boldsymbol{x}} = g(\boldsymbol{x}) \frac{\partial f(\boldsymbol{x})}{\partial \boldsymbol{x}} + f(\boldsymbol{x}) \frac{\partial g(\boldsymbol{x})}{\partial \boldsymbol{x}}.$$

④ Quotient principle: If $g(\boldsymbol{x}) \neq 0$, then

$$\frac{\partial f(\boldsymbol{x})/g(\boldsymbol{x})}{\partial \boldsymbol{x}} = \frac{1}{g^2(\boldsymbol{x})} \left[g(\boldsymbol{x}) \frac{\partial f(\boldsymbol{x})}{\partial \boldsymbol{x}} - f(\boldsymbol{x}) \frac{\partial g(\boldsymbol{x})}{\partial \boldsymbol{x}} \right].$$

⑤ Chain principle: If $\boldsymbol{y}(\boldsymbol{x})$ is a vector-valued function of $\boldsymbol{x}$, then

$$\frac{\partial f(\boldsymbol{y}(\boldsymbol{x}))}{\partial \boldsymbol{x}} = \frac{\partial \boldsymbol{y}^T(\boldsymbol{x})}{\partial \boldsymbol{x}} \frac{\partial f(\boldsymbol{y})}{\partial \boldsymbol{y}},$$

where $\frac{\partial \boldsymbol{y}^{\mathrm{T}}(\boldsymbol{x})}{\partial \boldsymbol{x}}$ is an $n \times n$ matrix.

⑥ If $\boldsymbol{a}$ is an $n \times 1$ constant vector, then

$$\frac{\partial \boldsymbol{a}^{\mathrm{T}} \boldsymbol{x}}{\partial \boldsymbol{x}} = \boldsymbol{a}, \quad \frac{\partial \boldsymbol{x}^{\mathrm{T}} \boldsymbol{a}}{\partial \boldsymbol{x}} = \boldsymbol{a}$$

⑦ If $\boldsymbol{A}$ and $\boldsymbol{y}$ are both independent of $\boldsymbol{x}$, then

$$\frac{\partial \boldsymbol{x}^{\mathrm{T}} \boldsymbol{A} \boldsymbol{y}}{\partial \boldsymbol{x}} = \boldsymbol{A} \boldsymbol{y}, \quad \frac{\partial \boldsymbol{y}^{\mathrm{T}} \boldsymbol{A} \boldsymbol{x}}{\partial \boldsymbol{x}} = \boldsymbol{A}^{\mathrm{T}} \boldsymbol{y}.$$

⑧ If $\boldsymbol{A}$ is a matrix independent of $\boldsymbol{x}$, then

$$\frac{\partial \boldsymbol{x}^{\mathrm{T}} \boldsymbol{A}}{\partial \boldsymbol{x}} = \boldsymbol{A}, \quad \frac{\partial \boldsymbol{x}^{\mathrm{T}} \boldsymbol{A} \boldsymbol{x}}{\partial \boldsymbol{x}} = \boldsymbol{A} \boldsymbol{x} + \boldsymbol{A}^{\mathrm{T}} \boldsymbol{x} = \left(\boldsymbol{A} + \boldsymbol{A}^{\mathrm{T}} \right) \boldsymbol{x}.$$

Especially, if $\boldsymbol{A}$ is a symmetric matrix, then $\frac{\partial \boldsymbol{x}^{\mathrm{T}} \boldsymbol{A} \boldsymbol{x}}{\partial \boldsymbol{x}} = 2\boldsymbol{A}\boldsymbol{x}$.

2.5.3 Gradient Matrix of Real Function

The gradient of a real function $f(\boldsymbol{A})$ with respect to an $m \times n$ real matrix $\boldsymbol{A}$ is an $m \times n$ matrix, called as gradient matrix, defined as

$$\frac{\partial f(\boldsymbol{A})}{\partial \boldsymbol{A}} = \begin{bmatrix} \frac{\partial f(\boldsymbol{A})}{\partial A_{11}} & \frac{\partial f(\boldsymbol{A})}{\partial A_{12}} & \cdots & \frac{\partial f(\boldsymbol{A})}{\partial A_{1n}} \\ \frac{\partial f(\boldsymbol{A})}{\partial A_{21}} & \frac{\partial f(\boldsymbol{A})}{\partial A_{22}} & \cdots & \frac{\partial f(\boldsymbol{A})}{\partial A_{2n}} \\ \vdots & \vdots & & \vdots \\ \frac{\partial f(\boldsymbol{A})}{\partial A_{m1}} & \frac{\partial f(\boldsymbol{A})}{\partial A_{m2}} & \cdots & \frac{\partial f(\boldsymbol{A})}{\partial A_{mn}} \end{bmatrix} = \nabla_{\boldsymbol{A}} f(\boldsymbol{A}),$$

where A_{ij} is the element of matrix $\boldsymbol{A}$ on its ith row and jth column.

Some properties of the gradient of a real function with respect to a matrix can be summarized as follows:

① If $f(\boldsymbol{A}) = c$ is a constant, where$\boldsymbol{A}$ is an $m \times n$ matrix, then $\frac{\partial c}{\partial \boldsymbol{A}} = \boldsymbol{O}_{m\times n}$.

② Linear principle: If $f(\boldsymbol{A})$ and $g(\boldsymbol{A})$ are real functions of matrix $\boldsymbol{A}$, and c_1 and c_2 are real constants, then

$$\frac{\partial[c_1 f(\boldsymbol{A}) + c_2 g(\boldsymbol{A})]}{\partial \boldsymbol{A}} = c_1 \frac{\partial f(\boldsymbol{A})}{\partial \boldsymbol{A}} + c_2 \frac{\partial g(\boldsymbol{A})}{\partial \boldsymbol{A}}.$$

③ Product principle: If $f(\boldsymbol{A})$ and $g(\boldsymbol{A})$ are real functions of matrix $\boldsymbol{A}$, then

$$\frac{\partial f(\boldsymbol{A}) g(\boldsymbol{A})}{\partial \boldsymbol{A}} = g(\boldsymbol{A}) \frac{\partial f(\boldsymbol{A})}{\partial \boldsymbol{A}} + f(\boldsymbol{A}) \frac{\partial g(\boldsymbol{A})}{\partial \boldsymbol{A}}.$$

④ Quotient principle: If $g(\boldsymbol{A}) \neq 0$,then

$$\frac{\partial f(\boldsymbol{A})/g(\boldsymbol{A})}{\partial(\boldsymbol{A})} = \frac{1}{g^2(\boldsymbol{A})}\left[g(\boldsymbol{A}) \frac{\partial f(\boldsymbol{A})}{\partial \boldsymbol{A}} - f(\boldsymbol{A}) \frac{\partial g(\boldsymbol{A})}{\partial \boldsymbol{A}}\right].$$

⑤ Chain principle: Let $\boldsymbol{A}$ be an $m \times n$ matrix, and $y = f(\boldsymbol{A})$ and $g(y)$ are real functions of matrix $\boldsymbol{A}$ and scalar y, respectively. Then

$$\frac{\partial g(f(\boldsymbol{A}))}{\partial \boldsymbol{A}} = \frac{dg(y)}{dy} \frac{\partial f(\boldsymbol{A})}{\partial \boldsymbol{A}}.$$

⑥ If $\boldsymbol{A} \in \Re^{m\times n}$, $\boldsymbol{x} \in \Re^{m\times 1}$, $\boldsymbol{y} \in \Re^{n\times 1}$, then

$$\frac{\partial \boldsymbol{x}^{\mathrm{T}} \boldsymbol{A} \boldsymbol{y}}{\partial \boldsymbol{A}} = \boldsymbol{A}\boldsymbol{y}^{\mathrm{T}}.$$

⑦ If $\boldsymbol{A} \in \Re^{n\times n}$ is nonsingular $\boldsymbol{x} \in \Re^{n\times 1}$,$\boldsymbol{y} \in \Re^{n\times 1}$, then

$$\frac{\partial \boldsymbol{x}^{\mathrm{T}} \boldsymbol{A}^{-1} \boldsymbol{y}}{\partial \boldsymbol{A}} = -\boldsymbol{A}^{-\mathrm{T}} \boldsymbol{A} \boldsymbol{y}^{\mathrm{T}} \boldsymbol{A}^{-\mathrm{T}}.$$

⑧ If $A \in \Re^{m \times n}, x, y \in \Re^{n \times 1}$, then

$$\frac{\partial x^{\mathrm{T}} A^{\mathrm{T}} A y}{\partial A} = A\left(xy^{\mathrm{T}} + yx^{\mathrm{T}}\right).$$

⑨ If $A \in \Re^{m \times n}, x, y \in \Re^{m \times 1}$, then

$$\frac{\partial x^{\mathrm{T}} A A^{\mathrm{T}} y}{\partial A} = \left(xy^{\mathrm{T}} + yx^{\mathrm{T}}\right)A.$$

2.5.4 Gradient Matrix of Trace Function

Here, we summarize some properties of gradient matrix of trace functions.

①–③ are gradient matrices of the trace of a single matrix.

① If W is an $m \times m$ matrix, then

$$\frac{\partial \mathrm{tr}(W)}{\partial W} = I_m.$$

② If an $m \times m$ matrix W is invertible, then

$$\frac{\partial \mathrm{tr}\left(W^{-1}\right)}{\partial W} = -\left(W^{-2}\right)^{\mathrm{T}}.$$

③ For the outer product of two vectors, it holds that

$$\frac{\partial \mathrm{tr}(xy^{\mathrm{T}})}{\partial x} = \frac{\partial tr(yx^{\mathrm{T}})}{\partial x} = y.$$

④–⑦ are gradient matrices of the trace of the product of two matrices.

④ If $W \in \Re^{m \times n}$, $A \in \Re^{n \times m}$, then

$$\frac{\partial \mathrm{tr}(WA)}{\partial W} = \frac{\partial \mathrm{tr}(AW)}{\partial W} = A^{\mathrm{T}}.$$

⑤ If $W \in \Re^{m \times n}, A \in \Re^{m \times n}$, then

$$\frac{\partial \mathrm{tr}\left(W^{\mathrm{T}} A\right)}{\partial W} = \frac{\partial \mathrm{tr}\left(A W^{\mathrm{T}}\right)}{\partial W} = A.$$

⑥ If $W \in \Re^{m\times n}$, then

$$\frac{\partial \mathrm{tr}\left(WW^{\mathrm{T}}\right)}{\partial W} = \frac{\partial \mathrm{tr}\left(W^{\mathrm{T}}W\right)}{\partial W} = 2W.$$

⑦ If $W \in \Re^{m\times n}$, then

$$\frac{\partial \mathrm{tr}(W^2)}{\partial W} = \frac{\partial \mathrm{tr}(WW)}{\partial W} = 2W^{\mathrm{T}}.$$

⑧ If $W, A \in \Re^{m\times m}$ and W is nonsingular, then

$$\frac{\partial \mathrm{tr}\left(AW^{-1}\right)}{\partial W} = -\left(W^{-1}AW^{-1}\right)^{\mathrm{T}}.$$

⑨–⑪ are gradient matrices of the trace of the product of three matrices.

⑨ If $W \in \Re^{m\times n}, A \in \Re^{m\times m}$, then

$$\frac{\partial \mathrm{tr}\left(W^{\mathrm{T}}AW\right)}{\partial W} = \left(A + A^{\mathrm{T}}\right)W.$$

Especially, if A is a symmetric matrix, then $\frac{\partial tr\left(W^{\mathrm{T}}AW\right)}{\partial W} = 2AW$

⑩ If $W \in \Re^{m\times n}, A \in \Re^{n\times n}$, then

$$\frac{\partial \mathrm{tr}\left(WAW^{\mathrm{T}}\right)}{\partial W} = W\left(A + A^{\mathrm{T}}\right).$$

Especially, if A is a symmetric matrix, then $\frac{\partial \mathrm{tr}\left(WAW^{\mathrm{T}}\right)}{\partial W} = 2WA$

⑪ If $W, A, B \in \Re^{m\times m}$ and W is nonsingular, then

$$\frac{\partial \mathrm{tr}\left(AW^{-1}B\right)}{\partial W} = -\left(W^{-1}BAW^{-1}\right)^{\mathrm{T}}.$$

2.5.5 Gradient Matrix of Determinant

Some properties of the gradient of the determinant of a matrix can be summarized as follows:

① Gradient of the determinant of a single nonsingular matrix

$$\frac{\partial|\boldsymbol{W}|}{\partial \boldsymbol{W}} = |\boldsymbol{W}|\left(\boldsymbol{W}^{-1}\right)^{\mathrm{T}} = \left(\boldsymbol{W}^{\#}\right)^{\mathrm{T}}$$

$$\frac{\partial|\boldsymbol{W}^{-1}|}{\partial \boldsymbol{W}} = -|\boldsymbol{W}|^{-1}\left(\boldsymbol{W}^{-1}\right)^{\mathrm{T}},$$

where $\boldsymbol{W}^{\#}$ is the adjoint matrix $\boldsymbol{A}$.

② Gradient of the logarithm of a determinant

$$\frac{\partial}{\partial \boldsymbol{W}}\log|\boldsymbol{W}| = \frac{1}{|\boldsymbol{W}|}\frac{\partial|\boldsymbol{W}|}{\partial \boldsymbol{W}},$$

$\boldsymbol{W}$ is nonsingular.

$$\frac{\partial}{\partial \boldsymbol{W}}\log|\boldsymbol{W}| = \left(\boldsymbol{W}^{-1}\right)^{\mathrm{T}},$$

the elements are independent to each other.

$$\frac{\partial}{\partial \boldsymbol{W}}\log|\boldsymbol{W}| = 2\boldsymbol{W}^{-1} - \mathrm{diag}\left(\boldsymbol{W}^{-1}\right),$$

$\boldsymbol{W}$ is symmetric matrix.

③ Gradient of the determinant of a two-matrix product

$$\frac{\partial|\boldsymbol{W}\boldsymbol{W}^{\mathrm{T}}|}{\partial \boldsymbol{W}} = 2|\boldsymbol{W}\boldsymbol{W}^{\mathrm{T}}|\left(\boldsymbol{W}\boldsymbol{W}^{\mathrm{T}}\right)^{-1}\boldsymbol{W}, \quad \mathrm{rank}(\boldsymbol{W}_{m\times n}) = m.$$

$$\frac{\partial|\boldsymbol{W}\boldsymbol{W}^{\mathrm{T}}|}{\partial \boldsymbol{W}} = 2|\boldsymbol{W}^{\mathrm{T}}\boldsymbol{W}|\boldsymbol{W}\left(\boldsymbol{W}^{\mathrm{T}}\boldsymbol{W}\right)^{-1}, \quad \mathrm{rank}(\boldsymbol{W}_{m\times n}) = n.$$

$$\frac{\partial|\boldsymbol{W}^2|}{\partial \boldsymbol{W}} = 2|\boldsymbol{W}|^2\left(\boldsymbol{W}^{-1}\right)^{\mathrm{T}}, \quad \mathrm{rank}(\boldsymbol{W}_{m\times m}) = m.$$

④ Gradient of the determinant of a three-matrix product

$$\frac{\partial|\boldsymbol{A}\boldsymbol{W}\boldsymbol{B}|}{\partial \boldsymbol{W}} = |\boldsymbol{A}\boldsymbol{W}\boldsymbol{B}|\boldsymbol{A}^{\mathrm{T}}\left(\boldsymbol{B}^{\mathrm{T}}\boldsymbol{W}^{\mathrm{T}}\boldsymbol{A}^{\mathrm{T}}\right)^{-1}\boldsymbol{B}^{\mathrm{T}}.$$

$$\frac{\partial|\boldsymbol{W}^{\mathrm{T}}\boldsymbol{A}\boldsymbol{W}|}{\partial \boldsymbol{W}} = 2\boldsymbol{A}\boldsymbol{W}\left(\boldsymbol{W}^{\mathrm{T}}\boldsymbol{A}\boldsymbol{W}\right)^{-1}, |\boldsymbol{W}^{\mathrm{T}}\boldsymbol{A}\boldsymbol{W}| > 0.$$

$$\frac{\partial|\boldsymbol{W}\boldsymbol{A}\boldsymbol{W}^{\mathrm{T}}|}{\partial \boldsymbol{W}} = \left[\left(\boldsymbol{W}\boldsymbol{A}\boldsymbol{W}^{\mathrm{T}}\right)^{-1}\right]^{\mathrm{T}}\boldsymbol{W}\left(\boldsymbol{A}^{\mathrm{T}} + \boldsymbol{A}\right).$$

2.5.6 Hessian Matrix

The Hessian matrix is defined as

$$\frac{\partial^2 f(\boldsymbol{x})}{\partial \boldsymbol{x}\partial \boldsymbol{x}^{\mathrm{T}}} = \frac{\partial}{\partial \boldsymbol{x}^{\mathrm{T}}}\left[\frac{\partial f(\boldsymbol{x})}{\partial \boldsymbol{x}}\right] = \begin{bmatrix} \frac{\partial^2 f}{\partial x_1 \partial x_1} & \frac{\partial^2 f}{\partial x_1 \partial x_2} & \cdots & \frac{\partial^2 f}{\partial x_1 \partial x_n} \\ \frac{\partial^2 f}{\partial x_2 \partial x_1} & \frac{\partial^2 f}{\partial x_2 \partial x_2} & \cdots & \frac{\partial^2 f}{\partial x_2 \partial x_n} \\ \vdots & \vdots & & \vdots \\ \frac{\partial^2 f}{\partial x_n \partial x_1} & \frac{\partial^2 f}{\partial x_n \partial x_2} & \cdots & \frac{\partial^2 f}{\partial x_n \partial x_n} \end{bmatrix}$$

and it can also be written as the gradient of gradient, i.e., $\nabla_x^2 f(\boldsymbol{x}) = \nabla_x(\nabla_x f(\boldsymbol{x}))$.

Here are some properties of Hessian matrix.

① For an $n \times 1$ constant vector $\boldsymbol{a}$, it holds that

$$\frac{\partial^2 \boldsymbol{a}^{\mathrm{T}}\boldsymbol{x}}{\partial \boldsymbol{x}\partial \boldsymbol{x}^{\mathrm{T}}} = \boldsymbol{O}_{n\times n}.$$

② If $\boldsymbol{A}$ is an $n \times n$ matrix, then

$$\frac{\partial^2 \boldsymbol{x}^{\mathrm{T}}\boldsymbol{A}\boldsymbol{x}}{\partial \boldsymbol{x}\partial \boldsymbol{x}^{\mathrm{T}}} = \boldsymbol{A} + \boldsymbol{A}^{\mathrm{T}}.$$

③ If $\boldsymbol{x}$ is an $n \times 1$ vector, $\boldsymbol{a}$ is an $m \times 1$ constant vector, $\boldsymbol{A}$ and $\boldsymbol{B}$, respectively, are $m \times n$ and $m \times m$ constant matrices, and $\boldsymbol{B}$ is symmetric, then

$$\frac{\partial^2 (\boldsymbol{a} - \boldsymbol{A}\boldsymbol{x})^{\mathrm{T}}\boldsymbol{B}(\boldsymbol{a} - \boldsymbol{A}\boldsymbol{x})}{\partial \boldsymbol{x}\partial \boldsymbol{x}^{\mathrm{T}}} = 2\boldsymbol{A}^{\mathrm{T}}\boldsymbol{B}\boldsymbol{A}.$$

2.6 Summary

The singular value decomposition, eigenvalue decomposition, Rayleigh quotient, and gradient and differentials of a matrix have been reviewed in a tutorial style in this chapter. The materials presented in this chapter are useful for the understanding of latter chapters, particularly for the chapters except 3 and 6.

References

1. Beltrami, E. (1873). Sulle funzioni bilineari, Giomale di Mathematiche ad Uso studenti Delle Uninersita. 11, 98–106. (An English translation by D Boley is available as University of Minnesota, Department of Computer Science), Technical Report 90–37, 1990.
2. Jordan, C. (1874). Memoire sur les formes bilineaires. *Journal of Mathematical Pures Application Eeuxieme Series, 19*, 35–54.
3. Autonne, L. (1902). Sur les groupes lineaires, reelles et orthogonaus. *Bulletin Social Math France, 30*, 121–134.
4. Eckart, C., & Young, G. (1939). A principal axis transformation for non-Hermitian matrices. *Null American Mathematics Society, 45*(2), 118–121.
5. Klema, V. C., & Laub, A. J. (1980). The singular value decomposition: Its computation and some application. *IEEE Transactions on Automatic Control, 25*(2), 164–176.
6. Zhang, X. D. (2004). *Matrix analysis and application*. Tsinghua University Press.
7. Jennings, A., & McKeown, J. J. (1992). *Matrix computations*. New York: Wiley.
8. Saad, Y. (1992). *Numerical methods for large eigenvalue problem*. New York: Machester University Press.
9. Dramc, Z. (1998). A tangent algorithm for computing the generalized singular value decomposition. *SIAM Journal of Numerical Analysis, 35*(5), 1804–1832.
10. Nour-Omid, B., Parlett, B. N., Ericsson, T., & Jensen, P. S. (1987). How to implement the spectral transformation. *Mathematics Computation, 48*(178), 663–673.
11. Yang, X., Sarkar, T. K., Yang, X., & Arvas, E. (1989). A survey of conjugate gradient algorithms for solution of extreme Eigen-problems of a symmetric matrix. *IEEE Transactions on Acoustic Speech and Signal Processing, 37*(10), 1550–1556.
12. Rayleigh, L. (1937). *The theory of sound* (2nd ed.). New York: Macmillian.
13. Parlett, B. N. (1974). The Rayleigh quotient iteration and some genelarizations of no normal matrices. *Mathematics of Computation, 28*(127), 679–693.
14. Parlett, B. N. (1980). *The symmetric Eigenvalue problem*. Englewood Cliffs, NJ: Prentice-Hall.
15. Chatelin, F. (1993). *Eigenvalues of matrices*. New York: Wiley.
16. Helmke, U., & Moore, J. B. (1994). *Optimization and dynamical systems*. London, UK: Springer-Verlag.
17. Golub, G. H., & Van Loan, C. F. (1989). *Matrix computation* (2nd ed.). Baltimore: The John Hopkins University Press.
18. Cirrincione, G., Cirrincione, M., Herault, J., & Van Huffel, S. (2002). The MCA EXIN neuron for the minor component analysis. *IEEE Transactions on Neural Networks, 13*(1), 160–187.
19. Lancaster, P., & Tismenetsky, M. (1985). *The theory of Matrices with Applications* (2nd ed.). New York: Academic.
20. Shavitt, I., Bender, C. F., Pipano, A., & Hosteny, R. P. (1973). The iterative calculation of several of the lowest or highest eigenvalues and corresponding eigenvectors of very large symmetric matrices. *Journal of Computation Physics, 11*(1), 90–108.
21. Chen, H., Sarkar, T. K., Brule, J., & Dianat, S. A. (1986). Adaptive spectral estimation by the conjugate gradient method. *IEEE Transactions on Acoustic Speech and Signal Processing, 34* (2), 271–284.
22. Huang, L. (1984). *The linear Algebra in system and control theory*. Beijing: Science Press.
23. Pease, M. C. (1965). *Methods of matrix algebra*. New York: Academic Press.
24. Magnus, J. R., & Neudecker, H. (1999). *Matrix differential calculus with application in statistics and econometrics, revised*. Chichester: Wiley.
25. Lutkepohl, H. (1996). *Handbook of matrices*. New York: Wiley.
26. Searle, S. R. (1982). *Matrix algebra useful for statistics*. New York: Wiley.

Chapter 3
Neural Networks for Principal Component Analysis

3.1 Introduction

PCA is a statistical method, which is directly related to EVD and SVD. Neural networks-based PCA method estimates PC online from the input data sequences, which especially suits for high-dimensional data due to the avoidance of the computation of large covariance matrix, and for the tracking of nonstationary data, where the covariance matrix changes slowly over time. Neural networks and algorithms for PCA will be described in this chapter, and algorithms given in this chapter are typically unsupervised learning methods.

PCA has been widely used in engineering and scientific disciplines, such as pattern recognition, data compression and coding, image processing, high-resolution spectrum analysis, and adaptive beamforming. PCA is based on the spectral analysis of the second moment matrix that statistically characterizes a random vector. PCA is directly related to SVD, and the most common way to perform PCA is via the SVD of a data matrix. However, the capability of SVD is limited for very large data sets.

It is well known that preprocessing usually maps a high-dimensional space to a low-dimensional space with the least information loss, which is known as feature extraction. PCA is a well-known feature extraction method, and it allows the removal of the second-order correlation among given random processes. By calculating the eigenvectors of the covariance matrix of the input vector, PCA linearly transforms a high-dimensional input vector into a low-dimensional one whose components are uncorrelated.

© Science Press, Beijing and Springer Nature Singapore Pte Ltd. 2017

X. Kong et al., *Principal Component Analysis Networks and Algorithms*,
DOI 10.1007/978-981-10-2915-8_3

PCA is often based on the optimization of some information criterion, such as the maximization of the variance of the projected data or the minimization of the reconstruction error. The aim of PCA is to extract m orthonormal directions $\bar{\bar{w}}_i \in \Re^n, i = 1, 2, \ldots, m, m < n$, in the input space that account for as much of the data's variance as possible. Subsequently, an input vector $\boldsymbol{x} \in \Re^n$ may be transformed into a lower m-dimensional space without losing essential intrinsic information. The vector x can be represented by being projected onto the m-dimensional subspace spanned by w_i using the inner products $\boldsymbol{x}^{\mathrm{T}}\bar{\bar{w}}_i$. This achieves dimensionality reduction.

PCA finds those unitary directions $\bar{\bar{w}} \in \Re^n$, along which the projections of the input vectors, known as the principal components (PCs), $y = \boldsymbol{x}^{\mathrm{T}}\bar{\bar{w}}$, have the largest variance $E_{\mathrm{PCA}}(\boldsymbol{w}) = E[y^2] = \bar{\bar{w}}^{\mathrm{T}}\boldsymbol{C}\bar{\bar{w}} = \bar{\bar{w}}^{\mathrm{T}}\boldsymbol{C}\bar{\bar{w}}/\|\boldsymbol{w}\|^2$, where $\bar{\bar{w}} = \boldsymbol{w}/\|\boldsymbol{w}\|$. When $\boldsymbol{w} = \alpha\boldsymbol{c}_1$, $E_{\mathrm{PCA}}(\boldsymbol{w})$ take its maximum value, where α is a scalar. When $\alpha = 1, \boldsymbol{w}$ becomes a unit vector. By repeating maximization of $E_{\mathrm{PCA}}(\boldsymbol{w})$ but limiting w to be orthogonal to $\boldsymbol{c}_1$, the maximization of $E_{\mathrm{PCA}}(\boldsymbol{w})$ is equal to λ_2 at $\boldsymbol{w} = \alpha\boldsymbol{c}_2$. Following this deflation procedure, all the m principal directions $\bar{\bar{w}}_i$ can be derived. The projections $y_i = \boldsymbol{x}^{\mathrm{T}}\bar{\bar{w}}_i, i = 1, 2, \ldots, m$ are the PCs of $\boldsymbol{x}$. A linear least square (LS) estimate $\hat{\boldsymbol{x}}$ can be constructed for the original input $\boldsymbol{x}$ as $\hat{\boldsymbol{x}} = \sum_{i=1}^{m} a_i(t)\bar{\bar{w}}_i$. As to other interpretations or analyses of PCA, see [1–4] for more details.

3.2 Review of Neural-Based PCA Algorithms

Neural networks on PCA pursue an effective "online" approach to update the eigen direction after each presentation of a data point, which are especially suitable for high-dimensional data and for the tracking of nonstationary data. In the last decades, many neural network-based PCA learning algorithms were proposed, among which, the Hebbian and Oja's learning rules are the bases. Overall, the existing neural network-based PCA algorithms can be grouped into the following classes: the Hebbian rule-based PCA algorithms, least mean squared error-based PCA algorithms, other optimization-based PCA algorithms, anti-Hebbian rule-based PCA algorithms, nonlinear PCA algorithms, constrained PCA algorithms, localized PCA algorithms, and other generalizations of the PCA. These algorithms will be analyzed and discussed in the above order.

3.3 Neural-Based PCA Algorithms Foundation

3.3.1 Hebbian Learning Rule

The classical Hebbian synaptic modification rule was first introduced in [5]. In Hebbian learning rule, the biological synaptic weights change in proportion to the correlation between the presynaptic and postsynaptic signals. For a single neuron, the Hebbian rule can be written as

$$\boldsymbol{w}(t+1) = \boldsymbol{w}(t) + \eta y(t)\boldsymbol{x}(t), \tag{3.1}$$

where the learning rate $\eta > 0$, $\boldsymbol{w} \in \Re^n$ is the weight vector, $\boldsymbol{x}(t) \in \Re^n$ is an input vector at time t, $y(t)$ is the output of the neuron defined by $y(t) = \boldsymbol{w}^{\mathrm{T}}(t)\boldsymbol{x}(t)$.

The convergence of Hebbian rule can be briefly analyzed as follows.

For a stochastic input vector $\boldsymbol{x}$, assuming that $\boldsymbol{x}$ and $\boldsymbol{w}$ are uncorrelated, the expected weight change is given by

$$\mathrm{E}[\Delta\boldsymbol{w}] = \eta\mathrm{E}[y\boldsymbol{x}] = \eta\mathrm{E}\left[\boldsymbol{x}\boldsymbol{x}^{\mathrm{T}}\boldsymbol{w}\right] = \eta\boldsymbol{C}\mathrm{E}[\boldsymbol{w}], \tag{3.2}$$

where $\mathrm{E}[\cdot]$ is the expectation operator, and $\boldsymbol{C} = \mathrm{E}[\boldsymbol{x}\boldsymbol{x}^{\mathrm{T}}]$ is the autocorrelation matrix of $\boldsymbol{x}$.

At equilibrium, $\mathrm{E}[\Delta\boldsymbol{w}] = 0$, and hence, it holds that the deterministic equation $\boldsymbol{C}\boldsymbol{w} = 0$. Due to the effect of noise terms, C is a full-rank positive-definite Hermitian matrix with positive eigenvalues $\lambda_i, i = 1, 2, \ldots, n$, and the associated orthogonal eigenvectors $\boldsymbol{c}_i$, where n = rank(C). Thus, $\boldsymbol{w} = 0$ is the only equilibrium state.

Equation (3.1) can be further represented in the continuous-time form

$$\dot{\boldsymbol{w}} = y\boldsymbol{x}. \tag{3.3}$$

Taking expectation on both sides, it holds that

$$\mathrm{E}[\dot{\boldsymbol{w}}] = \mathrm{E}[y\boldsymbol{x}] = \mathrm{E}\left[\boldsymbol{x}\boldsymbol{x}^{\mathrm{T}}\boldsymbol{w}\right] = \boldsymbol{C}\mathrm{E}[\boldsymbol{w}]. \tag{3.4}$$

This can be derived by minimizing the average instantaneous criterion function [6]

$$\mathrm{E}[E_{\mathrm{Hebb}}] = -\frac{1}{2}\mathrm{E}\left[y^2\right] = -\frac{1}{2}\mathrm{E}\left[\boldsymbol{w}^{\mathrm{T}}\boldsymbol{x}\boldsymbol{x}^{\mathrm{T}}\boldsymbol{w}\right] = -\frac{1}{2}\mathrm{E}\left[\boldsymbol{w}^{\mathrm{T}}\right]\boldsymbol{C}\mathrm{E}[\boldsymbol{w}], \tag{3.5}$$

where E_{Hebb} is the instantaneous criterion function. At equilibrium, $\mathrm{E}\left[\frac{\partial E_{\mathrm{Hebb}}}{\partial \boldsymbol{w}}\right] = -\boldsymbol{C}\mathrm{E}[\boldsymbol{w}] = \boldsymbol{0}$, thus $\boldsymbol{w} = \boldsymbol{0}$. Since $\mathrm{E}[\mathrm{H}(\boldsymbol{w})] = \mathrm{E}\left[\frac{\partial E^2_{\mathrm{Hebb}}}{\partial^2 \boldsymbol{w}}\right] = -\boldsymbol{C}$ is nonpositive for all $\mathrm{E}[\boldsymbol{w}]$, the solution $\boldsymbol{w} = \boldsymbol{0}$ is unstable, which drives $\boldsymbol{w}$ to infinite magnitude, with a direction parallel to that of the eigenvector of $\boldsymbol{C}$ associated with the largest eigenvalue [6]. Thus, the Hebbian rule is divergent.

To prevent the divergence of the Hebbian rule, one can normalize $\|\boldsymbol{w}\|$ to unity after each iteration [7]. This leads to the normalized Hebbian rule. Several other methods such as Oja's rule [8], Yuille's rule [9], Linsker's rule [10, 11], and Hassoun's rule [12] add a weight-decay term to the Hebbian rile to stabilize the algorithm.

3.3.2 Oja's Learning Rule

By adding a weight decay term into the Hebbian rule, Oja's learning rule was proposed in [8] and given by

$$\boldsymbol{w}(t+1) = \boldsymbol{w}(t) + \eta y(t)\boldsymbol{x}(t) - \eta y^2(t)\boldsymbol{w}(t). \tag{3.6}$$

Oja's rule converges to a state that minimizes (3.5) subject to $\|\boldsymbol{w}\| = 1$. The solution is the principal eigenvector of $\mathbf{C}$. For small η, Oja's rule is proved to be equivalent to the normalized Hebbian rule [8].

Using the stochastic learning theory, the continuous-time version of Oja's rule is given by a nonlinear stochastic differential equation

$$\dot{\boldsymbol{w}} = \eta(y\boldsymbol{x} - y^2\boldsymbol{w}). \tag{3.7}$$

The corresponding deterministic equation based on statistical average is thus derived as

$$\dot{\boldsymbol{w}} = \eta\left[\boldsymbol{C}\boldsymbol{w} - \left(\boldsymbol{w}^{\mathrm{T}}\boldsymbol{C}\boldsymbol{w}\right)\boldsymbol{w}\right]. \tag{3.8}$$

At equilibrium, it holds that

$$\boldsymbol{C}\boldsymbol{w} = \left(\boldsymbol{w}^{\mathrm{T}}\boldsymbol{C}\boldsymbol{w}\right)\boldsymbol{w}. \tag{3.9}$$

It can be easily seen that the solutions are $\boldsymbol{w} = \pm\boldsymbol{c}_i, i = 1, 2, \ldots, n$, whose associated eigenvalues λ_i are arranged in a descending order as $\lambda_1 \geq \lambda_2 \geq \cdots \geq \lambda_n \geq 0$.

Note that the average Hessian

$$\mathrm{H}(\boldsymbol{w}) = \frac{\partial}{\partial \boldsymbol{w}}\left[-\boldsymbol{C}\boldsymbol{w} + \left(\boldsymbol{w}^{\mathrm{T}}\boldsymbol{C}\boldsymbol{w}\right)\boldsymbol{w}\right] = -\boldsymbol{C} + \boldsymbol{w}^{\mathrm{T}}\boldsymbol{C}\boldsymbol{w}\boldsymbol{I} + 2\boldsymbol{w}\boldsymbol{w}^{\mathrm{T}}\boldsymbol{C} \tag{3.10}$$

is positive-definite only at $\boldsymbol{w} = \pm\boldsymbol{c}_1$, if $\lambda_1 \neq \lambda_2$ [12], where I is an $n \times n$ identity matrix. This can be seen from

$$\begin{aligned}\mathrm{H}(\boldsymbol{c}_i)\boldsymbol{c}_j &= (\lambda_i - \lambda_j)\boldsymbol{c}_j + 2\lambda_j\boldsymbol{c}_i\boldsymbol{c}_i^{\mathrm{T}}\boldsymbol{c}_j \\ &= \begin{cases} 2\lambda_i 2\boldsymbol{c}_i & i = j \\ (\lambda_i - \lambda_j)\boldsymbol{c}_j & i \neq j \end{cases}.\end{aligned} \tag{3.11}$$

Thus, Oja's rule always converges to the principal component of $\boldsymbol{C}$.

The convergence analysis of the stochastic discrete-time algorithms such as the gradient descent method is conventionally based on the stochastic approximation theory [13]. A stochastic discrete-time algorithm is first converted into deterministic continuous-time ODEs, and then analyzed by using Lyapunov's second theorem.

This conversion is based on the Robbins–Monro conditions, which require the learning rate to gradually approach zero as $t \to \infty$. This limitation is not practical for implementation, especially for the learning of nonstationary data. In [14], Zufiria proposed to convert the stochastic discrete-time algorithms into their deterministic discrete-time formulations that characterize their average evolution from a conditional expectation perspective. This method has been applied to Oja's rule and the dynamics have been analyzed, and chaotic behavior has been observed in some invariant subspaces. Such analysis can guarantee the convergence of the Oja's rule by selecting some constant learning rate. A constant learning rate for fast convergence has also been suggested as $\eta = 0.618\ \lambda_1$ [15]. Recently, the convergence of many PCA algorithms of Oja's rule type have been analyzed by using deterministic discrete-time methods, the details of which will be discussed in Chap. 6.

3.4 Hebbian/Anti-Hebbian Rule-Based Principal Component Analysis

Hebbian rule-based PCA algorithms include the single PCA algorithm, multiple PCA algorithms and principal subspace analysis algorithm. These neural PCA algorithms originate from the seminal work by Oja [8]. The output of the neuron is updated by $\boldsymbol{y} = \boldsymbol{w}^{\mathrm{T}}\boldsymbol{x}$, where $\boldsymbol{w} = (w_1, w_2, \ldots, w_{J1})^{\mathrm{T}}$. Here the activation function is the linear function $\varphi(x) = x$. The PCA turns out to be closely related to the Hebbian rule.

The PCA algorithms discussed in this section are based on the Hebbian rule. The network model was first proposed by Oja [16], where a J_1–J_2 FNN is used to extract the first J_2 PCs. The architecture of the PCA network is shown in Fig. 3.1, which is

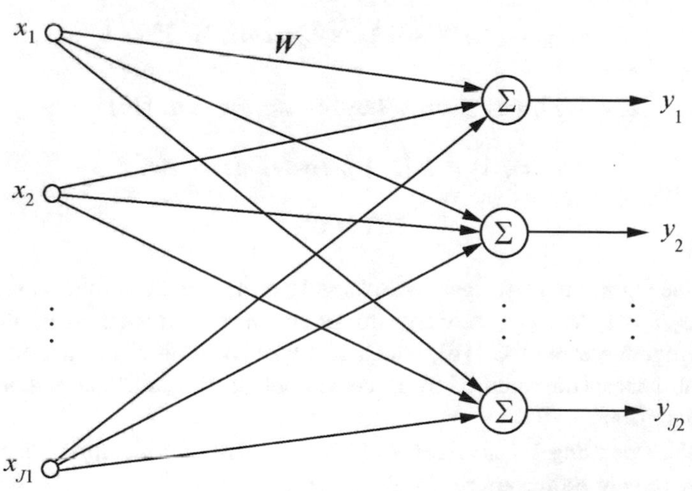

Fig. 3.1 Architecture of the PCA network

a simple expansion of the single-neuron PCA model. The output of the network is given by $\boldsymbol{y} = \boldsymbol{W}^{\mathrm{T}}\boldsymbol{x}$, where $\boldsymbol{y} = (y_1, y_2, \ldots, y_{J2})^{\mathrm{T}}$, $\boldsymbol{x} = (x_1, x_2, \ldots, x_{J1})^{\mathrm{T}}$, $\boldsymbol{W} = [\boldsymbol{w}_1, \boldsymbol{w}_2, \ldots, \boldsymbol{w}_{J2}]$, $\boldsymbol{w}_i = (w_{1i}, w_{2i}, \ldots, w_{J1i})^{\mathrm{T}}$.

3.4.1 *Subspace Learning Algorithms*

By using Oja's learning rule, $\boldsymbol{w}$ will converge to a unit eigenvector of the correlation matrix $\boldsymbol{C}$, and the variance of the output y is maximized. For zero-mean input data, this extracts the first PC. Here Oja's learning rule can be rewritten for the convenience of presentation as

$$\boldsymbol{w}(t+1) = \boldsymbol{w}(t) + \eta y(t)\boldsymbol{x}(t) - \eta y^2(t)\boldsymbol{w}(t), \tag{3.12}$$

where the term $y(t)\boldsymbol{x}(t)$ is the Hebbian term, and $-y^2(t)\boldsymbol{w}(t)$ is a decaying term, which is used to prevent instability. In order to keep the algorithm convergent, it is proved that $0 < \eta(t) < 1/1.2\lambda_1$ is required [16], where λ_1 is the largest eigenvalue of $\mathbf{C}$. If $\eta(t) \geq 1/\lambda_1$, $\mathbf{w}$ will not converge to $\pm\mathbf{c1}$ even if it is initially close to the target [17].

3.4.1.1 Symmetrical Subspace Learning Algorithm

Oja proposed a learning algorithm for the PCA network, referred to as the symmetrical subspace learning algorithm (SLA) [16]. The SLA can be derived by maximizing

$$E_{\mathrm{SLA}} = \frac{1}{2}\mathrm{tr}\left(\boldsymbol{W}^{\mathrm{T}}\boldsymbol{R}\boldsymbol{W}\right) \quad \text{subject to} \quad \boldsymbol{W}^{\mathrm{T}}\boldsymbol{W} = \boldsymbol{I}, \tag{3.13}$$

where $\boldsymbol{I}$ is a $\mathrm{J}_2 \times \mathrm{J}_2$ identity matrix. The SLA is given as [16]

$$\boldsymbol{w}_i(t+1) = \boldsymbol{w}_i(t) + \eta(t)y_i(t)[\boldsymbol{x}(t) - \hat{\boldsymbol{x}}(t)], \tag{3.14}$$

$$\hat{\boldsymbol{x}}(t) = \boldsymbol{W}\boldsymbol{y}. \tag{3.15}$$

After the algorithm converges, $\boldsymbol{W}$ is roughly orthonormal and the columns of $\boldsymbol{W}$, namely $\boldsymbol{w}_i, i = 1, 2, \ldots, J_2$, converge to some linear combination of the first J_2 principal eigenvectors of $\boldsymbol{C}$ [16], which is a rotated basis of the dominant eigenvector subspace. The value of $\boldsymbol{w}_i$ is dependent on the initial condition and the training samples.

The corresponding eigenvalues $\lambda_i, i = 1, 2, \ldots, J_2$, which approximate $E\left[y_i^2\right]$, can be adaptively estimated by

$$\hat{\lambda}_i(t+1) = \left(1 - \frac{1}{t+1}\right)\hat{\lambda}_i(t) + \frac{1}{t+1}y_i^2(t+1). \tag{3.16}$$

The PCA performs optimally when there is no noise process involved.

3.4.1.2 Weighted Subspace Learning Algorithm

The weighted SLA can be derived by maximizing the same criterion (3.13), with the constraint changed to $\boldsymbol{W}^{\mathrm{T}}\boldsymbol{W} = \boldsymbol{\alpha}$, where $\boldsymbol{\alpha} = diag(\alpha_1, \alpha_2, \ldots, \alpha_{J_2})$, is an arbitrary diagonal matrix with $\alpha_1 > \alpha_2 > \cdots > \alpha_{J_2} > 0$.

The weighted SLA is given by [18, 19]

$$\boldsymbol{w}_i(t+1) = \boldsymbol{w}_i(t) + \eta(t)y_i(t)[\boldsymbol{x}(t) - \gamma_i\hat{\boldsymbol{x}}(t)], \tag{3.17}$$

$$\hat{\boldsymbol{x}}(t) = \boldsymbol{W}\boldsymbol{y}, \tag{3.18}$$

for $i = 1, 2, \ldots, J_2$, where $\gamma_i, i = 1, 2, \ldots, J_2$, are coefficients satisfying $0 < \gamma_1 < \gamma_2 < \cdots < \gamma_{J_2}$.

Due to the asymmetry introduced by γ_i, $\boldsymbol{w}_i$ almost surely converges to the eigenvectors of $\boldsymbol{C}$. The weighted subspace algorithm can perform the PCA, however, norms of the weight vectors are not equal to unity.

The subspace and weighted subspace algorithms are nonlocal algorithms relying on the calculation of the errors and the backward propagation of the values between the layers [3]. Several algorithms converting PSA into PCA have been proposed, the details can be found in [3].

3.4.2 Generalized Hebbian Algorithm

By combining Oja's rule and the GSO procedure, Sanger proposed the GHA for extracting the first J_2 PCs [20]. The GHA can extract the first J_2 eigenvectors in the order of decreasing eigenvalues.

The GHA is given by [20]

$$\boldsymbol{w}_i(t+1) = \boldsymbol{w}_i(t) + \eta_i(t)y_i(t)[\boldsymbol{x}(t) - \hat{\boldsymbol{x}}_i(t)], \tag{3.19}$$

$$\hat{\boldsymbol{x}}_i(t) = \sum\nolimits_{j=1}^{i} \boldsymbol{w}_j(t)y_j(t), \tag{3.20}$$

for $i = 1, 2, \ldots, J_2$. The GHA becomes a local algorithm by solving the summation term in (3.20) in a recursive form

$$\hat{\boldsymbol{x}}_i(t) = \hat{\boldsymbol{x}}_{i-1}(t) + \boldsymbol{w}_i(t)y_i(t), \tag{3.21}$$

for $i = 1, 2, \ldots, J_2$, where $\hat{\boldsymbol{x}}_0(t) = 0$. $\eta_i(t)$ is usually selected the same for all neurons. When $\eta_i = \eta$ for all i, the algorithm can be written in a matrix form

$$\boldsymbol{W}(t+1) = \boldsymbol{W}(t) - \eta\boldsymbol{W}(t)\mathrm{LT}\left[\boldsymbol{y}(t)\boldsymbol{y}^T(t)\right] + \eta\boldsymbol{x}(t)\boldsymbol{y}^T(t), \tag{3.22}$$

where the operator LT[·] selects the lower triangle of input matrix. In the GHA, the mth neuron converges to the mth PC, and all the neurons tend to converge together. $\boldsymbol{w}_i$ and $E\left[y_i^2\right]$ approach $\boldsymbol{c}_i$ and λ_i, respectively, as $t \to \infty$.

Both the SLA and GHA algorithms employ implicit or explicit GSO to decorrelate the connection weights from one another. The weighted SLA algorithm performs well for extracting less-dominant components.

3.4.3 Learning Machine for Adaptive Feature Extraction via PCA

Learning machine for adaptive feature extraction via principal component analysis is called LEAP algorithm, and it is another local PCA algorithm for extracting all the J_2 PCs and their corresponding eigenvectors. The LEAP is given by

$$\boldsymbol{w}_i(t+1) = \boldsymbol{w}_i(t) + \eta\{\boldsymbol{B}_i(t)y_i(t)[\boldsymbol{x}(t) - \boldsymbol{w}_i(t)y_i(t)] - \boldsymbol{A}_i(t)\boldsymbol{w}_i(t)\}, \tag{3.23}$$

for $i = 1, 2, \ldots, J_2$, where η is the learning rate, $y_i(t)\boldsymbol{x}(t)$ is a Hebbian term, and

$$\boldsymbol{A}_i(t) = \begin{cases} 0, & i = 1 \\ \boldsymbol{A}_{i-1}(t) + \boldsymbol{w}_{i-1}(t)\boldsymbol{w}_{i-1}^T(t), & i = 2, \ldots, J_2 \end{cases}, \tag{3.24}$$

$$\boldsymbol{B}_i(t) = \boldsymbol{I} - \boldsymbol{A}_i(t), \quad i = 1, 2, \cdots, J_2. \tag{3.25}$$

The $J_1 \times J_1$ matrices $\boldsymbol{A}_i$ and $\boldsymbol{B}_i$ are important decorrelating terms for performing the GSO among all weights at each iteration. Unlike the SLA [16] and GHA [20] algorithms, whose stability analyses are based on the stochastic approximation theory [13], the stability analysis of the LEAP algorithm is based on Lyapunov's first theorem, and η can be selected as a small positive constant. Due to the use of a constant learning rate, the LEAP is capable of tracking nonstationary processes. The LEAP can satisfactorily extract PCs even for ill-conditioned autocorrelation matrices.

3.4.4 The Dot-Product-Decorrelation Algorithm (DPD)

The DPD algorithm is a nonlocal PCA algorithm, and it moves $\boldsymbol{w}_i, i = 1, 2, \ldots, J_2$, toward the J_2 principal eigenvectors c_i, ordered arbitrarily

$$\boldsymbol{w}_i(t+1) = \boldsymbol{w}_i(t) + \eta(t)\left[\boldsymbol{x}(t)y_i(t) - \left(\sum_{j=1}^{J_2} \boldsymbol{w}_j(t)\boldsymbol{w}_j^{\mathrm{T}}(t)\right)\frac{\boldsymbol{w}_i(t)}{\|\boldsymbol{w}_i(t)\|}\right], \tag{3.26}$$

where $\eta(t)$ satisfies the Robbins–Monro conditions. The algorithm induces the norms of the weight vectors toward the corresponding eigenvalues, i.e., $\|\boldsymbol{w}_i(t)\| \to \lambda_i(t)$, as $t \to \infty$. The algorithm is as fast as the GHA [20], weighted SLA [18, 19], and least mean squared error reconstruction (LMSER) [21] algorithms.

3.4.5 Anti-Hebbian Rule-Based Principal Component Analysis

When the update of a synaptic weight is proportional to the correlation of the presynaptic and postsynaptic activities, and the direction of the change is opposite to that in the Hebbian rule, the learning rule is called an anti-Hebbian learning rule [3]. The anti- Hebbian rule can be used to remove correlations between units receiving correlated inputs [22, 23], and it is inherently stable.

Anti-Hebbian rule-based PCA algorithms can be derived by using a network architecture of the J_1–J_2 FNN with lateral connections among the output units [22, 23]. The lateral connections can be in a symmetrical or hierarchical topology. A hierarchical lateral connection topology is illustrated in Fig. 3.2, based on which the Rubner–Tavan PCA algorithm [22, 23] and the APEX [24] were proposed. In [25], the local PCA algorithm is based on a full lateral connection topology. The feedforward weight matrix $\boldsymbol{W}$ is described in the preceding sections, and the

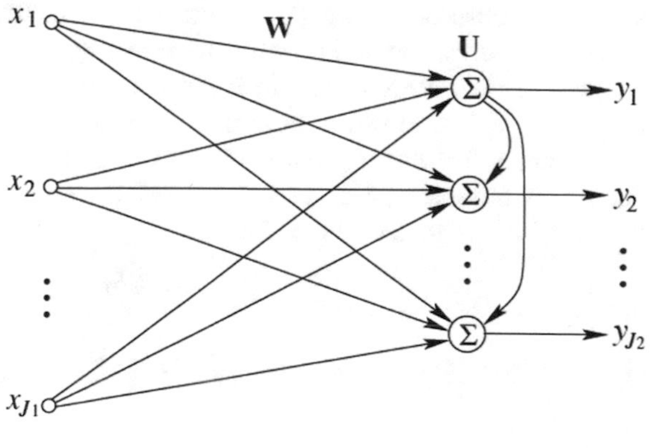

Fig. 3.2 Architecture of the PCA network with hierarchical lateral connections. The lateral weight matrix $\boldsymbol{U}$ is an upper triangular matrix with the diagonal elements being zero

lateral weight matrix $\boldsymbol{U} = [\boldsymbol{u}_1 \ldots \boldsymbol{u}_{J2}]$ is a $J_2 \times J_2$ matrix, where $\boldsymbol{u}_i = (u_{1i}, u_{2i}, \ldots, u_{J2i})^{\mathrm{T}}$ includes all the lateral weights connected to neuron i and u_{ji} denotes the lateral weight from neuron j to neuron i.

3.4.5.1 Rubner-Tavan PCA Algorithm

The Rubner-Tavan PCA algorithm is based on the PCA network with hierarchical lateral connection topology [22, 23]. The algorithm extracts the first J_2 PCs in a decreasing order of the eigenvalues. The output of the network is given by [22, 23]

$$y_i = \boldsymbol{w}_i^{\mathrm{T}}\boldsymbol{x} + \boldsymbol{u}_i^{\mathrm{T}}\boldsymbol{y}, \quad i = 1, 2, \ldots, J_2. \tag{3.27}$$

Note that $u_{ji} = 0$ for $j \geq i$ and $\boldsymbol{U}$ is a $J_2 \times J_2$ upper triangular matrix.

The weights $\boldsymbol{w}_i$ are trained by Oja's rule, and the lateral weights $\boldsymbol{u}_i$ are updated by the anti-Hebbian rule

$$\boldsymbol{w}_i(t+1) = \boldsymbol{w}_i(t) + \eta_1(t)y_i(t)[\boldsymbol{x}(t) - \hat{\boldsymbol{x}}(t)], \tag{3.28}$$

$$\hat{\boldsymbol{x}} = \boldsymbol{W}^{\mathrm{T}}\boldsymbol{y}, \tag{3.29}$$

$$\boldsymbol{u}_i(t+1) = \boldsymbol{u}_i(t) - \eta_2 y_i(t)\boldsymbol{y}(t). \tag{3.30}$$

This is a nonlocal algorithm. Typically, the learning rate $\eta_1 = \eta_2 > 0$ is selected as a small number between 0.001 and 0.1 or according to a heuristic derived from the Robbins–Monro conditions. During the training process, the outputs of the neurons are gradually uncorrelated and the lateral weights approach zero. The network should be trained until the lateral weights $\boldsymbol{u}_i$ are below a specified level.

3.4.5.2 APEX Algorithm

The APEX algorithm is used to adaptively extract the PCs [24]. The algorithm is recursive and adaptive, namely, given $i - 1$ PCs, it can produce the ith PC iteratively. The hierarchical structure of lateral connections among the output units serves the purpose of weight orthogonalization. This structure also allows the network to grow or shrink without retraining the old units. The convergence analysis of the APEX algorithm is based on the stochastic approximation theory, and the APEX is proved to have the property of exponential convergence.

Assuming that the correlation matrix $\boldsymbol{C}$ has distinct eigenvalues arranged in the decreasing order as $\lambda_1 > \lambda_2 > \cdots > \lambda_{J2}$ with the associated eigenvectors $\boldsymbol{w}_1, \ldots, \boldsymbol{w}_{J2}$, the algorithm is given by [24, 26]

$$\boldsymbol{y} = \boldsymbol{W}^{\mathrm{T}}\boldsymbol{x}, \tag{3.31}$$

$$y_i = \boldsymbol{w}_i^{\mathrm{T}}\boldsymbol{x} + \boldsymbol{u}^{\mathrm{T}}\boldsymbol{y}, \tag{3.32}$$

where $\boldsymbol{y} = (y_1, \ldots, y_{i-1})^{\mathrm{T}}$ is the output vector, $\boldsymbol{u} = (u_{1i}, u_{2i}, \ldots, u_{(i-1)i})^{\mathrm{T}}$, and $\boldsymbol{W} = [\boldsymbol{w}_1 \ldots \boldsymbol{w}_{i-1}]$ is the weight matrix of the first $i-1$ neurons. These definitions are for the first i neurons, which are different from their respective definitions given in the preceding sections. The iteration is given as [24, 26]

$$\boldsymbol{w}_i(t+1) = \boldsymbol{w}_i(t) + \eta_i(t)\left[y_i(t)\boldsymbol{x}(t) - y_i^2(t)\boldsymbol{w}_i(t)\right], \tag{3.33}$$

$$\boldsymbol{u}(t+1) = \boldsymbol{u}(t) - \eta_i(t)\left[y_i(t)\boldsymbol{y}(t) + y_i^2(t)\boldsymbol{u}(t)\right]. \tag{3.34}$$

Equations (3.33) and (3.34) are respectively the Hebbian and anti-Hebbian parts of the algorithm. y_i tends to be orthogonal to all the previous components due to the anti-Hebbian rule, also called the orthogonalization rule.

Both sequential and parallel APEX algorithms have been presented in [26]. In the parallel APEX, all J_2 output neurons work simultaneously. In the sequential APEX, the output neurons are added one by one. The sequential APEX is more attractive in practical applications, since one can decide a desirable number of neurons during the learning process. The APEX algorithm is especially useful when the number of required PCs is not known a priori. When the environment is changing over time, a new PC can be added to compensate for the change without affecting the previously computed principal components. Thus, the network structure can be expanded if necessary.

The stopping criterion can be that for each i the changes in $\boldsymbol{w}_i$ and $\boldsymbol{u}$ are below a threshold. At this time, $\boldsymbol{w}_i$ converges to the eigenvector of the correlation matrix $\boldsymbol{C}$ associated with the ith largest eigenvalue, and $\boldsymbol{u}$ converges to zero. The stopping criterion can also be that the change of the average output variance $\sigma_i^2(t)$ is sufficiently small.

Most existing linear complexity methods including the GHA [20], the SLA [16], and the PCA with the lateral connections require a computational complexity of $O(J_1J_2)$ per iteration. For the recursive computation of each additional PC, the APEX requires $O(J_1)$ operations per iteration, while the GHA utilizes $O(J_1J_2)$ per iteration. In contrast to the heuristic derivation of the APEX, a class of learning algorithms, called the *ψ-APEX*, is presented based on criterion optimization [27]. ψ can be selected as any function that guarantees the stability of the network. Some members in the class have better numerical performance and require less computational effort compared to that of both the GHA and the APEX.

3.5 Least Mean Squared Error-Based Principal Component Analysis

Existing PCA algorithms including the Hebbian rule-based algorithms can be derived by optimizing an objective function using the gradient descent method. The least mean squared error (LMSE)-based methods are derived from the modified MSE function

$$E(\boldsymbol{W}) = \sum_{t_1=1}^{t} \mu^{t-t_1} \left\| \boldsymbol{x}_{t_1} - \boldsymbol{W}\boldsymbol{W}^T \boldsymbol{x}_{t_1} \right\|^2, \tag{3.35}$$

where $0 < \mu \leq 1$ is a forgetting factor used for nonstationary observation sequences, and t is the current time instant. Many adaptive PCA algorithms actually optimize (3.35) by using the gradient descent method [21, 28] and the RLS method [28–32].

The gradient descent or Hebbian rule-based algorithms are highly sensitive to parameters such as η. It is difficult to choose proper parameters guaranteeing both a small misadjustment and a fast convergence. To overcome these drawbacks, applying the RLS to the minimization of (3.35) yields the RLS-based algorithms such as the adaptive principal components extraction (APEX) [24, 26], the Kalman-type RLS [29], the projection approximation subspace tracking (PAST) [28], the PAST with deflation (PASTd) [28], and the robust RLS algorithm (RRLSA) [31].

All RLS-based PCA algorithms exhibit fast convergence and high tracking accuracy and are suitable for slow changing nonstationary vector stochastic processes. All these algorithms correspond to a three-layer J_1-J_2-J_1 linear autoassociative network model, and they can extract all the J_2 PCs in a descending order of the eigenvalues, where a GSO-like orthonormalization procedure is used.

3.5.1 *Least Mean Square Error Reconstruction Algorithm (LMSER)*

The LMSER algorithm was derived based on the MSE criterion using the gradient descent method [21]. The LMSER algorithm can be written as

$$\boldsymbol{w}_i(t+1) = \boldsymbol{w}_i(t) + \eta(t)\{2\boldsymbol{A}(t) - \boldsymbol{C}_i(t)\boldsymbol{A}(t) - \boldsymbol{A}(t)\boldsymbol{C}_i(t) - \gamma[\boldsymbol{B}_i(t)\boldsymbol{A}(t) + \boldsymbol{A}(t)\boldsymbol{B}_i(t)]\}\boldsymbol{w}_i(t), \tag{3.36}$$

for $i = 1, 2, \ldots, J_2$, where $\boldsymbol{A}(t) = \boldsymbol{x}(t)\boldsymbol{x}^{\mathrm{T}}(t)$, $\boldsymbol{C}_i(t) = \boldsymbol{w}_i(t)\boldsymbol{w}_i^{\mathrm{T}}(t), i = 1, 2, \ldots, J_2$, $\boldsymbol{B}_i(t) = \boldsymbol{B}_{i-1}(t) + \boldsymbol{C}_{i-1}(t)$, $i = 2, \ldots, J_2$, and $\boldsymbol{B}_1(t) = \boldsymbol{0}$. The selection of $\eta(t)$ is based on the Robbins–Monro conditions and $\gamma \geq 1$.

The LMSER reduces to Oja's algorithm when $\boldsymbol{W}(t)$ is orthonormal, namely $\boldsymbol{W}^{\mathrm{T}}(t)\boldsymbol{W}(t) = \boldsymbol{I}$. Because of this, Oja's algorithm can be treated as an approximate stochastic gradient rule to minimize the MSE. Increasing the values of γ and δ results in a larger asymptotic MSE but faster convergence and vice versa, namely the stability speed problem. The LMSER uses nearly twice as much computation as the weighted SLA [18, 19] and the GHA [20], for each update of the weight. However, it leads to a smaller asymptotic and faster convergence for the minor eigenvectors [33].

3.5.2 Projection Approximation Subspace Tracking Algorithm (PAST)

The PASTd [28] is a well-known subspace tracking algorithm updating the signal eigenvectors and eigenvalues. The PASTd is based on the PAST. Both the PAST and the PASTd are derived for complex-valued signals, which are very common in signal processing area. At iteration t, the PASTd algorithm is given as [28]

$$y_i(t) = \boldsymbol{w}_i^H(t-1)\boldsymbol{x}_i(t), \tag{3.37}$$

$$\delta_i(t) = \mu\delta_i(t-1) + |y_i(t)|^2, \tag{3.38}$$

$$\hat{\boldsymbol{x}}_i(t) = \boldsymbol{w}_i(t-1)y_i(t), \tag{3.39}$$

$$\boldsymbol{w}_i(t) = \boldsymbol{w}_i(t-1) + [\boldsymbol{x}_i(t) - \hat{\boldsymbol{x}}_i(t)]\frac{y_i^*(t)}{\delta_i(t)}, \tag{3.40}$$

$$\boldsymbol{x}_{i+1}(t) = \boldsymbol{x}_i(t) - \boldsymbol{w}_i(t)y_i(t), \tag{3.41}$$

for $i = 1, \ldots, J_2$, where $\boldsymbol{x}_1(t) = \boldsymbol{x}_t$, and the superscript * denotes the conjugate operator.

$\boldsymbol{w}_i(0)$ and $\delta_i(0)$ should be suitably selected. $\boldsymbol{W}(0)$ should contain J_2 orthonormal vectors, which can be calculated from an initial block of data or from arbitrary initial data. A simple way is to set $\boldsymbol{W}(0)$ as the J_2 leading unit vectors of the $J_1 \times J_1$ identity matrix. $\delta_i(0)$ can be set as unity. The choice of these initial values affects the transient behavior, but not the steady-state performance of the algorithm. $\boldsymbol{w}_i(t)$ provides an estimate of the ith eigenvector, and $\delta_i(t)$ is an exponentially weighted estimate of the associated eigenvalue.

Both the PAST and the PASTd have linear computational complexity, that is, $O(J_1J_2)$ operations in every update, as in the cases of the SLA [16], the GHA [20], the LMSER [21], and the novel information criterion (NIC) algorithm [30]. The PAST computes an arbitrary basis of the signal subspace, while the PASTd is able to update the signal eigenvectors and eigenvalues. Both algorithms produce nearly orthonormal, but not exactly orthonormal, subspace basis or eigenvector estimates. If perfectly orthonormal eigenvector estimates are required, an orthonormalization procedure is necessary. The Kalman-type RLS [29] combines the basic RLS algorithm with the GSO procedure in a manner similar to that of the GHA. The Kalman-type RLS and the PASTd are exactly identical if the inverse of the covariance of the output of the ith neuron, $P_i(t)$, in the Kalman-type RLSA is set as $1/\delta_i(t)$ in the PASTd.

In the one-unit case, both the PAST and PASTd are identical to Oja's learning rule except that the PAST and the PASTd have a self-tuning learning rate $1/\delta_1(t)$. Both the PAST and the PASTd provide much more robust estimates than the EVD and converge much faster than the SLA [16]. The PASTd has been extended for the tracking of both the rank and the subspace by using information theoretic criteria such as the AIC and the MDL [34].

3.5.3 Robust RLS Algorithm (RRLSA)

The RRLSA [31] is more robust than the PASTd [28]. The RRLSA can be implemented in a sequential or parallel manner. Given the ith neuron, the sequential algorithm is given for all patterns as [31]

$$\overline{\overline{\boldsymbol{w}}}_i(t-1) = \frac{\boldsymbol{w}_i(t-1)}{\|\boldsymbol{w}_i(t-1)\|}, \tag{3.42}$$

$$y_i(t) = \overline{\overline{\boldsymbol{w}}}_i^{\mathrm{T}}(t-1)\boldsymbol{x}(t), \tag{3.43}$$

$$\hat{\boldsymbol{x}}_i(t) = \sum_{j=1}^{i-1} y_j(t)\overline{\overline{\boldsymbol{w}}}_j(t-1), \tag{3.44}$$

$$\boldsymbol{w}_i(t) = \mu\boldsymbol{w}_i(t-1) + [\boldsymbol{x}_i(t) - \hat{\boldsymbol{x}}_i(t)]y_i(t), \tag{3.45}$$

$$\hat{\lambda}_i(t) = \frac{\|\boldsymbol{w}_i(t)\|}{t}, \tag{3.46}$$

for $i = 1, \ldots, J_2$, where y_i is the output of the ith hidden unit, and $\boldsymbol{w}_i(0)$ is initialized as a small random value. By changing (3.44) into a recursive form, the RRLSA becomes a local algorithm.

The RRLSA has the same flexibility as the Kalman-type RLS [29], the PASTd, and the APEX, in that increasing the number of neurons does not affect the previously extracted principal components. The RRLSA naturally selects the inverse of the output energy as the adaptive learning rate for the Hebbian rule. The Hebbian and Oja rules are closely related to the RRLSA algorithm by suitable selection of the learning rates [31].

The RRLSA is also robust to the error accumulation from the previous components, which exists in the sequential PCA algorithms such as the Kalman-type RLS and the PASTd. The RRLSA converges rapidly, even if the eigenvalues extend over several orders of magnitude. According to the empirical results [31], the RRLSA provides the best performance in terms of convergence speed as well as steady-state error, whereas the Kalman-type RLS and the PASTd have similar performance, which is inferior to that of the RRLSA.

3.6 Optimization-Based Principal Component Analysis

The PCA can be derived by many optimization methods based on a properly defined objective function. This leads to many other algorithms, including gradient descent-based algorithms [9–11, 35], the CG method [36], and the quasi-Newton method [37, 38]. The gradient descent method usually converges to a local minimum. Second-order algorithms such as the CG and quasi-Newton methods typically converge much faster than first-order methods but have a computational complexity of $O(J_1^2 J_2)$ per iteration.

The infomax principle [10, 11] was first proposed by Linsker to describe a neural network algorithm. The principal subspace is derived by maximizing the mutual information criterion. Other examples of information criterion-based algorithms are the NIC algorithm [30] and the coupled PCA [39].

3.6.1 Novel Information Criterion (NIC) Algorithm

The NIC algorithm [30] is obtained by applying the gradient descent method to maximize the NIC. The NIC is a cost function very similar to the mutual information criterion [10, 11] but integrates a soft constraint on the weight orthogonalization

$$E_{\mathrm{NIC}} = \frac{1}{2}\left\{\ln\left(\det(\boldsymbol{W}^{\mathrm{T}}\boldsymbol{R}\boldsymbol{W})\right) - \mathrm{tr}(\boldsymbol{W}^{\mathrm{T}}\boldsymbol{W})\right\}. \tag{3.47}$$

Unlike the MSE, the NIC has a steep landscape along the trajectory from a small weight matrix to the optimum one. E_{NIC} has a single global maximum, and all the other stationary points are unstable saddle points. At the global maximum

$$E_{\mathrm{NIC}}^{*} = \frac{1}{2}\left(\sum_{i=1}^{J_2} \ln\,\lambda_i - J_2\right), \tag{3.48}$$

$\boldsymbol{W}$ yields an arbitrary orthonormal basis of the principal subspace.

The NIC algorithm was derived from E_{NIC} by using the gradient descent method, and the algorithm is given as

$$\boldsymbol{W}(t+1) = (1-\eta)\boldsymbol{W}(t) + \eta\hat{\boldsymbol{C}}(t+1)\boldsymbol{W}(t)\left[\boldsymbol{W}^{\mathrm{T}}(t)\hat{\boldsymbol{C}}(t+1)\boldsymbol{W}(t)\right]^{-1}, \tag{3.49}$$

where $\hat{\boldsymbol{C}}(t)$ is the estimate of the covariance matrix $\boldsymbol{C}(t)$

$$\hat{\boldsymbol{C}}(t) = \frac{1}{t}\sum_{i=1}^{t} \mu^{t-i}\boldsymbol{x}_i\boldsymbol{x}_i^{\mathrm{T}} = \mu\frac{t-1}{t}\hat{\boldsymbol{C}}(t-1) + \frac{1}{t}\boldsymbol{x}_i\boldsymbol{x}_i^{\mathrm{T}} \tag{3.50}$$

and $\mu \in (0, 1]$ is a forgetting factor. The NIC algorithm has a computational complexity of $O(J_1^2 J_2)$ per iteration.

Like the PAST algorithm [28], the NIC algorithm is a PSA method. It can extract the principal eigenvectors when the deflation technique is incorporated. The NIC algorithm converges much faster than the SLA and the LMSER and can globally converge to the PSA solution from almost any weight initialization. Reorthormalization can be applied so as to perform true PCA [30].

By selecting a well-defined adaptive learning rate, the NIC algorithm can also generalize some well-known PSA/PCA algorithms. For online implementation, an RLS version of the NIC algorithm has also been given in [30]. The PAST algorithm [28] is a special case of the NIC algorithm when η is unity, and the NIC algorithm essentially represents a robust improvement of the PAST.

In order to break the symmetry in the NIC, the weighted information criterion (WINC) [32] was proposed by adding a weight to the NIC. Two WINC algorithms are, respectively, derived by using the gradient ascent and the RLS. The gradient ascent-based WINC algorithm can be viewed as an extended weighted SLA with an adaptive step size, leading to a much faster convergence speed. The RLS-based WINC algorithm has not only fast convergence and high accuracy, but also a low computational complexity.

3.6.2 Coupled Principal Component Analysis

The most popular PCA or MCA algorithms do not consider eigenvalue estimates in the update of the weights, and they suffer from the stability speed problem because the eigen motion depends on the eigenvalues of the covariance matrix [39]. The convergence speed of a system depends on the eigenvalues of its Jacobian. In PCA algorithms, the eigen motion depends on the principal eigenvalue of the covariance matrix, while in MCA algorithms it depends on all eigenvalues [39].

Coupled learning rules can be derived by applying the Newton method to a common information criterion. In coupled PCA/MCA algorithms, both the eigenvalues and eigenvectors are simultaneously adapted. The Newton method yields averaged systems with identical speed of convergence in all eigen directions. The Newton descent-based PCA and MCA algorithms, respectively called nPCA and nMCA, are derived by using the information criterion [39]:

$$E_{\text{coupled}}(\boldsymbol{w}, \lambda) = \frac{\boldsymbol{w}^{\mathrm{T}} \boldsymbol{C} \boldsymbol{w}}{\lambda} - \boldsymbol{w}^{\mathrm{T}} \boldsymbol{w} + \ln \lambda, \tag{3.51}$$

where λ is an estimate of the eigenvalue.

By approximation $\boldsymbol{w}^{\mathrm{T}}\boldsymbol{w} \approx 1$, the nPCA is reduced to the ALA [17]. Further approximating the ALA by $\boldsymbol{w}^{\mathrm{T}}\boldsymbol{C}\boldsymbol{w} \approx \lambda$ leads to an algorithm called cPCA. The cPCA is a stable PCA algorithm, but there may be fluctuation in the weight vector length in the iteration process. This problem can be avoided by explicitly renormalizing the weight vector at every iteration, and this leads to the following robust PCA (rPCA) algorithm [39]:

$$\boldsymbol{w}(t+1)=\boldsymbol{w}(t)+\eta(t)\left(\frac{\boldsymbol{x}(t)y(t)}{\lambda(t)}-\boldsymbol{w}(t)\right), \tag{3.52}$$

$$\boldsymbol{w}(t+1)=\frac{\boldsymbol{w}(t+1)}{\|\boldsymbol{w}(t+1)\|}, \tag{3.53}$$

$$\lambda(t+1)=\lambda(t)+\eta(t)(y^2(t)-\lambda(t)), \tag{3.54}$$

where $\eta(t)$ is a small positive number and can be selected according to the Robbins–Monro conditions. The rPCA is shown to be closely related to the RRLSA algorithm [31] by applying the first-order Taylor approximation on the rPCA. The RRLSA can also be derived from the ALA algorithm by using the first-order Taylor approximation.

In order to extract multiple PCs, one has to apply an orthonormalization procedure, e.g., the GSO, or its first-order approximation as used in the SLA, or deflation as in the GHA. In the coupled learning rules, multiple PCs are simultaneously estimated by a coupled system of equations. It has been reported in [40] that in the coupled learning rules a first-order approximation of the GSO is superior to the standard deflation procedure in terms of orthonormality error and the quality of the eigenvectors and eigenvalues generated. An additional normalization step that enforces unit length of the eigenvectors further improves the orthonormality of the weight vectors [40].

3.7 Nonlinear Principal Component Analysis

The aforementioned PCA algorithms apply a linear transform to the input data. The PCA is based on the Gaussian assumption for data distribution, and the optimality of the PCA results from taking into account only the second-order statistics, namely the covariances. For non-Gaussian data distributions, the PCA is not able to capture complex nonlinear correlations, and nonlinear processing of the data is usually more efficient. Nonlinearities introduce higher-order statistics into the computation in an implicit way. Higher-order statistics, defined by cumulants or higher-than-second moments, are needed for a good characterization of non-Gaussian data.

The Gaussian distribution is only one of the canonical exponential distributions, and it is suitable for describing real-valued data. In the case of binary-valued, integer-valued, or non-negative data, the Gaussian assumption is inappropriate, and a family of exponential distributions can be used. For example, the Poisson distribution is better suited for integer data and the Bernoulli distribution to binary data, and an exponential distribution to nonnegative data. All these distributions belong to the exponential family. The PCA can be generalized to distributions of the exponential family. This generalization is based on a generalized linear model and criterion functions using the Bregman distance. This approach permits hybrid dimensionality reduction in which different distributions are used for different attributes of the data.

When the feature space is nonlinearly related to the input space, we need to use nonlinear PCA. The outputs of nonlinear PCA networks are usually more independent than their respective linear cases. For non-Gaussian input data, the PCA may fail to provide an adequate representation, while a nonlinear PCA permits the extraction of higher-order components and provides a sufficient representation. Nonlinear PCA networks and learning algorithms can be classified into symmetric and hierarchical ones similar to those for the PCA networks. After training, the lateral connections between output units are not needed, and the network becomes purely feedforward. In the following, we discuss the kernel PCA, robust PCA, and nonlinear PCA.

3.7.1 Kernel Principal Component Analysis

Kernel PCA [41, 42] is a special, linear algebra-based nonlinear PCA, which introduces kernel functions into the PCA. The kernel PCA first maps the original input data into a high-dimensional feature space using the kernel method and then calculates the PCA in the high-dimensional feature space. The linear PCA in the high-dimensional feature space corresponds to a nonlinear PCA in the original input space.

Given an input pattern set $\{\boldsymbol{x}_i \in \Re^{J_1} | i = 1, 2, \ldots, N\}, \varphi : \Re^{J_1} \rightarrow \Re^{J_2}$ is a nonlinear map from the J_1-dimensional input to the J_2-dimensional feature space. A J_2-by-J_2 correlation matrix in the feature space is defined by

$$\boldsymbol{C}_1 = \frac{1}{N} \sum_{i=1}^{N} \varphi(\boldsymbol{x}_i) \varphi^{\mathrm{T}}(\boldsymbol{x}_i). \tag{3.55}$$

Like the PCA, the set of feature vectors is limited to zero mean

$$\frac{1}{N} \sum_{i=1}^{N} \varphi(\boldsymbol{x}_i) = 0. \tag{3.56}$$

A procedure to select $\boldsymbol{\phi}$ satisfying (3.56) is given in [41, 42]. The PCs can then be computed by solving the eigenvalue problem [41, 42]

$$\lambda \boldsymbol{v} = \boldsymbol{C}_1 \boldsymbol{v} = \frac{1}{N} \sum_{j=1}^{N} \left(\varphi(\boldsymbol{x}_i)^{\mathrm{T}} v \right) \varphi(\boldsymbol{x}_i). \tag{3.57}$$

Thus, $\boldsymbol{v}$ must be in the span of the mapped data

$$\boldsymbol{v} = \sum_{i=1}^{N} \alpha_i \varphi(\boldsymbol{x}_i). \tag{3.58}$$

After premultiplying both sides of (3.58) by $\boldsymbol{\phi}(\mathbf{x}_j)$ and performing mathematical manipulations, the kernel PCA problem reduces to

$$\boldsymbol{K\alpha} = \lambda \boldsymbol{\alpha}, \tag{3.59}$$

where λ and $\boldsymbol{\alpha} = (\alpha_1, \ldots, \alpha_N)^{\mathrm{T}}$ are, respectively, the eigenvalues and the associated eigenvectors of $\boldsymbol{K}$, and $\boldsymbol{K}$ is an $N \times N$ kernel matrix with

$$\boldsymbol{K}_{ij} = \kappa(\boldsymbol{x}_i, \boldsymbol{x}_j) = \varphi^T(\boldsymbol{x}_i)\varphi(\boldsymbol{x}_j), \tag{3.60}$$

where $\kappa(\cdot)$ is a kernel function.

Popular kernel functions used in the kernel method are the polynomial, Gaussian kernel, and sigmoidal kernels, which are, respectively, given by

$$\kappa(\boldsymbol{x}_i, \boldsymbol{x}_j) = \left(\boldsymbol{x}_i^{\mathrm{T}}\boldsymbol{x}_j + \theta\right)^{a0}, \tag{3.61}$$

$$\kappa(\boldsymbol{x}_i, \boldsymbol{x}_j) = \mathrm{e}^{-\frac{\left\|x_i - x_j\right\|^2}{2\sigma^2}}, \tag{3.62}$$

$$\kappa(\boldsymbol{x}_i, \boldsymbol{x}_j) = \tanh\left(c_0\left(\boldsymbol{x}_i^{\mathrm{T}}\boldsymbol{x}_j\right) + \theta\right), \tag{3.63}$$

where a_0 is a positive integer, $\sigma > 0$, and c_0, $\theta \in R$. Even if the exact form of $\phi(\cdot)$ does not exist, any symmetric function $\kappa\ (\boldsymbol{x}_i, \boldsymbol{x}_j)$ satisfying Mercer's theorem can be used as a kernel function.

Arrange the eigenvalues in the descending order $\lambda_1 \geq \lambda_2 \geq \cdots \geq \lambda_{J2} > 0$ and denote their associated eigenvectors as $\boldsymbol{\alpha}_1, \ldots\ \boldsymbol{\alpha}_{J2}$. The eigenvectors are further normalized as $\boldsymbol{\alpha}_k^T \boldsymbol{\alpha}_k = 1/\lambda_k$.

The nonlinear PCs of $\boldsymbol{x}$ can be extracted by projecting the mapped pattern $\boldsymbol{\phi}(\boldsymbol{x})$ onto $\boldsymbol{v}_k$

$$\boldsymbol{v}_k^{\mathrm{T}} \varphi(\boldsymbol{x}) = \sum_{j=1}^{N} \alpha_{k,j} \kappa(\boldsymbol{x}_j, \boldsymbol{x}), \tag{3.64}$$

for $k = 1, 2, \ldots, J_2$, where $\alpha_{k,\ j}$ is the jth element of $\boldsymbol{\alpha}_k$.

The kernel PCA algorithm is much more complicated and may sometimes be trapped more easily into local minima. The PCA needs to deal with an eigenvalue problem of a $J_1 \times J_1$ matrix, while the kernel PCA needs to solve an eigenvalue problem of an $N \times N$ matrix. Sparse approximation methods can be applied to reduce the computational cost.

3.7.2 Robust/Nonlinear Principal Component Analysis

In order to increase the robustness of the PCA against outliers, a simple way is to eliminate the outliers or replace them by more appropriate values. A better alternative is to use a robust version of the covariance matrix based on the *M*-estimator. The data from which the covariance matrix is constructed may be weighted such that the samples far from the mean have less importance.

Several popular PCA algorithms have been generalized into robust versions by applying a statistical physics approach [43], where the defined objective function can be regarded as a soft generalization of the *M*-estimator. In this subsection, robust PCA algorithms are defined so that the optimization criterion grows less than quadratically and the constraint conditions are the same as for the PCA algorithms [44], which are based on a quadratic criterion. The robust PCA problem usually leads to mildly nonlinear algorithms, in which the nonlinearities appear at selected places only and at least one neuron produces the linear response $y_i = \boldsymbol{x}^{\mathrm{T}}\boldsymbol{w}_i$. When all neurons generate nonlinear responses $y_i = \varphi(\boldsymbol{x}^{\mathrm{T}}\boldsymbol{w}_i)$, the algorithm is referred to as the nonlinear PCA.

Variance Maximization-based Robust Principal Component Analysis:

The PCA is to maximize the output variances $E\left[y_i^2\right] = E\left[\left(\boldsymbol{w}_i^{\mathrm{T}}\boldsymbol{x}\right)^2\right] = \boldsymbol{w}_i^{\mathrm{T}}\boldsymbol{C}\boldsymbol{w}_i$ of the linear network under orthonormality constraints. In the hierarchical case, the constraints take the form $\boldsymbol{w}_i^{\mathrm{T}}\boldsymbol{w}_j = \delta_{ij}, j \leq i, \delta_{ij}$ being the Kronecker delta. In the symmetric case, symmetric orthonormality constraints $\boldsymbol{w}_i^{\mathrm{T}}\boldsymbol{w}_j = \delta_{ij}$ are applied. The SLA and GHA algorithms correspond to the symmetric and hierarchical network structures, respectively.

To derive robust PCA algorithms, the variance maximization criterion is generalized as $E\left[\sigma\left(\boldsymbol{w}_i^{\mathrm{T}}\boldsymbol{x}\right)\right]$ for the *i*th neuron, subject to hierarchical or symmetric orthonormality constraints, where $\sigma(x)$ is the *M*-estimator assumed to be a valid differentiable cost function that grows less than quadratically, at least for large x. Examples of such functions are $\sigma(x) = \ln\cosh(x)$ and $\sigma(x) = |x|$. The robust PCA in general does not coincide with the corresponding PCA solution, although it can be close to it. The robust PCA is derived by applying the gradient descent method [21, 44]

$$\boldsymbol{w}_i(t+1) = \boldsymbol{w}_i(t) + \eta(t)\varphi(y_i(t))\boldsymbol{e}_i(t), \tag{3.65}$$

$$\boldsymbol{e}_i(t) = \boldsymbol{x}(t) - \hat{\boldsymbol{x}}_i(t), \tag{3.66}$$

$$\hat{\boldsymbol{x}}_i(t) = \sum_{j=1}^{I(i)} y_j(t)\boldsymbol{w}_j(t), \tag{3.67}$$

where $\boldsymbol{e}_i(t)$ is the instantaneous representation error vector, and the influence function $\varphi(\boldsymbol{x}) = \mathrm{d}\sigma(\boldsymbol{x})/\mathrm{d}\boldsymbol{x}$.

In the symmetric case, $I(i) = J_2$ and the errors $\boldsymbol{e}_i(t) = \boldsymbol{e}(t)$, $i = 1, \ldots, J_2$. When $\phi(x) = x$, the algorithm is simplified to the SLA. Otherwise, it defines a robust generalization of Oja's rule, first proposed quite heuristically. In the hierarchical case, $I(i) = i$, $i = 1, \ldots, J_2$. If $\phi(x) = x$, the algorithm coincides exactly with the GHA; Otherwise, it defines a robust generalization of the GHA. In the hierarchical case, $\mathbf{e}_i(t)$ can be calculated in a recursive form $\boldsymbol{e}_i(t) = \boldsymbol{e}_{i-1}(t) - y_i(t)\boldsymbol{w}_i(t)$, with $\boldsymbol{e}_0(t) = \boldsymbol{x}(t)$.

Mean Squared Error Minimization-based Robust Principal Component Analysis:

PCA algorithms can also be derived by minimizing the MSE $E\left[\|\boldsymbol{e}_i\|^2\right]$, where $\boldsymbol{e}_i(t) = \boldsymbol{x}(t) - \hat{\boldsymbol{x}}_i(t)$. Accordingly, robust PCA can be obtained by minimizing $\boldsymbol{1}^{\mathrm{T}}E[\sigma(\boldsymbol{e}_i)] = E\left[\|h(\boldsymbol{e}_i)\|^2\right]$, where $\boldsymbol{1}$ is a J_2-dimensional vector, all of whose entries are unity, and $\sigma(\cdot)$ and $h(\cdot)$ are applied componentwise on the input vector. Here, $h(x) = \sqrt{\sigma(x)}$. When $\sigma(x) = x^2$, it corresponds to the MSE. A robust PCA is defined if $\sigma(x)$ grows less than quadratically. Using the gradient descent method leads to

$$\boldsymbol{w}_i(t+1) = \boldsymbol{w}_i(t) + \eta(t)\left[\boldsymbol{w}_i(t)^{\mathrm{T}}\varphi(\boldsymbol{e}_i(t))\boldsymbol{x}(t) + \boldsymbol{x}^{\mathrm{T}}(t)\boldsymbol{w}_i(t)\varphi(\boldsymbol{e}_i(t))\right], \tag{3.68}$$

where $\boldsymbol{w}_i$ estimates the robust counterparts of the principal eigenvectors $\boldsymbol{c}_i$. The first term in the bracket is very small and can be neglected, and thus we can get a simplified algorithm

$$\boldsymbol{w}_i(t+1) = \boldsymbol{w}_i(t) + \eta(t)\boldsymbol{x}^{\mathrm{T}}(t)\boldsymbol{w}_i(t)\varphi(\boldsymbol{e}_i(t)) = \boldsymbol{w}_i(t) + \eta(t)y_i(t)\varphi(\boldsymbol{e}_i(t)). \tag{3.69}$$

Algorithms (3.69) and (3.65) resemble each other. However, Algorithm (3.69) generates a linear final input–output mapping, while in Algorithm (3.65) the input–output mapping is nonlinear. When $\phi(x) = x$, algorithms (3.69) and (3.65) are the same as the SLA in the symmetric case, and the same as the GHA in the hierarchical case.

Another Nonlinear Extension to Principal Component Analysis:

A nonlinear PCA algorithm may be derived by the gradient descent method for minimizing the MSE $E\left[\|\boldsymbol{\varepsilon}_i\|^2\right]$, where the error vector $\boldsymbol{\varepsilon}_i$ is a nonlinear extension to $\boldsymbol{e}_i(t) = \boldsymbol{x}(t) - \hat{\boldsymbol{x}}_i(t)$. The nonlinear PCA so obtained has a form similar to the robust PCA given by (3.65) through (3.67)

$$\boldsymbol{w}_i(t+1) = \boldsymbol{w}_i(t) + \eta(t)\varphi(y_i(t))\boldsymbol{\varepsilon}_i(t), \tag{3.70}$$

$$\boldsymbol{\varepsilon}_i(t) = \boldsymbol{x}(t) - \sum_{j=1}^{I(i)} \varphi(y_j(t))\boldsymbol{w}_j(t), \tag{3.71}$$

for $i = 1, \ldots, J_2$.

In this case, $I(i) = J_2$ and all $\varepsilon_i(t)$ are the same. The nonlinear PCA in the hierarchical case is a direct nonlinear generalization of the GHA. In the hierarchical case, $I(i) = i$ and (3.71) can be computed recursively

$$\varepsilon_i(t) = \varepsilon_{i-1}(t) - \varphi(y_i(t))\boldsymbol{w}_i(t), \tag{3.72}$$

with $\varepsilon_0(t) = \boldsymbol{x}(t)$.

It has been pointed out in [44] that robust and nonlinear PCA algorithms have better stability than the corresponding PCA algorithms if the (odd) nonlinearity $\phi(x)$ grows less than linearly, namely $|\phi(x)| < |x|$. On the contrary, nonlinearities growing faster than linearly cause stability problems easily and therefore are not recommended.

3.7.3 Autoassociative Network-Based Nonlinear PCA

The MLP can be used to perform nonlinear dimension reduction and hence nonlinear PCA. Both the input and output layers of the MLP have J_1 units, and one of its hidden layers, known as the bottleneck or representation layer, has J_2 units, $J_2 < J_1$. The network is trained to reproduce its input vectors. This kind of network is called the autoassociative MLP. After the network is trained, it performs a projection onto the J_2-dimensional subspace spanned by the first J_2 principal components of the data. The vectors of weights leading to the hidden units form a basis set that spans the principal subspace, and data compression therefore occurs in the bottleneck layer. Many applications of the MLP in autoassociative mode for PCA are available in the literature [45, 46].

The three-layer autoassociative J_1-J_2-J_1 feedforward network or MLP network can also be used to extract the first J_2 principal components of J_1-dimensional data. If nonlinear activation functions are applied in the hidden layer, the network performs as a nonlinear PCA network. In the case of nonlinear units, local minima certainly appear. However, if linear units are used in the output layer, nonlinearity in the hidden layer is theoretically meaningless [45]. This is due to the fact that the network tries to approximate a linear mapping.

3.8 Other PCA or Extensions of PCA

Besides the algorithms reviewed in the preceding parts, there exist lots of other PCAs or their extensions. For example, there are minor component analysis, constrained PCA, localized PCA, incremental PCA, supervised PCA, complex-valued PCA, two-dimensional PCA, generalized eigenvalue decomposition, singular value decomposition, canonical correlation analysis, etc. Among these algorithms, minor component analysis (MCA), generalized eigenvalue decomposition, and singular value decomposition are important PCA algorithms or extensions. So we will have separate chapters to study them, respectively. See Chaps. 4, 8 and 9 for more details. Here, we only discuss the remaining algorithms.

Constrained PCA: When certain subspaces are less preferred than others, this yields the constrained PCA [47]. The optimality criterion for constrained PCA is variance maximization, as in PCA, but with an external subspace orthogonality constraint that extracts principal components orthogonal to some undesired subspace [3]. Constrained PCA first decomposes the data matrix by projecting the data matrix onto the spaces spanned by matrices of external information and then applies PCA to the decomposed matrices, which involves generalized SVD. APEX can be applied to recursively solve the constrained PCA problem [26].

Localized PCA: The nonlinear PCA problem can be overcome using localized PCA [3]. First, the data space is partitioned into a number of disjunctive regions, followed by the estimation of the principal subspace within each partition by linear PCA. The distribution is then collectively modeled by a collection of linear PCA models, each characterizing a partition. It should be noted that the localized PCA is different from local PCA. In the latter, the update at each node makes use of only local information. VQ-PCA [48] is a locally linear model that uses vector quantization to define the Voronoi regions for localized PCA. An online localized PCA algorithm [49] was developed by extending the neural gas method. ASSOM is another localized PCA for unsupervised extraction of invariant local features from the input data. Localized PCA provides an efficient means to decompose high-dimensional data compression problems into low-dimensional ones [3].

Incremental PCA: Incremental PCA algorithm can update eigenvectors and eigenvalues incrementally. It is applied to a single training sample at a time, and the intermediate eigen problem must be solved repeatedly for every training sample [50]. Chunk incremental PCA [51] processes a chunk of training samples at a time. It can reduce the training time effectively and obtain major eigenvectors with fairly good approximation. In Chunk incremental PCA, the update of an eigen space is completed by performing single eigenvalue decomposition. The SVD updating-based incremental PCA algorithm [52] gives a close approximation to the batch-mode PCA method, and the approximation error is proved to be bounded. Candid covariance-free IPCA [53] is a fast incremental PCA algorithm, which is used to compute the principal components of a sequence of samples incrementally without estimating the covariance matrix.

Supervised PCA: Like supervised clustering, supervised PCA [54] is achieved by augmenting the input of PCA with the class label of the data set. Class-augmented PCA [55] is a supervised feature extraction method, which is composed of processes for encoding the class information, augmenting the encoded information to data, and extracting features from class-augmented data by applying PCA.

Complex-valued PCA: Complex PCA is a generalization of PCA in complex-valued data sets [56], and it employs the same neural network architecture as for PCA, but with complex weights. Complex-domain GHA [57] extends GHA for complex principal component extraction, and it is very similar to GHA except that complex notations are introduced. In [58], a complex-valued neural network, model is developed for nonlinear complex PCA, and it uses the architecture of Kramer's nonlinear PCA network, but with complex weights and biases. The algorithm can extract nonlinear features missed by PCA. Both PAST and PASTd are, respectively, the PSA and PCA algorithms derived for complex-valued signals [28]. Complex-valued APEX [59] actually allows for extracting a number of principal components from a complex-valued signal. The robust complex PCA algorithms have also been derived in [60] for hierarchically extracting principal components of complex-valued signals using a robust statistics-based loss function.

Two-dimensional PCA: Because of the small-sample-size problem for image representation, PCA is prone to be overfitted to the training set. Two-dimensional PCA can address these problems. In two-dimensional PCA, an image covariance matrix is constructed directly using the original image matrices instead of the transformed vectors, and its eigenvectors are derived for image feature extraction.

2DPCA [61] evaluates the covariance (scatter) matrix more accurately than PCA does, since it only reflects the information between rows and is a row-based PCA. Diagonal PCA [62] improves 2DPCA by defining the image scatter matrix as the covariances between the variations of the rows and those of the columns of the images and is more accurate than PCA and 2DPCA. In modular PCA [63], an image is divided into n_1 subimages and PCA is performed on all these subimages. 2DPCA and modular PCA both solve the overfitting problems by reducing the dimension and by increasing the training vectors yet introduce the high feature dimension problem.

Bidirectional PCA [64] reduces the dimension in both column and row directions for image feature extraction, whose feature dimension is much less than that of 2DPCA. It has to be performed in batch mode. PCA-L_1 [65] is a fast and robust L_1-norm-based PCA. L_1-norm-based two-dimensional PCA (2DPCA-L_1) [66] is a two-dimensional generalization of PCA-L_1 [65]. It avoids the eigen decomposition process, and its iteration step is easy to perform. The uncorrelated multilinear PCA algorithm [67] is used for unsupervised subspace learning of tensorial data. It not only obtains features that maximize the variance captured, but also enforces a zero-correlation constraint, thus extracting uncorrelated features.

3.9 Summary

An overview of a variety of neural network-based principal component analysis algorithms has been presented in this chapter. Many new adaptive PCA algorithms are being added to this field, indicating a consistent interest in this direction. Nevertheless, neural network-based PCA algorithms have been considered a matured subject. Many problems and current research interest lie in performance analysis of PCA algorithms, minor component analysis, generalization or extensions of PCA algorithms, etc., which will be discussed in the next chapters.

References

1. Diamantaras, K. I., & Kung, S. Y. (1996). *Principal component neural networks: Theory and applications*. New York: Wiley.
2. Jolliffe, I. T. (2002). *Principal component analysis* (2nd ed.). Berlin: Springer.
3. Du, K. L., & Swamy, M. N. S. (2013). *Neural networks and statistical learning*. Berlin: Springer.
4. Liu J. L, Wang H., Lu J. B., Zhang B. B., & Du K. L. (2012). Neural network implementations for PCA and its extensions. *Artificial Intelligence*. doi:10.5402/2012/847305
5. Hebb, D. O. (1949). *The organization of behavior*. New York: Wiley.
6. Zafeiriou, S., & Petrou, M. (2010). Nonlinear non-negative component analysis algorithms. *IEEE Transactions on Image Processing, 19*(4), 1050–1066.
7. Rubner, J., & Tavan, P. (1989). A self-organizing network for principal-component analysis. *Europhysics Letters, 10*(7), 693–698.
8. Oja, E. (1982). A simplified neuron model as a principal component analyzer. *Journal of Mathematical Biology, 15*(3), 267–273.
9. Yuille, A. L., Kammen, D. M., & Cohen, D. S. (1989). Quadrature and development of orientation selective cortical cells by Hebb rules. *Biological Cybernetics, 61*(3), 183–194.
10. Linsker, R. (1986). From basic network principles to neural architecture. *Proceedings of the National Academy of Sciences of the USA* (Vol. 83, pp. 7508–7512), 8390-8394, 9779-8783.
11. Linsker, R. (1988). Self-organization in a perceptual network. *IEEE Computer, 21*(3), 105–117.
12. Hassoun, M. H. (1995). *Fundamentals of artificial neural networks*. Cambridge, MA: MIT Press.
13. Ljung, L. (1977). Analysis of recursive stochastic algorithm. *IEEE Transactions on Automatic Control, 22*(4), 551–575.
14. Zufiria, P. J. (2002). On the discrete-time dynamics of the basic Hebbian neural-network node. *IEEE Transactions on Neural Networks, 13*(6), 1342–1352.
15. Zhang, Y., Ye, M., Lv, J. C., & Tan, K. K. (2005). Convergence analysis of a deterministic discrete system of Oja's PCA learning algorithm. *IEEE Transactions on Neural Networks, 16*(6), 1318–1328.
16. Oja, E., & Karhunen, J. (1985). On stochastic approximation of the eigenvectors and eigenvalues of the expectation of a random matrix. *Journal of Mathematical Analysis and Applications, 106*(1), 69–84.
17. Chen, L. H., & Chang, S. (1995). An adaptive learning algorithm for principal component analysis. *IEEE Transactions on Neural Networks, 6*(5), 1255–1263.
18. Oja, E. (1992). Principal components, minor components, and linear neural networks. *Neural Networks, 5*(6), 929–935.
19. Oja, E., Ogawa, H., & Wangviwattana, J. (1992). Principal component analysis by homogeneous neural networks. *IEICE Transactions on Information and Systems, 75*(3), 366–382.
20. Sanger, T. D. (1989). Optimal unsupervised learning in a single-layer linear feedforward neural network. *Neural Networks, 2*(6), 459–473.
21. Xu, L. (1993). Least mean square error reconstruction principle for self-organizing neural-nets. *Neural Networks, 6*(5), 627–648.
22. Rubner, J., & Schulten, K. (1990). Development of feature detectors by self-organization. *Biological Cybernetics, 62*(62), 193–199.
23. Rubner, J., & Tavan, P. (1989). A self-organizing network for principal-component analysis. *Europhysics Letters, 10*(7), 693–698.
24. Kung, S. Y., & Diamantaras, K. I. (1990). A neural network learning algorithm for adaptive principal components extraction (APEX). *Proceedings of IEEE ICCASSP* (pp. 861–864). Albuquerque, NM.

25. Foldiak, P. (1989). Adaptive network for optimal linear feature extraction. *Proceedings of International Joint Conference Neural Networks (IJCNN)* (Vol. 1, pp. 401–405). Washington, DC.
26. Kung, S. Y., Diamantaras, K. I., & Taur, J. S. (1994). Adaptive principal components extraction (APEX) and applications. *IEEE Transactions on Signal Processing, 42*(5), 1202–1217.
27. Fiori, S., & Piazza, F. (1998). A general class of ψ-APEX PCA neural algorithms. *IEEE Transactions on Circuits and Systems I, 47*(9), 1394–1397.
28. Yang, B. (1995). Projection approximation subspace tracking. *IEEE Transactions on Signal Processing, 43*(1), 95–107.
29. Bannour, S., & Azimi-Sadjadi, M. R. (1995). Principal component extraction using recursive least squares learning. *IEEE Transactions on Neural Networks, 6*(2), 457–469.
30. Miao, Y., & Hua, Y. (1998). Fast subspace tracking and neural network learning by a novel information criterion. *IEEE Transactions on Signal Processing, 46*(7), 1967–1979.
31. Ouyang, S., Bao, Z., & Liao, G. (2000). Robust recursive least squares learning algorithm for principal component analysis. *IEEE Transactions on Neural Networks, 11*(1), 215–221.
32. Ouyang, S., & Bao, Z. (2002). Fast principal component extraction by a weighted information criterion. *IEEE Transactions on Signal Processing, 50*(8), 1994–2002.
33. Chatterjee, C., Roychowdhury, V. P., & Chong, E. K. P. (1998). On relative convergence properties of principal component analysis algorithms. *IEEE Transactions on Neural Networks, 9*(2), 319–329.
34. Yang, B. (1995). An extension of the PASTd algorithm to both rank and subspace tracking. *IEEE Signal Processing Letters, 2*(9), 179–182.
35. Chauvin, Y. (1989). Principal component analysis by gradient descent on a constrained linear Hebbian cell. *Proceedings of the International Joint Conference on Neural Networks* (pp. 373–380). Washington, DC.
36. Fu, Z., & Dowling, E. M. (1995). Conjugate gradient eigenstructure tracking for adaptive spectral estimation. *IEEE Transactions on Signal Processing, 43*(5), 1151–1160.
37. Kang, Z., Chatterjee, C., & Roychowdhury, V. P. (2000). An adaptive quasi-Newton algorithm for eigensubspace estimation. *IEEE Transactions on Signal Processing, 48*(12), 3328–3333.
38. Ouyang, S., Ching, P. C., & Lee, T. (2003). Robust adaptive quasi-Newton algorithms for eigensubspace estimation. *IEEE Proceedings—Vision, Image and Signal Processing, 150*(5), 321–330.
39. Moller, R., & Konies, A. (2004). Coupled principal component analysis. *IEEE Transactions on Neural Network, 15*(1), 214–222.
40. Moller, R. (2006). First-order approximation of Gram-Schmidt orthonormalization beats deflation in coupled PCA learning rules. *Neurocomputing, 69*(13–15), 1582–1590.
41. Schölkopf, B., Smola, A., & Müller, K. R. (1998). Nonlinear component analysis as a kernel eigenvalue problem. *Neural Computations, 10*(5), 1299–1319.
42. Schölkopf, B., Mika, S., Burges, C., Knirsch, P., Müller, K. R., Rätsch, G., et al. (1970). Input space vs. feature space in kernel-based methods. *IEEE Transactions on Neural Networks, 10*(5), 1000–1017.
43. Xu, L., & Yuille, A. L. (1995). Robust principal component analysis by self-organizing rules based on statistical physics approach. *IEEE Transactions on Neural Networks, 6*(1), 131–143.
44. Karhunen, J., & Joutsensalo, J. (1995). Generalizations of principal component analysis, optimization problems, and neural networks. *Neural Networks, 8*(4), 549–562.
45. Bourlard, H., & Kamp, Y. (1988). Auto-association by multilayer perceptrons and singular value decomposition. *Biological Cybernetics, 59*(4–5), 291–294.
46. Kramer, M. A. (1991). Nonlinear principal component analysis using autoassociative neural networks. *AIChE Journal, 37*(2), 233–243.
47. Kung, S. Y. (1990). Constrained principal component analysis via an orthogonal learning network. *Proceedings of the IEEE International Symposium on Circuits and Systems* (Vol. 1, pp. 719–722). New Orleans, LA.

48. Kambhatla, N., & Leen, T. K. (1993). Fast non-linear dimension reduction. *Proceedings of IEEE International Conference on Neural Networks* (Vol. 3, pp. 1213–1218). San Francisco, CA.
49. Moller, R., & Hoffmann, H. (2004). An extension of neural gas to local PCA. *Neurocomputing, 62*(1), 305–326.
50. Hall, P., & Martin, R. (1998). Incremental eigenanalysis for classification. *Proceedings of British Machine Vision Conference* (Vol. 1, pp. 286–295).
51. Ozawa, S., Pang, S., & Kasabov, N. (2008). Incremental learning of chunk data for online pattern classification systems. *IEEE Transactions on Neural Networks, 19*(6), 1061–1074.
52. Zhao, H., Yuen, P. C., & Kwok, J. T. (2006). A novel incremental principal component analysis and its application for face recognition. *IEEE Transactions on Systems, Man, and Cybernetics, 36*(4), 873–886.
53. Weng, J., Zhang, Y., & Hwang, W. S. (2003). Candid covariance-free incremental principal component analysis. *IEEE Transactions on Pattern Analysis and Machine Intelligence, 25*(8), 1034–1040.
54. Chen, S., & Sun, T. (2005). Class-information-incorporated principal component analysis. *Neurocomputing, 69*(1–3), 216–223.
55. Park, M. S., & Choi, J. Y. (2009). Theoretical analysis on feature extraction capability of class-augmented PCA. *Pattern Recognition, 42*(11), 2353–2362.
56. Horel, J. D. (1984). Complex principal component analysis: Theory and examples. *Journal of Climate and Applied Meteorology, 23*(12), 1660–1673.
57. Zhang, Y., & Ma, Y. (1997). CGHAfor principal component extraction in the complex domain. *IEEE Transactions on Neural Networks, 8*(5), 1031–1036.
58. Rattan, S. S. P., & Hsieh, W. W. (2005). Complex-valued neural networks for nonlinear complex principal component analysis. *Neural Networks, 18*(1), 61–69.
59. Chen, Y., & Hou, C. (1992). High resolution adaptive bearing estimation using a complex-weighted neural network. *Proceedings of IEEE International Conference on Acoustics, Speech, and Signal Processing (ICASSP)* (Vol. 2, pp. 317–320). San Francisco, CA.
60. Cichocki, A., Swiniarski, R. W., & Bogner, R. E. (2010). Hierarchical neural network forrobust PCA computation of complex valued signals. In *World Congress on Neural Networks* (pp. 818–821).
61. Yang, J., Zhang, D., Frangi, A. F., & Yang, J. Y. (2004). Two-dimensional PCA: A new approach to appearance-based face representation and recognition. *IEEE Transactions on Pattern Analysis Machine Intelligence, 26*(1), 131–137.
62. Zhang, D., Zhou, Z. H., & Chen, S. (2006). Diagonal principal component analysis for face recognition. *Pattern Recognition, 39*(1), 140–142.
63. Gottumukkal, R., & Asari, V. K. (2004). An improved face recognition technique based on modular PCA approach. *Pattern Recognition Letters, 25*(4), 429–436.
64. Zuo, W., Zhang, D., & Wang, K. (2006). Bidirectional PCA with assembled matrix distance metric for image recognition. *IEEE Transactions on Systems, Man, and Cybernetics B, 36*(4), 863–872.
65. Kwak, N. (2008). Principal component analysis based on L1-norm maximization. *IEEE Transactions on Pattern Analysis and Machine Intelligence, 30*(9), 1672–1680.
66. Li, X., Pang, Y., & Yuan, Y. (2010). L1-norm-based 2DPCA. *IEEE Transactions on Systems, Man, and Cybernetics B, 40*(4), 1170–1175.
67. Lu, H., Plataniotis, K. N. K., & Venetsanopoulos, A. N. (2009). Uncorrelated multilinear principal component analysis for unsupervised multilinear subspace learning. *IEEE Transactions on Neural Networks, 20*(11), 1820–1836.

Chapter 4
Neural Networks for Minor Component Analysis

4.1 Introduction

The minor subspace (MS) is a subspace spanned by all the eigenvectors associated with the minor eigenvalues of the autocorrelation matrix of a high-dimensional vector sequence. The MS, also called the noise subspace (NS), has been extensively used in array signal processing. The NS tracking is a primary requirement in many real-time signal processing applications such as the adaptive direction-of-arrival (DOA) estimation, the data compression in data communications, the solution of a total least squares problem in adaptive signal processing, and the feature extraction technique for a high-dimensional data sequence. Although the MS can be efficiently obtained by the algebraic approaches such as the QR decomposition, such approaches usually have the computational complexity of $O(N^2 r)$ per data update, where N and r are the dimensions of the high-dimensional vector sequence and the MS, respectively. Hence, it is of great interest to find some learning algorithms with lower computational complexity for adaptive signal processing applications.

The minor component analysis (MCA) deals with the recovery of the eigenvector associated with the smallest eigenvalue of the autocorrelation matrix of the input data, and it is an important statistical method for extracting minor component. To solve the MCA problem, many neural learning algorithms have been proposed for over 30 years [1–7, 8, 9–11]. These learning algorithms can be used to extract minor component from input data without calculating the correlation matrix in advance, which makes neural networks method more suitable for real-time applications. In neural network algorithms for MCA, the only nonlinear network is the Hopfield network by Mathew and Reddy [12, 13], in which a constrained energy function was proposed, using a penalty function, to minimize the RQ. The neurons use sigmoidal activation functions; however, the structure of the network is problem-dependent (the number of neurons is equal to the dimension of the eigenvectors). In addition, it is necessary to estimate the trace of the covariance

© Science Press, Beijing and Springer Nature Singapore Pte Ltd. 2017
X. Kong et al., *Principal Component Analysis Networks and Algorithms*,
DOI 10.1007/978-981-10-2915-8_4

matrix for selecting appropriate penalty factors. All other existing neural networks are made up of one simple linear neuron.

Linear neurons are the simplest units to build a neural network. They are often considered uninteresting because linear functions can be computed with linear networks and a network with several layers of linear units can always be collapsed into a linear network without any hidden layer by multiplying the weights in a proper fashion [4]. On the contrary, there are very important advantages. Oja [14] has found that a simple linear neuron with an unsupervised constrained Hebbian learning rule can extract the principal component from stationary input data. Contrary to the nonlinear neural networks which are seriously plagued by the problem of local minima of their cost function, the linear networks have simple cost landscapes [4, 15].

The adaptive algorithms for tracking one minor component (MC) have been proposed in [2–4], all resulting in adaptive implementation of Pisarenko's harmonic retrieve estimator [16]. Thompson [17] proposed an adaptive algorithm for extracting the smallest eigenvector from a high-dimensional vector stream. Yang and Kaveh [18] extended Thompson's algorithm [17] to estimate multiple MCs with the inflation procedure. However, Yang and Kaveh's algorithm requires normalization operation. Oja [2] and Xu et al. [1] proposed several efficient algorithms that can avoid the normalized operation. Luo et al. [3] presented a minor component analysis (MCA) algorithm that does not need any normalization operation. Recently, some modifications for Oja's MS tracking algorithms have been proposed in [4, 19, 8, 9, 10]. Chiang and Chen [20] showed that a learning algorithm can extract multiple MCs in parallel with the appropriate initialization instead of the inflation method. On the basis of an information criterion and by extending and modifying the total least mean squares (TLMS) algorithm [5], Ouyang et al. [21] developed an adaptive MC tracker that can automatically find the MS without using the inflation method. Recently, Cirrincione et al. [4], [22] proposed a learning algorithm called MCA EXIN that may have satisfactory convergence. Zhang and Leung [23] proposed a much more general model for the MC and provided an efficient technique for analyzing the convergence properties of these algorithms. Interestingly, Cirrincione et al. analyzed these algorithms in detail.

4.2 Review of Neural Network-Based MCA Algorithms

The neural network-based MCA algorithms in the literature can be roughly classified into the following classes: nonstabilizing algorithm, self-stabilizing algorithm, fast algorithms, etc. In [4], Cirrincione et al. distinguished between two classes of learning algorithms according to the time course of the length of the eigenvector estimate. In the algorithms of the first class, e.g., OJAn, Luo, and MCA EXIN, the length of the weight vector in the fixed point is undetermined. In an exact solution of the differential equation, the weight vector length would not deviate from its initial value. However, when a numerical procedure (like Euler's method) is applied, all these rules are plagued by "sudden divergence" of the weight vector length. In the rules of the second class, e.g., Doug, Chen, XU*, OJA*, OJA +, FENG, AMEX, ZLlog, ZLdiff, and Wang, the weight vector length converges toward a fixed value. Within the second class, some rules are self-stabilizing with respect to the time course of the weight vector length. In these algorithms, the weight vector length converges toward a fixed value independent of the presented input vector. The Doug and Chen rules are strictly self-stabilizing, while OJA+, OJA*, and XU* show this property in the vicinity of the fixed point. All rules lacking self-stabilization are potentially prone to fluctuations and divergence in the weight vector length [10]. Besides, there still exists a need to develop fast minor subspace tracking algorithms. Due to the direct use of Gram–Schmidt orthonormalization, the computational complexity of DPM algorithm [18] is $O(np^2)$. The Fast Rayleigh's quotient-based Adaptive Noise Subspace (FRANS) [24] algorithm reduces its dominant complexity to $3np$. Afterward, several algorithms such as FDPM [25], [26], HFRANS [27], FOOJA [28], and YAST [29] were proposed. In the following, we will discuss some well-known MCA/MSA algorithms.

4.2.1 Extracting the First Minor Component

The well-known algorithms for extracting the first minor component include the anti-Hebbian learning rule (OJA) [1], the normalized anti-Hebbian learning rule (OJAn) [1], the constrained anti-Hebbian learning algorithm [30], OJA+ algorithm [2], Luo algorithm [3], the total least mean squares algorithm (TLMS) [5], and the MCA EXIN algorithm [4]. The OJA algorithm tends rapidly to infinite magnitudes of the weights, and the OJAn algorithm leads to better convergence, but it may also lead to infinite magnitudes of weights before the algorithm converges [4]. The constrained anti-Hebbian learning algorithm has a simple structure and requires a low computational complexity per update, which has been applied to adaptive FIR and IIR filtering. It can be used to solve the TLS parameter estimation and has been extended for complex-valued TLS problem. The TLMS algorithm is a random adaptive algorithm for extracting the MC, based on which an adaptive step-size learning algorithm [31] has been derived for extracting the MC by introducing information criterion. The algorithm outperforms the TLMS in terms of both convergence speed and estimation accuracy.

Consider a linear unit with input $\boldsymbol{x}(t) = [x_1(t), x_2(t), \ldots, x_n(t)]^{\mathrm{T}}$ and $y(t) = \sum_{i=1}^{N} w_i(t)x_i(t) = \boldsymbol{w}^{\mathrm{T}}(t)\boldsymbol{x}(t)$, where $\boldsymbol{w}(t) = [w_1(t), w_2(t), \ldots, w_n(t)]^{\mathrm{T}}$ is the weight vector. In the MCA analysis, $\boldsymbol{x}(t)$ is a bounded continuous-valued stationary ergodic data vector with finite second moments. The existing well-known learning laws for the MCA of the autocorrelation matrix $\boldsymbol{R} = E[\boldsymbol{x}(t)\boldsymbol{x}^{\mathrm{T}}(t)]$ of the input vector $\boldsymbol{x}(t)$ are listed below.

Oja's MCA algorithm: Changing the Oja's learning law for PCA into a constrained anti-Hebbian rule and by reversing the sign, Oja's MCA algorithm is given by

$$\boldsymbol{w}(t+1) = \boldsymbol{w}(t) - \alpha(t)y(t)[\boldsymbol{x}(t) - y(t)\boldsymbol{w}(t)], \tag{4.1}$$

where $\alpha(t)$ is a positive learning rate. The OJAn algorithm is

$$\boldsymbol{w}(t+1) = \boldsymbol{w}(t) - \alpha(t)y(t)\left[\boldsymbol{x}(t) - \frac{y(t)\boldsymbol{w}(t)}{\boldsymbol{w}^{\mathrm{T}}(t)\boldsymbol{w}(t)}\right]. \tag{4.2}$$

Another Oja's learning rule (OJA+) is as follows:

$$\boldsymbol{w}(t+1) = \boldsymbol{w}(t) - \alpha(t)\left[y(t)\boldsymbol{x}(t) - \left(y^2(t) + 1 - \|\boldsymbol{w}(t)\|_2^2\right)\boldsymbol{w}(t)\right]. \tag{4.3}$$

In [3], Luo et al. proposed the following rule:

$$\boldsymbol{w}(t+1) = \boldsymbol{w}(t) - \alpha(t)\left[\boldsymbol{w}^{\mathrm{T}}(t)\boldsymbol{w}(t)y(t)\boldsymbol{x}(t) - y^2(t)\boldsymbol{w}(t)\right]. \tag{4.4}$$

In [5], TLMS algorithm was given by:

$$\boldsymbol{w}(t+1) = \boldsymbol{w}(t) - \alpha(t)[\boldsymbol{w}^{\mathrm{T}}(t)\boldsymbol{w}(t)y(t)\boldsymbol{x}(t) - \boldsymbol{w}(t)]. \tag{4.5}$$

In [31], by defining an information criterion: $\max_{\boldsymbol{w}}\{J(\boldsymbol{w}) = \frac{1}{2}(\boldsymbol{w}^{\mathrm{T}}\boldsymbol{R}\boldsymbol{w} - \log\|\boldsymbol{w}\|^2)\}$, an adaptive step-size learning algorithm has been derived for extracting the MC as follows:

$$\boldsymbol{w}(t+1) = \boldsymbol{w}(t) - \frac{\eta}{\|\boldsymbol{w}(t)\|^2}\left[\|\boldsymbol{w}(t)\|^2\boldsymbol{x}(t)\boldsymbol{x}^{\mathrm{T}}(t)\boldsymbol{w}(t) - \boldsymbol{w}(t)\right]. \tag{4.6}$$

In order to improve the performance of MCA algorithms, a novel algorithm called MCA EXIN [4] was proposed as follows:

$$\boldsymbol{w}(t+1) = \boldsymbol{w}(t) - \frac{\alpha(t)y(t)}{\|\boldsymbol{w}(t)\|_2^2}\left[\boldsymbol{x}(t) - \frac{y(t)\boldsymbol{w}(t)}{\|\boldsymbol{w}(t)\|_2^2}\right]. \tag{4.7}$$

The convergence of the above MCA algorithms is indirectly proven by the convergence of their corresponding averaging ODE. In [4], the analysis of the temporal behavior of all the above MCA neurons is analyzed by using not only the ODE approximation, but especially the stochastic discrete laws. Using only the ODE approximation does not reveal some of the most important features of these algorithms. For instance, it will be shown that the constancy of the weight modulus for OJAn and Luo, which is the consequence of the use of the ODE, is not valid, except, as a very first approximation, in approaching the minor component. This study will also lead to the very important problem of sudden divergence [4].

4.2.2 Oja's Minor Subspace Analysis

Oja's minor subspace analysis (MSA) algorithm can be formulated by reversing the sign of the learning rate of SLA for PSA [32]

$$\boldsymbol{W}(t+1) = \boldsymbol{W}(t) - \eta[\mathbf{x}(t) - \boldsymbol{W}(t)\boldsymbol{y}(t)]\boldsymbol{y}^{\mathrm{T}}(t), \tag{4.8}$$

where $\boldsymbol{y}(t) = \boldsymbol{W}^{\mathrm{T}}(t)\boldsymbol{x}(t)$, $\eta > 0$ is the learning rate. This algorithm requires the assumption that the smallest eigenvalue of the autocorrelation matrix $\boldsymbol{C}$ is less than unity. However, Oja's MSA algorithm is known to diverge [2]. The bigradient PSA algorithm is a modification to the SLA and is obtained by introducing an additional bigradient term embodying the orthonormal constraints of the weights, and it can be used for MSA by reversing the sign of η.

4.2.3 Self-stabilizing MCA

The concept of self-stability was presented in [10]. If the norm of the weight vector in the algorithm converges toward a fixed value independent of the presented input vector, then this algorithm is called self-stabilizing. Since all algorithms lacking self-stability are prone to fluctuations and divergence in the weight vector norm, the self-stability is an indispensable property for adaptive algorithms.

The MCA algorithm proposed in [33] can be written as

$$\boldsymbol{W}(t+1) = \boldsymbol{W}(t) - \eta[\boldsymbol{x}(t)\boldsymbol{y}^{\mathrm{T}}(t)\boldsymbol{W}^{\mathrm{T}}(t)\boldsymbol{W}(t) - \boldsymbol{W}(t)\boldsymbol{y}(t)\boldsymbol{y}^{\mathrm{T}}(t)]. \tag{4.9}$$

During initialization, $\boldsymbol{W}^{\mathrm{T}}(0)\boldsymbol{W}(0)$ is required to be diagonal. Algorithm (4.9) suffers from a marginal instability, and thus, it requires intermittent normalization such that $\|\boldsymbol{w}_i\| = 1$.

A self-stabilizing MCA algorithm was given in [8] as

$$\boldsymbol{W}(t+1) = \boldsymbol{W}(t) - \eta\left[\boldsymbol{x}(t)\boldsymbol{y}^{\mathrm{T}}(t)\boldsymbol{W}^{\mathrm{T}}(t)\boldsymbol{W}(t)\boldsymbol{W}^{\mathrm{T}}(t)\boldsymbol{W}(t) - \boldsymbol{W}(t)\boldsymbol{y}(t)\boldsymbol{y}^{\mathrm{T}}(t)\right]. \tag{4.10}$$

Algorithm (4.10) is self-stabilizing, such that none of $\|\boldsymbol{w}_i(t)\| = 1$ deviates significantly from unity. Algorithm (4.10) diverges for PCA when $-\eta$ is changed to $+\eta$. Both Algorithms (4.9) and (4.10) have complexities of $O(J_1 J_2)$.

4.2.4 Orthogonal Oja Algorithm

The orthogonal Oja (OOja) algorithm consists of Oja's MSA plus an orthogonalization of $\boldsymbol{W}(t)$ at each iteration [34] $\boldsymbol{W}^{\mathrm{T}}(t)\boldsymbol{W}(t) = \boldsymbol{I}$. A Householder transform-based implementation of the MCA algorithm was given as [27]

$$\hat{\boldsymbol{x}}(t) = \boldsymbol{W}(t)\boldsymbol{y}(t) \tag{4.11}$$

$$\boldsymbol{e}(t) = \boldsymbol{x}(t) - \hat{\boldsymbol{x}}(t) \tag{4.12}$$

$$\vartheta(t) = \frac{1}{\sqrt{1 + \eta^2 \|\boldsymbol{e}(t)\|^2 \|\boldsymbol{y}(t)\|^2}} \tag{4.13}$$

$$\boldsymbol{u}(t) = \frac{1 - \vartheta(t)}{\eta \|\boldsymbol{y}(t)\|^2} \hat{\boldsymbol{x}}(t) + \vartheta(t)\boldsymbol{e}(t) \tag{4.14}$$

$$\bar{\bar{\boldsymbol{u}}}(t) = \frac{\boldsymbol{u}(t)}{\|\boldsymbol{u}(t)\|} \tag{4.15}$$

$$\boldsymbol{v}(t) = \boldsymbol{W}^{\mathrm{T}}(t)\bar{\bar{\boldsymbol{u}}}(t) \tag{4.16}$$

$$\boldsymbol{W}(t+1) = \boldsymbol{W}(t) - 2\bar{\bar{\boldsymbol{u}}}(t)\boldsymbol{v}^{\mathrm{T}}(t), \tag{4.17}$$

where $\boldsymbol{W}$ is initialized as an arbitrary orthogonal matrix and $\boldsymbol{y}$ is given by $\boldsymbol{y}(t) = \boldsymbol{W}^{\mathrm{T}}(t)\boldsymbol{x}(t)$. The OOja is numerically very stable. By reversing the sign of η, we extract J_2 PCs.

The normalized Oja (NOja) was derived by optimizing the MSE criterion subject to an approximation to the orthonormal constraint $\boldsymbol{W}^{\mathrm{T}}(t)\boldsymbol{W}(t) = \boldsymbol{I}$ [35]. This leads to the optimal learning rate. The normalized orthogonal Oja (NOOja) is an orthogonal version of the NOja such that $\boldsymbol{W}^{T}(t)\boldsymbol{W}(t) = \boldsymbol{I}$ is perfectly satisfied [35]. Both algorithms offer, as compared to Oja's SLA, a faster convergence, orthogonality, and a better numerical stability with a slight increase in the computational complexity. By switching the sign of η in the given learning algorithms, both NOja and NOOja can be used for the estimation of minor and principal subspaces of a vector sequence. All the Algorithms (4.8), (4.9), (4.10), OOja, NOja, and NOOja have a complexity of $O(J_1 J_2)$ [34]. OOja, NOja, and NOOja require less computation load than Algorithms (4.9) and (4.10) [34, 35].

4.2.5 Other MCA Algorithm

By using the Rayleigh quotient as an energy function, the invariant-norm MCA [3] was analytically proved to converge to the first MC of the input signals. The MCA algorithm has been extended to sequentially extract multiple MCs in the ascending order of the eigenvalues by using the idea of sequential elimination in [36]. However, the invariant-norm MCA [3] leads to divergence in finite time [4], and this drawback can be eliminated by renormalizing the weight vector at each iteration. In [11, 33], an alternative MCA algorithm for extracting multiple MCs was described by using the idea of sequential addition, and a conversion method between the MCA and the PCA was also discussed.

Based on a generalized differential equation for the generalized eigenvalue problem, a class of algorithms can be obtained for extracting the first PC or MC by selecting different parameters and functions [23]. Many existing PCA algorithms, e.g., the ones in [14, 37, 38], and MCA algorithms, e.g., the one in [37], are special cases of this class. All the algorithms of this class have the same order of convergence speed and are robust to implementation error.

A rapidly convergent quasi-Newton method has been applied to extract multiple MCs in [13]. This algorithm has a complexity of $O(J_2 J_1^2)$, but with a quadratic convergence. It makes use of an implicit orthogonalization procedure that is built into it through an inflation technique.

4.3 MCA EXIN Linear Neuron

In [4], Cirrincione et al. proposed a MCA EXIN neuron, and its algorithm is as follows:

$$w(t+1) = w(t) - \frac{\alpha(t)y(t)}{\|w(t)\|_2^2}\left[x(t) - \frac{y(t)w(t)}{\|w(t)\|_2^2}\right]. \tag{4.7}$$

The convergence of MCA EXIN is indirectly proven by the convergence of their corresponding averaging ODE. Moreover, its temporal behavior is analyzed by the stochastic discrete laws. This study leads to the very important problem of sudden divergence. In this subsection, we will briefly analyze the MCA EXIN neuron and its algorithm [4].

4.3.1 The Sudden Divergence

The squared modulus of the weight vector at an instant is given by

$$\|w(t+1)\|_2^2 = \|w(t)\|_2^2 + \frac{\alpha^2(t)}{4}\|w(t)\|_2^{-2}\|x(t)\|_2^4 \sin^2 2\theta_{xw}, \tag{4.18}$$

where θ_{xw} is the angle between the directions of $\boldsymbol{x}(t)$ and $\boldsymbol{w}(t)$. From (4.18), the dependence of the squared modulus on the square of the learning rate is apparent. As a consequence, for low learning rates (as near convergence), the increase in the weight modulus is less significant.

The following observations can be made [4]:

(1) Except for particular conditions, the weight modulus always increases

$$\|\boldsymbol{w}(t+1)\|_2^2 > \|\boldsymbol{w}(t)\|_2^2. \tag{4.19}$$

These particular conditions, i.e., all data are in exact particular directions, are too rare to be met in a noisy environment.

(2) $\sin^2 2\theta_{xw}$ is a positive function with four peaks within the interval $(-\pi, \pi]$. This is one of the possible interpretations of the oscillatory behavior of the weight modulus.

In summary, the property of constant modulus for MCA EXIN is not correct [4]. In the following, the divergence will be studied in further detail.

Averaging the behavior of the weight vector after the first critical point has been reached $(t \geq t_0)$, it then follows that $\boldsymbol{w}(t) = \|\boldsymbol{w}(t)\|_2 z_n,\ \forall t \geq t_0$, where z_n is the unit eigenvector associated with the smallest eigenvalue of the autocorrelation matrix of the input vector $\boldsymbol{x}(t)$. From (4.7), the discrete law can be easily obtained for the update of the weight modulus. This discrete law can be regarded as a discretization of the following ODE:

$$\frac{\mathrm{d}\boldsymbol{w}^{\mathrm{T}}\boldsymbol{w}}{\mathrm{d}t} = \frac{1}{(\boldsymbol{w}^{\mathrm{T}}\boldsymbol{w})^2} E\left[\left\| y\boldsymbol{x} - \frac{y^2}{\boldsymbol{w}^{\mathrm{T}}\boldsymbol{w}} \right\|_2^2\right]. \tag{4.20}$$

Without loss of generality, the input data are considered Gaussian. After some matrix manipulations, which can be found in [4], it holds that:

$$\frac{\mathrm{d}p}{\mathrm{d}t} = \frac{1}{p^2} E\left[-\lambda_n^2 + \lambda_n \mathrm{tr}(R)\right] p, \tag{4.21}$$

where $p = \|\boldsymbol{w}(t)\|_2^2$, λ_n is its smallest eigenvalue of $\boldsymbol{R}$, i.e., autocorrelation matrix of the input data. Solving this differential equation with $p(0) = 1$ for the sake of simplicity yields

$$p(t) = \sqrt{1 + 2\left[-\lambda_n^2 + \lambda_n \mathrm{tr}(R)\right] t} \tag{4.22}$$

and since the quantity in brackets is never negative, it follows that $p(t) \to \infty$, as $t \to \infty$. Here, the norm of $\boldsymbol{w}(t)$ diverges.

For Luo MCA algorithm, it holds that

$$p(t) = \frac{1}{\sqrt{1 - 2\left[-\lambda_n^2 + \lambda_n \mathrm{tr}(R)\right]t}}. \tag{4.23}$$

In this case, the divergence happens in a *finite* time (called sudden divergence), i.e.,

$$p(t) \to \infty \quad \text{when} \quad t \to t_\infty = \frac{1}{2\sqrt{-\lambda_n^2 + \lambda_n \mathrm{tr}(R)}}. \tag{4.24}$$

Hence, t_∞ depends on the spreading of the eigenvalue spectrum of R. If the eigenvalues of R are clustered, the sudden divergence appears late. Furthermore, t_∞ is proportional to the inverse of λ_n (high λ_n means noisy data) [4].

4.3.2 The Instability Divergence

In [4], the instability divergence was defined, and it is related to the dynamic stability of algorithm and learning rate. We know that in an iterative algorithm, the learning rate $\alpha(t)$ must be very small to avoid the instability and consequent divergence of the learning law. This implies some problems [4]: (1) A small learning rate gives a low learning speed; (2) it is difficult to find a good learning rate to prevent learning divergence; and (3) the transient and accuracy in the solution are both affected by the choice of the learning rate.

The analysis of instability divergence of MCA EXIN linear neuron is very complex, and the details can be found in [4]. In Sect. 4.4, the analysis method will be discussed in detail.

4.3.3 The Numerical Divergence

The MCA learning laws are iterative algorithms and have a different computational cost per iteration. The limited precision (quantization) errors can degrade the solution of the gradient-based algorithms with regard to the performance achievable in infinite precision. These errors accumulate in time without bound, leading, in the long term, to an eventual overflow. This kind of divergence is here called numerical divergence. There are two sources of quantization errors [4]: (1) The analog-to-digital conversion used to obtain the discrete time series input; for a uniform quantization characteristics, the quantization is zero mean. (2) The finite word length used to store all internal algorithmic quantities; this error is not zero mean. This mean is the result of the use of multiplication schemes that either truncate or round products to fit the given fixed word length.

It is found that the degradation of the solution is proportional to the conditioning of the input, i.e., to the spread of the eigenvalue spectrum of the input autocorrelation matrix. Hence, this problem is important for near singular matrices, e.g., in the application of MCA for the computation of translation in computer vision [4]. Obviously, decreasing the learning rate in the infinite precision algorithm leads to improved performance. Nevertheless, this decrease increases the deviation from infinite precision performance. However, increasing the learning rate can also magnify numerical errors, so there is a trade-off in terms of numerical effects on the size of the learning rate [4].

4.4 A Novel Self-stabilizing MCA Linear Neurons

As mentioned in Sects. 4.2 and 4.3, several adaptive algorithms for tracking one minor component have been proposed. The dynamics of many MCA algorithms have been studied, and a divergence problem of the weight vector norm has been found in some existing MCA algorithms, e.g., OJAn algorithm [1] and Luo algorithm [3]. Also, sudden divergence has been found in some existing algorithms, e.g., Luo algorithm, OJA algorithm [1], and OJA+ algorithms [2] on some condition. In order to guarantee convergence, several self-stabilizing MCA learning algorithms have been proposed [8–10]. In these algorithms, the weight vector of the neuron can be guaranteed to converge to a normalized minor component.

The objective of this section is to develop more satisfactory learning algorithm for the adaptive tracking of MS. For neural network-based learning algorithms, the convergence is crucial to their practical applications. Usually, MSA (or MCA) learning algorithms are described by stochastic discrete time systems. Traditionally, convergence of MSA algorithms is analyzed via a corresponding DCT system, but some restrictive conditions must be satisfied in this method. It is realized that using only DCT method does not reveal some of the most important features of these algorithms. The SDT method uses directly the stochastic discrete learning laws to analyze the temporal behavior of MCA algorithms and has been given more and more attention [4]. In this section, we will introduce a self-stabilizing MCA algorithm and extended it for the tracking of MS [39], which has a more satisfactory numerical stability compared to some existing MSA algorithms, and the dynamics of this algorithm will be analyzed via DCT and SDT methods [39].

4.4.1 A Self-stabilizing Algorithm for Tracking One MC

(1) A self-stabilizing MCA algorithm

Let us consider a single linear neuron with the following input–output relation:

$$y(t) = \boldsymbol{W}^{\mathrm{T}}(t)\boldsymbol{X}(t), (t = 0, 1, 2, \ldots),$$

where $y(t)$ is the neuron output, the input sequence $\{\boldsymbol{X}(t)|\boldsymbol{X}(t) \in \boldsymbol{R}^n(t = 0, 1, 2, \ldots)\}$ is a zero-mean stationary stochastic process, and $\boldsymbol{W}(t) \in \boldsymbol{R}^n(t = 0, 1, 2, \ldots)$ is the weight vector of the neuron. Let $\boldsymbol{R} = E[\boldsymbol{X}(t)\boldsymbol{X}^{\mathrm{T}}(t)]$ denote the autocorrelation matrix of the input sequence $\boldsymbol{X}(t)$, and let λ_i and $\boldsymbol{v}_{\mathbf{i}}$ $(i = 1, 2, \ldots, N)$ denote the eigenvalues and the associated orthonormal eigenvectors of $\boldsymbol{R}$, respectively. We can arrange the orthonormal eigenvectors $\boldsymbol{v}_1, \boldsymbol{v}_2, \cdots, \boldsymbol{v}_N$ such that the associated eigenvalues are in a nondecreasing order: $0 < \lambda_1 \leq \lambda_2 \leq \cdots \leq \lambda_N$.

The dynamics of some major MCA algorithms, e.g., OJA, OJAn, OJA+, Luo, and OJAm algorithms, have been studied, and a MCA EXIN algorithm based on the gradient flow of the Rayleigh quotient of the autocorrelation matrix $\boldsymbol{R}(= E[\boldsymbol{X}(t)\boldsymbol{X}^{\mathrm{T}}(t)])$ on $\boldsymbol{R}^n - \{0\}$ was presented as follows [4]:

$$\boldsymbol{W}(t+1) = \boldsymbol{W}(t) - \alpha(t)\left(\boldsymbol{W}^{\mathrm{T}}(t)\boldsymbol{W}(t)\right)^{-1}\left[(y(t)\boldsymbol{X}(t) - y^2(t)\boldsymbol{W}(t)\left(\boldsymbol{W}^{\mathrm{T}}(t)\boldsymbol{W}(t)\right)^{-1}\right], \tag{4.25}$$

where $\alpha(t)$ is the learning rate, which controls the stability and rate of convergence of the algorithm. MCA EXIN algorithm is analyzed in detail, and it is concluded that the algorithm is the best MCA neuron in terms of stability (no finite time divergence), speed, and accuracy. However, by using the same analytical approach, it is easy to show that it is possible that MCA EXIN converges to infinity. In order to avoid the possible divergence and preserve the good performance of MCA EXIN as much as possible, we propose a modified algorithm as follows:

$$\boldsymbol{W}(t+1) = \boldsymbol{W}(t) - \alpha(t)\left(\boldsymbol{W}^{\mathrm{T}}(t)\boldsymbol{W}(t)\right)^{-1}\left[y(t)\boldsymbol{X}(t) - \left(y^2(t) + 1 - \|\boldsymbol{W}(t)\|^4\right)\left(\boldsymbol{W}^{\mathrm{T}}(t)\boldsymbol{W}(t)\right)^{-1}\boldsymbol{W}(t)\right]. \tag{4.26}$$

The difference between MCA EXIN algorithm and the modified one is that the latter refers to OJA+ algorithm, and adds a term $(1 - \|\boldsymbol{W}(t)\|^4)\boldsymbol{W}(t)$. Then, this renders our algorithm to have satisfactory one-tending property (which will be explained later), and it outperforms some existing MCA algorithms, e.g., OJAm algorithm.

(2) The convergence analysis via DCT method

Usually, MCA learning algorithms are described by SDT systems. It is very difficult to study the convergence of the SDT system directly. So far, dynamics of most of MCA algorithms are indirectly proved via a corresponding DCT system. According to the stochastic approximation theory [40, 41], it can be shown that if some conditions are satisfied, then the asymptotic limit of the discrete learning algorithm of (4.26) can be obtained by solving the following continuous time differential equations:

$$\frac{\mathrm{d}\boldsymbol{W}(t)}{\mathrm{d}t}=-\left(\boldsymbol{W}^{\mathrm{T}}(t)\boldsymbol{W}(t)\right)^{-1}\left[y(t)\boldsymbol{X}(t)-\left(y^{2}(t)+1-\|\boldsymbol{W}(t)\|^{4}\right)\left(\boldsymbol{W}^{\mathrm{T}}(t)\boldsymbol{W}(t)\right)^{-1}\boldsymbol{W}(t)\right]. \tag{4.27}$$

Assume $\boldsymbol{X}(t)$ is stationary, and $\boldsymbol{X}(t)$ is not correlated with $\boldsymbol{W}(t)$, and taking expectation on both sides of Eq. (4.27), then Eq. (4.27) can be approximated by the following ordinary differential equation (ODE):

$$\frac{\mathrm{d}\boldsymbol{W}(t)}{\mathrm{d}t}=-\left(\boldsymbol{W}^{\mathrm{T}}(t)\boldsymbol{W}(t)\right)^{-1}\left[\boldsymbol{R}\boldsymbol{W}(t)-\left(\boldsymbol{W}^{\mathrm{T}}(t)\boldsymbol{R}\boldsymbol{W}(t)+1-\|\boldsymbol{W}(t)\|^{4}\right)\left(\boldsymbol{W}^{\mathrm{T}}(t)\boldsymbol{W}(t)\right)^{-1}\boldsymbol{W}(t)\right]. \tag{4.28}$$

The asymptotic property of (4.28) approximates that of (4.27), and the asymptotic property of (4.28) can be ensured by the following theorem.

Theorem 4.1 *Let* $\boldsymbol{R}$ *be a positive semi-definite matrix,* λ_1 *and* $\boldsymbol{v}_1$ *be respectively its smallest eigenvalue and the associated normalized eigenvector with nonzero first component. If the initial weight vector* $\boldsymbol{W}(0)$ *satisfies* $\boldsymbol{W}^{\mathrm{T}}(0)\boldsymbol{v}_1 \neq 0$ *and* λ_1 *is single, then* $\lim_{t\to\infty}\boldsymbol{W}(t)=\pm\boldsymbol{v}_1$*, i.e.* $\boldsymbol{W}(t)$ *tends to* $\pm\boldsymbol{v}_1$ *asymptotically as* $t\to\infty$.

Proof Denote N eigenvalues of $\boldsymbol{R}$ by $\lambda_1,\lambda_2,\ldots,\lambda_N$, where λ_1 is the smallest eigenvalue and denotes a set of associated normalized eigenvectors by $\boldsymbol{v}_1,\boldsymbol{v}_2,\ldots,\boldsymbol{v}_N$. So $\boldsymbol{R}$ and $\boldsymbol{W}(t)$ can be written as

$$\boldsymbol{R}=\sum_{i=1}^{N}\lambda_i\boldsymbol{v}_i\boldsymbol{v}_i^{\mathrm{T}},\ \boldsymbol{W}(t)=\sum_{i=1}^{N}f_i(t)\boldsymbol{v}_i. \tag{4.29}$$

Then, it holds that

$$\begin{aligned}\frac{\mathrm{d}\boldsymbol{W}(t)}{\mathrm{d}t}&=\sum_{i=1}^{N}\frac{\mathrm{d}f_i(t)}{\mathrm{d}t}\boldsymbol{v}_i\\&=\frac{-\sum_{i=1}^{N}\|\boldsymbol{W}(t)\|^{2}(\lambda_i f_i(t)\boldsymbol{v}_i)+\left(\boldsymbol{W}^{\mathrm{T}}(t)\boldsymbol{R}\boldsymbol{W}(t)+1-\|\boldsymbol{W}(t)\|^{4}\right)\sum_{i=1}^{N}(f_i(t)\boldsymbol{v}_i)}{\|\boldsymbol{W}(t)\|^{4}}\\&=\sum_{i=1}^{N}\left(\left(-\lambda_i\|\boldsymbol{W}(t)\|^{2}+(\boldsymbol{W}^{\mathrm{T}}(t)\boldsymbol{R}\boldsymbol{W}(t)+1-\|\boldsymbol{W}(t)\|^{4})\right)f_i(t)\boldsymbol{v}_i\right)\|\boldsymbol{W}(t)\|^{-4}.\end{aligned} \tag{4.30}$$

and

$$\frac{\mathrm{d}f_i(t)}{\mathrm{d}t}=\left(-\lambda_i\|\boldsymbol{W}(t)\|^{2}+\left(\boldsymbol{W}^{\mathrm{T}}(t)\boldsymbol{R}\boldsymbol{W}(t)+1-\|\boldsymbol{W}(t)\|^{4}\right)\right)f_i(t)\|\boldsymbol{W}(t)\|^{-4} \tag{4.31}$$
$\forall\, i=1,2,\ldots,N.$

Since $f_i(t) = \boldsymbol{W}^{\mathrm{T}}(t)\boldsymbol{v}_i$ and $\boldsymbol{W}^{\mathrm{T}}(0)\boldsymbol{v}_1 \neq 0$, we have $f_1(t) \neq 0$ $(\forall t \geq 0)$. Define

$$\gamma_i(t) = \frac{f_i(t)}{f_1(t)} \ (i = 2, \ldots, N). \tag{4.32}$$

And then it follows that

$$\frac{\mathrm{d}\gamma_i(t)}{\mathrm{d}t} = \frac{(\lambda_1 - \lambda_i)\|\boldsymbol{W}(t)\|^2 f_i(t) f_1(t)}{f_1^2(t)\|\boldsymbol{W}(t)\|^4} = \frac{(\lambda_1 - \lambda_i)\gamma_i(t)}{\|\boldsymbol{W}(t)\|^2}, \tag{4.33}$$

whose solution on $[0, \infty]$ is

$$\gamma_i(t) = \exp\left(\frac{(\lambda_1 - \lambda_i)}{\|\boldsymbol{W}(t)\|^2} \int_0^t \mathrm{d}\tau\right) \ \forall i = 2, \ldots, N. \tag{4.34}$$

If $\lambda_i > \lambda_1$ (i.e., the smallest eigenvalue is single but not multiple), then $\gamma_i(t)$ tends to zero as $t \to \infty$ $(\forall i = 2, \ldots, N)$. Consequently, $\lim_{t \to \infty} f_i(t) = 0$ $(\forall i = 2, \ldots, N)$.

So we have

$$\lim_{t \to \infty} \boldsymbol{W}(t) = \lim_{t \to \infty}\left(\sum_{i=1}^{N} f_i(t)\boldsymbol{v}_i\right) = f_1(t)\boldsymbol{v}_1. \tag{4.35}$$

From (4.35), it follows that

$$\lim_{t \to \infty}\|\boldsymbol{W}(t)\| = \lim_{t \to \infty}\|f_1(t)\boldsymbol{v}_1\| = \lim_{t \to \infty}\|f_1(t)\|. \tag{4.36}$$

However, by differentiating $\boldsymbol{W}^{\mathrm{T}}\boldsymbol{W}$ along the solution of (4.28), it holds that

$$\begin{aligned}
\frac{\mathrm{d}\boldsymbol{W}^{\mathrm{T}}(t)\boldsymbol{W}(t)}{\mathrm{d}t} &= -2\left(\boldsymbol{W}^{\mathrm{T}}(t)\boldsymbol{W}(t)\right)^{-2}\Big[\|\boldsymbol{W}(t)\|^2\boldsymbol{W}^{\mathrm{T}}(t)\boldsymbol{R}\boldsymbol{W}(t) \\
&\quad -\left(\boldsymbol{W}^{\mathrm{T}}(t)\boldsymbol{R}\boldsymbol{W}(t) + 1 - \|\boldsymbol{W}(t)\|^4\right)\|\boldsymbol{W}(t)\|^2\Big] \\
&= -2\left(\boldsymbol{W}^{\mathrm{T}}(t)\boldsymbol{W}(t)\right)^{-1}\Big[\boldsymbol{W}^{\mathrm{T}}(t)\boldsymbol{R}\boldsymbol{W}(t) - \left(\boldsymbol{W}^{\mathrm{T}}(t)\boldsymbol{R}\boldsymbol{W}(t) + 1 - \|\boldsymbol{W}(t)\|^4\right)\Big] \\
&= 2\left(\boldsymbol{W}^{\mathrm{T}}(t)\boldsymbol{W}(t)\right)^{-1}\left[1 - \|\boldsymbol{W}(t)\|^4\right] \\
&= \begin{cases} > 0 & \text{for} \quad \|\boldsymbol{W}(t)\| < 1 \\ < 0 & \text{for} \quad \|\boldsymbol{W}(t)\| > 1 \\ = 0 & \text{for} \quad \|\boldsymbol{W}(t)\| = 1. \end{cases}
\end{aligned} \tag{4.37}$$

This means that $\lim_{t\to\infty}\|\boldsymbol{W}(t)\| = 1$. Thus, we have $\lim_{t\to\infty} f_1(t) = \pm 1$, which gives $\lim_{t\to\infty} \boldsymbol{W}(t) = \pm \boldsymbol{v}_1$.

This completes the proof.

The asymptotic behavior of MCA ODE can only be considered in the limits of validity of this asymptotic theory. Therefore, the above theorem's result is only approximately valid in the first part of the time evolution of the MCA learning law, i.e., in approaching the minor component.

(3) The divergence analysis

Cirrincione et al. found that in an exact solution of the differential equation associated with some MCA algorithms, e.g., Luo, OJAn, and MCA EXIN algorithms, the weight vector length would not deviate from its initial value. However, when a numerical procedure (like Euler's method) is applied, all these rules are plagued by "divergence or sudden divergence" of the weight vector length. Obviously, only from the analysis of the ordinary differential equation, it is not sufficient to determine the convergence of the weight vector length for MCA algorithms. Thus, it is necessary and important to analyze the temporal behavior of MCA algorithms via the stochastic discrete law.

The purpose of this section is to analyze the temporal behavior of the proposed algorithm by using not only the ODE approximation, but especially, the stochastic discrete laws. Cirrincione found a sudden divergence for Luo algorithm (OJA and OJA+ also have this phenomenon on some condition). Sudden divergence is very adverse for practical application. Does our proposed algorithm have a sudden divergence? In this section, we will study the proposed algorithm in detail.

Averaging the behavior of the weight vector after the first critical point has been reached $(t \geq t_0)$, it follows that:

$$\boldsymbol{W}(t) = \|\boldsymbol{W}(t)\|\boldsymbol{v}_1 \quad \forall t \geq t_0, \tag{4.38}$$

where $\boldsymbol{v}_1$ is the unit eigenvector associated with the smallest eigenvalue of $\boldsymbol{R}$.

From (4.26), it holds that

$$\|\boldsymbol{W}(t+1)\|^2 = \|\boldsymbol{W}(t)\|^2 + \|\Delta\boldsymbol{W}(t)\|^2 + 2\boldsymbol{W}^{\mathrm{T}}(t)\Delta\boldsymbol{W}(t).$$

Neglecting the second-order term in $\alpha(t)$, the above equation can be regarded as a discretization of the following ODE:

$$\begin{aligned}
\frac{\mathrm{d}\|\boldsymbol{W}(t)\|^2}{\mathrm{d}t} &= E\{2\boldsymbol{W}^{\mathrm{T}}(t)\Delta\boldsymbol{W}(t)\} \\
&= E\left\{2\boldsymbol{W}^{\mathrm{T}}(t)\cdot\left\{-(\boldsymbol{W}^{\mathrm{T}}(t)\boldsymbol{W}(t))^{-1}\left[y(t)\boldsymbol{X}(t) - \left(y^2(t) + 1 - \|\boldsymbol{W}(t)\|^4\right)(\boldsymbol{W}^{\mathrm{T}}(t)\boldsymbol{W}(t))^{-1}\boldsymbol{W}(t)\right]\right\}\right\} \\
&= -2(\boldsymbol{W}^{\mathrm{T}}(t)\boldsymbol{W}(t))^{-2}\left[\|\boldsymbol{W}(t)\|^2\boldsymbol{W}^{\mathrm{T}}(t)\boldsymbol{R}\boldsymbol{W}(t) - \left(\boldsymbol{W}^{\mathrm{T}}(t)\boldsymbol{R}\boldsymbol{W}(t) + 1 - \|\boldsymbol{W}(t)\|^4\right)\|\boldsymbol{W}(t)\|^2\right] \\
&= 2(\boldsymbol{W}^{\mathrm{T}}(t)\boldsymbol{W}(t))^{-1}\left[1 - \|\boldsymbol{W}(t)\|^4\right].
\end{aligned} \tag{4.39}$$

The above equation can be approximated as

$$\frac{\mathrm{d}p}{\mathrm{d}t} = \frac{2}{p}(1 - p^2), \tag{4.40}$$

where $p = \|\boldsymbol{W}(t)\|^2$. Denote the time instant at which the MC direction is approached as t_0 and the corresponding value of the squared weight modulus as p_0. The solution of (4.40) is given by

$$\begin{cases} |1 - p^2| = |1 - p_0^2|\mathrm{e}^{-4(t-t_0)} & \text{if } p_0 \neq 1 \\ p = p_0 & \text{if } p_0 = 1. \end{cases} \tag{4.41}$$

Figure 4.1 shows these results for different values of p_0.

From the above results, it can be seen that the norm of the weight increases or decreases to one according to the initial weight modulus and the sudden divergence does not happen in a finite time. From (4.41), it is obvious that the rate of increase or decrease in the weight modulus depends only on the initial weight modulus and is not relevant to the eigenvalue of the autocorrelation matrix of the input vector.

(4) The convergence analysis via SDT method

The above analysis is based on a fundamental theorem of stochastic approximation theory [40, 41]. The obtained result is then an approximation on some conditions. The use of the stochastic discrete laws is a direct analytical method. The purpose of this section is to analyze the temporal behavior of our MCA neurons and the

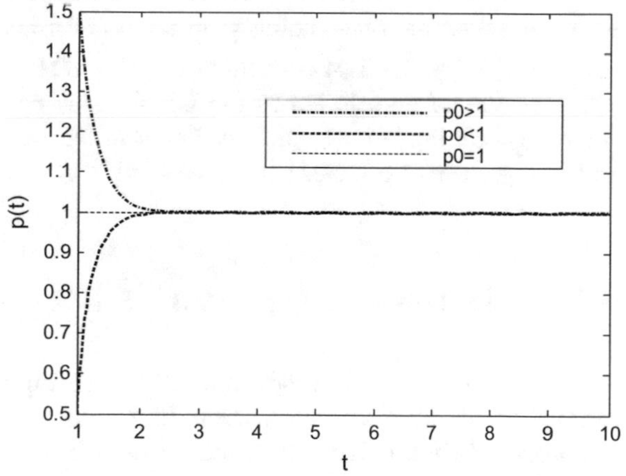

Fig. 4.1 Asymptotic behavior of the ODE for different values of the initial conditions

relation between the dynamic stability and learning rate, by using mainly the SDT system following the approach in [4].

From (4.26), it holds that

$$
\begin{aligned}
\|\boldsymbol{W}(t+1)\|^2 &= \boldsymbol{W}^{\mathrm{T}}(t+1)\boldsymbol{W}(t+1) \\
&= \|\boldsymbol{W}(t)\|^2 - 2\alpha(t)\left(\boldsymbol{W}^{\mathrm{T}}(t)\boldsymbol{W}(t)\right)^{-2}\left(\|\boldsymbol{W}(t)\|^2 y^2(t) - \left(y^2(t)+1-\|\boldsymbol{W}(t)\|^4\right)\|\boldsymbol{W}(t)\|^2\right) \\
&\quad + \alpha^2(t)\left(\boldsymbol{W}^{\mathrm{T}}(t)\boldsymbol{W}(t)\right)^{-4}\left(\|\boldsymbol{W}(t)\|^4 y^2(t)\|\boldsymbol{X}(t)\|^2\right. \\
&\quad \left. -2\|\boldsymbol{W}(t)\|^2 y^2(t)\left(y^2(t)+1-\|\boldsymbol{W}(t)\|^4\right) + \left(y^2(t)+1-\|\boldsymbol{W}(t)\|^4\right)^2\|\boldsymbol{W}(t)\|^2\right) \\
&= \|\boldsymbol{W}(t)\|^2 + 2\alpha(t)\left(\boldsymbol{W}^{\mathrm{T}}(t)\boldsymbol{W}(t)\right)^{-1}\left(1-\|\boldsymbol{W}(t)\|^4\right) + O\left(\alpha^2(t)\right) \\
&\approx \|\boldsymbol{W}(t)\|^2 + 2\alpha(t)\left(\boldsymbol{W}^{\mathrm{T}}(t)\boldsymbol{W}(t)\right)^{-1}\left(1-\|\boldsymbol{W}(t)\|^4\right).
\end{aligned}
\tag{4.42}
$$

Hence, if the learning factor is small enough and the input vector is bounded, we can make such analysis as follows by neglecting the second-order terms of the $\alpha(t)$.

$$
\begin{aligned}
\frac{\|\boldsymbol{W}(t+1)\|^2}{\|\boldsymbol{W}(t)\|^2} &\approx 1 + 2\alpha(t)\left(\boldsymbol{W}^{\mathrm{T}}(t)\boldsymbol{W}(t)\right)^{-2}\left(1-\|\boldsymbol{W}(t)\|^4\right) \\
&= \begin{cases} >1 & \text{for} \quad \|\boldsymbol{W}(0)\| < 1 \\ <1 & \text{for} \quad \|\boldsymbol{W}(0)\| > 1 \\ =1 & \text{for} \quad \|\boldsymbol{W}(0)\| = 1. \end{cases}
\end{aligned}
\tag{4.43}
$$

This shows that $\|\boldsymbol{W}(t+1)\|^2$ tends to one whether $\|\boldsymbol{W}(t)\|$ is equal to one or not, which is called the one-tending property (OTP); i.e. the weight modulus remains constant ($\|\boldsymbol{W}(t)\|^2 \rightarrow 1$ at convergence). The OTP indicates that $\boldsymbol{W}(0)$ with modulus one should be selected as the initial value of the proposed algorithm; thus, some practical limitations which may be resulted from an inappropriate initial value and a larger learning factor can be avoided.

4.4.2 MS Tracking Algorithm

In this section, we will introduce a self-stabilizing neural network learning algorithm for tracking minor subspace in high-dimensional data stream. Dynamics of the proposed algorithm are analyzed via a DCT and a SDT systems. The proposed algorithm provides an efficient online learning for tracking the MS and can track an orthonormal basis of the MS. Computer simulations are carried out to confirm the theoretical results.

The MCA learning algorithm given in Sect. 4.4.1 extracts only one component. We can easily extend the algorithm for tracking multiple MCs or MS. Let $\boldsymbol{U} = [\boldsymbol{u}_1, \boldsymbol{u}_2, \ldots, \boldsymbol{u}_r] \in \boldsymbol{R}^{N\times r}$ denote the weight matrix, where $\boldsymbol{u}_i \in \boldsymbol{R}^{N\times 1}$ represents the ith column vector of $\boldsymbol{U}$ and also denote the weight vector of the ith neuron of a multiple-input–multiple-output (MIMO) linear neural network. The input–output relation of the MIMO linear neural network is described by

$$\boldsymbol{y}(t) = \boldsymbol{U}^{\mathrm{T}}(t)\boldsymbol{x}(t). \tag{4.44}$$

The extended learning algorithm for training the weight matrix is given by

$$\begin{aligned}\boldsymbol{U}(t+1) = \boldsymbol{U}(t) - \mu(t)[\boldsymbol{x}(t)\boldsymbol{y}^{\mathrm{T}}(t) \\ - \boldsymbol{U}(t)\{\boldsymbol{U}^{\mathrm{T}}(t)\boldsymbol{U}(t)\}^{-1}(\boldsymbol{y}(t)\boldsymbol{y}^{T}(t) + \boldsymbol{I} - \{\boldsymbol{U}^{\mathrm{T}}(t)\boldsymbol{U}(t)\}^{2})]\{\boldsymbol{U}^{\mathrm{T}}(t)\boldsymbol{U}(t)\}^{-1}.\end{aligned} \tag{4.45}$$

It should be noted that (4.45) is not a trivial extension of (4.26). Although (4.26) has many extended forms, it may be difficult to find their corresponding Lyapunov functions in order to analyze their stability.

(1) Convergence analysis

Under similar conditions as those defined in [42], using the techniques of stochastic approximation theory [40, 41], we can obtain the following averaging differential equation

$$\begin{aligned}\frac{\mathrm{d}\boldsymbol{U}(t)}{\mathrm{d}t} = -[\boldsymbol{R}\boldsymbol{U}(t) - \boldsymbol{U}(t)\{\boldsymbol{U}^{\mathrm{T}}(t)\boldsymbol{U}(t)\}^{-1}(\boldsymbol{U}^{\mathrm{T}}(t)\boldsymbol{R}\boldsymbol{U}(t) \\ + \boldsymbol{I} - \{\boldsymbol{U}^{\mathrm{T}}(t)\boldsymbol{U}(t)\}^{2})]\{\boldsymbol{U}^{\mathrm{T}}(t)\boldsymbol{U}(t)\}^{-1}.\end{aligned} \tag{4.46}$$

The energy function associated with (4.46) is given by

$$E(\boldsymbol{U}) = \frac{1}{2}\mathrm{tr}\left\{(\boldsymbol{U}^{\mathrm{T}}\boldsymbol{R}\boldsymbol{U})(\boldsymbol{U}^{\mathrm{T}}\boldsymbol{U})^{-1}\right\} + \frac{1}{2}\mathrm{tr}\left\{\boldsymbol{U}^{\mathrm{T}}\boldsymbol{U} + (\boldsymbol{U}^{\mathrm{T}}\boldsymbol{U})^{-1}\right\}. \tag{4.47}$$

The gradient of $E(\boldsymbol{U})$ with respect to $\boldsymbol{U}$ is

$$\begin{aligned}\nabla E(\boldsymbol{U}) &= \boldsymbol{R}\boldsymbol{U}(\boldsymbol{U}^{\mathrm{T}}\boldsymbol{U})^{-1} - \boldsymbol{U}^{\mathrm{T}}\boldsymbol{R}\boldsymbol{U}\boldsymbol{U}(\boldsymbol{U}^{\mathrm{T}}\boldsymbol{U})^{-2} + \boldsymbol{U}[\boldsymbol{I} - (\boldsymbol{U}^{\mathrm{T}}\boldsymbol{U})^{-2}] \\ &= \left\{\boldsymbol{R}\boldsymbol{U}\boldsymbol{U}^{\mathrm{T}}\boldsymbol{U} - \boldsymbol{U}\left(\boldsymbol{U}^{\mathrm{T}}\boldsymbol{R}\boldsymbol{U} + \boldsymbol{I} - (\boldsymbol{U}^{\mathrm{T}}\boldsymbol{U})^{2}\right)\right\}(\boldsymbol{U}^{\mathrm{T}}\boldsymbol{U})^{-2} \\ &= \left\{\boldsymbol{R}\boldsymbol{U} - \boldsymbol{U}(\boldsymbol{U}^{\mathrm{T}}\boldsymbol{U})^{-1}\left(\boldsymbol{U}^{\mathrm{T}}\boldsymbol{R}\boldsymbol{U} + \boldsymbol{I} - (\boldsymbol{U}^{\mathrm{T}}\boldsymbol{U})^{2}\right)\right\}(\boldsymbol{U}^{\mathrm{T}}\boldsymbol{U})^{-1}.\end{aligned} \tag{4.48}$$

Clearly, (4.46) is equivalent to the following equation:

$$\frac{\mathrm{d}\boldsymbol{U}}{\mathrm{d}t} = -\nabla E(\boldsymbol{U}). \tag{4.49}$$

Differentiating $E(\boldsymbol{U})$ along the solution of (4.46) yields

$$\begin{aligned} \frac{\mathrm{d}E(\boldsymbol{U})}{\mathrm{d}t} &= \frac{\mathrm{d}\boldsymbol{U}^{\mathrm{T}}}{\mathrm{d}t}\nabla E(\boldsymbol{U}) \\ &= -\frac{\mathrm{d}\boldsymbol{U}^{\mathrm{T}}}{\mathrm{d}t}\frac{\mathrm{d}\boldsymbol{U}}{\mathrm{d}t}. \end{aligned} \tag{4.50}$$

Since the extended form of Algorithm (4.45) has a Lyapunov function $E(\boldsymbol{U})$ only with a lower bound [43], the corresponding averaging equation converges to the common invariance set $P = \{\boldsymbol{U}|\nabla E(\boldsymbol{U}) = 0\}$ from any initial value $\boldsymbol{U}(0)$.

(2) Divergence analysis

Theorem 4.2 *If the learning factor $\mu(t)$ is small enough and the input vector is bounded, then the state flows in Algorithm (4.45) for tracking the MS are bounded.*

Proof Since the learning factor $\mu(t)$ is small enough and the input vector is bounded, we have

$$\begin{aligned} \left\|\boldsymbol{U}^{\mathrm{T}}(t+1)\boldsymbol{U}(t+1)\right\|_F^2 &= \mathrm{tr}[\boldsymbol{U}^{\mathrm{T}}(t+1)\boldsymbol{U}(t+1)] \\ &= \mathrm{tr}\{\{\boldsymbol{U}(t) - \mu(t)[\boldsymbol{x}(t)\boldsymbol{y}^{\mathrm{T}}(t) - \boldsymbol{U}(t)\{\boldsymbol{U}^{\mathrm{T}}(t)\boldsymbol{U}(t)\}^{-1}(\boldsymbol{y}(t)\boldsymbol{y}^{\mathrm{T}}(t) \\ &\quad + \boldsymbol{I} - \{\boldsymbol{U}^{\mathrm{T}}(t)\boldsymbol{U}(t)\}^2)]\{\boldsymbol{U}^{\mathrm{T}}(t)\boldsymbol{U}(t)\}^{-1}\}^{\mathrm{T}} \\ &\quad \times \{\boldsymbol{U}(t) - \mu(t)[\boldsymbol{x}(t)\boldsymbol{y}^{\mathrm{T}}(t) - \boldsymbol{U}(t)\{\boldsymbol{U}^{\mathrm{T}}(t)\boldsymbol{U}(t)\}^{-1}(\boldsymbol{y}(t)\boldsymbol{y}^{\mathrm{T}}(t) \\ &\quad + \boldsymbol{I} - \{\boldsymbol{U}^{\mathrm{T}}(t)\boldsymbol{U}(t)\}^2)]\{\boldsymbol{U}^{\mathrm{T}}(t)\boldsymbol{U}(t)\}^{-1}\}\} \\ &\approx \mathrm{tr}[\boldsymbol{U}^{\mathrm{T}}(t)\boldsymbol{U}(t)] - 2\mu(t)\mathrm{tr}[(\{\boldsymbol{U}^{\mathrm{T}}(t)\boldsymbol{U}(t)\}^2 - \boldsymbol{I})\{\boldsymbol{U}^{\mathrm{T}}(t)\boldsymbol{U}(t)\}^{-1}] \\ &= \mathrm{tr}[\boldsymbol{U}^{\mathrm{T}}(t)\boldsymbol{U}(t)] - 2\mu(t)[\mathrm{tr}\{\boldsymbol{U}^{\mathrm{T}}(t)\boldsymbol{U}(t)\} - \mathrm{tr}\{\boldsymbol{U}^{\mathrm{T}}(t)\boldsymbol{U}(t)\}^{-1}]. \end{aligned} \tag{4.51}$$

Notice that in (4.51), the second-order terms associated with the learning factor have been neglected. It holds that

$$\begin{aligned} &\left\|\boldsymbol{U}^T(t+1)\boldsymbol{U}(t+1)\right\|_F^2/\left\|\boldsymbol{U}^T(t)\boldsymbol{U}(t)\right\|_F^2 \\ &\approx 1 - 2\mu(t)[\mathrm{tr}\{\boldsymbol{U}^T(t)\boldsymbol{U}(t)\} - \mathrm{tr}\{\boldsymbol{U}^T(t)\boldsymbol{U}(t)\}^{-1}]/\left\|\boldsymbol{U}^T(t)\boldsymbol{U}(t)\right\|_F^2 \\ &= 1 - 2\mu(t)\left(1 - \frac{\mathrm{tr}\{\boldsymbol{U}^T(t)\boldsymbol{U}(t)\}^{-1}}{\mathrm{tr}\{\boldsymbol{U}^T(t)\boldsymbol{U}(t)\}}\right). \end{aligned} \tag{4.52}$$

It is obvious that there exists a $\mathrm{tr}\{\boldsymbol{U}^{\mathrm{T}}(t)\boldsymbol{U}(t)\}$ large enough such that $(1-\mathrm{tr}\{\boldsymbol{U}^{\mathrm{T}}(t)\boldsymbol{U}(t)\}^{-1}/\mathrm{tr}\{\boldsymbol{U}^{\mathrm{T}}(t)\boldsymbol{U}(t)\})>0$, which results in $\|\boldsymbol{U}^{\mathrm{T}}(t+1)\boldsymbol{U}(t+1)\|_F^2/\|\boldsymbol{U}^{\mathrm{T}}(t)\boldsymbol{U}(t)\|_F^2<1$. Thus, the state flow in the proposed algorithm is bounded. This completes the proof.

(3) Landscape of nonquadratic criteria and global asymptotical convergence

Given $\boldsymbol{U}\in R^{N\times r}$ in the domain $\Omega=\{\boldsymbol{U}|0<\boldsymbol{U}^{\mathrm{T}}\boldsymbol{R}\boldsymbol{U}<\infty,\ \boldsymbol{U}^{\mathrm{T}}\boldsymbol{U}\neq 0\}$, we analyze the following nonquadratic criterion (NQC) for tracking the MS:

$$\min_{\boldsymbol{U}} E(\boldsymbol{U})=\frac{1}{2}\mathrm{tr}\{(\boldsymbol{U}^{\mathrm{T}}\boldsymbol{R}\boldsymbol{U})(\boldsymbol{U}^{\mathrm{T}}\boldsymbol{U})^{-1}\}+\frac{1}{2}\mathrm{tr}\{\boldsymbol{U}^{\mathrm{T}}\boldsymbol{U}+(\boldsymbol{U}^{\mathrm{T}}\boldsymbol{U})^{-1}\}. \tag{4.53}$$

Feng et al. [9] analyzed the landscape of nonquadratic criteria for the OJAm algorithm in detail. We can refer to the analysis method of OJAm algorithm to analyze Algorithm (4.45). Here, we only give the resulting theorems. The landscape of $E(\boldsymbol{U})$ is depicted by the following Theorems 4.3 and 4.4.

Theorem 4.3 *$\boldsymbol{U}$ is a stationary point of $E(\boldsymbol{U})$ in the domain Ω if and only if $\boldsymbol{U}=\boldsymbol{L}_r\boldsymbol{Q}$, where $\boldsymbol{L}_r\in\boldsymbol{R}^{N\times r}$ consists of the r eigenvectors of $\boldsymbol{R}$, and $\boldsymbol{Q}$ is a $r\times r$ orthogonal matrix.*

Theorem 4.4 *In the domain Ω, $E(\boldsymbol{U})$ has a global minimum that is achieved if and only if $\boldsymbol{U}=\boldsymbol{L}_{(n)}\boldsymbol{Q}$, where $\boldsymbol{L}_{(n)}=[\boldsymbol{v}_1,\boldsymbol{v}_2,\ldots,\boldsymbol{v}_r]$. At a global minimum, $E(\boldsymbol{U})=\frac{1}{2}\sum_{i=1}^{r}\lambda_i+r$. All the other stationary points of $\boldsymbol{U}=\boldsymbol{L}_r\boldsymbol{Q}$ $(\boldsymbol{L}_r\neq\boldsymbol{L}_{(n)})$ are saddle (unstable) points of $E(\boldsymbol{U})$.*

The proofs of Theorems 4.3 and 4.4 can be referred to Section IV in [9]. They are similar in most parts. From the previous theorems, it is obvious that the minimum of $E(\boldsymbol{U})$ automatically orthonormalizes the columns of $\boldsymbol{U}$, and at the minimum of $E(\boldsymbol{U})$, $\boldsymbol{U}$ only produces an arbitrary orthonormal basis of the MS but not the multiple MC.

The global asymptotical convergence of Algorithm (4.45) by considering its gradient rule (4.46) can be given by the following theorem.

Theorem 4.5 *Given the ordinary differential Eq. (4.46) and an initial value $\boldsymbol{U}(0)\in\Omega$, then $\boldsymbol{U}(t)$ globally asymptotically converges to a point in the set $\boldsymbol{U}=\boldsymbol{L}_{(n)}\boldsymbol{Q}$ as $t\to\infty$,where $\boldsymbol{L}_{(n)}=[\boldsymbol{v}_1,\boldsymbol{v}_2,\ldots,\boldsymbol{v}_r]$ and $\boldsymbol{Q}$ denotes a $r\times r$ unitary orthogonal matrix.*

Remark 4.1 OJAn, Luo, MCA EXIN, FENGm, and OJAm algorithms have been extended for tracking MS as in Eq. (4.45), and simulations have been performed [9]. It is concluded that the state matrices in OJAn, Luo, MCA EXIN, and FENGm do not converge to an orthonormal basis of the MS, but OJAm can. From the previous analysis, we can conclude that Algorithm (4.26) can be extended for tracking MS and can converge to an orthonormal basis of the MS.

4.4.3 Computer Simulations

(1) Simulation experiment on MCA algorithm

In this section, we provide simulation results to illustrate the convergence and stability of MCA algorithm (4.26). Since the OJAm algorithm and Douglas algorithm are self-stabilizing and have better performance than other MCA algorithms, we compare the performance of the proposed MCA algorithm with these algorithms in the following simulations. In the simulations, we use above three algorithms to extract the minor component from the input data sequence which is generated by $\boldsymbol{X}(t) = \boldsymbol{C} \cdot \boldsymbol{y}(t)$, where $\boldsymbol{C}$ = rand*n*(5, 5)/5 and $\boldsymbol{y}(t) \in \boldsymbol{R}^{5 \times 1}$ is Gaussian and randomly generated. In order to measure the convergence speed of learning algorithm, we compute the norm of $\boldsymbol{W}(t)$ and the direction cosine at the *t*th update:

$$\text{Direction Cosine}(t) = \frac{\left|\boldsymbol{W}^{\mathrm{T}}(t) \cdot \boldsymbol{v}_1\right|}{\|\boldsymbol{W}(t)\| \cdot \|\boldsymbol{v}_1\|},$$

where $\boldsymbol{v}_1$ is the unit eigenvector associated with the smallest eigenvalue of $\boldsymbol{R}$. If direction cosine(*t*) converges to 1, $\boldsymbol{W}(t)$ must converge to the direction of minor component $\boldsymbol{v}_1$. Figures 4.2 and 4.3 show the simulation curves about the convergence of $\|\boldsymbol{W}(t)\|$ and direction cosine(*t*) (DC), respectively. The learning constant in the OJAm and the proposed algorithm is 0.3, while the learning constant in the Douglas is 0.1. All the algorithms start from the same initial value that is randomly generated and normalized to modulus one.

From the simulation results, we can see easily that when the weight norm and the direction cosine in the OJAm, Douglas, and Algorithm (4.26) all converge, and the

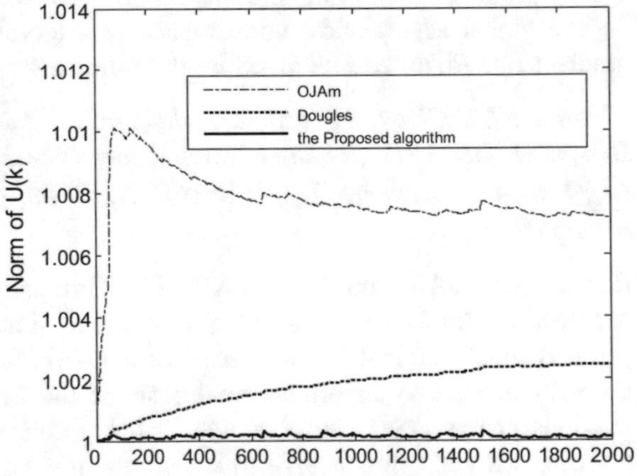

Fig. 4.2 Convergence of $\|\boldsymbol{W}(t)\|$

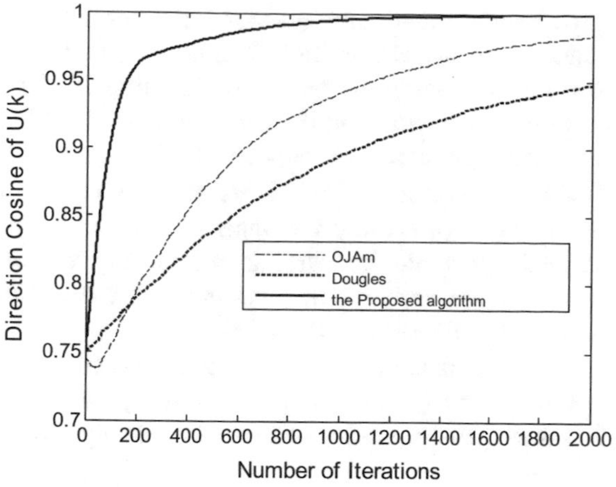

Fig. 4.3 Convergence of direction cosine(t)

convergence precision in Algorithm (4.26) is the best. Since the OJAn, MCA EXIN, and FENGm are divergent and the OJAm outperforms these algorithms [9], it seems that Algorithm (4.26) for tracking one MC works more satisfactorily than most existing MCA algorithms.

(2) Simulation experiment on MSA algorithm

In this section, we provide simulation results to illustrate the convergence and stability of MSA algorithm (4.45). The self-stabilizing Douglas algorithm is extended for tracking MS as in Eq. (4.45). Since the OJAm algorithm has better performance than other MSA algorithms, we compare performance of the proposed MSA algorithm with the OJAm and Douglas algorithms in the following simulations. Here, an MS of dimension 5 is tracked. The vector data sequence is generated by $\boldsymbol{X}(t) = \boldsymbol{B} \cdot \boldsymbol{y}(t)$, where $\boldsymbol{B}$ is randomly generated. In order to measure the convergence speed and precision of learning algorithm, we compute the norm of a state matrix at the tth update:

$$\rho(\boldsymbol{U}(t)) = \left\| \boldsymbol{U}^{\mathrm{T}}(t)\boldsymbol{U}(t) \right\|_2,$$

and the deviation of a state matrix from the orthogonality at the tth update, which is defined as:

$$\mathrm{dist}(\boldsymbol{U}(t)) = \left\| \boldsymbol{U}^{\mathrm{T}}(t)\boldsymbol{U}(t)[\mathrm{diag}(\boldsymbol{U}^{\mathrm{T}}(t)\boldsymbol{U}(t))]^{-1} - \boldsymbol{I}_r \right\|_F.$$

This simulation can be divided into two parts. In the first part, let $\boldsymbol{B}$ = (1/11) rand$n$(11, 11), and $\boldsymbol{y}(t) \in \boldsymbol{R}^{11\times 1}$ be Gaussian, spatially temporally white, and randomly generated. We simulate the algorithms starting from the same initial value

$\boldsymbol{U}(0)$ which is randomly generated and normalized to modulus one. The learning constants in the OJAm and Douglas are 0.02 and 0.01, respectively, while the learning constant in the proposed algorithm is 0.01. Figures 4.4 and 4.5 show the norm of state matrix and deviation of a state matrix from the orthogonality versus the number of iterations, respectively. In the second part, let $\boldsymbol{B}$ = (1/31) rand*n*(31, 31), and $\boldsymbol{y}(t) \in \boldsymbol{R}^{31\times 1}$ be Gaussian, spatially temporally white, and randomly generated. The learning constants in the OJAm and Douglas are 0.04 and 0.02, respectively, while the learning constant in the proposed algorithm is 0.02. Other conditions are similar to the ones in the first part. We can get the simulation results as shown in Figs. 4.6 and 4.7.

From the simulation results, we can see easily that the state matrices in the OJAm, Douglas, and Algorithm (4.45) all converge to an orthonormal basis of the MS, and the convergence precision of state matrix in algorithm (4.45) is the best, and there are

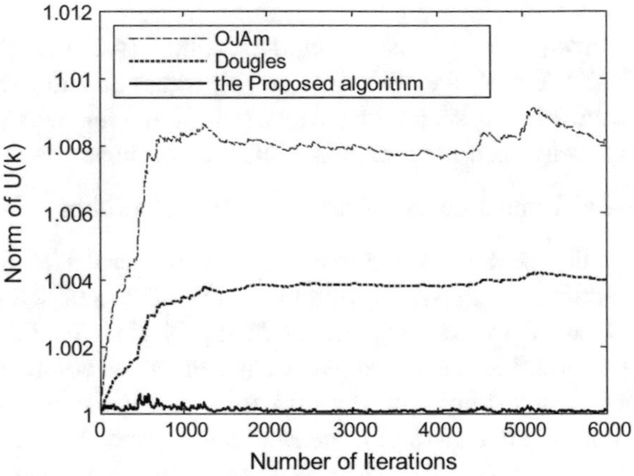

Fig. 4.4 Evolution curves of the norm of state matrix

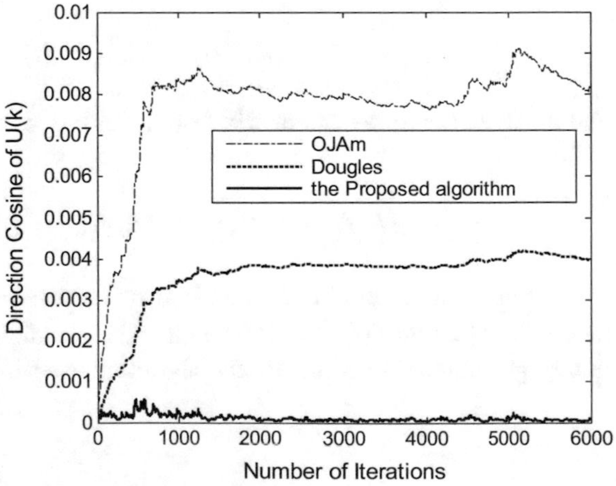

Fig. 4.5 Deviation from the orthogonality

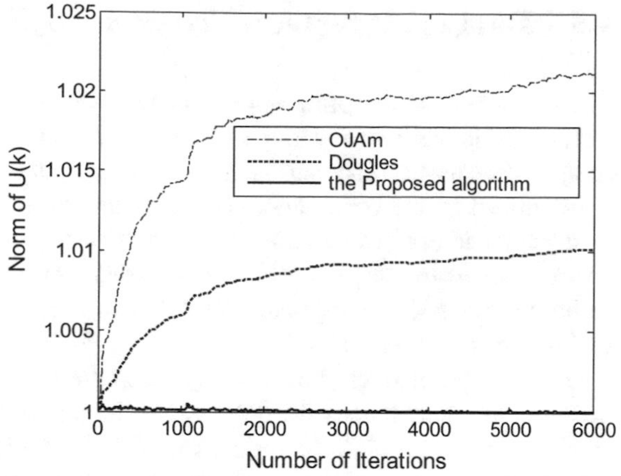

Fig. 4.6 Evolution curves of the norm of state matrix

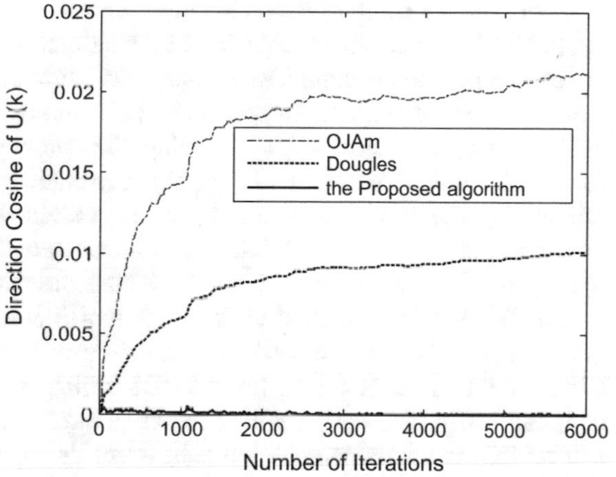

Fig. 4.7 Deviation from the orthogonality

residual deviations of state matrices from the orthogonality in the OJAm and Douglas. Since the OJAm is superior to OJAn, MCA EXIN, and FENGm, algorithm (4.45) seems to work more satisfactorily than most existing MSA algorithms.

In summary, a self-stabilizing MCA learning algorithm is presented in this section. The algorithm has been extended for tracking MS, and a self-stabilizing MSA algorithm is developed. The theoretical analysis of the proposed MCA algorithm is given via a corresponding DCT system and a SDT system. The globally asymptotic stability of the averaging equation of the proposed MSA algorithm has been studied. Simulation experiments have shown that the proposed MCA algorithm can efficiently extract one MC and works satisfactorily, the proposed MSA algorithm makes the corresponding state matrix tend to column-orthonormal basis of the MS, and the performance is superior to that of other MSA algorithms for high-dimensional data stream.

4.5 Total Least Squares Problem Application

The problem of *linear parameter estimation* gives rise to an overdetermined set of linear equations $\boldsymbol{Ax} \approx \boldsymbol{b}$ where $\boldsymbol{A}$ is the *data matrix* and $\boldsymbol{b}$ is the *observation vector*. In the least squares (LS) approach, there is an underlying assumption that all errors are confined to the observation vector. This assumption is often unrealistic: The data matrix is not error-free because of sampling errors, human errors, modeling errors, and instrumental errors. The method of *total least squares* (TLS) is a technique devised to compensate for data errors, and a complete analysis of the TLS problem can be seen [44–48].

The TLS problem can be solved by using direct and iterative methods. The direct methods compute directly the SVD. Since the number of multiplications in SVD for an $N \times N$ matrix is $6N^3$, the application of TLS problems is very limited in practice. Among the iterative methods, which are good for slow-changing set of equations, the most important are the inverse iteration, the ordinary and inverse Chebyshev iteration, the Rayleigh quotient iteration, and the Lanczos methods (for a survey, see [48]). The neural approaches can be considered iterative methods, and they have lower computational complexity compared with other iterative methods, which make them more suitable for real-time applications.

There are two neural ways of solving TLS problem: (1) One is a linear neuron for MCA, which finds the MC of the correlation matrix of the input data by minimizing the Rayleigh quotient with a gradient learning, where a subsequent normalization is needed. (2) Another is linear neural network acting directly on a hyperplane (TLS NN). The existing TLS NNs are the Hopfield-like neural network of Luo [7, 49]. The principal limit of it is that it is linked to the dimensions of the data matrix and cannot be used without structural changes for other TLS problems. Others are the linear neurons [30, 50–52], which is correct enough for small gains and, especially, for weight norms much smaller than one. The TLS EXIN linear neuron [53] is a new neural approach, which is superior in performance.

The objective of this section is to develop more satisfactory self-stabilizing TLS neural approach, which is applied to the parameter estimation of an adaptive FIR filters for system identification in the presence of additive noise in both input and output signals.

4.5.1 *A Novel Neural Algorithm for Total Least Squares Filtering*

In this section, we present a neural approach for solving the TLS problem. It is based on a linear neuron with a self-stabilizing neural algorithm, capable of resolving the TLS problem present in the parameter estimation of an adaptive FIR filters for system identification, where noisy errors affect not only the observation vector but also the data matrix. The learning rule is analyzed mathematically, and the condition to guarantee the stability of algorithm is obtained. The computer simulations are given to illustrate the effectiveness of this neural approach.

(1) TLS linear neuron with a self-stabilizing algorithm

The TLS method approximately solves a set of m linear equations in $n \times 1$ unknowns x represented in matrix form by

$$\boldsymbol{A}\boldsymbol{x} \approx \boldsymbol{b}, \tag{4.54}$$

where $\boldsymbol{A}$ is the *data matrix* composed of elements a_{ij} and columns a_i, and $\boldsymbol{b}$ is the *observation vector* composed of elements b_i and $m > n$ (*overdetermined system*). The TLS solution $\boldsymbol{X}_{\mathrm{TLS}}$ can be obtained as [5]

$$\begin{pmatrix} \boldsymbol{X}_{\mathrm{TLS}} \\ -1 \end{pmatrix} = -\frac{\boldsymbol{V}_{n+1}}{v_{n+1,n+1}}, \tag{4.55}$$

where $\boldsymbol{V}_{n+1}$ is the right singular vector associated with the smallest singular of $\boldsymbol{C} = [\boldsymbol{A}|\boldsymbol{b}]$, and $v_{n+1,n+1}$ is the last component of $\boldsymbol{V}_{n+1}$. Vector $\boldsymbol{V}_{n+1}$ is equivalent to the eigenvector associated with the smallest eigenvalue of the autocorrelation matrix of the input data. Thus, an adaptive algorithm for extracting eigenvector associated with the smallest eigenvalue of the autocorrelation matrix of the input data can be used to solve the TLS problem.

Consider the adaptive filter when the input and output observation data are both corrupted by noise, and denote the input and output observation series at k as $\{[\tilde{x}(k), \tilde{d}(k)] | k = 1, 2, \ldots, N\}$, $\tilde{x}(k) = x(k) + n_i(k)$, $\tilde{d}(k) = d(k) + n_o(k)$, where $n_i(k)$ and $n_o(k)$ are additive noises. Denote the filter weight vector as $\boldsymbol{H}(k) = [h_1, \ldots, h_n]^{\mathrm{T}}$ and the input vector at k as $\tilde{\boldsymbol{X}}(k)$. Then, the filter output at k can be written as $y(k) = \widetilde{\boldsymbol{X}}^{\mathrm{T}}(k)\boldsymbol{H}(k)$ and the output error is $\varepsilon(k) = y(k) - \tilde{d}(k)$. Denote the augmented input vector as $\boldsymbol{Z}(k) = [\widetilde{\boldsymbol{X}}^{\mathrm{T}}(k), \tilde{d}(k)]^{\mathrm{T}}$ and the augmented weighted vector as $\boldsymbol{W}(k) = [\boldsymbol{H}^{\mathrm{T}}(k), -1]^{\mathrm{T}}$. Then, the output error can be written as $\varepsilon(k) = \boldsymbol{Z}^{\mathrm{T}}(k)\boldsymbol{W}(k)$.

Let the Rayleigh quotient of the augmented weighted vector $\boldsymbol{W}(k)$ be the TLS cost function. An adaptive algorithm can then be obtained as follows:

$$\boldsymbol{W}(k+1) = \boldsymbol{W}(k) - \alpha(k)\varepsilon(k)\frac{\boldsymbol{Z}(k)\|\boldsymbol{W}(k)\|_2^2 - \varepsilon(k)\boldsymbol{W}(k)}{\|\boldsymbol{W}(k)\|_2^4}, \tag{4.56}$$

where $\alpha(k)$ is the learning rate, which controls the stability and rate of convergence of the algorithm. Equation (4.56) is the MCA EXIN algorithm for the TLS problem [4]. In [4] and [22], Algorithm (4.56) is analyzed in detail and it is concluded that the above algorithm is the best TLS neuron in terms of stability (no finite time divergence), speed, and accuracy. However, the MCA or TLS EXIN algorithm are divergent and do not have self-stabilization property [43].

In order to avoid the possible divergence and preserve the good performance as much as possible, based on the MCA EXIN algorithm and by adding a restriction on norm of weight, we proposed a modified algorithm as follows:

$$\boldsymbol{W}(k+1) = \boldsymbol{W}(k) - \frac{\alpha(k)}{\|\boldsymbol{W}(k)\|^2}\left[\varepsilon(k)\boldsymbol{Z}(k) - \frac{(\varepsilon^2(k) + 1 - \|\boldsymbol{W}(k)\|^2)}{\|\boldsymbol{W}(k)\|^2}\boldsymbol{W}(k)\right]. \tag{4.57}$$

From (4.57), the temporal derivative $D(k)$ of the squared weight vector norm is:

$$\begin{aligned} D(k) &= \frac{1}{2}\frac{\mathrm{d}\boldsymbol{W}^{\mathrm{T}}(t)\boldsymbol{W}(t)}{\mathrm{d}t} = \boldsymbol{W}^{\mathrm{T}}(k)\Delta\boldsymbol{W}(k) = \boldsymbol{W}^{\mathrm{T}}(k)(\boldsymbol{W}(k+1) - \boldsymbol{W}(k)) \\ &= -\frac{1}{\|\boldsymbol{W}(t)\|^2}\left[\varepsilon^2(k) - (\varepsilon^2(k) + 1 - \|\boldsymbol{W}(t)\|^2)\right] \\ &= \frac{1}{\|\boldsymbol{W}(t)\|^2}[1 - \|\boldsymbol{W}(t)\|^2] = \begin{cases} > 0 & \|\boldsymbol{W}(t)\| < 1 \\ < 0 & \|\boldsymbol{W}(t)\| > 1 \\ = 0 & \|\boldsymbol{W}(t)\| = 1. \end{cases} \end{aligned}$$

Clearly, the sign of the temporal derivative $D(k)$ is independent of $\varepsilon(k)$, but depends only on the sign of $\left(1 - \|\boldsymbol{W}(k)\|^2\right)$. This means that Algorithm (4.57) has self-stabilizing property. Thus, the norm of weight vector may stably approach the equilibrium point 1, as $k \to \infty$. This also renders our algorithm to have satisfactory one-tending property, and it outperforms OJA, OJAn, OJA+, and other existing TLS algorithms. The presented TLS neuron is a linear unit with n inputs, n weights, one output, and one training error, and the presented algorithm is a modified gradient algorithm, which can be used for the NN weights estimation where the input and output observation data are both corrupted by noise.

(2) Performance analysis via a DCT method

According to the stochastic approximation theory [40, 41], it can be shown that if some conditions are satisfied [4], Eq. (4.57) can be effectively represented by Eq. (4.58), i.e., their asymptotic paths are close with a large probability, and eventually, the solution of Eq. (4.57) tends with probability one to the uniformly asymptotically stable solution of the ODE.

$$\frac{\mathrm{d}\boldsymbol{W}(t)}{\mathrm{d}t} = -\frac{1}{\|\boldsymbol{W}(t)\|^2}\left[\boldsymbol{R}\boldsymbol{W}(t) - \frac{(\boldsymbol{W}^{\mathrm{T}}(t)\boldsymbol{R}\boldsymbol{W}(t) + 1 - \|\boldsymbol{W}_2(t)\|^2)}{\|\boldsymbol{W}(t)\|^2}\boldsymbol{W}(t)\right]. \tag{4.58}$$

From a computational point of view, the most important conditions are as follows:

(1) $\boldsymbol{Z}(t)$ is zero mean, stationary, and bounded with probability one;

(2) $\alpha(t)$ is a decreasing sequence of positive scalars;
(3) $\sum_t \alpha(t) = \infty$;
(4) $\sum_t \alpha^p(t) < \infty$ for some p;
(5) $\sum_{t\to\infty} \sup[(1/\alpha(t)) - (1/\alpha(t-1))] < \infty$.

The asymptotic property of (4.58) can be ensured by the following theorem.

Theorem 4.6 *Let* $\boldsymbol{R}$ *be a positive semi-definite matrix,* λ_n *and* $\boldsymbol{V}_n$ *be respectively its smallest eigenvalue and associated normalized eigenvector with nonzero last component. If the initial weight vector* $\boldsymbol{W}(0)$ *satisfies* $\boldsymbol{W}^{\mathrm{T}}(0)\boldsymbol{V}_n \neq 0$ *and* λ_n *is single, then* $\lim_{t\to\infty} \boldsymbol{W}(t) = \pm \boldsymbol{V}_n$ *holds.*

The proof of Theorem 4.6 is similar to that of Theorem 4.1 in [54]. For details, refer to [54]. Only the difference is given below.

From a series of consequence, we have

$$\lim_{t\to\infty} \boldsymbol{W}(t) = \lim_{t\to\infty}\left[\sum_{i=1}^{n} f_i(t)\boldsymbol{V}_i\right] = f_n(t)\boldsymbol{V}_n. \tag{4.59}$$

Equation (4.59) shows that $\boldsymbol{W}(t)$ tends to the direction of the eigenvector associated with the smallest eigenvalue of the autocorrelation matrix of the input data. From (4.59), it follows that

$$\lim_{t\to\infty} \|\boldsymbol{W}(t)\| = \lim_{t\to\infty} \|f_n(t)\boldsymbol{V}_n\| = \lim_{t\to\infty} \|f_n(t)\|. \tag{4.60}$$

Furthermore, from (4.60), it holds that

$$\begin{aligned}
\frac{\mathrm{d}\boldsymbol{W}^{\mathrm{T}}(t)\boldsymbol{W}(t)}{\mathrm{d}t} &= -\frac{2}{\|\boldsymbol{W}(t)\|^4}\left[\|\boldsymbol{W}(t)\|^2\boldsymbol{W}^{\mathrm{T}}(t)\boldsymbol{R}\boldsymbol{W}(t) - (\boldsymbol{W}^{\mathrm{T}}(t)\boldsymbol{R}\boldsymbol{W}(t) + 1 - \|\boldsymbol{W}(t)\|^2)\|\boldsymbol{W}(t)\|^2\right] \\
&= -\frac{2}{\|\boldsymbol{W}(t)\|^2}\left[\boldsymbol{W}^{\mathrm{T}}(t)\boldsymbol{R}\boldsymbol{W}(t) - (\boldsymbol{W}^{\mathrm{T}}(t)\boldsymbol{R}\boldsymbol{W}(t) + 1 - \|\boldsymbol{W}(t)\|^2)\right] = 2\left[\frac{1}{\|\boldsymbol{W}(t)\|^2} - 1\right] \\
&= \begin{cases} > 0 & \|\boldsymbol{W}(t)\| < 1 \\ < 0 & \|\boldsymbol{W}(t)\| > 1 \\ = 0 & \|\boldsymbol{W}(t)\| = 1. \end{cases}
\end{aligned} \tag{4.61}$$

Because the initial value of the squared modulus is larger than or equal to one in the TLS problem, and from (4.61), we have $\lim_{t\to\infty} \|\boldsymbol{W}(t)\| = 1$. Thus, we have $\lim_{t\to\infty} f_n(t) = \pm 1$, which gives $\lim_{t\to\infty} \boldsymbol{W}(t) = \pm \boldsymbol{V}_n$.

This completes the proof.

(3) Divergence analysis

In [4] and [22], Cirrincione et al. found a sudden divergence for Luo algorithm. Sudden divergence is very adverse for practical applications. Does our proposed algorithm have a sudden divergence? In this section, we will study Algorithm (4.57) in detail.

In [4], Corollary 19 gave simple convergence behavior for the solutions of a gradient flow as follows. The solutions of a gradient flow have a particularly simple convergence behavior; i.e., there are no periodic solutions, strange attractors, or any chaotic behaviors. Based on the above corollary, dynamic behaviors of the weight vector for Luo, OJAn, and MCA EXIN are described, and MCA divergences for Luo, OJAn, MCA EXIN, OJA, and OJA+ learning laws are proved. Following the above analysis method, the divergence analysis of the proposed algorithm is performed as follows.

Averaging the behavior of the weight vector after the first critical point has been reached $(t \geq t_0)$, it follows that [4]:

$$\boldsymbol{W}(t) = \|\boldsymbol{W}(t)\|\boldsymbol{V}_n, \ \forall t \geq t_0 \tag{4.62}$$

where $\mathbf{V}_n$ is the unit eigenvector associated with the smallest eigenvalue of $\boldsymbol{R}$.

From (4.61), we can obtain that $\mathrm{d}\|\boldsymbol{W}(t)\|^2/\mathrm{d}t = 2(1/\|\boldsymbol{W}(t)\|^2 - 1)$. Assuming that the MC direction has been approached, the equation can be approximated as

$$\frac{\mathrm{d}p}{\mathrm{d}t} = 2\left(\frac{1}{p} - 1\right), \tag{4.63}$$

where $p = \|\boldsymbol{W}(t)\|^2$. Define the time instant at which the MC direction is approached as t_0 and the corresponding value of the squared weight modulus as p_0. The solution of (4.11) is given by

$$\begin{cases} p + \ln|p - 1| = p_0 + \ln|p_0 - 1| - 2(t - t_0) & \text{if } p_0 \neq 1 \\ p = p_0 & \text{if } p_0 = 1. \end{cases} \tag{4.64}$$

We have simulated the asymptotic behavior of (4.63) for different initial squared weight modulus p, which is analogous with Fig. 4.1 and is not drawn here. Thus, the norm of the weight vector of proposed algorithm tends to one and the sudden divergence does not happen in a finite time.

(4) The temporal behavior analysis via a SDT method

The above analysis is based on a fundamental theorem of stochastic approximation theory. The obtained results are then approximations on some condition. The purpose of this section is to analyze the temporal behavior of our TLS neurons and the relation between the dynamic stability and learning rate, by using mainly the stochastic discrete time system following the approach in [41, 22, 54].

Define

$$r'(k) = \frac{\left|\boldsymbol{W}^{\mathrm{T}}(k+1)\boldsymbol{Z}(k)\right|^2}{\|\boldsymbol{W}(k+1)\|^2},\ r(k) = \frac{\left|\boldsymbol{W}^{\mathrm{T}}(k)\boldsymbol{Z}(k)\right|^2}{\|\boldsymbol{W}(k)\|^2}, \tag{4.65}$$

$$\rho(\alpha) = \frac{r'}{r} \geq 0,\ p = \|\boldsymbol{W}(k)\|^2,\ u = \varepsilon^2(k). \tag{4.66}$$

The two scalars r' and r represent, respectively, the squared perpendicular distance from the input $\boldsymbol{Z}(k)$ to the data fitting hyperplane whose normal is given by the weight and passes through the origin, after and before the weight increment. Recalling the definition of MC, we should have $r' \leq r$. If this inequality is not valid, it means that the learning law increases the estimation error due to disturbances caused by noisy data. When this disturbance is too large, it will make $\boldsymbol{W}(k)$ deviate drastically from the normal learning, which may result in divergence or fluctuations (implying an increased learning time).

Next, we will prove a theorem which provides the stability condition for Algorithm (4.57).

Theorem 4.7 *Under definitions* (4.65)–(4.66), *if*

$$0 < \alpha(k) < \frac{2p^2}{p\|\boldsymbol{Z}(k)\|^2 - 2(u+1-p)} \wedge p\|\boldsymbol{Z}(k)\|^2 - 2(u+1-p) > 0,$$

then $r' \leq r$, *which implies algorithm* (4.57) *is stable.*

Proof From Eq. (4.57), we have

$$\begin{aligned}
\boldsymbol{W}^{\mathrm{T}}(k+1)\boldsymbol{Z}(k) &= \varepsilon(k) - \frac{\alpha(k)}{\|\boldsymbol{W}(k)\|^4}\left[\|\boldsymbol{W}(k)\|^2\varepsilon(k)\|\boldsymbol{Z}(k)\|^2 - (\varepsilon^2(k) + 1 - \|\boldsymbol{W}(k)\|^2)\varepsilon(k)\right] \\
&= \varepsilon(k)\left(1 - \frac{\alpha(k)}{\|\boldsymbol{W}(k)\|^4}\left[\|\boldsymbol{W}(k)\|^2\|\boldsymbol{Z}(k)\|^2 - (\varepsilon^2(k) + 1 - \|\boldsymbol{W}(k)\|^2)\right]\right)
\end{aligned} \tag{4.67}$$

From (4.57), it holds that

$$\begin{aligned}
\|\boldsymbol{W}(k+1)\|^2 &= \boldsymbol{W}^{\mathrm{T}}(k+1)\boldsymbol{W}(k+1) \\
&= \|\boldsymbol{W}(k)\|^2 - \frac{2\alpha(k)}{\|\boldsymbol{W}(k)\|^4}(\|\boldsymbol{W}(k)\|^2\varepsilon^2(k) - (\varepsilon^2(k) + 1 - \|\boldsymbol{W}(k)\|^2)\|\boldsymbol{W}(k)\|^2) \\
&\quad + \frac{\alpha^2(k)}{\|\boldsymbol{W}(k)\|^8}\Big(\|\boldsymbol{W}(k)\|^4\varepsilon^2(k)\|\boldsymbol{Z}(k)\|^2 - 2\|\boldsymbol{W}(k)\|^2\varepsilon^2(k)(\varepsilon^2(k) + 1 - \|\boldsymbol{W}(k)\|^2) \\
&\quad + (\varepsilon^2(k) + 1 - \|\boldsymbol{W}(k)\|^2)^2\|\boldsymbol{W}(k)\|^2\Big).
\end{aligned} \tag{4.68}$$

Therefore,

$$\begin{aligned}\rho(\alpha) &= \frac{r'(k)}{r(k)} = \frac{(\mathbf{W}^{\mathrm{T}}(k+1)\mathbf{Z}(k))^2}{\|\mathbf{W}(k+1)\|^2}\frac{\|\mathbf{W}(k)\|^2}{(\varepsilon(k))^2} \\ &= \frac{\left(1 - \frac{\alpha(k)}{\|\mathbf{W}(k)\|^4}[\|\mathbf{W}(k)\|^2\|\mathbf{Z}(k)\|^2 - (\varepsilon^2(k) + 1 - \|\mathbf{W}(k)\|^2)]\right)^2}{1 - \frac{2\alpha(k)}{\|\mathbf{W}(k)\|^2}\left(1 - \frac{1}{\|\mathbf{W}(k)\|^2}\right) + \frac{\alpha^2(k)}{\|\mathbf{W}(k)\|^2}E} \\ &= \frac{p^2(1-\alpha(k)q)^2}{p^2 - 2\alpha(k)(p-1) + \alpha^2(k)pE},\end{aligned} \tag{4.69}$$

where $q = (1/p^2)(\|\mathbf{Z}(k)\|^2 p - (u+1-p))$ and $E = (1/p^3)(up\|\mathbf{Z}(k)\|^2 - 2u(u+1-p) + (u+1-p)^2)$.

Then, $\rho(\alpha) < 1$ (dynamic stability) if and only if

$$\begin{aligned}p^2(1-\alpha(k)q)^2 &< p^2 - 2\alpha(k)(p-1) + \alpha^2(k)\frac{1}{p^2}(up\|\mathbf{Z}(k)\|^2 - 2u(u+1 \\ &\quad - p) + (u+1-p)^2).\end{aligned} \tag{4.70}$$

Notice that $u/p = \|\mathbf{Z}(k)\|^2\cos^2\theta_{\mathbf{ZW}}$, where $\theta_{\mathbf{ZW}}$ is the angle between the augmented input vector and the augmented weight vector.

From (4.70), it holds that

$$\alpha(k)\|\mathbf{Z}(k)\|^2 p\sin^2\theta_{\mathbf{ZW}}(\alpha(k)(2u+2-2p-p\|\mathbf{Z}(k)\|^2) + 2p^2) > 0. \tag{4.71}$$

The dynamic stability condition is then

$$0 < \alpha(k) < \frac{2p^2}{p\|\mathbf{Z}(k)\|^2 - 2(u+1-p)} \wedge p\|\mathbf{Z}(k)\|^2 - 2(u+1-p) > 0 \tag{4.72}$$

The second condition implies the absence of the negative instability. It can be rewritten as

$$\cos^2\theta_{\mathbf{ZW}} \leq \frac{1}{2} + \frac{1}{\|\mathbf{Z}(k)\|^2}\frac{p-1}{p}. \tag{4.73}$$

In reality, the second condition is included in the first one. For the case $0 < \alpha_b \leq \gamma < 1$, it holds

$$\begin{aligned}\cos^2\theta_{\mathbf{ZW}} &\leq \frac{1}{2} + \frac{1}{\|\mathbf{Z}(k)\|^2}\frac{p-1}{p} - \frac{p}{\gamma\|\mathbf{Z}(k)\|^2} \\ &\leq \frac{1}{2} + \frac{1}{\|\mathbf{Z}(k)\|^2} - \frac{2}{\|\mathbf{Z}(k)\|^2\sqrt{\gamma}} = \Upsilon,\end{aligned} \tag{4.74}$$

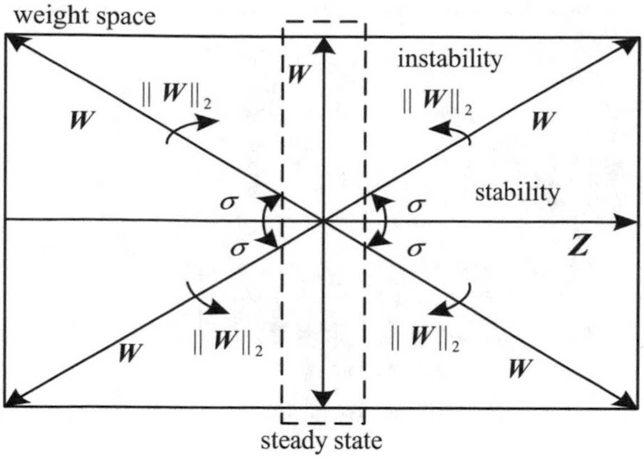

Fig. 4.8 Stability subspace of the weight space with regard to the input vector (two-dimensional space) of proposed algorithm

which is more restrictive than (4.73). Figure 4.8 shows this condition, where $\sigma = \arccos\sqrt{\Upsilon}$. From (4.74), we can see that the decrease of γ increases the stability and the angle $\theta_{\boldsymbol{ZW}}$ is no relevant to the change in weight vector modulus. From Fig. 4.8, it is apparent that in the transient (in general low $\theta_{\boldsymbol{ZW}}$), there are less fluctuations and this is beneficial to the stability. From (4.74), it can be concluded that the angle $\theta_{\boldsymbol{ZW}}$ between the weight vector $\boldsymbol{W}(k)$ and the input vector $\boldsymbol{Z}(k)$ is equal to or smaller than 45° when the proposed algorithm is convergent.

This completes the proof.

4.5.2 *Computer Simulations*

In this section, we provide simulation results to illustrate the convergence and stability of Algorithm (4.57) in a total least squares filtering. Since the OJAm [9] and MCA EXIN algorithms [4] have better performance than other TLS algorithms, we compare the performance of the proposed TLS algorithm of (4.57) with these algorithms in the following simulations.

The additive noise is independent zero-mean white noise. The input signal is zero-mean white Gaussian noise with variance 1. In the simulation, the above algorithms are used for a linear system identification problem. The linear system is given by $\boldsymbol{H} = [-0.3, -0.9, 0.8, -0.7, 0.6, 0.2, -0.5, 1.0, -0.7, 0.9, -0.4]^{\mathrm{T}}$, and its normalized one is $\overline{\boldsymbol{H}} = \boldsymbol{H}/\|\boldsymbol{H}\|$. Their convergence is compared under different noise environment and by using different learning factors. Define the learning error of the weight vector as: $\mathrm{Error}(k) = 10\log\left(\left\|\overline{\boldsymbol{H}} - \widehat{\boldsymbol{H}}(k)\right\|^2\right)$, where $\widehat{\boldsymbol{H}}(k)$ is the estimated weight vector. Figure 4.9 shows the learning curves for SNR=10dB with

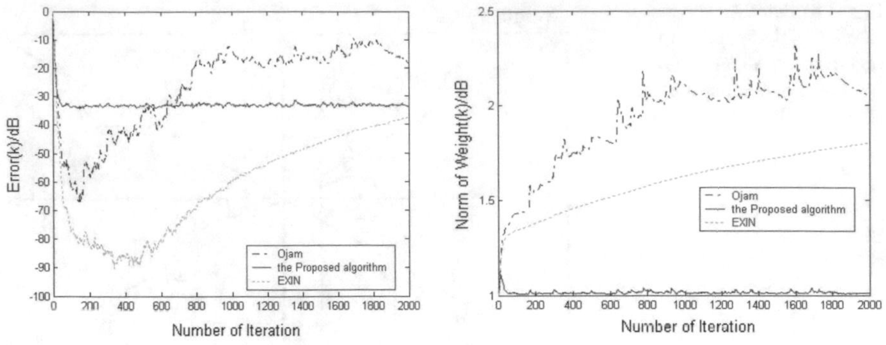

(a) Error of the weight vector (b) norm of the weight vector

Fig. 4.9 Learning curve at $\mu = 0.1$, 10 dB

learning factors equaling to 0.1. All the learning curves below are obtained by averaging over 30 independent experiments. Assume that the SNR of the input signal is equal to the rate of the output signal.

These learning curves indicate that for FIR filters with larger number of coefficients, Algorithm (4.57) has excellent convergence and stable accuracy, and the norm of weight of the algorithm converges to one. From Fig. 4.9(a), we can see that the convergence and stable accuracy are obviously better than other two algorithms when a larger learning factor is used or the SNR is smaller in the linear system. From Fig. 4.9(b), we can see that the norm of the weight of algorithm (4.57) converges to one, and the norms of the weight of the other algorithms are divergent. The above learning curves indicate that Algorithm (4.57) is good for larger learning factor or under higher noise environments.

In this section, a TLS neuron with a self-stabilizing algorithm has been presented for the parameter estimation of adaptive FIR filters for system identification where noisy errors affect not only the observation vector but also the data matrix. Compared with other TLS algorithms, the neural approach is self-stabilizing and considerably outperforms the existing TLS methods when a larger learning factor is used or the signal-to-noise rate (SNR) is lower.

4.6 Summary

In this chapter, several neural network-based MCA algorithms, e.g., extracting the first minor component analysis algorithm, Oja's minor subspace analysis algorithm, the self-stabilizing MCA algorithm, the MCA EXIN algorithm, and orthogonal Oja algorithm, have been reviewed. Three types of divergence for MCA algorithms, i.e., the sudden divergence, the instability divergence, and the numerical divergence, have been analyzed and discussed. Finally, two algorithms we proposed, i.e., a novel self-stabilizing MCA linear neuron and a novel neural algorithm for TLS filtering, are presented, and their performances, e.g., convergence and self-stabilizing, are mathematically proved and numerically verified.

References

1. Xu, L., Oja, E., & Suen, C. (1992). Modified Hebbian learning for curve and surface fitting. *Neural Networks, 5*(3), 441–457.
2. Oja, E. (1992). Principal components, minor components and linear neural networks. *Neural Networks, 5*(6), 927–935.
3. Luo, F., Unbehauen, R., & Cichocki, A. (1997). A minor component analysis algorithm. *Neural Networks, 10*(2), 291–297.
4. Cirrincione, G., & Cirrincione, M. (2002). The MCA EXIN neuron for the minor component analysis. *IEEE Transactions on On Neural Networks, 13*(1), 160–187.
5. Feng, D. Z., Bao, Z., & Jiao, L. C. (1998). Total least mean squares algorithm. *IEEE Transactions on Signal Processing, 46*(8), 2122–2130.
6. Dunne, B. E., & Williamson, G. A. (2000). Stable simplified gradient algorithms for total least squares filtering. In *Proceedings of 32nd Ann. Asilomar Conference on Signals, Systems, and Computers, Pacific Grove*, CA (pp. 1762–1766).
7. Luo, F., Li, Y., & He, C. (1996). Neural network approach to the TLS linear prediction frequency estimation problem. *Neurocomputing, 11*(1), 31–42.
8. Douglas, S. C., Kung, S. Y., & Amari, S. (1998). A self-stabilized minor subspace rule. *IEEE Signal Processing Letter, 5*(12), 328–330.
9. Feng, D. Z., Zheng, W. X., & Jia, Y. (2005). Neural network Learning algorithms for tracking minor subspace in high-dimensional data stream. *IEEE Transactions on Neural Networks, 16* (3), 513–521.
10. Möller, R. (2004). A self-stabilizing learning rule for minor component analysis. *Internal Journal of Neural Systems, 14*(1), 1–8.
11. Chen, T., & Amari, S. (2001). Unified stabilization approach to principal and minor components extraction algorithms. *Neural Networks, 14*(10), 1377–1387.
12. Mathew, G., & Reddy, V. U. (1994). Orthogonal eigensubspace estimation using neural networks. *IEEE Transactions on Signal Processing, 42*(7), 1803–1811.
13. Mathew, G., Reddy, V. U., & Dasgupta, S. (1995). Adaptive estimation of eigensubspace. *IEEE Transactions on Signal Processing, 43*(2), 401–411.
14. Oja, E. (1982). Simplified neuron model as a principal component analyzer. *Journal of Mathematics Biology, 15*(3), 267–273.
15. Baldi, P., & Hornik, K. (1989). Neural networks and principal component analysis: Learning from examples without local minima. *Neural Networks, 2*(1), 52–58.
16. Pisarenko, V. F. (1973). The retrieval of harmonics from a covariance function. *Geophysics Journal of Astronomic Society, 33*(33), 347–366.
17. Thompson P. A. (1979). An adaptive spectral analysis technique for unbiased estimation in the presence of white noise. In *Proceedings of 13th Asilomar Conference on Circuits, Systems Computation, Pacific Grove, CA* (pp. 529–533).
18. Yang, J. F., & Kaveh, M. (1988). Adaptive eigensubspace algorithms for direction or frequency estimation and tracking. *IEEE Transactions on Acoustic, Speech, and Signal Processing, 36*(2), 241–251.
19. Taleb, A., & Cirrincione, G. (1999). Against the convergence of the minor component analysis neurons. *IEEE Transactions on Neural Networks, 10*(1), 207–210.
20. Chiang, C. T., & Chen, Y. H. (1999). On the inflation method in adaptive noise subspace estimator. *IEEE Transactions on Signal Processing, 47*(4), 1125–1129.

21. Ouyang, S., Bao, Z., Liao, G. S., & Ching, P. C. (2001). Adaptive minor component extraction with modular structure. *IEEE Transactions on Signal Processing, 49*(9), 2127–2137.
22. Cirrincione, G. (1998). *A neural approach to the structure from motion problem*, Ph.D. dissertation, LIS INPG Grenoble, Grenoble, France.
23. Zhang, Q., & Leung, Y. W. (2000). A class of learning algorithms for principal component analysis and minor component analysis. *IEEE Transactions on Neural Networks, 11*(2), 529–533.
24. Attallah, S., & Abed-Meraim, K. (2002). Low-cost adaptive algorithm for noise subspace estimation. *Electronic Letter, 38*(12), 609–611.
25. Doukopoulos, X. G., & Moustakides, G. V. (2008). Fast and stable subspace tracking. *IEEE Transactions on Signal Processing, 56*(4), 1452–1465.
26. Doukopoulos, X. G., & Moustakides, G. V. (2006). Blind adaptive channel estimation in OFDM systems. *IEEE Transactions on Wireless Communication, 5*(7), 1716–1725.
27. Attallah, S. (2006). The generalized Rayleigh's quotient adaptive noise subspace algorithm: A Householder transformation-based implementation. *IEEE Transactions on Circuits Systems II, 53*(1), 3–7.
28. Bartelmaos S. & Abed-Meraim K. (2006). Principal and minor subspace tracking: Algorithms and stability analysis. In *Proceeding of IEEE International Conference on Acoustic, Speech, and Signal Processing* (Vol. 3, pp. 560–563).
29. Badeau, R., Richard, G., & David, B. (2008). Fast and stable YAST algorithm for principal and minor subspace tracking. *IEEE Transactions on Signal Processing, 56*(8), 3437–3446.
30. Gao, K. Q., Ahmad, M. O., & Swamy, M. N. (1994). A constrained anti-Hebbian learning algorithm for total least-squares estimation with applications to adaptive FIR and IIR filtering. *IEEE Transactions on Circuits and Systems II: Analog and Digital Signal Processing, 41*(11), 718–729.
31. Ouyang, S., Bao, Z., & Liao, G. (1999). Adaptive step-size minor component extraction algorithm. *Electronics Letter, 35*(6), 443–444.
32. Oja, E., & Karhunen, J. (1985). On stochastic approximation of the eigenvectors and eigenvalues of the expectation of a random matrix. *Journal of Mathematical Analysis and Applications, 106*(1), 69–84.
33. Chen, T., Amari, S., & Lin, Q. (1998). A unified algorithm for principal and minor components extraction. *Neural Networks, 11*(3), 385–390.
34. Abed-Meraim, K., Attallah, S., Chkeif, A., & Hua, Y. (2000). Orthogonal Oja algorithm. *IEEE Signal Processing Letter, 7*(5), 116–119.
35. Attallah, S., & Abed-Meraim, K. (2001). Fast algorithms for subspace tracking. *IEEE Signal Processing Letter, 8*(7), 3–206.
36. Chen, T. P., Amari, S. I., & Murata, N. (2001). Sequential extraction of minor components. *Neural Processing Letter, 13*(3), 195–201.
37. Wang, L., & Karhunen, J. (1996). A simplified neural bigradient algorithm for robust PCA and MCA. *International Journal of Neural Systems, 7*(1), 53–67.
38. Yuille, A. L., Kammen, D. M., & Cohen, D. S. (1989). Quadrature and development of orientation selective cortical cells by Hebb rules. *Biological Cybernetics, 61*(3), 183–194.
39. Kong, X. Y., Hu, C. H., & Han, C. Z. (2010). A self-stabilizing MSA algorithm in high-dimension data stream. *Neural Networks, 23*(7), 865–871.
40. Kushner, H., & Clark, D. (1978). *Stochastic approximation methods for constrained and unconstrained systems*. New York: Springer-Verlag.
41. Ljung, L. (1977). Analysis of recursive stochastic algorithms. *IEEE Transactions on Automatic Control, 22*(4), 551–575.
42. Zufiria, P. J. (2002). On the discrete-time dynamics of the basic Hebbian neural network node. *IEEE Transactions on Neural Networks, 13*(6), 1342–1352.
43. LaSalle, J. P. (1976). *The stability of dynamical systems*. Philadelphia, PA: SIAM.
44. Golub, G. H., & Van Loan, C. F. (1980). An analysis of the total least squares problem. *SIAM Journal of Numerical Analysis, 17*(6), 883–893.

45. Golub, G. H., & Reinsch, C. (1970). Singular value decomposition and least squares solutions. *Numerical Mathematics, 14*(5), 403–420.
46. Golub, G. H. (1973). Some modified eigenvalue problems. *SIAM Review, 15*, 318–344.
47. Gleser, L. J. (1981). Estimation in a multivariate "errors in variables" regression model: Large sample results. *Annals Statistic, 9*(1), 24–44.
48. Van Huffel, S., & Vandewalle, J. (1991). *The total least squares problems: Computational aspects and analysis*. SIAM, Philadelphia: Frontiers in Applied Mathematics.
49. Luo, F., & Unbehauen, R. (1997). *Applied neural networks for signal processing*. Cambridge: Cambridge University Press.
50. Gao, K. Q., Ahmad, M. O., & Swamy, M. N. (1992). Learning algorithm for total least-squares adaptive signal processing. *Electronic Letter, 28*(4), 430–432.
51. Cichocki, A., & Unbehauen, R. (1993). *Neural networks for optimization and signal processing*. New York: Wiley.
52. Cichocki, A., & Unbehauen, R. (1994). Simplified neural networks for solving linear least squares and total least squares problems in real time. *IEEE Transactions on Neural Networks, 5*(6), 910–923.
53. Cirrincione, G., & Cirrincione, M. (1999). Linear system identification using the TLS EXIN neuron. *Neurocomputing, 28*(1), 53–74.
54. Kong, X. Y., Han, C. Z., & Wei, R. X. (2006). Modified gradient algorithm for total least squares filtering. *Neurocomputing, 70*(1), 568–576.
55. Peng, D., & Zhang, Y. (2005). Convergence analysis of an effective MCA learning algorithm. In *Proceedings of International Conference on Neural Networks and Brain*, 2003–2008.
56. Peng, D., & Zhang, Y. (2007). Dynamics of generalized PCA and MCA learning algorithms. *IEEE Transactions on Neural Networks, 18*(6), 1777–1784.
57. Peng, D., Zhang, Y., & Luo, W. (2007). Convergence analysis of a simple minor component analysis algorithm. *Neural Networks, 20*(7), 842–850.
58. Peng, D., Zhang, Y., Lv, J., & Xiang, Y. (2007). A neural networks learning algorithm for minor component analysis and its convergence analysis. *Neurocomputing, 71*(7–9), 1748–1752.
59. Peng, D., Zhang, Y., & Xiang, Y. (2008). On the discrete time dynamics of a self-stabilizing MCA learning algorithm. *Mathematical and Computer Modeling, 47*(9–10), 903–916.
60. Peng, D., Zhang, Y., Lv, J., & Xiang, Y. (2008). A stable MCA learning algorithm. *Computers and Mathematics with Application, 56*(4), 847–860.
61. Ye, M., Fan, X., & Li, X. (2006). A class of self-stabilizing MCA learning algorithms. *IEEE Transactions on Neural Networks, 17*(6), 1634–1638.
62. Kong, X. Y., Hu, C. H., & Han, C. Z. (2010). On the discrete-time dynamics of a class of self-stabilizing MCA extraction algorithms. *IEEE Transactions on Neural Networks, 21*(1), 175–181.
63. Magnus, J. R., & Neudecker, H. (1991). *Matrix differential calculus with applications in statistics and econometrics* (2nd ed.). New York: Wiley.
64. Golub, G. H., & Van Loan, C. F. (1989). *Matrix computations* (2nd ed.). Baltimore, MD: The Johns Hopkins University. Press.
65. Abed-Meraim, K., Chkeif, A., & Hua, Y. (2000). Fast orthonormal PAST algorithm. *IEEE Signal Processing Letter, 7*(3), 60–62.
66. Badeau, R., David, B., & Richard, G. (2005). Fast approximated power iteration subspace tracking. *IEEE Transactions on Signal Processing, 53*(8), 2931–2941.
67. Strobach, P. (2009). The fast recursive row-Householder subspace tracking algorithm. *Signal Processing, 89*(12), 2514–2528.
68. Yang, L., Attallah, S., Mathew, G., & Abed-Meraim, K. (2008). Analysis of orthogonality error propagation for FRANS and HFRANS algorithms. *IEEE Transactions on Signal Processing, 56*(9), 4515–4521.
69. Strobach, P. (2000). Square-root QR inverse iteration for tracking the minor subspace. *IEEE Transactions on Signal Processing, 48*(1), 2994–2999.

70. Yang, L., & Attallah, S. (2009). Adaptive noise subspace estimation algorithm suitable for VLSI implementation. *IEEE Signal Processing Letter, 16*(12), 1075–1078.
71. Strobach, P. (2009). The QS-Householder sliding window Bi-SVD subspace tracker. *IEEE Transactions on Signal Processing, 57*(11), 4260–4268.
72. Badeau, R., David, B., & Richard, G. (2005). Yet another subspace tracker. In *Proceeding of IEEE International Conference on Acoustic, Speech, and Signal Processing* (Vol. 4, pp. 329–332).
73. Badeau, R., David, B., & Richard, G. (2006). YAST algorithm for minor subspace tracking. In *Proceedings of IEEE International Conference on Acoustic, Speech, and Signal Processing* (Vol. 3, pp. 552–555).
74. Bartelmaos, S., & Abed-Meraim, K. (2007). An efficient & stable algorithm for minor subspace tracking and stability analysis. In *Proceedings of IEEE International Conference on Acoustic, Speech, and Signal Processing* (Vol. 3, pp. 1301–1304).
75. Wang, R., Yao, M. L., Zhang, D. M., & Zou, H. X. (2012). A novel orthonormalization matrix based fast and stable dpm algorithm for principal and minor subspace tracking. *IEEE Transactions on Signal Processing, 60*(1), 466–472.

Chapter 5
Dual Purpose for Principal and Minor Component Analysis

5.1 Introduction

The PS is a subspace spanned by all eigenvectors associated with the principal eigenvalues of the autocorrelation matrix of a high-dimensional vector sequence, and the subspace spanned by all eigenvectors associated with the minor eigenvalues is called the MS. PSA is a powerful technique in many information processing fields, e.g., feature extraction and data compression. Whereas in many real-time signal processing applications, e.g., the adaptive direction-of-arrival (DOA) estimation, the adaptive solution of a total least squares problem in adaptive signal processing, and curve and surface fitting, minor subspace tracking is a primary requirement. Neural networks can be used to handle PSA and MSA, which possess many obvious advantages, e.g., lower computational complexity compared with the traditional algebraic approaches [1]. Hence, it is interesting to find some learning algorithms with lower computational complexity for adaptive signal processing applications.

In the past decades, many neural network learning algorithms have been proposed to extract PS [2–9] or MS [1, 10–18]. In the class of PS tracking, lots of learning algorithms, e.g., Oja's subspace algorithm [19], the symmetric error correction algorithm [20], and the symmetric back propagation algorithm [21] were proposed based on some heuristic reasoning [22]. Afterward, some information criteria were proposed, and the corresponding algorithms, e.g., LMSER algorithm [9], the projection approximation subspace tracking (PAST) algorithm [23], the conjugate gradient method [24], the Gauss-Newton method [25], and the novel information criterion (NIC) algorithm were developed [22]. These gradient-type algorithms are claimed to be globally convergent. In the class of MS tracking, many algorithms [10–18] have been proposed on the basis of feedforward neural network models. Recently, an information criterion was proposed in [1], and the corresponding globally convergent gradient algorithm, called OJAm, was developed. The OJAm provided an efficient online learning for tracking the MS.

© Science Press, Beijing and Springer Nature Singapore Pte Ltd. 2017

X. Kong et al., *Principal Component Analysis Networks and Algorithms*,
DOI 10.1007/978-981-10-2915-8_5

However, the above algorithms are either for PS tracking or for MS tracking. Are there such algorithms as dual-purpose subspace tracking algorithm, which is capable of both PS and MS tracking by simply switching the sign in the same learning rule? The objective of this chapter is to introduce such a dual-purpose subspace tracking algorithm. There are several reasons to develop dual-purpose algorithms. First, in many information processing systems, it is necessary to extract the main features inherent in complex, high-dimensional input data streams, and PSA is a technique to extract principal components simultaneously. On the other hand, the subspace spanned by the eigenvectors associated with the smallest eigenvalues is the minor subspace, which, in signal processing, represents the noise subspace and is also an important topic to be investigated [26]. Second, the only difference between PCA/PSA and MCA/MSA rule in the dual-purpose algorithm is the sign of the right side of the iterative equation. This simplicity is of practical significance in the implementation of algorithms, and it can reduce the complexity and cost of hardware realization [27–29]. Third, it is found that many PCA/PSA algorithms do not have the corresponding MCA/MSA algorithms dual to each other [27]. It is of significance to elucidate the fundamental reason for this. Moreover, by the stability analysis of the dual-purpose algorithms, we can better understand the dynamics of some PCA/PSA or MCA/MSA algorithms and may find some valuable phenomena. In summary, we think it is necessary to develop dual-purpose learning algorithms, which are of significance in both practical applications and theory.

In fact, many pioneering works have been done by Chen et al. [27, 26, 30, 31] in this research area. In [26, 31], a conversion mechanism between PSA/PCA and MSA/MCA was given, and it is shown that every PCA algorithm accompanies an MCA algorithm and vice versa. However, by the conversion mechanism, the resultant dual-purpose algorithms have different structures. In [30], a unified algorithm capable of both PCA and MCA by simply switching the sign of the same learning rule was proposed. However, if the discrete-time update analysis was used, the MCA rule of unified algorithm in [30] would suffer from a marginal instability [27]. In [27], the unified algorithms in [30] were modified by adding a penalty term, and then a unified stabilization approach for principal and minor component extraction algorithms was proposed and the dynamical behaviors of several PCA/MCA algorithms were investigated in detail. [27] laid sound theoretical foundations for dual-purpose algorithm research. Afterward, a dual-purpose principal and minor component flow was proposed in [32]. Recently, several self-normalizing dual systems for minor and principal component extraction were also proposed, and their stability was widely analyzed [28]. Through our analysis, we can see that most self-normalizing dual systems in [28] can be viewed as generalizations of the unified stabilization approach.

However, of these dual systems for PS and MS tracking, most were developed on some heuristic reasoning. There are few algorithms that possess a unified information criterion (UIC) formulation and a globally convergent gradient update equation derived from the UIC. It is well known that a properly chosen criterion is very important for developing any learning algorithm [22]. Different from the derivation of the above dual-purpose algorithms in [26–31], we first proposed a novel UIC formulation, and then based on this information criterion we derived our dual-purpose learning algorithm for adaptively tracking PS and MS. Our dual-purpose algorithms are exact subspace gradient flow and both are self-stabilizing. Compared with some existing unified algorithms, our dual-purpose algorithms have faster convergence for the weight matrix, and the computational complexity do not increase, which will be seen in the latter simulations. In this chapter, the landscape of the UIC formulation, self-stabilizing property, and the global asymptotical convergence of our dual-purpose subspace gradient flow will be analyzed in detail. These theoretical results will lay a solid foundation for the applications of this algorithm.

5.2 Review of Neural Network-Based Dual-Purpose Methods

The dual-purpose principal and minor subspace gradient flow can be used to track PS. If altered simply by the sign, it can also serve as a MS tracker. This is of practical significance in the implementation of algorithms.

5.2.1 Chen's Unified Stabilization Approach

In many information processing systems, it is necessary to extract the main features or eliminate the noise inherent in complex, high-dimensional input data streams. Two of the most general purpose feature extraction techniques are PCA and MCA. However, there appeared two intriguing puzzles in PCA and MCA. The first one is why it is more difficult to find p principal components (eigenvectors) than to find the principal subspace. Based on the subspace method, Xu used $\boldsymbol{D} = \mathrm{diag}(d_1^2, \ldots, d_p^2), d_1 > d_2 > \cdots > d_p > 0$ to solve the first puzzle. The second puzzle is concerned with MCA algorithms. Since PCA algorithms use the gradient method, the corresponding MCA algorithms seem to be derived from the same idea by just changing the cost function. This idea suggests to change the sign of the right-hand side of the PCA algorithms. However, this idea does not work in general. Most MCA algorithms derived from changing the sign of its PCA algorithm suffer from marginal instability in discrete-time updating.

In [27], Chen proposed a unified stabilization approach for principal and minor component extraction algorithms as follows:

$$\boldsymbol{W}(t) = \{\boldsymbol{C}\boldsymbol{W}(t)\boldsymbol{W}^{\mathrm{T}}(t)\boldsymbol{W}(t) - \boldsymbol{W}^{\mathrm{T}}(t)\boldsymbol{W}(t)\boldsymbol{C}\boldsymbol{W}(t)\} + \boldsymbol{W}(t)\{\boldsymbol{E} - \boldsymbol{W}^{\mathrm{T}}(t)\boldsymbol{W}(t)\}, \tag{5.1}$$

$$\boldsymbol{W}(t) = -\{\boldsymbol{C}\boldsymbol{W}(t)\boldsymbol{W}^{\mathrm{T}}(t)\boldsymbol{W}(t) - \boldsymbol{W}^{\mathrm{T}}(t)\boldsymbol{W}(t)\boldsymbol{C}\boldsymbol{W}(t)\} + \boldsymbol{W}(t)\{\boldsymbol{E} - \boldsymbol{W}^{\mathrm{T}}(t)\boldsymbol{W}(t)\}, \tag{5.2}$$

where $\boldsymbol{W}(0) = \boldsymbol{U}(0)\boldsymbol{D}(0)\boldsymbol{V}(0)^{\mathrm{T}}$ is the singular value decomposition of the initial matrix $\boldsymbol{W}(0)$ and $\boldsymbol{E} = \boldsymbol{W}^{\mathrm{T}}(0)\boldsymbol{W}(0) = \boldsymbol{V}(0)\boldsymbol{D}(0)\boldsymbol{V}^{\mathrm{T}}(0)$. The first term is to extract principal or minor components, and the second term is used to recover deviation $\boldsymbol{W}^{\mathrm{T}}(t)\boldsymbol{W}(t)$ from *E*. The rigorous stability analysis shows that Chen's dual-purpose algorithms to principal and minor component extraction are both self-stabilizing [27].

5.2.2 Hasan's Self-normalizing Dual Systems

As indicated in [32], the task of developing an MCA flow is perceived as being more complicated than that for a PCA flow. The Hasan's work in [28] shows that perhaps there are as many MCA/MSA dynamical flows as there are PCA/PSA flows. A common method for converting a PCA/PSA flow into an MCA/MSA one is to change the sign of the given matrix, or by using the inverse of the original matrix [28, 32]. However, inverting a large matrix is a costly task, and changing the sign of the original matrix does not always generate a stable system unless frequent orthonormalization is employed during the numerical implementation. In [28], Hasan proposed a framework to generate classes of stable dynamic systems that can be easily converted from PCA flow into MCA flow and vice versa.

First, based on the Rayleigh quotient $\mathrm{tr}\left\{(\boldsymbol{W}^{\mathrm{T}}\boldsymbol{C}\boldsymbol{W})(\boldsymbol{W}^{\mathrm{T}}\boldsymbol{W})^{-1}\right\}$ and the inverse Rayleigh quotient $\mathrm{tr}\left\{(\boldsymbol{W}^{\mathrm{T}}\boldsymbol{W})(\boldsymbol{W}^{\mathrm{T}}\boldsymbol{C}\boldsymbol{W})^{-1}\right\}$, the following two systems can be obtained

$$\dot{\boldsymbol{W}} = \boldsymbol{W}\boldsymbol{W}^{\mathrm{T}}\boldsymbol{C}\boldsymbol{W} - \boldsymbol{C}\boldsymbol{W}\boldsymbol{W}^{\mathrm{T}}\boldsymbol{W}, \tag{5.3}$$

$$\dot{\boldsymbol{W}} = \boldsymbol{C}\boldsymbol{W}\boldsymbol{W}^{\mathrm{T}}\boldsymbol{W} - \boldsymbol{W}\boldsymbol{W}^{\mathrm{T}}\boldsymbol{C}\boldsymbol{W}. \tag{5.4}$$

These two systems both appear in the numerator of the gradient of the Rayleigh and the inverse Rayleigh quotients. For (5.3) and (5.4), the following theorem holds.

Theorem 5.1 *Systems* (5.3) *and* (5.4) *are stable and if* $\boldsymbol{W}(t)$ *is a solution of either systems for* $t \geq 0$, *then* $\boldsymbol{W}(t)^{\mathrm{T}}\boldsymbol{W}(t) = \boldsymbol{W}(0)^{\mathrm{T}}\boldsymbol{W}(0)$ *for any* $t \geq 0$.

Proof To examine the critical points of the Rayleigh quotient and the inverse Rayleigh quotient for the simple case where $\boldsymbol{W} \in \mathrm{R}^{n \times 1}$, let $\boldsymbol{A}$ be symmetric and $\boldsymbol{W}$ be one-dimensional. System (5.3) can be rewritten as

$$\dot{W} = \nabla_W \left(\frac{W^T W}{W^T C W} \right) (W^T C W)^2, \tag{5.5}$$

and system (5.4) can be rewritten as

$$\dot{W} = -\nabla_W \left(\frac{W^T C W}{W^T W} \right) (W^T W)^2. \tag{5.6}$$

This means that $WW^T CW - CWW^T W = \nabla_W \left(\frac{W^T W}{W^T CW} \right) (W^T CW)^2 = -\nabla_W \left(\frac{W^T CW}{W^T W} \right) (W^T W)^2$. Thus, both systems (5.3) and (5.4) are gradient-like systems since $(W^T CW)^2$ and $(W^T W)^2$ are positive definite.

To understand the behavior of the Rayleigh quotient along the trajectory of (5.3) and (5.4), let $f(W) = \frac{W^T CW}{W^T W}$, $C^T = C$, Then, $\dot{f} = \nabla_W (f(W))^T \dot{W} = -\nabla_W (f(W))^T \nabla_W (f(W)) (W^T CW)^2 \leq 0$. Consequently, $f(W(t))$ is a decreasing function of $t \geq 0$. Furthermore, since it is bounded from below, $\lim_{t \to \infty} f(W(t))$ exists. Also, note that $V(W) = W^T W$ remains constant along the trajectory of system (5.3), and the function $V = W^T CW$ is decreasing since $\dot{V} = (W^T CW)^2 - W^T C^2 WW^T W < 0$. This implies that $\mathbf{W}^T(t) C \mathbf{W}(t) \leq \mathbf{W}_0^T C \mathbf{W}_0$ and $\mathbf{W}^T(t) \mathbf{W}(t) \leq \mathbf{W}_0^T \mathbf{W}_0$ for $t \geq 0$.

This completes the proof [28].

In the following, several generalizations of systems (5.3) and (5.4) are provided.

Theorem 5.2 *Consider the dynamic systems*

$$\dot{W} = W\mathrm{K}(W) - CWW^T W, \tag{5.7}$$

$$\dot{W} = CWW^T W - W\mathrm{K}(W), \tag{5.8}$$

where $\mathrm{K}(W) : \Re^{n \times p} \to \Re^{p \times p}, p \leq n$ *is a continuously differentiable function. If* $\mathrm{K} + \mathrm{K}^T = W^T CW + W^T C^T W + \alpha(I - W^T W) B(W)$, *where* $\alpha \geq 0$ *and* $B(W) + B(W)^T$ *is positive definite, then systems* (5.7) *and* (5.8) *are stable.*

Proof By considering a Lyapunov function of the form $V(W) = (1/4)\mathrm{tr}\left((W^T W - I)^2 \right)$, it can be shown that the time derivative of V along the trajectory $W(t)$ of system (5.7) is

$$\begin{aligned} \dot{V} &= \mathrm{tr}\{ (W^T W - I)(\mathrm{K}^T - W^T CW) W^T W \} \\ &= \frac{1}{2} \mathrm{tr}\{ (W^T W - I)(\mathrm{K}^T + \mathrm{K} - W^T CW - W^T C^T W) W^T W \} \\ &= -\frac{\alpha}{4} \mathrm{tr}\left\{ (W^T W - I)^2 (B(W) + B(W)^T) \right\} \leq 0. \end{aligned}$$

Since $V(W) \to \infty$ as $\|W\| \to \infty$, Theorem 5.2 guarantees that system (5.7) is stable. Similarly, system (5.8) is stable.

This completes the proof [28].

Special Case: Based on the above theorem, several variations of (5.3) and (5.4) may be derived. For example, assume that in Theorem 5.2, let $\mathrm{K} - \boldsymbol{W}^{\mathrm{T}}\boldsymbol{C}\boldsymbol{W} = \alpha B(\boldsymbol{W})\left(\boldsymbol{I} - \boldsymbol{W}^{\mathrm{T}}\boldsymbol{W}\right)$, where $B(\boldsymbol{W}) + B(\boldsymbol{W})^{\mathrm{T}}$ is positive definite, and $\alpha \geq 0$. Then, systems (5.7) and (5.8) are simplified to

$$\dot{\boldsymbol{W}} = \boldsymbol{W}\boldsymbol{W}^{\mathrm{T}}\boldsymbol{C}\boldsymbol{W} - \boldsymbol{C}\boldsymbol{W}\boldsymbol{W}^{\mathrm{T}}\boldsymbol{W} - \alpha \boldsymbol{W} B(\boldsymbol{W})\left(\boldsymbol{W}^{\mathrm{T}}\boldsymbol{W} - \boldsymbol{I}\right), \tag{5.9}$$

$$\dot{\boldsymbol{W}} = \boldsymbol{C}\boldsymbol{W}\boldsymbol{W}^{\mathrm{T}}\boldsymbol{W} - \boldsymbol{W}\boldsymbol{W}^{\mathrm{T}}\boldsymbol{C}\boldsymbol{W} - \alpha \boldsymbol{W} B(\boldsymbol{W})\left(\boldsymbol{W}^{\mathrm{T}}\boldsymbol{W} - \boldsymbol{I}\right). \tag{5.10}$$

In particular, when $B(\boldsymbol{W}) = \boldsymbol{I}$, the following MSA/PSA systems hold:

$$\dot{\boldsymbol{W}} = \boldsymbol{W}\boldsymbol{W}^{\mathrm{T}}\boldsymbol{C}\boldsymbol{W} - \mathbf{C}\boldsymbol{W}\boldsymbol{W}^{\mathrm{T}}\boldsymbol{W} - \alpha \boldsymbol{W}\left(\boldsymbol{W}^{\mathrm{T}}\boldsymbol{W} - \boldsymbol{I}\right), \tag{5.9}$$

$$\dot{\boldsymbol{W}} = \boldsymbol{C}\boldsymbol{W}\boldsymbol{W}^{\mathrm{T}}\boldsymbol{W} - \boldsymbol{W}\boldsymbol{W}^{\mathrm{T}}\boldsymbol{C}\boldsymbol{W} - \alpha \boldsymbol{W}\left(\boldsymbol{W}^{\mathrm{T}}\boldsymbol{W} - \boldsymbol{I}\right). \tag{5.10}$$

When C is symmetric, other variations follow by incorporating the term $-\alpha \boldsymbol{C}^k \boldsymbol{W}\left(\boldsymbol{W}^{\mathrm{T}}\boldsymbol{W} - \boldsymbol{I}\right)$ into systems (5.3) and (5.4):

$$\dot{\boldsymbol{W}} = \pm\left\{\boldsymbol{W}\boldsymbol{W}^{\mathrm{T}}\boldsymbol{C}\boldsymbol{W} - \boldsymbol{C}\boldsymbol{W}\boldsymbol{W}^{\mathrm{T}}\boldsymbol{W}\right\} - \alpha \boldsymbol{C}^k \boldsymbol{W}\left(\boldsymbol{W}^{\mathrm{T}}\boldsymbol{W} - \boldsymbol{I}\right), \tag{5.11}$$

where k is an integer and $\alpha \geq 0$. Here, the choices of the + and − signs yield MSA and PSA systems, respectively. If the − sign is chosen and $\alpha = k = 1$, it follows from (5.11) that

$$\dot{\boldsymbol{W}} = \boldsymbol{C}\boldsymbol{W} - \boldsymbol{W}\boldsymbol{W}^{\mathrm{T}}\boldsymbol{C}\boldsymbol{W}, \tag{5.12}$$

which is one form of Oja's subspace system. When $\alpha = k = 1$, the system (5.11) with the + sign reduces to

$$\dot{\boldsymbol{W}} = \boldsymbol{W}\boldsymbol{W}^{\mathrm{T}}\boldsymbol{C}\boldsymbol{W} - \boldsymbol{C}\boldsymbol{W}\left(2\boldsymbol{W}^{\mathrm{T}}\boldsymbol{W} - \boldsymbol{I}\right). \tag{5.13}$$

This MSA system is known in the literature and was analyzed in [17].

To convert MSA/PSA systems into MCA/PCA learning rules, a diagonal matrix D may be incorporated in the above systems. For example, if I in system (5.11) is replaced by $\boldsymbol{D}$, the following systems can be obtained

$$\dot{\boldsymbol{W}} = \pm\left\{\boldsymbol{W}\boldsymbol{W}^{\mathrm{T}}\boldsymbol{C}\boldsymbol{W} - \boldsymbol{C}\boldsymbol{W}\boldsymbol{W}^{\mathrm{T}}\boldsymbol{W}\right\} - \alpha \boldsymbol{C}^k \boldsymbol{W}\left(\boldsymbol{W}^{\mathrm{T}}\boldsymbol{W} - \boldsymbol{D}\right). \tag{5.14}$$

If $\alpha = k = 1$, the two systems in (5.14) are simplified to

$$\dot{\boldsymbol{W}} = \boldsymbol{C}\boldsymbol{W}\boldsymbol{D} - \boldsymbol{W}\boldsymbol{W}^{\mathrm{T}}\boldsymbol{C}\boldsymbol{W}, \tag{5.15}$$

$$\dot{\boldsymbol{W}} = \boldsymbol{W}\boldsymbol{W}^{\mathrm{T}}\boldsymbol{C}\boldsymbol{W} - \boldsymbol{C}\boldsymbol{W}\left(2\boldsymbol{W}^{\mathrm{T}}\boldsymbol{W} - \boldsymbol{D}\right). \tag{5.16}$$

System (5.15) is sometimes called Xu's weighted PCA rule [9]. Also, system (5.16) is an MCA version of the minor subspace system (5.13). For other more dual-purpose PCA/MCA systems derived using logarithmic cost functions, see [28]. In [28], the PCA/PSA and MCA/MSA learning differential equations of the previous sections may be modified to obtain PCA/PSA and MCA/MSA learning differential equations for the generalized eigenvalue problem involving two matrices $\boldsymbol{A}$ and $\boldsymbol{B}$, where $\boldsymbol{B}$ is positive definite and $\boldsymbol{A}^{\mathrm{T}} = \boldsymbol{A}$.

5.2.3 Peng's Unified Learning Algorithm to Extract Principal and Minor Components

In [29], Peng proposed a unified PCA and MCA algorithms as follows:

$$\begin{aligned} \boldsymbol{w}(k+1) &= \boldsymbol{w}(k) \\ &\pm \eta\left\{ y(k)\boldsymbol{x}(k) - y^2(k)\boldsymbol{w}(k) + \frac{\boldsymbol{w}_0^{\mathrm{T}}\boldsymbol{x}(k)\boldsymbol{x}^{\mathrm{T}}(k)\boldsymbol{w}_0}{\boldsymbol{w}_0^{\mathrm{T}}\boldsymbol{w}_0}\left[\|\boldsymbol{w}(k)\|^2 - 1\right]\boldsymbol{w}(k)\right\}, \end{aligned} \tag{5.17}$$

where $\eta > 0$ is a constant learning rate, $\boldsymbol{w}_0$ is an n-dimensional randomly selected nonzero vector. Here, the choices of the + and − signs yield PCA and MCA algorithms, respectively.

In order to find a sufficient condition to guarantee the convergence of (5.17), Peng analyzed the dynamics of (5.17) via the DDT approach. A DDT system of (5.17) is be given by

$$\boldsymbol{w}(k+1) = \boldsymbol{w}(k) \pm \eta\left\{ \boldsymbol{R}\boldsymbol{w}(k) - \boldsymbol{w}^{\mathrm{T}}(k)\boldsymbol{R}\boldsymbol{w}(k)\boldsymbol{w}(k) + \frac{\boldsymbol{w}_0^{\mathrm{T}}\boldsymbol{R}\boldsymbol{w}_0}{\boldsymbol{w}_0^{\mathrm{T}}\boldsymbol{w}_0}\left[\|\boldsymbol{w}(k)\|^2 - 1\right]\boldsymbol{w}(k)\right\}, \tag{5.18}$$

where $\boldsymbol{R} = \mathrm{E}[\boldsymbol{x}(k)\boldsymbol{x}^{\mathrm{T}}(k)]$ is the correlation matrix of $\boldsymbol{x}(k)$. By denoting $\mu = \left(\boldsymbol{w}_0^{\mathrm{T}}\boldsymbol{R}\boldsymbol{w}_0\right)/\left(\boldsymbol{w}_0^{\mathrm{T}}\boldsymbol{w}_0\right)$, (5.18) can be simplified to

$$\boldsymbol{w}(k+1) = \boldsymbol{w}(k) + \eta\boldsymbol{C}\boldsymbol{w}(k) - \eta\boldsymbol{w}^{\mathrm{T}}(k)\boldsymbol{C}\boldsymbol{w}(k)\boldsymbol{w}(k) \tag{5.19}$$

where $\boldsymbol{C} = \pm(\boldsymbol{R} - \mu\boldsymbol{I})$ and $\boldsymbol{I}$ is an $n \times n$ unitary matrix. In [29], the dynamics of system (5.19) is analyzed, and some sufficient conditions to guarantee its convergence were obtained.

5.2.4 Manton's Dual-Purpose Principal and Minor Component Flow

It is known that principal component flows are flows converging to the eigenvectors associated with the largest eigenvalues of a given symmetric matrix, and minor component flows converge to the eigenvectors associated with the smallest eigenvalues. Traditional flows require the matrix to be positive definite. Moreover, finding well-behaved minor component flows appeared to be harder and unrelated to the principal component case. Manton derived a flow which can be used to extract either the principal or the minor components and which does not require the matrix to be positive definite. The flow is shown to be a generalization of the Oja–Brockett flow.

In [32], Manton used the following cost function:

$$f(\boldsymbol{W}) = \frac{1}{2}\operatorname{tr}\{\boldsymbol{CWNW}^{\mathrm{T}}\} + \frac{1}{4}\mu\|\boldsymbol{N} - \boldsymbol{W}^{\mathrm{T}}\boldsymbol{W}\|^2, \tag{5.20}$$

where the following assumptions are made, i.e., (1) scalar $\mu \in \Re$ is strictly positive, matrix $\boldsymbol{C} \in \Re^{n\times n}$ is symmetric, and matrix $\boldsymbol{N} \in \Re^{p\times p}$ is diagonal with distinct positive eigenvalues.

Then, the critical points of (5.20) were derived. The local stability analysis of these critical points was conducted, and local minima were established. It was proved that the essentially unique local minimum of the cost function (5.20) corresponds to the minor components of $\boldsymbol{C}$. It is therefore natural to consider the corresponding gradient flow. The gradient flow can be written as

$$\dot{\boldsymbol{W}} = -\boldsymbol{CWN} + \mu\boldsymbol{W}(\boldsymbol{N} - \boldsymbol{W}^{\mathrm{T}}\boldsymbol{W}), \tag{5.21}$$

which is the minor component flow. For the detailed theorem and its proof, see [32].

For convenience, the Oja–Brockett flow is restated here but using different variables,

$$\dot{\boldsymbol{Z}} = \boldsymbol{CZN} - \boldsymbol{ZNZ}^{\mathrm{T}}\boldsymbol{CZ},\ \boldsymbol{Z} \in \Re^{n\times p} \tag{5.22}$$

where $\boldsymbol{C} \in \Re^{n\times n}$ and $\boldsymbol{N} \in \Re^{p\times p}$ are positive definite symmetric matrices with distinct eigenvalues. The columns of $\boldsymbol{Z}$ converge to the eigenvectors associated with the p largest eigenvalues of $\boldsymbol{C}$ in an order determined by $\boldsymbol{N}$ [32].

Using the linear coordinate transformation $\boldsymbol{W} = \lambda^{-1/2}\boldsymbol{C}^{1/2}\boldsymbol{ZN}^{1/2}$, which is only defined if both $\boldsymbol{C}$ and $\boldsymbol{N}$ are positive definite symmetric matrices, the Oja–Brockett flow (5.22) becomes

$$\dot{\boldsymbol{W}} = (\boldsymbol{C} - \lambda\boldsymbol{I})\boldsymbol{WN} + \lambda\boldsymbol{W}(\boldsymbol{N} - \boldsymbol{W}^{\mathrm{T}}\boldsymbol{W}), \tag{5.23}$$

which is the minor component flow (5.21) with $\boldsymbol{A} = \lambda\boldsymbol{I} - \boldsymbol{C}$ and $\mu = \lambda$.

Since the Oja–Brockett flow requires $\boldsymbol{C}$ and $\boldsymbol{N}$ to be positive definite and symmetric, the transformation from the Oja–Brockett flow (5.22) to the minor component flow (5.21) is always valid. However, the reverse transformation from (5.21) to (5.22) is not always possible, meaning that the minor component flow (5.21) is a strict generalization of the Oja–Brockett flow. The reverse of $\boldsymbol{A} = \lambda\boldsymbol{I} - \boldsymbol{C}$ and $\mu = \lambda$ is $\boldsymbol{C} = \lambda\boldsymbol{I} - \boldsymbol{A}$ and $\lambda = \mu$. Since $\boldsymbol{C}$ must be positive definite for $\boldsymbol{W} = \lambda^{-1/2}\boldsymbol{C}^{1/2}\boldsymbol{Z}\boldsymbol{N}^{1/2}$ to be defined (and for (5.22) to be stable), the reverse transformation is only valid if μ is larger than the largest eigenvalue of $\boldsymbol{A}$. Since (5.21) is a generalization of (5.22), it is natural to consider using (5.21) with $\boldsymbol{A} = \lambda\boldsymbol{I} - \boldsymbol{C}$ to find the principal components of $\boldsymbol{C}$. Since $\boldsymbol{A}$ is not required to be positive definite in (5.21), the choice of λ is relatively unimportant. Thus, Manton's dual-purpose principal and minor component flow can be rewritten as follows:

$$\dot{\boldsymbol{W}} = \pm\boldsymbol{C}\boldsymbol{W}\boldsymbol{N} + \mu\boldsymbol{W}\left(\boldsymbol{N} - \boldsymbol{W}^{\mathrm{T}}\boldsymbol{W}\right), \tag{5.24}$$

where the choices of the + and − signs yield principal and minor component flow, respectively.

5.3 A Novel Dual-Purpose Method for Principal and Minor Subspace Tracking

In this section, a UIC will be introduced, and a dual-purpose principal and minor subspace gradient flow will be derived based on this information criterion. In this dual-purpose gradient flow, the weight matrix length is self-stabilizing. The dual-purpose gradient flow can efficiently track an orthonormal basis of the PS or MS.

5.3.1 Preliminaries

5.3.1.1 Definitions and Properties

Definition 5.1 For a $r \times r$ matrix $\boldsymbol{B}$, its EVD is represented as $\boldsymbol{B} = \boldsymbol{\Phi}\boldsymbol{\Psi}\boldsymbol{\Phi}^{-1}$, where $\boldsymbol{\Phi}$ denotes a $r \times r$ matrix formed by all its eigenvectors, and $\boldsymbol{\Psi} = \mathrm{diag}(\lambda_1, \ldots, \lambda_r) > 0$ is a diagonal matrix formed by all its eigenvalues.

Property 5.1 *If* $\boldsymbol{A}$ *is an* $m \times n$ *matrix and* $\boldsymbol{B}$ *is an* $n \times m$ *matrix, then it holds that* $\mathrm{tr}(\boldsymbol{A}\boldsymbol{B}) = \mathrm{tr}(\boldsymbol{B}\boldsymbol{A})$.

5.3.1.2 Conventional Formulation for PSA or MSA

Suppose that the vector sequence $\boldsymbol{x}_k, k = 1, 2, \ldots$ is a stationary stochastic process with zero mean and covariance matrix $\boldsymbol{R} = E\left[\boldsymbol{x}_k \boldsymbol{x}_k^{\mathrm{T}}\right] \in \mathbb{R}^{N \times N}$. Let λ_i and $\boldsymbol{v}_i (i = 1, 2, \ldots, N)$ denote the eigenvalues and the associated orthonormal eigenvectors of $\boldsymbol{R}$, respectively. We can arrange the orthonormal eigenvectors $\boldsymbol{v}_1, \boldsymbol{v}_2, \ldots, \boldsymbol{v}_N$ in such a way that the associated eigenvalues are in a nondecreasing order: $0 < \lambda_1 \leq \lambda_2 \leq \cdots \leq \lambda_N$. The eigenvalue decomposition (EVD) of $\boldsymbol{R}$ is represented as $\boldsymbol{R} = \sum_{i=1}^{N} \lambda_i \boldsymbol{v}_i \boldsymbol{v}_i^{\mathrm{T}} = \boldsymbol{L} \boldsymbol{\Lambda} \boldsymbol{L}^{\mathrm{T}}$, where $\boldsymbol{\Lambda} = \mathrm{diag}(\lambda_1, \ldots, \lambda_N)$ and $\boldsymbol{L} = [\boldsymbol{v}_1, \boldsymbol{v}_2, \ldots, \boldsymbol{v}_N]$.

For some applications in information processing fields, usually it is not necessary to perform true PCA, and the PSA is sufficient to yield the optimal solution by an arbitrary orthonormal basis spanning the principal subspace [22]. Similarly in some applications, we require only to find the MS spanned by $\boldsymbol{v}_1, \boldsymbol{v}_2, \ldots, \boldsymbol{v}_r$, where r is the dimension of the MS [1]. Only for PS tracking or MS tracking, lots of algorithms can be used. Here, our objective is to find such algorithms as dual-purpose subspace tracking, which are capable of both PS and MS tracking by simply switching the sign in the same learning rule and have also self-stabilizing property in both PS and MS tracking.

Let $\boldsymbol{W} = [\boldsymbol{u}_1, \boldsymbol{u}_2, \ldots, \boldsymbol{u}_r] \in R^{N \times r}$ denote the weight matrix, where $\boldsymbol{u}_i \in \Re^{N \times 1}$ represents the ith column vector of $\boldsymbol{W} = \boldsymbol{L}_{(n)} \boldsymbol{Q}$ and also represents the weight vector of the ith neuron of a multiple-input–multiple-output (MIMO) linear neural network, and r is the dimension of the subspace. The input–output relation of the MIMO linear neural network is described by

$$\boldsymbol{y}_k = \boldsymbol{W}^{\mathrm{T}} \boldsymbol{x}_k \tag{5.25}$$

where $\boldsymbol{y}_k$ is a low-dimensional representation of $\boldsymbol{x}_k$. By minimizing the MSE between $\boldsymbol{x}_k$ and its reconstruction or maximizing the variance of $\boldsymbol{y}_k$ and using the exact gradient descent rule, a few PS tracking algorithms [9, 33] have been derived. The two frameworks mentioned above are in fact equivalent [22]. In [22], a novel information criterion was proposed, and a fast PS tracking algorithm was derived by using the gradient method. Recently, a novel random gradient-based algorithm was proposed for online tracking the MS, and a corresponding energy function was used to analyze the globally asymptotical convergence [1]. However, our analysis indicates that the above information criteria could not derive dual-purpose subspace tracking algorithms by only changing the sign of the given matrix, which will be analyzed briefly in what follows. For the convenience of analysis, here we only take single component extraction algorithm into account.

In [9], by minimizing the MSE of reconstructing $\boldsymbol{x}_k$ from $\boldsymbol{y}_k$, i.e.,

$$\begin{aligned} \min_{\boldsymbol{W}} \{J_{\mathrm{MSE}}(\boldsymbol{W})\} &= \frac{1}{2} E\left\{ \|\boldsymbol{x}_k - \boldsymbol{W} \boldsymbol{y}_k\|^2 \right\} \\ &= \frac{1}{2} \left[\mathrm{tr}(\boldsymbol{R}) - \mathrm{tr}\left(2\boldsymbol{W}^{\mathrm{T}} \boldsymbol{R} \boldsymbol{W} - \boldsymbol{W}^{\mathrm{T}} \boldsymbol{R} \boldsymbol{W} \boldsymbol{W}^{\mathrm{T}} \boldsymbol{W}\right) \right], \end{aligned} \tag{5.26}$$

Following the exact gradient descent rule to minimize $J_{\text{MSE}}(\boldsymbol{W})$, the well-known LMSER algorithm for principal subspace analysis was derived as follows:

$$\boldsymbol{W}_{k+1} = \boldsymbol{W}_k + \eta\left(2\boldsymbol{R} - \boldsymbol{R}\boldsymbol{W}_k\boldsymbol{W}_k^{\mathrm{T}} - \boldsymbol{W}_k\boldsymbol{W}_k^{\mathrm{T}}\boldsymbol{R}\right)\boldsymbol{W}_k. \tag{5.27}$$

For one-dimensional case, (5.27) can be written as $\boldsymbol{w}_{k+1} = \boldsymbol{w}_k + \eta(2\boldsymbol{R} - \boldsymbol{R}\boldsymbol{w}_k\boldsymbol{w}_k^{\mathrm{T}} - \boldsymbol{w}_k\boldsymbol{w}_k^{\mathrm{T}}\boldsymbol{R})\boldsymbol{w}_k$. From this equation, it is easy to obtain that:

$$\begin{aligned}\frac{\boldsymbol{w}_{k+1}^{\mathrm{T}}\boldsymbol{w}_{k+1}}{\boldsymbol{w}_k^{\mathrm{T}}\boldsymbol{w}_k} &\approx 1 + \frac{2\eta(2\boldsymbol{w}_k^{\mathrm{T}}\boldsymbol{R}\boldsymbol{w}_k - 2\boldsymbol{w}_k^{\mathrm{T}}\boldsymbol{R}\boldsymbol{w}_k\boldsymbol{w}_k^{\mathrm{T}}\boldsymbol{w}_k)}{\boldsymbol{w}_k^{\mathrm{T}}\boldsymbol{w}_k} = 1 + \frac{4\eta\boldsymbol{w}_k^{\mathrm{T}}\boldsymbol{R}\boldsymbol{w}_k(1 - \boldsymbol{w}_k^{\mathrm{T}}\boldsymbol{w}_k)}{\boldsymbol{w}_k^{\mathrm{T}}\boldsymbol{w}_k} \\ &= \begin{cases} > 1 & \text{for} \quad \boldsymbol{w}_k^{\mathrm{T}}\boldsymbol{w}_k < 1 \\ < 1 & \text{for} \quad \boldsymbol{w}_k^{\mathrm{T}}\boldsymbol{w}_k > 1 \\ = 1 & \text{for} \quad \boldsymbol{w}_k^{\mathrm{T}}\boldsymbol{w}_k = 1. \end{cases}\end{aligned} \tag{5.28}$$

Obviously, it shows that the LMSER algorithm for principal component analysis has self-stabilizing property, which means that the weight vector length is self-stabilizing, i.e., moving toward unit length at each learning step [17]. If we change the sign of matrix R in (5.26) and use the exact gradient descent rule, we can obtain the LMSER algorithm for minor component analysis as follows:

$$\boldsymbol{w}_{k+1} = \boldsymbol{w}_k - \eta\left(2\boldsymbol{R} - \boldsymbol{R}\boldsymbol{w}_k\boldsymbol{w}_k^{\mathrm{T}} - \boldsymbol{w}_k\boldsymbol{w}_k^{\mathrm{T}}\boldsymbol{R}\right)\boldsymbol{w}_k. \tag{5.29}$$

From (5.29), it follows that:

$$\begin{aligned}\frac{\boldsymbol{w}_{k+1}^{\mathrm{T}}\boldsymbol{w}_{k+1}}{\boldsymbol{w}_k^{\mathrm{T}}\boldsymbol{w}_k} &\approx 1 - \frac{2\eta(2\boldsymbol{w}_k^{\mathrm{T}}\boldsymbol{R}\boldsymbol{w}_k - 2\boldsymbol{w}_k^{\mathrm{T}}\boldsymbol{R}\boldsymbol{w}_k\boldsymbol{w}_k^{\mathrm{T}}\boldsymbol{w}_k)}{\boldsymbol{w}_k^{\mathrm{T}}\boldsymbol{w}_k} = 1 - \frac{4\eta\boldsymbol{w}_k^{\mathrm{T}}\boldsymbol{R}\boldsymbol{w}_k(1 - \boldsymbol{w}_k^{\mathrm{T}}\boldsymbol{w}_k)}{\boldsymbol{w}_k^{T}\boldsymbol{w}_k} \\ &= \begin{cases} < 1 & \text{for} \quad \boldsymbol{w}_k^{\mathrm{T}}\boldsymbol{w}_k < 1 \\ > 1 & \text{for} \quad \boldsymbol{w}_k^{\mathrm{T}}\boldsymbol{w}_k > 1 \\ = 1 & \text{for} \quad \boldsymbol{w}_k^{\mathrm{T}}\boldsymbol{w}_k = 1. \end{cases}\end{aligned} \tag{5.30}$$

It is clear that the LMSER algorithm for minor component analysis does not have self-stabilizing property. We can perform the above analyses on some algorithms in [9, 1, 22, 33, 34], etc., and conclude that the dual-purpose self-stabilizing algorithm for PCA and MCA cannot be derived from their objective functions by only changing the sign of the given matrix.

Recently, considerable interest has been given to the construction and analysis of dual systems for minor and principal component extraction (or subspace analysis) [26–31, 35]. The above analysis has shown that an appropriately selected objective function is the key in deriving dual-purpose self-stabilizing systems. In order to obtain the desirable dual-purpose algorithm for PS and MS tracking, it is necessary to develop more novel information criterion.

5.3.2 A Novel Information Criterion and Its Landscape

5.3.2.1 A Novel Criterion for PSA and MSA

Given $\boldsymbol{W} \in R^{N\times r}$ in the domain $\Omega = \left\{\boldsymbol{W}|0 < \boldsymbol{W}^{\mathrm{T}}\boldsymbol{R}\boldsymbol{W} < \infty, \boldsymbol{W}^{\mathrm{T}}\boldsymbol{W} \neq 0\right\}$, we present a novel unified nonquadratic criterion (NQC) for PSA and MSA as follows:

$$\begin{aligned}&\min_{W} J_{\mathrm{NQC}}(\boldsymbol{W},\boldsymbol{R})\\ &J_{\mathrm{NQC}}(\boldsymbol{W},\boldsymbol{R}) = \pm\frac{1}{2}\mathrm{tr}\left\{\left(\boldsymbol{W}^{T}\boldsymbol{R}\boldsymbol{W}\right)\left(\boldsymbol{W}^{T}\boldsymbol{W}\right)^{-1}\right\} + \frac{1}{2}\mathrm{tr}\left\{\left[\boldsymbol{I} - \left(\boldsymbol{W}^{\mathrm{T}}\boldsymbol{W}\right)\right]^{2}\right\},\end{aligned} \tag{5.31}$$

where "+" is for MS tracking, "−" is for PS tracking, and $\boldsymbol{I}$ is Identity matrix. From (5.31), we can see that $J_{\mathrm{NQC}}(\boldsymbol{W},\boldsymbol{R})$ has a lower bound and approaches infinity from the above as $\boldsymbol{W}^{\mathrm{T}}\boldsymbol{W} \to \infty$. Obviously, the gradient searching algorithm can be derived based on the above unified NQC. This criterion is referred to as novel because it is different from all existing PSA or MSA criteria. It is worth noting that if we replace the Rayleigh quotient $\left(\boldsymbol{W}^{\mathrm{T}}\boldsymbol{R}\boldsymbol{W}\right)\left(\boldsymbol{W}^{\mathrm{T}}\boldsymbol{W}\right)^{-1}$ by $\left(\boldsymbol{W}^{\mathrm{T}}\boldsymbol{R}\boldsymbol{W}\right)$ in (5.31) when "−" is used, then the objective function can be used to derive a few PCA algorithms for one-dimensional case [25, 36]. However, following the analysis method in Sect. 5.3.1.2, it can be easily seen that the objective function in [25, 36] cannot derive dual-purpose self-stabilizing gradient learning algorithms by only changing the sign of the given matrix. The landscape of this novel criterion is depicted by the following four theorems. Since the matrix differential method will be used extensively, interested readers may refer to [37] for more details.

5.3.2.2 Landscape of Nonquadratic Criteria

Given $\boldsymbol{W} \in R^{N\times r}$ in the domain $\Omega = \left\{\boldsymbol{W}|0 < \boldsymbol{W}^{\mathrm{T}}\boldsymbol{R}\boldsymbol{W} < \infty, \boldsymbol{W}^{\mathrm{T}}\boldsymbol{W} \neq 0\right\}$, we analyze the following NQC for tracking the MS:

$$\min_{\boldsymbol{W}} E1(\boldsymbol{W}) = \frac{1}{2}\mathrm{tr}\left\{\left(\boldsymbol{W}^{\mathrm{T}}\boldsymbol{R}\boldsymbol{W}\right)\left(\boldsymbol{W}^{\mathrm{T}}\boldsymbol{W}\right)^{-1}\right\} + \frac{1}{2}\mathrm{tr}\left\{\left[\boldsymbol{I} - \left(\boldsymbol{W}^{\mathrm{T}}\boldsymbol{W}\right)\right]^{2}\right\}. \tag{5.32}$$

The landscape of $E1(\boldsymbol{W})$ is depicted by Theorems 5.3 and 5.4.

Theorem 5.3 *$\boldsymbol{W}$ is a stationary point of $E1(\boldsymbol{W})$ in the domain Ω if and only if $\boldsymbol{W} = \boldsymbol{L}_r\boldsymbol{Q}$, where $\boldsymbol{L}_r \in R^{N\times r}$ consist of the r eigenvectors of $\boldsymbol{R}$, and $\boldsymbol{Q}$ is a $r \times r$ orthogonal matrix.*

If $\boldsymbol{W}$ is expanded by the eigenvector basis into $\boldsymbol{W} = \boldsymbol{L}^{\mathrm{T}}\tilde{\boldsymbol{W}}$, then we can write the NQC for the expanded coefficient matrix from (5.32) as

$$\min_{\tilde{W}} \tilde{E}1(\tilde{W}) = \frac{1}{2}\mathrm{tr}\left\{ (\tilde{W}^{\mathrm{T}} \Lambda \tilde{W})(\tilde{W}^{\mathrm{T}}\tilde{W})^{-1} \right\} + \frac{1}{2}\mathrm{tr}\left\{ [I - (\tilde{W}^{\mathrm{T}}\tilde{W})]^{2} \right\}, \tag{5.33}$$

where $\tilde{W} \in R^{N \times r}$ is an expanded coefficient matrix, and Λ is a $N \times N$ diagonal matrix given by $\mathrm{diag}(\lambda_1, \ldots \lambda_N)$. Obviously, (5.33) represents an equivalent form of (5.32). Thus, Theorem 5.3 is equivalent to the following Corollary 5.1. Therefore, we will only provide the proof of Corollary 5.1.

Corollary 5.1 $\tilde{W}$ *is a stationary point of* $\tilde{E}1(\tilde{W})$ *in the domain* $\tilde{\Omega} = \{\tilde{W} | 0 < \tilde{W}^{\mathrm{T}} \Lambda \tilde{W} < \infty, \tilde{W}^{\mathrm{T}}\tilde{W} \neq 0\}$ *if and only if* $\tilde{W} = P_r Q$, *where* P_r *is a* $N \times r$ *permutation matrix in which each column has exactly one nonzero element equal to 1, and each row has, at most, one nonzero element, where* $N \geq r$.

Proof The gradient of $\tilde{E}1(\tilde{W})$ with respect to $\tilde{W}$ can be written as

$$\nabla_{\tilde{W}} \tilde{E}1(\tilde{W}) = \Lambda \tilde{W}(\tilde{W}^{\mathrm{T}}\tilde{W})^{-1} - \tilde{W}(\tilde{W}^{\mathrm{T}}\tilde{W})^{-2}\tilde{W}^{\mathrm{T}}\Lambda\tilde{W} + \tilde{W}[I - (\tilde{W}^{\mathrm{T}}\tilde{W})], \tag{5.34}$$

where the notation $\nabla_{\tilde{W}}$ denotes $\partial\tilde{E}1/\partial\tilde{W}$, so does for (5.35).

Given a point in $\{\tilde{W} | \tilde{W} = P_r Q$ for any unitary orthonormal $Q\}$, we have

$$\begin{aligned} \nabla_{\tilde{W}} \tilde{E}1(P_r Q) &= \Lambda P_r Q (Q^{\mathrm{T}} P_r^{\mathrm{T}} P_r Q)^{-1} - P_r Q (Q^{\mathrm{T}} P_r^{\mathrm{T}} P_r Q)^{-2} Q^{\mathrm{T}} P_r^{\mathrm{T}} \Lambda P_r Q + P_r Q [I - (Q^{\mathrm{T}} P_r^{\mathrm{T}} P_r Q)] \\ &= \Lambda P_r Q - P_r^{\mathrm{T}} Q Q^{\mathrm{T}} P_r^{\mathrm{T}} \Lambda P_r Q. \end{aligned} \tag{5.35}$$

Conversely, $\tilde{E}1(\tilde{W})$ at a stationary point should satisfy $\nabla_{\tilde{W}} \tilde{E}1(\tilde{W}) = 0$, which yields

$$\left\{ \Lambda \tilde{W}(\tilde{W}^{\mathrm{T}}\tilde{W})^{-1} - \tilde{W}(\tilde{W}^{\mathrm{T}}\tilde{W})^{-2}\tilde{W}^{\mathrm{T}}\Lambda\tilde{W} + \tilde{W}[I - (\tilde{W}^{\mathrm{T}}\tilde{W})] \right\} = 0. \tag{5.36}$$

Premultiplying both sides of (5.36) by $\tilde{W}^{\mathrm{T}}$, we have

$$\begin{aligned} &\left\{ \tilde{W}^{\mathrm{T}} \Lambda \tilde{W}(\tilde{W}^{\mathrm{T}}\tilde{W})^{-1} - \tilde{W}^{\mathrm{T}}\tilde{W}(\tilde{W}^{\mathrm{T}}\tilde{W})^{-2}\tilde{W}^{\mathrm{T}}\Lambda\tilde{W} + \tilde{W}T\tilde{W}[I - (\tilde{W}^{\mathrm{T}}\tilde{W})] \right\} \\ &= \left\{ \tilde{W}^{\mathrm{T}} \Lambda \tilde{W}(\tilde{W}^{\mathrm{T}}\tilde{W})^{-1} - (\tilde{W}^{\mathrm{T}}\tilde{W})^{-1}\tilde{W}^{\mathrm{T}}\Lambda\tilde{W} + \tilde{W}T\tilde{W}[I - (\tilde{W}^{\mathrm{T}}\tilde{W})] \right\} \\ &= \left\{ \tilde{W}^{\mathrm{T}}\tilde{W}[I - (\tilde{W}^{\mathrm{T}}\tilde{W})] \right\} \\ &= \tilde{W}^{\mathrm{T}}\tilde{W} - (\tilde{W}^{\mathrm{T}}\tilde{W})^{2} = 0. \end{aligned}$$

From the above equation, we have

$$\tilde{W}^{\mathrm{T}}\tilde{W} = I_{\mathrm{r}} \tag{5.37}$$

which implies that the columns of $\tilde{\boldsymbol{W}} \in \boldsymbol{R}^{N\times r}$ are column-orthonormal at a stationary point of $\tilde{E}1(\tilde{\boldsymbol{W}})$. From (5.36) and (5.37), it holds that

$$\boldsymbol{\Lambda}\tilde{\boldsymbol{W}} = \tilde{\boldsymbol{W}}(\tilde{\boldsymbol{W}}^{\mathrm{T}}\boldsymbol{\Lambda}\tilde{\boldsymbol{W}}) \tag{5.38}$$

Let $\tilde{\boldsymbol{W}} = \left[\breve{u}_1^T, \ldots \breve{u}_N^T\right]^T$, where $\breve{u}_i(i = 1,\ldots,N)$ is a row vector, and $\boldsymbol{B} = \tilde{\boldsymbol{W}}^{\mathrm{T}}\boldsymbol{\Lambda}\tilde{\boldsymbol{W}}$ that is a $r \times r$ symmetric positive definite matrix. Then, an alternative form of (5.38) is

$$\sigma_i\breve{u}_i = \breve{u}_i\boldsymbol{B}(i = 1,\ldots,N) \tag{5.39}$$

Obviously, (5.39) shows the EVD for $\boldsymbol{B}$. Since $\boldsymbol{B}$ is a $r \times r$ symmetric positive definite matrix, it has only r orthonormal left row eigenvectors, which means that $\tilde{\boldsymbol{W}}$ has only r orthonormal row vectors. Moreover, all the r nonzero row vectors in $\tilde{\boldsymbol{W}}$ form an orthonormal matrix, which means that $\tilde{\boldsymbol{W}}$ can always be represented as $\tilde{\boldsymbol{W}} = \boldsymbol{P}_r\boldsymbol{Q}$.

This completes the proof.

Theorem 5.3 establishes the property for all the stationary points of $E1(\boldsymbol{W})$. The next theorem will distinguish the global minimum point set attained by $\boldsymbol{W}$ spanning the MS from the other stationary points that are saddle (unstable) points.

Theorem 5.4 *In the domain* Ω, $E1(\boldsymbol{W})$ *has a global minimum that is achieved if and only if* $\boldsymbol{W} = \boldsymbol{L}_{(n)}\boldsymbol{Q}$, *where* $\boldsymbol{L}_{(n)} = [\boldsymbol{v}_1, \boldsymbol{v}_2, \ldots, \boldsymbol{v}_r]$. *At the global minimum,* $E1(\boldsymbol{W}) = (1/2)\sum_{i=1}^{r}\lambda_i$. *All the other stationary points of* $\boldsymbol{W} = \boldsymbol{L}_r\boldsymbol{Q}(\boldsymbol{L}_r \neq \boldsymbol{L}_{(n)})$ *are saddle (unstable) points of* $E1(\boldsymbol{W})$.

Similarly, we can show that Theorem 5.4 is equivalent to the following Corollary 5.2, and Corollary 5.2 is proved to indirectly prove Theorem 5.4.

Corollary 5.2 *In the domain* Ω, $\tilde{E}1(\tilde{\boldsymbol{W}})$ *has a global minimum that is achieved if and only if* $\tilde{\boldsymbol{W}} = \overline{\boldsymbol{P}}\boldsymbol{Q}$, *where* $\overline{\boldsymbol{P}} = \begin{pmatrix}\tilde{\boldsymbol{P}} & \boldsymbol{0}\end{pmatrix}^{\mathrm{T}} \in R^{N\times r}$, *and* $\tilde{\boldsymbol{P}}$ *is a* $r \times r$ *permutation matrix. At the global minimum, we have* $\tilde{E}1(\tilde{\boldsymbol{W}}) = (1/2)\sum_{i=1}^{r}\lambda_i$. *All the other stationary points* $\tilde{\boldsymbol{W}} = \boldsymbol{P}_{\mathrm{r}}\tilde{\boldsymbol{Q}}(\boldsymbol{P}_r \neq \overline{\boldsymbol{P}})$ *are saddle (unstable) points of* $\tilde{E}1(\tilde{\boldsymbol{W}})$, *where* $\tilde{\boldsymbol{Q}}$ *is a* $r\times r$ *orthogonal matrix.*

Proof By computing $\tilde{E}1(\tilde{\boldsymbol{W}})$ in the stationary point set for the domain $\{\tilde{\boldsymbol{W}}|\tilde{\boldsymbol{W}}^{\mathrm{T}}\boldsymbol{\Lambda}\tilde{\boldsymbol{W}} > 0, \tilde{\boldsymbol{W}}^{\mathrm{T}}\tilde{\boldsymbol{W}} \neq 0\}$, we can directly verify that a global minimum of $\tilde{E}1(\tilde{\boldsymbol{W}})$ is achieved if and only if $\tilde{\boldsymbol{W}} \in \{\overline{\boldsymbol{P}}\boldsymbol{Q}|\boldsymbol{Q}^{\mathrm{T}}\overline{\boldsymbol{\Lambda}}\boldsymbol{Q} > 0\}$, where the first r row vectors of the permutation matrix $\overline{\boldsymbol{P}}$ contain all the nonzero elements of $\overline{\boldsymbol{P}}$. By substituting the above $\tilde{\boldsymbol{W}}$ into the gradient of $E1(\boldsymbol{W})$ with respect to $\boldsymbol{W}$ and performing some algebraic operations, we can get the global minimum of $\tilde{E}1(\tilde{\boldsymbol{W}})$ as follows:

$$\tilde{E}1(\tilde{\boldsymbol{W}}) = \frac{1}{2}\sum_{i=1}^{r}\lambda_i. \tag{5.40}$$

Moreover, we can determine whether a stationary point of $\tilde{E}1(\tilde{\boldsymbol{W}})$ is saddle (unstable) in such a way that within an infinitesimal neighborhood near the stationary point, there is a point $\boldsymbol{W}'$ such that its value $\tilde{E}1(\boldsymbol{W}')$ is less than $\tilde{E}1(\tilde{\boldsymbol{W}})$.

Let $\boldsymbol{P_r} \neq \overline{\boldsymbol{P}}$. There exists, at least, a nonzero element in the row vectors from $r+1$ to N for $\boldsymbol{P_r}$. Since $\overline{\boldsymbol{P}}$ and $\boldsymbol{P}_r$ are two permutation matrices, there exist certainly two diagonal matrices $\overline{\Lambda}$ and $\hat{\Lambda}$ such that $\overline{\boldsymbol{P}}^{\mathrm{T}}\boldsymbol{\Lambda}\overline{\boldsymbol{P}} = \overline{\boldsymbol{P}}^{\mathrm{T}}\overline{\boldsymbol{P}\Lambda}$ and $\boldsymbol{P}_r^{\mathrm{T}}\boldsymbol{\Lambda}\boldsymbol{P}_r = \boldsymbol{P}_r^{\mathrm{T}}\boldsymbol{P}_r\hat{\boldsymbol{\Lambda}}$, where $\hat{\Lambda}$ is a diagonal matrix $\mathrm{diag}(\lambda_{\hat{j}1},\ldots\lambda_{\hat{j}r})$ associated with $\boldsymbol{P}_{\mathrm{r}}$ and $\boldsymbol{\Lambda}$, in which $\hat{j}_i$ is an integer such that the permutation matrix $\boldsymbol{P}_{\mathrm{r}}$ has exactly the nonzero entry equal to 1 in row $\hat{j}_i$ and column i.

This yields

$$\overline{\boldsymbol{P}}^{\mathrm{T}}\boldsymbol{\Lambda}\overline{\boldsymbol{P}} = \overline{\boldsymbol{\Lambda}}, \tag{5.41}$$

$$\boldsymbol{P}_r^{\mathrm{T}}\boldsymbol{\Lambda}\boldsymbol{P}_r = \hat{\boldsymbol{\Lambda}}. \tag{5.42}$$

Thus, it holds that

$$\mathrm{tr}(\overline{\boldsymbol{P}}^{\mathrm{T}}\boldsymbol{\Lambda}\overline{\boldsymbol{P}}) = \sum_{i=1}^{r}\lambda_i, \tag{5.43}$$

$$\mathrm{tr}(\boldsymbol{P}_r^{\mathrm{T}}\boldsymbol{\Lambda}\boldsymbol{P}_r) = \sum_{i=1}^{r}\lambda_{\hat{j}_i}. \tag{5.44}$$

If $\lambda_{\hat{j}_i}(i=1,\ldots,r)$ are rearranged in a nondecreasing order $\tilde{\lambda}_1 \leq \tilde{\lambda}_2 \leq \cdots \leq \tilde{\lambda}_r$, then it follows that $\lambda_i \leq \tilde{\lambda}_i(i=1,\ldots,r-1)$, and $\lambda_r < \tilde{\lambda}_r$ for $\boldsymbol{P}_r \neq \overline{\boldsymbol{P}}$. That is,

$$\mathrm{tr}(\overline{\boldsymbol{P}}^{\mathrm{T}}\boldsymbol{\Lambda}\overline{\boldsymbol{P}}) < \mathrm{tr}(\boldsymbol{P}_r^{\mathrm{T}}\boldsymbol{\Lambda}\boldsymbol{P}_r). \tag{5.45}$$

Since

$$\begin{aligned}\tilde{E}1(\boldsymbol{P}_r\boldsymbol{Q}) &= \frac{1}{2}\mathrm{tr}\left\{(\boldsymbol{Q}^{\mathrm{T}}\boldsymbol{P}_r^{\mathrm{T}}\boldsymbol{\Lambda}\boldsymbol{P}_r\boldsymbol{Q})(\boldsymbol{Q}^{\mathrm{T}}\boldsymbol{P}_r^{\mathrm{T}}\boldsymbol{P}_r\boldsymbol{Q})^{-1}\right\} + \frac{1}{2}\mathrm{tr}\{\boldsymbol{I} - (\boldsymbol{Q}^{\mathrm{T}}\boldsymbol{P}_r^{\mathrm{T}}\boldsymbol{P}_r\boldsymbol{Q})\} \\ &= \frac{1}{2}\mathrm{tr}\{(\boldsymbol{Q}^{\mathrm{T}}\boldsymbol{P}_r^{\mathrm{T}}\boldsymbol{\Lambda}\boldsymbol{P}_r\boldsymbol{Q})\} = \frac{1}{2}\mathrm{tr}\{\boldsymbol{P}_r^{\mathrm{T}}\boldsymbol{\Lambda}\boldsymbol{P}_r\},\end{aligned} \tag{5.46}$$

$$\begin{aligned}\tilde{E}1(\overline{\boldsymbol{P}})\boldsymbol{Q} &= \frac{1}{2}\mathrm{tr}\left\{\left(\boldsymbol{Q}^{\mathrm{T}}\overline{\boldsymbol{P}}^{\mathrm{T}}\boldsymbol{\Lambda}\overline{\boldsymbol{P}}\boldsymbol{Q}\right)\left(\boldsymbol{Q}^{\mathrm{T}}\overline{\boldsymbol{P}}^{\mathrm{T}}\overline{\boldsymbol{P}}\boldsymbol{Q}\right)^{-1}\right\} + \frac{1}{2}\mathrm{tr}\left\{\boldsymbol{I} - \left(\boldsymbol{Q}^{\mathrm{T}}\overline{\boldsymbol{P}}^{\mathrm{T}}\overline{\boldsymbol{P}}\boldsymbol{Q}\right)\right\} \\ &= \frac{1}{2}\mathrm{tr}\left\{\left(\boldsymbol{Q}^{\mathrm{T}}\overline{\boldsymbol{P}}^{\mathrm{T}}\boldsymbol{\Lambda}\overline{\boldsymbol{P}}\boldsymbol{Q}\right)\right\} = \frac{1}{2}\mathrm{tr}\left\{\overline{\boldsymbol{P}}^{\mathrm{T}}\boldsymbol{\Lambda}\overline{\boldsymbol{P}}\right\}.\end{aligned} \tag{5.47}$$

Thus, (5.45) becomes

$$\tilde{E}1(\boldsymbol{P}_r\boldsymbol{Q}) > \tilde{E}1(\overline{\boldsymbol{P}}\boldsymbol{Q}), \tag{5.48}$$

which means that the set $\left\{\boldsymbol{P}_r\boldsymbol{Q}|\boldsymbol{Q}^{\mathrm{T}}\hat{\Lambda}\boldsymbol{Q} > 0 \text{ and } \boldsymbol{P}_r \neq \overline{\boldsymbol{P}}\right\}$ is not a global minimum point set.

Since $\boldsymbol{P}_r \neq \overline{\boldsymbol{P}}$, we can always select a column $\overline{\boldsymbol{P}}_i(1 \le i \le r)$ from $\overline{\boldsymbol{P}} = \left[\overline{\boldsymbol{P}}_1, \ldots, \overline{\boldsymbol{P}}_r\right]$, such that

$$\overline{\boldsymbol{P}}_i^{\mathrm{T}}\boldsymbol{P}_r = 0, \tag{5.49}$$

Otherwise, $\boldsymbol{P}_r \doteq \overline{\boldsymbol{P}}$. Moreover, we can always select a column $\overline{\boldsymbol{P}}_{r,j}(1 \le j \le r)$ from $\boldsymbol{P}_r = \left[\boldsymbol{P}_{r,1}, \ldots, \boldsymbol{P}_{r,r}\right]$ such that

$$\boldsymbol{P}_{r,j}^{\mathrm{T}}\overline{\boldsymbol{P}} = 0, \tag{5.50}$$

Otherwise, $\boldsymbol{P}_r \doteq \overline{\boldsymbol{P}}$. Let $\overline{\boldsymbol{P}}_i$ have nonzero element only in row $\bar{j}_i$ and $\boldsymbol{P}_{r,j}$ have nonzero entry only in row $\hat{j}_j$. Obviously,$\bar{j}_i < \hat{j}_j$ and $\lambda_{\hat{j}_j} > \lambda_{\bar{j}_i}$; otherwise, $\boldsymbol{P}_r \doteq \overline{\boldsymbol{P}}$. Define an orthonormal matrix as $\boldsymbol{B} = \left[\boldsymbol{P}_{r,1}, \ldots, (\boldsymbol{P}_{r,i} + \varepsilon\overline{\boldsymbol{P}}_i)/\sqrt{1+\varepsilon^2}, \ldots, \boldsymbol{P}_{r,r}\right]$, where ε is a positive infinitesimal. Considering that $\boldsymbol{P}_{r,j}$ and $\overline{\boldsymbol{P}}_i$ have one nonzero entry, it follows that

$$\boldsymbol{\Lambda B} = \left[\lambda_{\hat{j}_1}\boldsymbol{P}_{r,1}, \ldots, \left(\lambda_{\hat{j}_j}\boldsymbol{P}_{r,i} + \lambda_{\bar{j}_i}\varepsilon\bar{\boldsymbol{P}}_i\right)/\sqrt{1+\varepsilon^2}, \ldots, \lambda_{\hat{j}_r}\boldsymbol{P}_{r,r}\right]. \tag{5.51}$$

Considering (5.49), (5.50), and (5.51), we have

$$\boldsymbol{B}^{\mathrm{T}}\boldsymbol{\Lambda B} = \operatorname{diag}\left[\lambda_{\hat{j}_1}, \ldots, (\lambda_{\hat{j}_j} + \varepsilon^2\lambda_{\bar{j}_i})/(1+\varepsilon^2), \ldots, \lambda_{\hat{j}_r}\right]. \tag{5.52}$$

Since $\boldsymbol{P}_r$ is an $N \times r$ permutation matrix, we have

$$\boldsymbol{P}_r^{\mathrm{T}}\boldsymbol{\Lambda P}_r = \operatorname{diag}\left[\lambda_{\hat{j}_1}, \ldots, \lambda_{\hat{j}_j}, \ldots, \lambda_{\hat{j}_r}\right]. \tag{5.53}$$

Then,

$$\begin{aligned}
&\boldsymbol{B}^{\mathrm{T}}\boldsymbol{\Lambda B} - \boldsymbol{P}_r^{\mathrm{T}}\boldsymbol{\Lambda P}_r \\
&\quad = \operatorname{diag}\left[\lambda_{\hat{j}_1}, \ldots, (\lambda_{\hat{j}_j} + \varepsilon^2\lambda_{\bar{j}_i})/(1+\varepsilon^2), \ldots, \lambda_{\hat{j}_r}\right] - \operatorname{diag}\left[\lambda_{\hat{j}_1}, \ldots, \lambda_{\hat{j}_j}, \ldots, \lambda_{\hat{j}_r}\right] \\
&\quad = \operatorname{diag}\left[0, \ldots, 0, (-\lambda_{\hat{j}_j} + \lambda_{\bar{j}_i})\varepsilon^2/(1+\varepsilon^2), 0, \ldots, 0\right].
\end{aligned}$$

Since $\lambda_{\bar{j}_j} > \lambda_{\bar{j}_i}$, $\boldsymbol{B}^{\mathrm{T}}\boldsymbol{\Lambda}\boldsymbol{B} - \boldsymbol{P}_r^{\mathrm{T}}\boldsymbol{\Lambda}\boldsymbol{P}_r$ is a negative definite matrix. Thus, we have

$$\begin{aligned}\tilde{E}1(\boldsymbol{BQ}) &= \frac{1}{2}\mathrm{tr}\left\{\left(\boldsymbol{Q}^{\mathrm{T}}\boldsymbol{B}^{\mathrm{T}}\boldsymbol{\Lambda}\boldsymbol{BQ}\right)\left(\boldsymbol{Q}^{\mathrm{T}}\boldsymbol{B}^{\mathrm{T}}\boldsymbol{BQ}\right)^{-1}\right\} + \frac{1}{2}\mathrm{tr}\left\{\boldsymbol{I} - \left(\boldsymbol{Q}^{\mathrm{T}}\boldsymbol{B}^{\mathrm{T}}\boldsymbol{BQ}\right)\right\} \\ &= \frac{1}{2}\mathrm{tr}\left\{\boldsymbol{B}^{\mathrm{T}}\boldsymbol{\Lambda}\boldsymbol{B}\right\} < \tilde{E}1(\boldsymbol{P}_r\boldsymbol{Q}) \\ &= \frac{1}{2}\mathrm{tr}\left\{\boldsymbol{P}_r^{\mathrm{T}}\boldsymbol{\Lambda}\boldsymbol{P}_r\right\}.\end{aligned}$$

This means that $\left\{\boldsymbol{P}_r\boldsymbol{Q}|\boldsymbol{Q}^{\mathrm{T}}\hat{\Lambda}Q > 0 \text{ and } \boldsymbol{P}_r \neq \overline{\boldsymbol{P}}\right\}$ is a saddle (unstable) point set. This completes the proof.

The analysis about the landscape of NQC for tracking the PS is similar to Theorems 5.3 and 5.4. Here, we only give the resulting theorems, and the detailed proofs are omitted.

Obviously, the NQC for tracking the PS can be written as

$$\underset{\boldsymbol{W}}{\mathrm{Min}}\, E2(\boldsymbol{W}) = -\frac{1}{2}\mathrm{tr}\left\{\left(\boldsymbol{W}^{\mathrm{T}}\boldsymbol{RW}\right)\left(\boldsymbol{W}^{\mathrm{T}}\boldsymbol{W}\right)^{-1}\right\} + \frac{1}{2}\mathrm{tr}\left\{\left[\boldsymbol{I} - \left(\boldsymbol{W}^{\mathrm{T}}\boldsymbol{W}\right)\right]^{2}\right\}. \tag{5.54}$$

The landscape of $E2(\boldsymbol{W})$ is depicted by the following Theorems 5.5 and 5.6.

Theorem 5.5 *$\boldsymbol{W}$ is a stationary point of $E2(\boldsymbol{W})$ in the domain Ω if and only if $\boldsymbol{W} = \tilde{\boldsymbol{L}}_r\boldsymbol{Q}$, where $\tilde{\boldsymbol{L}}_r \in R^{N\times r}$ consist of the r eigenvectors of $\boldsymbol{R}$, and $\boldsymbol{Q}$ is a $r \times r$ orthogonal matrix.*

Theorem 5.6 *In the domain Ω, $E2(\boldsymbol{W})$ has a global minimum that is achieved if and only if $\boldsymbol{W} = \tilde{\boldsymbol{L}}_{(n)}\boldsymbol{Q}$, where $\tilde{\boldsymbol{L}}_{(n)} = [\boldsymbol{v}_{N-r+1}, \boldsymbol{v}_{N-r+2}, \ldots, \boldsymbol{v}_N]$. At the global minimum, $E2(\boldsymbol{W}) = -(1/2)\sum_{i=N-r+1}^{N}\lambda_i$. All the other stationary points of $\boldsymbol{W} = \tilde{\boldsymbol{L}}_r\boldsymbol{Q}(\tilde{\boldsymbol{L}}_r \neq \tilde{\boldsymbol{L}}_{(n)})$ are saddle (unstable) points of $E2(\boldsymbol{W})$.*

5.3.3 Dual-Purpose Subspace Gradient Flow

5.3.3.1 Dual Purpose Gradient Flow

Suppose that the vector sequence $\boldsymbol{x}_k, k = 1, 2, \ldots$ is a stationary stochastic process with zero mean and covariance matrix $\boldsymbol{R} = E\left[\boldsymbol{x}_k\boldsymbol{x}_k^{\mathrm{T}}\right] \in R^{N\times N}$. According to the stochastic learning law, the weight matrix $\boldsymbol{W}$ changes in random directions and is uncorrected with $\boldsymbol{x}_k$. Here, we take $J_{\mathrm{NQC}}(\boldsymbol{W}, \boldsymbol{R})$, which is the unified NQC for PSA and MSA in (5.31), as the cost or energy function. Then, up to a constant, the gradient flow of $J_{\mathrm{NQC}}(\boldsymbol{W}, \boldsymbol{R})$ is given by

$$\frac{\mathrm{d}\boldsymbol{W}(t)}{\mathrm{d}t} = \mp\left[\boldsymbol{R}\boldsymbol{W}(t) - \boldsymbol{W}(t)\left\{\boldsymbol{W}^{\mathrm{T}}(t)\boldsymbol{W}(t)\right\}^{-1}\boldsymbol{W}^{\mathrm{T}}(t)\boldsymbol{R}\boldsymbol{W}(t)\right]\left\{\boldsymbol{W}^{\mathrm{T}}(t)\boldsymbol{W}(t)\right\}^{-1} + \boldsymbol{W}(t)\left[\boldsymbol{I} - \left\{\boldsymbol{W}^{\mathrm{T}}(t)\boldsymbol{W}(t)\right\}\right], \tag{5.55}$$

which is the average version of the continuous-time differential equation

$$\frac{\mathrm{d}\boldsymbol{W}(t)}{\mathrm{d}t} = \mp\left[\boldsymbol{x}(t)\boldsymbol{x}^{T}(t)\boldsymbol{W}(t) - \boldsymbol{W}(t)\left\{\boldsymbol{W}^{\mathrm{T}}(t)\boldsymbol{W}(t)\right\}^{-1}\boldsymbol{W}^{\mathrm{T}}(t)\boldsymbol{x}(t)\boldsymbol{x}^{T}(t)\boldsymbol{W}(t)\right]\left\{\boldsymbol{W}^{\mathrm{T}}(t)\boldsymbol{W}(t)\right\}^{-1} + \boldsymbol{W}(t)\left[\boldsymbol{I} - \left\{\boldsymbol{W}^{\mathrm{T}}(t)\boldsymbol{W}(t)\right\}\right], \tag{5.56}$$

which, after discretization, gives a nonlinear stochastic learning rule

$$\boldsymbol{W}_{k+1} = \boldsymbol{W}_k \mp \mu\left[\boldsymbol{x}_k\boldsymbol{y}_k^{\mathrm{T}} - \boldsymbol{W}_k\left\{\boldsymbol{W}_k^{\mathrm{T}}\boldsymbol{W}_k\right\}^{-1}\boldsymbol{y}_k\boldsymbol{y}_k^{\mathrm{T}}\right]\left\{\boldsymbol{W}_k^{\mathrm{T}}\boldsymbol{W}_k\right\}^{-1} + \mu\boldsymbol{W}_k\left[\boldsymbol{I} - \left\{\boldsymbol{W}_k^{\mathrm{T}}\boldsymbol{W}_k\right\}\right], \tag{5.57}$$

where $0 < \mu < 1$ denotes the learning step size, and if "+" is used, then (5.57) is a PSA algorithm, and if "−" is used, then (5.57) is a MSA algorithm. (5.25) and (5.57) constitute our unified dual-purpose principal and minor subspace gradient flow. The gradient flow (5.57) has a computational complexity of $3Nr^2 + (4/3)r^3 + 4Nr + r^2$ flops per update, which is cheaper than $2N^2r + O(Nr^2)$ for algorithm in [22], $12N^2r + O(Nr^2)$ for algorithm in [24], and $8N^2r + O(Nr^2)$ for algorithm in [25]. The operations involved in (5.57) are simple matrix addition, multiplication, and inversion, which are easy for systolic array implementation [22].

5.3.3.2 Convergence Analysis

Under similar conditions to those defined in [38, 39], using the techniques of stochastic approximation theory [38, 39], (5.55) can be regarded as the corresponding averaging differential equation of algorithm (5.57). Next, we study the convergence of (5.55) via the Lyapunov function theory.

For MS tracking algorithm, we can give the energy function associated with (5.55) as follows:

$$E1(\boldsymbol{W}) = \frac{1}{2}\mathrm{tr}\left\{\left(\boldsymbol{W}^{\mathrm{T}}\boldsymbol{R}\boldsymbol{W}\right)\left(\boldsymbol{W}^{\mathrm{T}}\boldsymbol{W}\right)^{-1}\right\} + \frac{1}{2}\mathrm{tr}\left\{\left(\boldsymbol{I} - \left(\boldsymbol{W}^{T}\boldsymbol{W}\right)\right)^{2}\right\}. \tag{5.58}$$

The gradient of $E1(\boldsymbol{W})$ with respect to $\boldsymbol{W}$ is given by

$$\nabla E_1(\boldsymbol{W}) = \left[\boldsymbol{R}\boldsymbol{W} - \boldsymbol{W}\left\{\boldsymbol{W}^{\mathrm{T}}\boldsymbol{W}\right\}^{-1}\boldsymbol{W}^{\mathrm{T}}\boldsymbol{R}\boldsymbol{W}\right]\left\{\boldsymbol{W}^{\mathrm{T}}\boldsymbol{W}\right\}^{-1} + \left[\boldsymbol{W}\left\{\boldsymbol{W}^{\mathrm{T}}\boldsymbol{W}\right\}^{-1} - \boldsymbol{W}\right]. \tag{5.59}$$

Clearly, (5.55) for MS tracking is equivalent to the following:

$$\frac{\mathrm{d}\boldsymbol{W}}{\mathrm{d}t} = -\nabla E1(\boldsymbol{W}). \tag{5.60}$$

Differentiating $E1(\boldsymbol{W})$ along the solution of (5.55) for MS tracking algorithm yields

$$\frac{\mathrm{d}E1(\boldsymbol{W})}{\mathrm{d}t} = \frac{\mathrm{d}\boldsymbol{W}^{\mathrm{T}}}{\mathrm{d}t}\nabla E1(\boldsymbol{W}) = -\frac{\mathrm{d}\boldsymbol{W}^{\mathrm{T}}}{\mathrm{d}t}\frac{\mathrm{d}\boldsymbol{W}}{\mathrm{d}t}. \tag{5.61}$$

Since algorithm (5.55) for MS tracking has the Lyapunov function $E1(\boldsymbol{W})$ only with a lower bound, the corresponding averaging equation converges to the common invariance set $P = \{\boldsymbol{W}|\nabla E1(\boldsymbol{W}) = 0\}$ from any initial value $\boldsymbol{W}(0)$.

Similarly, for PS tracking algorithm, we can give the energy function associated with (5.55) as follows:

$$E2(\boldsymbol{W}) = -\frac{1}{2}\mathrm{tr}\left\{(\boldsymbol{W}^{\mathrm{T}}\boldsymbol{R}\boldsymbol{W})(\boldsymbol{W}^{\mathrm{T}}\boldsymbol{W})^{-1}\right\} + \frac{1}{2}\mathrm{tr}\left\{[\boldsymbol{I} - (\boldsymbol{W}^{\mathrm{T}}\boldsymbol{W})]^2\right\}. \tag{5.62}$$

According to the property of Rayleigh quotient, we can see that when $\|\boldsymbol{W}\| \to \infty, E2(\boldsymbol{W}) \to \infty$. Thus, it can be shown that $G_c = \{W; E2(\boldsymbol{W}) < c\}$ is bounded for each c.

The gradient of $E2(\boldsymbol{W})$ with respect to $\boldsymbol{W}$ is given by

$$\begin{aligned}\nabla E2(\boldsymbol{W}) &= -\left[\boldsymbol{R}\boldsymbol{W}(\boldsymbol{W}^{\mathrm{T}}\boldsymbol{W})^{-1} - \boldsymbol{W}(\boldsymbol{W}^{\mathrm{T}}\boldsymbol{W})^{-2}\boldsymbol{W}^{\mathrm{T}}\boldsymbol{R}\boldsymbol{W}\right] - \boldsymbol{W}[\boldsymbol{I} - (\boldsymbol{W}^{\mathrm{T}}\boldsymbol{W})] \\ &= -\left[\boldsymbol{R}\boldsymbol{W} - \boldsymbol{W}(\boldsymbol{W}^{\mathrm{T}}\boldsymbol{W})^{-1}\boldsymbol{W}^{\mathrm{T}}\boldsymbol{R}\boldsymbol{W}\right](\boldsymbol{W}^{\mathrm{T}}\boldsymbol{W})^{-1} - \boldsymbol{W}[\boldsymbol{I} - (\boldsymbol{W}^{\mathrm{T}}\boldsymbol{W})].\end{aligned} \tag{5.63}$$

Clearly, (5.55) for PS tracking is equivalent to the following:

$$\frac{\mathrm{d}\boldsymbol{W}}{\mathrm{d}t} = -\nabla E2(\boldsymbol{W}). \tag{5.64}$$

Differentiating $E2(\boldsymbol{W})$ along the solution of (5.55) for PS tracking algorithm yields

$$\frac{\mathrm{d}E2(\boldsymbol{W})}{\mathrm{d}t} = \frac{\mathrm{d}\boldsymbol{W}^{\mathrm{T}}}{\mathrm{d}t}\nabla E2(\boldsymbol{W}) = -\frac{\mathrm{d}\boldsymbol{W}^{\mathrm{T}}}{\mathrm{d}t}\frac{\mathrm{d}\boldsymbol{W}}{\mathrm{d}t}. \tag{5.65}$$

Since algorithm (5.55) for PS tracking has the Lyapunov function $E2(\boldsymbol{W})$ only with a lower bound, the corresponding averaging equation converges to the common invariance set $P = \{\boldsymbol{W}|\nabla E2(\boldsymbol{W}) = 0\}$ from any initial value $\boldsymbol{W}(0)$.

From the above analysis, we can conclude that the dual-purpose algorithm (5.57) for tracking PS and MS can converge to a common invariance set from any initial weight value.

5.3.3.3 Self-stability Property Analysis

Next, we will study the self-stability property of (5.57).

Theorem 5.7 *If the learning factor μ is small enough and the input vector is bounded, then the state flows in the unified learning algorithm* (5.57) *for tracking the MS and PS is bounded.*

Proof Since the learning factor μ is small enough and the input vector is bounded, we have

$$\begin{aligned}\|\boldsymbol{W}_{k+1}\|_F^2 &= \mathrm{tr}\left[\boldsymbol{W}_{k+1}^{\mathrm{T}}\boldsymbol{W}_{k+1}\right] = \mathrm{tr}\left\{\left\{\boldsymbol{W}_k \pm \mu\left[\boldsymbol{x}_k\boldsymbol{y}_k^{\mathrm{T}} - \boldsymbol{W}_k\{\boldsymbol{W}_k^{T}\boldsymbol{W}_k\}^{-1}\boldsymbol{y}_k\boldsymbol{y}_k^{\mathrm{T}}\right]\{\boldsymbol{W}_k^{\mathrm{T}}\boldsymbol{W}_k\}^{-1} + \mu\boldsymbol{W}_k[\boldsymbol{I} - \{\boldsymbol{W}_k^{\mathrm{T}}\boldsymbol{W}_k\}]\right\}^{\mathrm{T}}\right.\\&\qquad\left.\times\left\{\boldsymbol{W}_k \pm \mu\left[\boldsymbol{x}_k\boldsymbol{y}_k^{\mathrm{T}} - \boldsymbol{W}_k\{\boldsymbol{W}_k^{\mathrm{T}}\boldsymbol{W}_k\}^{-1}\boldsymbol{y}_k\boldsymbol{y}_k^{\mathrm{T}}\right]\{\boldsymbol{W}_k^{\mathrm{T}}\boldsymbol{W}_k\}^{-1} + \mu\boldsymbol{W}_k[\boldsymbol{I} - \{\boldsymbol{W}_k^{\mathrm{T}}\boldsymbol{W}_k\}]\right\}\right\}\\&\approx \mathrm{tr}\left[\boldsymbol{W}_k^{\mathrm{T}}\boldsymbol{W}_k\right] + 2\mu\cdot\mathrm{tr}[\{\boldsymbol{W}_k^{\mathrm{T}}\boldsymbol{W}_k\}(\boldsymbol{I} - \{\boldsymbol{W}_k^{\mathrm{T}}\boldsymbol{W}_k\})]\\&= \mathrm{tr}\left[\boldsymbol{W}_k^{\mathrm{T}}\boldsymbol{W}_k\right] - 2\mu\left[\mathrm{tr}\{\boldsymbol{W}_k^{\mathrm{T}}\boldsymbol{W}_k\}^2 - \mathrm{tr}\{\boldsymbol{W}_k^{\mathrm{T}}\boldsymbol{W}_k\}\right].\end{aligned} \tag{5.66}$$

Notice that in the previous formula the second-order terms associated with the learning factor have been neglected. It holds that

$$\begin{aligned}\|\boldsymbol{W}_{k+1}\|_F^2/\|\boldsymbol{W}_k\|_F^2 &\approx 1 - 2\mu\left[\mathrm{tr}\{\boldsymbol{W}_k^{\mathrm{T}}\boldsymbol{W}_k\}^2 - \mathrm{tr}\{\boldsymbol{W}_k^{\mathrm{T}}\boldsymbol{W}_k\}\right]/\|\boldsymbol{W}_k\|_F^2\\&= 1 - 2\mu\cdot\left(\frac{\mathrm{tr}\{\boldsymbol{W}_k^{\mathrm{T}}\boldsymbol{W}_k\}^2}{\mathrm{tr}\{\boldsymbol{W}_k^{\mathrm{T}}\boldsymbol{W}_k\}} - 1\right) = 1 - 2\mu\cdot\left(\frac{\mathrm{tr}\{\boldsymbol{M}^2\}}{\mathrm{tr}\{\boldsymbol{M}\}} - 1\right)\\&= 1 - 2\mu\cdot\left(\frac{\mathrm{tr}\{\boldsymbol{\Phi}\boldsymbol{\Psi}\boldsymbol{\Phi}^{-1}\boldsymbol{\Phi}\boldsymbol{\Psi}\boldsymbol{\Phi}^{-1}\}}{\mathrm{tr}\{\boldsymbol{\Phi}\boldsymbol{\Psi}\boldsymbol{\Phi}^{-1}\}} - 1\right) = 1 - 2\mu\cdot\left(\frac{\mathrm{tr}\{\boldsymbol{\Phi}\boldsymbol{\Psi}^2\boldsymbol{\Phi}^{-1}\}}{\mathrm{tr}\{\boldsymbol{\Phi}\boldsymbol{\Psi}\boldsymbol{\Phi}^{-1}\}} - 1\right)\\&= 1 - 2\mu\cdot\left(\frac{\mathrm{tr}\{\boldsymbol{\Phi}^{-1}\boldsymbol{\Phi}\boldsymbol{\Psi}^2\}}{\mathrm{tr}\{\boldsymbol{\Phi}^{-1}\boldsymbol{\Phi}\boldsymbol{\Psi}\}} - 1\right) = 1 - 2\mu\cdot\left(\frac{\mathrm{tr}\{\boldsymbol{\Psi}^2\}}{\mathrm{tr}\{\boldsymbol{\Psi}\}} - 1\right)\\&= 1 - 2\mu\cdot\left(\frac{\sum_{i=1}^{r}\zeta_i^2}{\sum_{i=1}^{r}\zeta_i} - 1\right) = 1 - 2\mu\cdot\left(\frac{\sum_{i=1}^{r}\zeta_i(\zeta_i - 1)}{\sum_{i=1}^{r}\zeta_i}\right)\\&= \begin{cases} > 1 & \text{for} \quad \zeta_i < 1, (i = 1, 2, \ldots, r)\\ = 1 & \text{for} \quad \zeta_i = 1, (i = 1, 2, \ldots, r)\\ < 1 & \text{for} \quad \zeta_i > 1, (i = 1, 2, \ldots, r),\end{cases}\end{aligned} \tag{5.67}$$

where $\boldsymbol{M} = \boldsymbol{W}_k^{\mathrm{T}}\boldsymbol{W}_k$ is a $r \times r$ matrix, and its EVD is represented as $\boldsymbol{M} = \boldsymbol{\Phi\Psi\Phi}^{-1}$, where $\boldsymbol{\Phi}$ denotes a $r \times r$ matrix formed by all its eigenvectors and $\boldsymbol{\Psi} = \mathrm{diag}(\zeta_1, \ldots, \zeta_r) > 0$ is a diagonal matrix formed by all its eigenvalues. It is obvious that the state flow in dual-purpose subspace tracking algorithm (5.57) approaches one. This completes the proof.

Next, we will study the globally asymptotical convergence of the dual-purpose algorithm (5.57) for tracking PS and MS following the approaches in [1].

5.3.4 Global Convergence Analysis

We now study the global convergence property of the dual-purpose algorithm by considering the gradient rule (5.57). Under the conditions that $\boldsymbol{x}_k$ is a stationary process and the step size μ is small enough, the discrete-time difference Eq. (5.57) approximates the continuous-time ODE (5.55). By analyzing the global convergence property of (5.55), we can establish the condition for the global convergence of (5.57). In particular, we will answer the following questions based on the Lyapunov function approach [31].

1. Is the dynamic system described by (5.55) able to globally converge to the principal subspace solution when "+" sign is used? Or if "−" sign is used, is the dynamic system (5.55) able to globally converge to the minor subspace solution?
2. What is the domain of attraction around the equilibrium attained at the PS (or at the MS), or equivalently, what is the initial condition to ensure the global convergence?

These questions can be answered by the following theorem.

Theorem 5.8 *Given the ODE(5.55) and an initial value $\boldsymbol{W}(0) \in \Omega$, then $\boldsymbol{W}(t)$ globally asymptotically converges to a point in the set $\boldsymbol{W} = \boldsymbol{L}_{(n)}\boldsymbol{Q}$ as $t \to \infty$ when "+" sign in (5.55) is used, where $\boldsymbol{L}_{(n)} = [\boldsymbol{v}_1, \boldsymbol{v}_2, \ldots, \boldsymbol{v}_r]$ and $\boldsymbol{Q}$ denotes a $r \times r$ unitary orthogonal matrix; and if "−" sign in (5.55) is used, $\boldsymbol{W}(t)$ globally asymptotically converges to a point in the set $\boldsymbol{W} = \tilde{\boldsymbol{L}}_{(n)}\boldsymbol{Q}$ as $t \to \infty$, where $\tilde{\boldsymbol{L}}_{(n)} = [\boldsymbol{v}_{N-r+1}, \boldsymbol{v}_{N-r+2}, \ldots, \boldsymbol{v}_N]$.*

Proof For algorithm (5.55) for MS tracking, "−" sign in (5.55) is used. Under this condition, it is known from (5.60) that $\boldsymbol{W}(t)$ globally asymptotically converges to a point in the invariance (stationary point) set of $E1(\boldsymbol{W})$. At a saddle (unstable) point, (5.55) is unstable. Thus, we can conclude that $\boldsymbol{W}(t)$ globally asymptotically converges to a point in the global minimum point set $\boldsymbol{W} = \boldsymbol{L}_{(n)}\boldsymbol{Q}$. For algorithm (5.55) for PS tracking, similarly we can conclude that $\boldsymbol{W}(t)$ globally asymptotically converges to a point in the global minimum point set $\boldsymbol{W} = \tilde{\boldsymbol{L}}_{(n)}\boldsymbol{Q}$. This completes the proof.

Remark 5.1 From the above analyses, it is concluded that $\boldsymbol{W}_k$ will asymptotically converge to matrix with orthonormal columns as $k \to \infty$ for stationary signals. When $\boldsymbol{W}_k^{\mathrm{T}}\boldsymbol{W}_k$ is approximated by $\boldsymbol{I}_r$, (5.57) is simplified to the stochastic Oja's algorithm for PS and MS tracking:

$$\boldsymbol{W}_{k+1} = \boldsymbol{W}_k \mp \mu\left[\boldsymbol{x}_k\boldsymbol{y}_k^{\mathrm{T}} - \boldsymbol{W}_k\boldsymbol{y}_k\boldsymbol{y}_k^{\mathrm{T}}\right]. \tag{5.68}$$

In this sense, Oja's algorithm for PS and MS tracking is derived as an approximate stochastic gradient rule to minimize the proposed NQC. However, it is worth noting that Oja's algorithm for MS tracking, where "−" is used in (5.68), does not have the self-stabilizing property, whereas the dual-purpose subspace tracking algorithm (5.57) is self-stabilizing whether for PS tracking or for MS tracking.

5.3.5 *Numerical Simulations*

In this section, we provide several interesting experiments to illustrate the performance of the dual-purpose principal and minor subspace flow. The first experiment mainly shows the self-stabilizing property of the dual-purpose gradient flow via simulations, the second experiments give some performance comparisons with other algorithms, and the third experiment provides some examples of practical applications.

5.3.5.1 Self-stabilizing Property and Convergence

Here, a PS or MS with dimension 5 is tracked. The vector data sequence is generated by $\boldsymbol{X}_k = \boldsymbol{B} \cdot \boldsymbol{y}_k$, where $\boldsymbol{B}$ is randomly generated. In order to measure the convergence speed and precision of learning algorithm, we compute the norm of a state matrix at the kth update $\rho(\boldsymbol{W}_k)$ and the index parameter $\mathrm{dist}(\boldsymbol{W}_k)$, which means the deviation of a state matrix from the orthogonality. Clearly, if $\mathrm{dist}(\boldsymbol{W}_k)$ converges to zero, then it means that $\boldsymbol{W}_k$ produces an orthonormal basis of the MS or PS.

In this simulation, let $\boldsymbol{B}$ =(1/31)randn(31, 31) and $\boldsymbol{y}_t \in \boldsymbol{R}^{31\times 1}$ be Gaussian, spatially and temporally white, and randomly generated. In order to show that the self-stabilizing property of the dual-purpose algorithm, let the initial weight value be randomly generated and normalized to modulus 2.5, which is larger than 1. Figures 5.1 and 5.2 are the simulation results for PSA and MSA on this condition, respectively. Figures 5.3 and 5.4 are, respectively, the simulation results for PSA and MSA with the initial weight modulus value normalized to 0.5, which is smaller than 1. In Figs. 5.1, 5.2, 5.3, and 5.4, all the learning curves are obtained by averaging over 30 independent experiments.

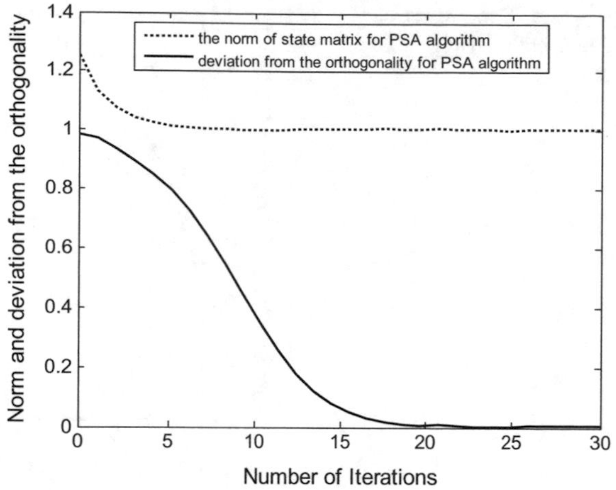

Fig. 5.1 Experiment on PSA with $\|\boldsymbol{W}_0\|_2 = 2.5$, $\mu = 0.2$

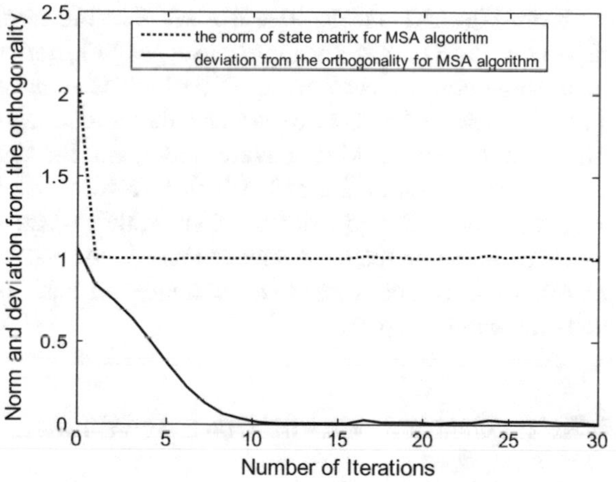

Fig. 5.2 Experiment on MSA with $\|\boldsymbol{W}_0\|_2 = 2.5$, $\mu - 0.3$

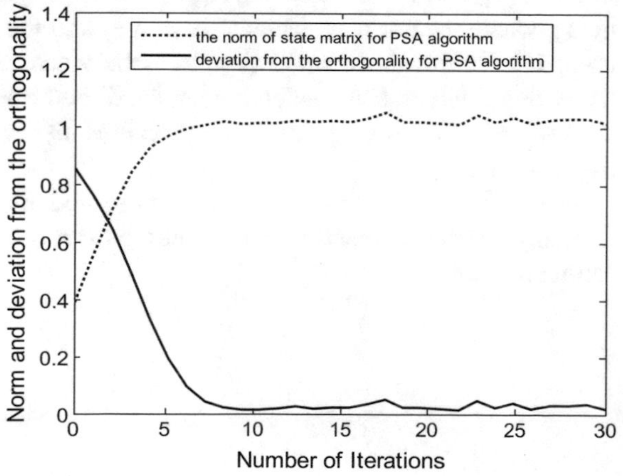

Fig. 5.3 Experiment on PSA with $\|\boldsymbol{W}_0\|_2 = 0.5$, $\mu = 0.3$

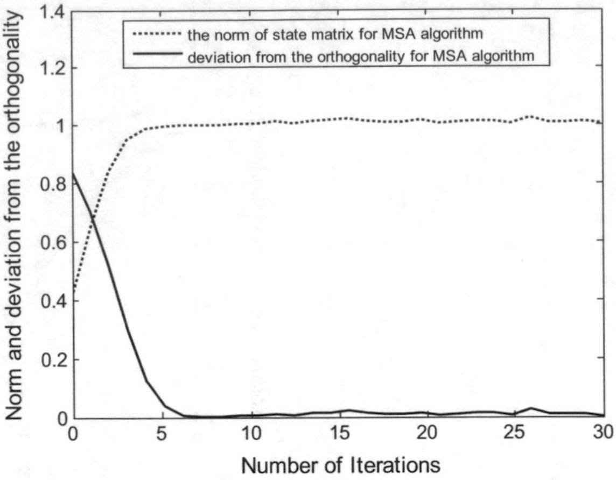

Fig. 5.4 Experiment on MSA with $\|\boldsymbol{W}_0\|_2 = 0.5$, $\mu = 0.4$

From Figs. 5.1 and 5.3, we can see that the norm of the state matrices in the algorithm (5.57) for PS tracking converges to 1, and the state matrices can converge to an orthonormal basis of the PS, which is consistent with Theorem 5.8. From Figs. 5.2 and 5.4, we also can see that the norm of the state matrices in the algorithm for MS tracking converges to 1, and the state matrices can converge to an orthonormal basis of the MS, which is also consistent with Theorem 5.8. From Figs. 5.1, 5.2, 5.3, and 5.4, it is obvious that whether the norm of the initial state matrices is larger than, or smaller than 1, the state matrices in the dual-purpose algorithm all can converge to an orthonormal basis of PS or MS, which shows the self-stabilizing property.

5.3.5.2 The Contrasts with Other Algorithms

In this simulation, the PS tracking algorithm is compared with the LMSER algorithm in [9], and the MS tracking algorithm is compared with the OJAm algorithm in [1], where the norm of state matrix $\rho(\boldsymbol{W}_k)$ and the index parameter $\text{dist}(\boldsymbol{W}_k)$ are used. Like in the above simulation, here the vector data sequence is generated by $\boldsymbol{X}_k = \boldsymbol{B} \cdot \boldsymbol{y}_k$, where $\boldsymbol{B}$ is randomly generated, and a PS or MS with dimension 5 is tracked. All the learning curves are obtained by averaging over 30 independent experiments.

From Figs. 5.5 and 5.6, it can be observed that our dual-purpose subspace tracking algorithm outperforms the others in terms of both convergence speed and estimation accuracy.

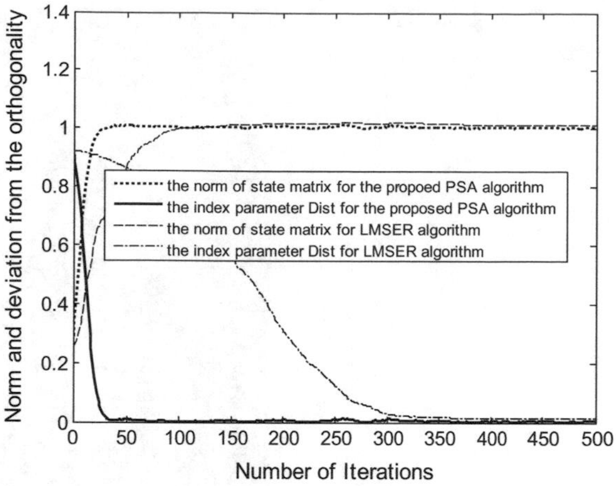

Fig. 5.5 Learning curves for PS tracking algorithm and LMSER algorithm, $\eta = 0.1$

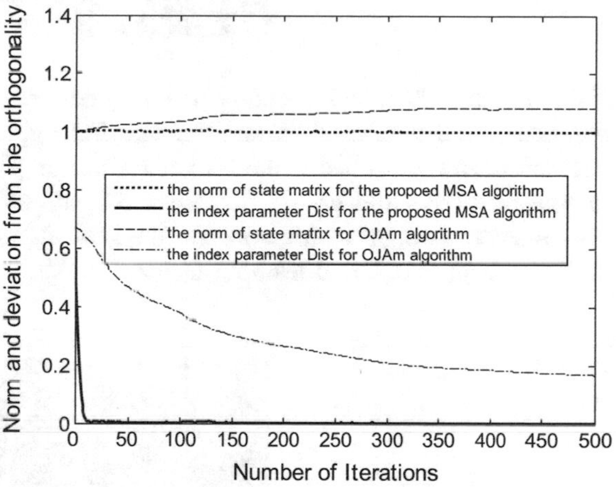

Fig. 5.6 Learning curves for MS tracking algorithm and OJAm algorithm, $\eta = 0.2$

5.3.5.3 Examples from Practical Applications of Our Unified Algorithm

Data compression is an important application of PS tracking analysis. In this simulation, we use the PSA algorithm to compress the well-known Lenna picture of 512×512 pixels as shown in Fig. 5.7. The original Lenna picture is decomposed into 8×8 nonoverlapping blocks, and from the block set, a 64-dimensional vector set $\boldsymbol{h} = \{x(k)|x(k) \in R^{64}\ (k = 1, 2,\ldots, 4096)\}$ can be constructed. After the removal of the mean and normalization, vectors randomly selected from this vector set form an input sequence for the PSA algorithm. In order to measure convergence and accuracy, we compute the norm of $\boldsymbol{W}(k)$ and the *direction cosine*(k) between the weight matrix $\boldsymbol{W}(k)$ and the true PS $\boldsymbol{V}$ of the vector set $\boldsymbol{h}$ at the kth update. Clearly,

Fig. 5.7 Original Lenna image

if the norm of $\boldsymbol{W}(k)$ and *direction cosine*(*k*) converge to 1, the weight matrix must approach the direction of PS and the algorithm gives the right result. The reconstructed Lenna image using the PSA algorithm is presented in Fig. 5.8, where the dimension of the subspace is 6. Figure 5.9 shows the convergence of direction cosine and the weight matrix norm for the PSA algorithm with initial weight vector norm $\|\boldsymbol{W}(0)\| = 0.2$, and learning factor $\mu = 0.2$.

Fig. 5.8 Reconstructed results

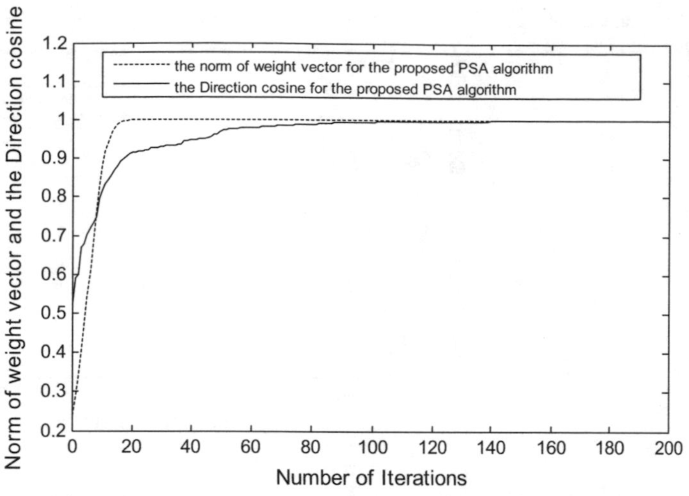

Fig. 5.9 Convergence curve of the proposed PSA algorithm

From Figs. 5.8and 5.9, we can see that our PSA algorithm achieves a satisfactory compression performance, and its convergence and accuracy are also satisfactory.

An important application of the minor component analysis is to solve the total least squares (TLS) problem. In this section, we will use our MSA algorithm, where the dimension is one, to conduct the line fitting under the TLS criterion. By adding Gaussian noises to the 400 sample points on the line $x_2 = -0.5x_1$, we can obtain a set of 2-dimensional data points $T = \left\{[x_1(t), x_2(t)]^{\mathrm{T}}, k = 1, 2, \ldots, 400\right\}$ as shown in Fig. 5.10. The problem of line fitting is to find a parameterized line model (e.g., $w_1x_1 + w_2x_2 = 0$) to fit the point set, such that the sum of the squared perpendicular

Fig. 5.10 The line fitting problem and results

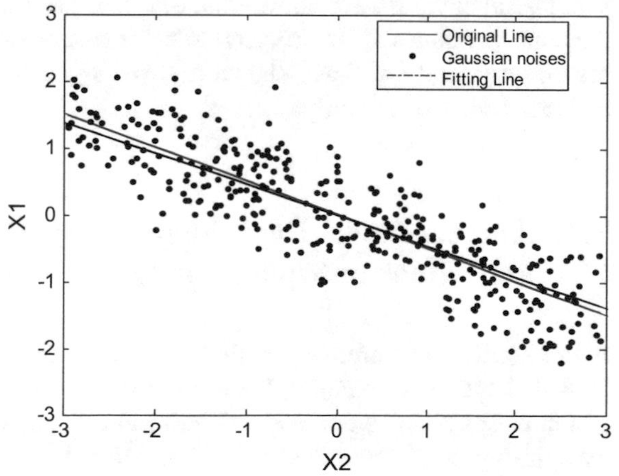

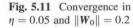

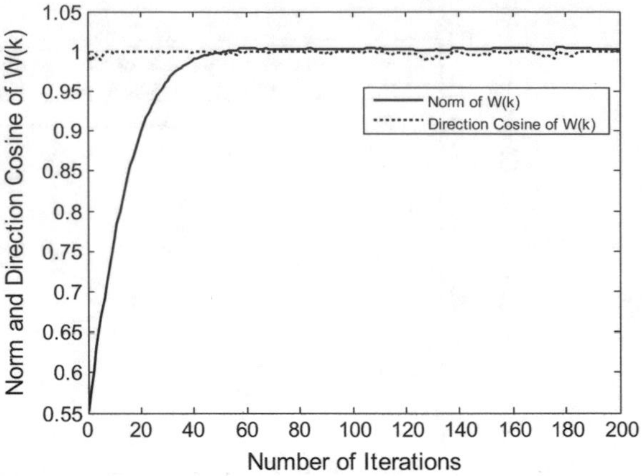

Fig. 5.11 Convergence in $\eta = 0.05$ and $\|\boldsymbol{W}_0\| = 0.2$

distances between the line and these data points is minimized. This line fitting problem can be reduced to a problem of estimating minor component of this data set T. After the removal of the mean and normalization, randomly select vectors from the data set T as the inputs for our MSA algorithm. Figure 5.11 shows the convergence of the weight vector $\boldsymbol{W}(k)$ with learning rate $\mu = 0.05$, where the direction cosine between the weight vector $\boldsymbol{W}(k)$ and the true minor component v_n is computed. After 150 iterations, the weight vector $\boldsymbol{W}(k)$ converges to $w^* = [0.41930.9100]^T$. Figure 4.10 gives the fitting results of $\boldsymbol{w}^*$, from which we can see that our MSA algorithm has a satisfactory convergence performance.

In summary, in this section, a novel unified NQC for PS and MS tracking has been introduced. Based on this criterion, a dual-purpose principal and minor subspace gradient flow is derived. Also, the landscape of nonquadratic criteria of the gradient flow is analyzed. The averaging equation of the dual-purpose algorithm for PS and MS tracking exhibits a single global minimum that is achievable if and only if its state matrix spans the PS or MS of the autocorrelation matrix of a vector data stream. Simulations have shown that our dual-purpose gradient flow can guarantee the corresponding state matrix tend to column-orthonormal basis of the PS and MS, respectively, and also have shown that the algorithm has fast convergence speed and can work satisfactorily.

5.4 Another Novel Dual-Purpose Algorithm for Principal and Minor Subspace Analysis

In this section, we introduce another novel unified information criterion (NUIC) for PSA and MSA by changing the second term of (5.31) into a nonquadratic form, which makes NUIC a nonquadratic criterion. Then, the NUIC algorithm is derived by applying gradient method to NUIC, which has fast convergence speed.

5.4.1 The Criterion for PSA and MSA and Its Landscape

The information criterion for PSA and MSA is:

$$\max_{W}\left\{J_{\mathrm{NUIC}}(\boldsymbol{W}) = \pm\frac{1}{2}\mathrm{tr}\left[(\boldsymbol{W}^{\mathrm{T}}\boldsymbol{R}\boldsymbol{W})(\boldsymbol{W}^{\mathrm{T}}\boldsymbol{W})^{-1}\right] + \frac{1}{2}\mathrm{tr}\left[\ln(\boldsymbol{W}^{\mathrm{T}}\boldsymbol{W}) - \boldsymbol{W}^{\mathrm{T}}\boldsymbol{W}\right]\right\}, \tag{5.69}$$

where "+" means the PSA information criterion and "−" means the MSA information criterion. The landscape of NUIC is depicted by the following four theorems.

Let λ_i and $\boldsymbol{u}_i$, $i = 1, 2, \ldots, n$ denote the eigenvalues and the associated orthonormal eigenvectors of $\boldsymbol{R}$. Arrange the orthonormal eigenvectors $\boldsymbol{u}_1, \boldsymbol{u}_2, \cdots, \boldsymbol{u}_n$ in such a way that the associated eigenvalues are in a nonascending order: $\lambda_1 \geq \lambda_2 \geq \cdots \geq \lambda_n$, then the eigenvalue decomposition (EVD) of $\boldsymbol{R}$ can be written as

$$\boldsymbol{R} = \boldsymbol{U}\boldsymbol{\Lambda}\boldsymbol{U}^{\mathrm{T}} = \boldsymbol{U}_1\boldsymbol{\Lambda}\boldsymbol{U}_1^{\mathrm{T}} + \boldsymbol{U}_2\boldsymbol{\Lambda}\boldsymbol{U}_2^{\mathrm{T}} \tag{5.70}$$

where $\boldsymbol{\Lambda} = \mathrm{diag}(\lambda_1, \ldots \lambda_r, \lambda_{r+1}, \ldots, \lambda_n), \boldsymbol{U}_1 = [\boldsymbol{u}_1, \ldots, \boldsymbol{u}_r]$ and $\boldsymbol{U}_2 = [\boldsymbol{u}_{r+1}, \ldots, \boldsymbol{u}_n]$. If the eigenvalues of $\boldsymbol{R}$ are allowed to be in an arbitrary order instead of the nonascending order, the EVD can also be represented by

$$\boldsymbol{R} = \boldsymbol{U}'\boldsymbol{\Lambda}'(\boldsymbol{U}')^{\mathrm{T}} = \boldsymbol{U}_r\boldsymbol{\Lambda}_r\boldsymbol{U}_r^{\mathrm{T}} + \boldsymbol{U}_{n-r}\boldsymbol{\Lambda}_{n-r}\boldsymbol{U}_{n-r}^{\mathrm{T}}, \tag{5.71}$$

where $\boldsymbol{U}' = [\boldsymbol{U}_r, \boldsymbol{U}_{n-r}]$, $\boldsymbol{\Lambda}' = \mathrm{diag}(\lambda_1', \ldots, \lambda_{\mathrm{r}}', \lambda_{\mathrm{r}+1}', \ldots, \lambda_{\mathrm{n}}')$.

Given $\boldsymbol{W}$ in the domain $\{\boldsymbol{W}|\boldsymbol{W}^{\mathrm{T}}\boldsymbol{R}\boldsymbol{W} > 0\}$, we analyze the following information criterion for tracking PS:

$$\max_{W}\left\{E_1(\boldsymbol{W}) = \frac{1}{2}\mathrm{tr}\left[(\boldsymbol{W}^{\mathrm{T}}\boldsymbol{R}\boldsymbol{W})(\boldsymbol{W}^{\mathrm{T}}\boldsymbol{W})^{-1}\right] + \frac{1}{2}\mathrm{tr}\left[\ln(\boldsymbol{W}^{\mathrm{T}}\boldsymbol{W}) - \boldsymbol{W}^{\mathrm{T}}\boldsymbol{W}\right]\right\} \tag{5.72}$$

Theorem 5.9 *$\boldsymbol{W}$ is a stationary point of $E_1(\boldsymbol{W})$ in the domain $\{\boldsymbol{W}|\boldsymbol{W}^{\mathrm{T}}\boldsymbol{R}\boldsymbol{W} > 0\}$ if and only if $\boldsymbol{W} = \boldsymbol{U}_r\boldsymbol{Q}$, where $\boldsymbol{U}_r \in \Re^{n\times r}$ consist any r distinct orthonormal eigenvectors of $\boldsymbol{R}$, and $\boldsymbol{Q}$ is an arbitrary orthogonal matrix.*

Proof Since $\boldsymbol{W}^{\mathrm{T}}\boldsymbol{R}\boldsymbol{W}$ and $\boldsymbol{W}^{\mathrm{T}}\boldsymbol{W}$ are positive definite, they are invertible. Thus, the gradient of $E_1(\boldsymbol{W})$ with respect to $\boldsymbol{W}$ exits and is given by

$$\nabla E_1(\boldsymbol{W}) = \left[\boldsymbol{R}\boldsymbol{W} - \boldsymbol{W}\{\boldsymbol{W}^{\mathrm{T}}\boldsymbol{W}\}^{-1}\boldsymbol{W}^{\mathrm{T}}\boldsymbol{R}\boldsymbol{W}\right]\{\boldsymbol{W}^{\mathrm{T}}\boldsymbol{W}\}^{-1} + \left[\boldsymbol{W}\{\boldsymbol{W}^{\mathrm{T}}\boldsymbol{W}\}^{-1} - \boldsymbol{W}\right]. \tag{5.73}$$

If $\boldsymbol{W} = \boldsymbol{U}_r\boldsymbol{Q}$, where $\boldsymbol{Q}$ is an arbitrary orthogonal matrix, we can obtain that $\nabla E_1(\boldsymbol{W}) = 0$. Conversely, by definition, the stationary point of $E_1(\boldsymbol{W})$ satisfies $\nabla E_1(\boldsymbol{W}) = 0$, which yields

$$\left[\boldsymbol{RW} - \boldsymbol{W}\{\boldsymbol{W}^T\boldsymbol{W}\}^{-1}\boldsymbol{W}^{\mathrm{T}}\boldsymbol{RW}\right]\{\boldsymbol{W}^{\mathrm{T}}\boldsymbol{W}\}^{-1} = -\left[\boldsymbol{W}\{\boldsymbol{W}^{\mathrm{T}}\boldsymbol{W}\}^{-1} - \boldsymbol{W}\right]. \tag{5.74}$$

Premultiplying both sides of (5.74) by $\boldsymbol{W}^{\mathrm{T}}$, we obtain $\boldsymbol{W}^{\mathrm{T}}\boldsymbol{W} = \boldsymbol{I}_r$, which implies that the columns of $\boldsymbol{W}$ are orthonormal at any stationary point $E_1(\boldsymbol{W})$. Let $\boldsymbol{W}^{\mathrm{T}}\boldsymbol{RW}=\boldsymbol{Q}^{\mathrm{T}}\boldsymbol{\Lambda}'_r\boldsymbol{Q}$ be the EVD and substitute it into (5.74). Then, we have $\boldsymbol{RU}'_r = \boldsymbol{U}'_r\boldsymbol{\Lambda}'_r$, where $\boldsymbol{U}'_r = \boldsymbol{WQ}^{\mathrm{T}}$ with $(\boldsymbol{U}'_r)\boldsymbol{U}'_r = \boldsymbol{I}_r$. Since $\boldsymbol{\Lambda}'_r$ is a diagonal matrix and $\boldsymbol{U}'_r$ has full rank, $\boldsymbol{U}'_r$ and $\boldsymbol{\Lambda}'_r$ must be the same as $\boldsymbol{U}_{\mathbf{r}}$ and $\boldsymbol{\Lambda}_r$.

Theorem 5.10 *In the domain* $\{\boldsymbol{W}|\boldsymbol{W}^{\mathrm{T}}\boldsymbol{RW} > 0\}$, $E_1(\boldsymbol{W})$ *has a global maximum, which is achieved if and only if* $\boldsymbol{W} = \boldsymbol{U}_1\boldsymbol{Q}$, *where* $\boldsymbol{U}_1 = [\boldsymbol{u}_1, \boldsymbol{u}_2, \ldots, \boldsymbol{u}_r]$ *and* $\boldsymbol{Q}$ *is an arbitrary orthogonal matrix. All the other stationary points are saddle points of* $E_1(\boldsymbol{W})$. *At this global maximizer,*

$$E_1(\boldsymbol{W}) = 1/2\sum_{i=1}^{r}\lambda_i + 1/2(\ln(r) - r). \tag{5.75}$$

Proof According to Theorem 5.9, any $\boldsymbol{W} = \boldsymbol{U}_r\boldsymbol{Q}$ is the stationary point of $E_1(\boldsymbol{W})$. Let $L_1 = \{i_1, i_2, \ldots, i_r\}$, whose elements are the indexes of the eigenvectors which make up the matrix $\boldsymbol{U}_r$. Similarly, let $L_2 = \{1, 2, \ldots, r\}$.

For any $L_1 (L_1 \neq L_2)$, there must exist j, which satisfies $j \in L_1$ and $j \notin L_2$. Then, replace the component $\boldsymbol{u}_j$ of matrix $\boldsymbol{U}_r$ by $\boldsymbol{u}_j + \varepsilon\boldsymbol{u}_k$, where $k \in L_1, k \notin L_2$ and $\forall\varepsilon > 0$. Let $\boldsymbol{U}'_r$ be the resultant new matrix, and $\boldsymbol{W}' = \boldsymbol{U}'_r\boldsymbol{Q}$. Then, we have

$$E_1(\boldsymbol{W}') - E_1(\boldsymbol{W}) = \frac{1}{2}(\lambda_k - \lambda_j)\varepsilon^2 + o(\varepsilon^2). \tag{5.76}$$

It can be easily seen that along the direction of the component $\boldsymbol{u}_k$, $E_1(\boldsymbol{W})$ will increase. In addition, if the component $\boldsymbol{u}_j$ is replaced by $\boldsymbol{u}_j + \varepsilon\boldsymbol{u}_j$, and let $\boldsymbol{U}''_r$ be the resultant new matrix, and $\boldsymbol{W}'' = \boldsymbol{U}''_r\boldsymbol{Q}$. Then, we have

$$E_1(\boldsymbol{W}'') - E_1(\boldsymbol{W}) = -2\varepsilon^2 + o(\varepsilon^2). \tag{5.77}$$

This means along the direction of the component $\boldsymbol{u}_j$, $E_1(\boldsymbol{W})$ will decrease. Therefore, $\boldsymbol{W} = \boldsymbol{U}_r\boldsymbol{Q}$ is a saddle point. Conversely, it can be shown that if any component of $\boldsymbol{U}_r$ is perturbed by $\boldsymbol{u}_k$ $(1 \le k \le r)$,$E_1(\boldsymbol{W})$ will decrease.

Therefore, $\tilde{\boldsymbol{W}} = \boldsymbol{U}_1\boldsymbol{Q}$ is the unique global maximizer. This means $E_1(\boldsymbol{W})$ has a global maximum without any other local maximum. It can be easily seen that at the global point

$$E_1(\boldsymbol{W}) = \frac{1}{2}\sum_{i=1}^{r}\lambda_i + \frac{1}{2}(\ln(r) - r). \tag{5.78}$$

The landscape of NUIC for tracking the MS is similar to Theorems 5.9 and 5.10. Due to limited space, here we only give the resulting theorems. Obviously, the information criterion for tracking the MS can be written as follows:

$$\max_{\boldsymbol{W}}\left\{E_2(\boldsymbol{W}) = -\frac{1}{2}\mathrm{tr}\left[\left(\boldsymbol{W}^{\mathrm{T}}\boldsymbol{R}\boldsymbol{W}\right)\left(\boldsymbol{W}^{\mathrm{T}}\boldsymbol{W}\right)^{-1}\right] + \frac{1}{2}\mathrm{tr}\left[\ln\left(\boldsymbol{W}^{\mathrm{T}}\boldsymbol{W}\right) - \boldsymbol{W}^{\mathrm{T}}\boldsymbol{W}\right]\right\}. \tag{5.79}$$

The landscape of $E_2(\boldsymbol{W})$ is depicted by Theorems 5.3 and 5.4.

Theorem 5.11 *$\boldsymbol{W}$ is a stationary point of $E_2(\boldsymbol{W})$ in the domain $\{\boldsymbol{W}|\boldsymbol{W}^{\mathrm{T}}\boldsymbol{R}\boldsymbol{W} > 0\}$ if and only if $\boldsymbol{W} = \boldsymbol{U}_r\boldsymbol{Q}$, where $\mathbf{U}_r \in R^{n\times r}$ consist any r distinct orthonormal eigenvectors of $\boldsymbol{R}$, and $\boldsymbol{Q}$ is an arbitrary orthogonal matrix.*

Theorem 5.12 *In the domain $\{\boldsymbol{W}|\boldsymbol{W}^{\mathrm{T}}\boldsymbol{R}\boldsymbol{W} > 0\}$, $E_2(\boldsymbol{W})$ has a global maximum, which is achieved if and only if $\boldsymbol{W} = \tilde{\boldsymbol{U}}_1\boldsymbol{Q}$, where $\tilde{\boldsymbol{U}}_1 = [\boldsymbol{u}_{n-r+1}, \boldsymbol{u}_{n-r+2}, \ldots, \boldsymbol{u}_n]$ and $\boldsymbol{Q}$ is an arbitrary orthogonal matrix. All the other stationary points are saddle points of $E_2(\boldsymbol{W})$. At this global maximum,*

$$E_2(\boldsymbol{W}) = -1/2\sum_{i=1}^{r}\lambda_i + 1/2(\ln(r) - r). \tag{5.80}$$

From Theorems 5.9 and 5.10, we can conclude that $E_1(\boldsymbol{W})$ has a global maximum and no local oncs. This means iterative methods like the gradient ascent search method can be used to scarch the global maximizer of $E_1(\boldsymbol{W})$. We can obtain similar conclusions for $E_2(\boldsymbol{W})$.

5.4.2 Dual-Purpose Algorithm for PSA and MSA

The gradient of $J_{\mathrm{NUIC}}(\boldsymbol{W})$ with respect to $\boldsymbol{W}$ is given by

$$\frac{\mathrm{d}\boldsymbol{W}(t)}{\mathrm{d}t} = \pm\left[\boldsymbol{R}\boldsymbol{W}(t) - \frac{\boldsymbol{W}(t)\boldsymbol{W}^{\mathrm{T}}(t)\boldsymbol{R}\boldsymbol{W}(t)}{\boldsymbol{W}^{T}(t)\boldsymbol{W}(t)}\right]\left\{\boldsymbol{W}^{\mathrm{T}}(t)\boldsymbol{W}(t)\right\}^{-1} + \left[\frac{\boldsymbol{W}(t)}{\boldsymbol{W}^{\mathrm{T}}(t)\boldsymbol{W}(t)} - \boldsymbol{W}(t)\right]. \tag{5.81}$$

By using the stochastic approximation theory, we have the following differential equation

$$\begin{aligned}\frac{\mathrm{d}\boldsymbol{W}(t)}{\mathrm{d}t} &= \pm\left[\boldsymbol{x}(t)\boldsymbol{x}^{\mathrm{T}}(t)\boldsymbol{W}(t) - \frac{\boldsymbol{W}(t)\boldsymbol{W}^{\mathrm{T}}(t)\boldsymbol{x}(t)\boldsymbol{x}^{\mathrm{T}}(t)\boldsymbol{W}(t)}{\boldsymbol{W}^{\mathrm{T}}(t)\boldsymbol{W}(t)}\right] \\ &\quad \times \left\{\boldsymbol{W}^{\mathrm{T}}(t)\boldsymbol{W}(t)\right\}^{-1} + \left[\boldsymbol{W}(t)\left\{\boldsymbol{W}^{\mathrm{T}}(t)\boldsymbol{W}(t)\right\}^{-1} - \boldsymbol{W}(t)\right].\end{aligned} \tag{5.82}$$

Discretizing (5.82), then we can get the following nonlinear stochastic algorithm:

$$\begin{aligned}\boldsymbol{W}_{k+1} &= \boldsymbol{W}_k \pm \mu\left[\boldsymbol{x}\boldsymbol{y}_k^{\mathrm{T}} - \boldsymbol{W}_k\left(\boldsymbol{W}_k^{\mathrm{T}}\boldsymbol{W}_k\right)^{-1}\boldsymbol{y}_k\boldsymbol{y}_k^{\mathrm{T}}\right]\left(\boldsymbol{W}_k^{\mathrm{T}}\boldsymbol{W}_k\right)^{-1} \\ &\quad + \mu\left[\boldsymbol{W}_k - \boldsymbol{W}_k\left(\boldsymbol{W}_k^{\mathrm{T}}\boldsymbol{W}_k\right)\right]\left(\boldsymbol{W}_k^{\mathrm{T}}\boldsymbol{W}_k\right)^{-1}.\end{aligned} \tag{5.83}$$

where $0 < \mu < 1$ is the learning rate. If "+" is used, then (5.83) is a PSA algorithm; and if "−" is used, then (5.83) is a MSA algorithm.

5.4.3 Experimental Results

5.4.3.1 Simulation Experiment

In this section, we compare the NUIC algorithm with the following algorithms, i.e., fast PSA algorithm proposed by Miao [22] (NIC), fast MSA algorithm proposed by Feng [1] (OJAm), unified algorithm for PSA, and MSA [40] (UIC). In order to measure the convergence speed and the precision of the algorithms, we compute the norm $\rho(\boldsymbol{W}_k)$ of the weight matrix and the index parameter $\mathrm{dist}(\boldsymbol{W}_k)$.

In this simulation, we adopt the method in [1] to generate the input sequences: The input vector is generated by $\boldsymbol{x}_k = \boldsymbol{B} \cdot z_k$, where $\boldsymbol{B}$ is a 31×31 matrix and is randomly generated and $z_k \in \Re^{31\times 1}$ is a zero-mean white noise with variance $\sigma^2 = 1$. In order to make a fair comparison, the same initial conditions are used in the algorithms, i.e., the same initial weight matrix and learning rates are used in all algorithms. The simulation results are shown in Figs. 5.12, 5.13, 5.14, and 5.15, and a PS or MS with dimension 16 is tracked. The processing time of these algorithms is listed in Table 5.1. The learning curves and the processing times are obtained by averaging over 100 independent experiments.

From Table 5.1, we can see that the NUIC algorithm and the algorithm in [40] have almost the same processing time and have less time than other algorithms. From the four figures, we can see that the norm of the weight matrices in the NUIC algorithm for PS or MS tracking converges to a constant, and the index parameters converge to zero, which means the algorithm can track the designated subspace (PS or MS). When compared with other algorithms, we can conclude that no matter for PS or MS tracking, the NUIC algorithm has faster convergence speed than other algorithms, which is mainly due to the complete nonquadratic criterion of NUIC and the time-varying step size of the NUIC algorithm. From the last several steps of

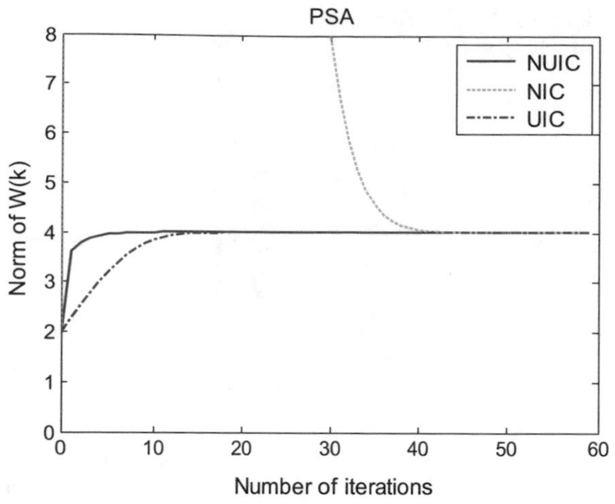

Fig. 5.12 Norm curves for PS with $\mu = 0.2$

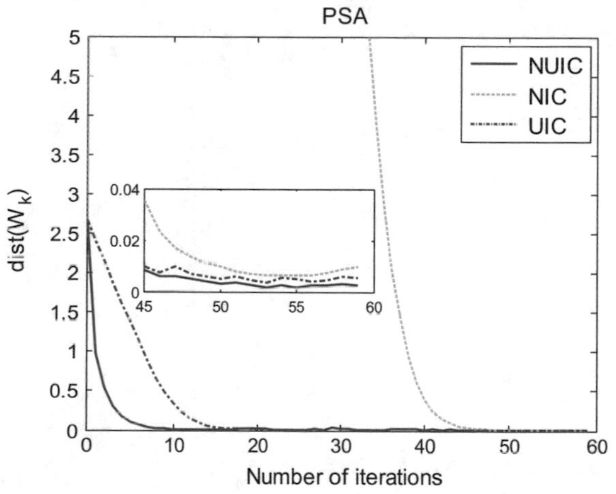

Fig. 5.13 Deviation curves for PS with $\mu = 0.2$

the simulation results shown in Figs. 5.13 and 5.15, we can see that although those algorithms can converge to the designated subspace, but the values convergence to are different. From Figs. 5.13 and 5.15, we can see that the index parameter $\mathrm{dist}(\boldsymbol{W}_k)$ of the NUIC algorithm has the smallest derivation from zero among these algorithms. In other words, the NUIC algorithm has the best estimation accuracy as shown in Fig. 5.14.

5.4.3.2 Real Application Experiment

The first experiment is direction-of-arrival (DOA) estimation. Consider the scenario where two equipower incoherent plane waves impinge on a uniform linear array

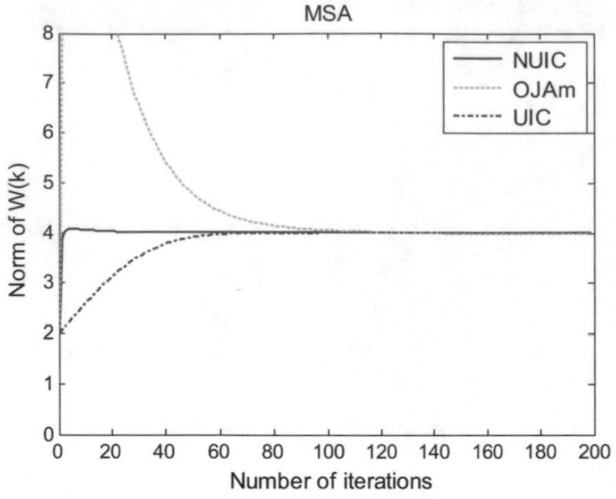

Fig. 5.14 Norm curves for MS with $\mu = 0.05$

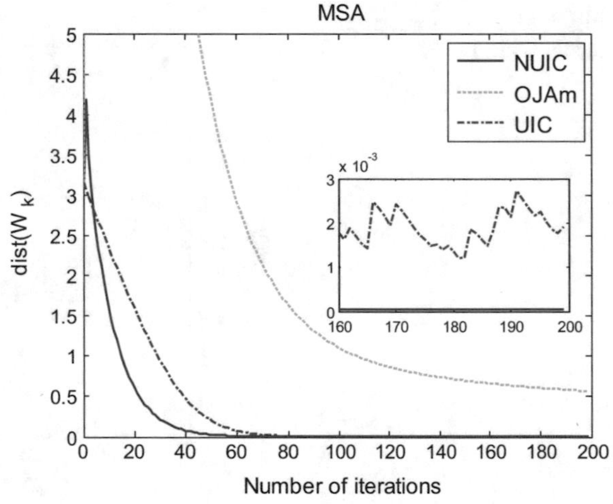

Fig. 5.15 Deviation curves for PS with $\mu = 0.05$

Table 5.1 Summary of processing time

Time (ms)	NUIC	UIC	NIC	OJAm
PSA	6.0	5.8	12.2	
MSA	5.6	5.2		7.0

with 13 sensors from 40° to 80°. The receiver noise is spatially white with unit variance $\sigma^2=1$, and the signal-to-noise ratio is 10 dB. Three algorithms (NIC, UIC, and NUIC) are used to estimate the signal (principal) subspace, and then the ESPRIT method is used to get the DOA estimates. Figures 5.16 and 5.17 show the DOA learning curves of the three algorithms. It can be seen that the three algorithms can provide satisfactory DOA estimates after some iterations. However, the NUIC algorithm has the fastest convergence speed. In order to show the estimation

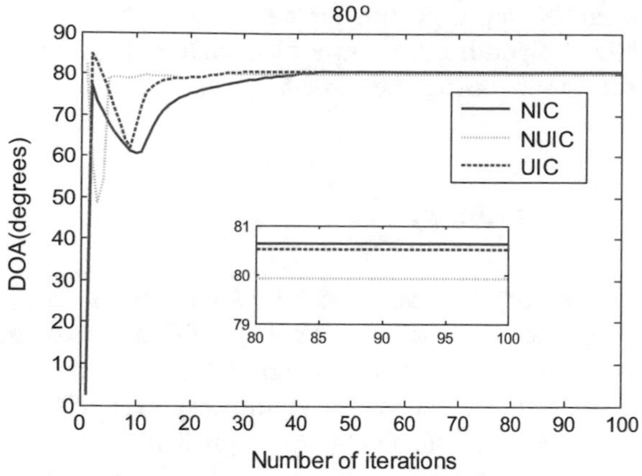

Fig. 5.16 DOA learning curves (80°)

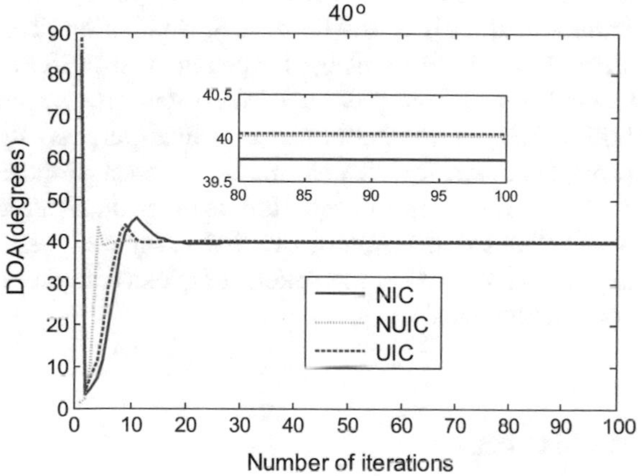

Fig. 5.17 DOA learning curves (40°)

Table 5.2 DOA estimation values

	80(°)			40(°)		
	NIC	UIC	NUIC	NIC	UIC	NUIC
Estimation(°)	80.6382	80.5008	79.9155	39.7476	40.0439	40.0253
Deviation(°)	0.6382	0.5008	0.0845	0.2524	0.0439	0.0253

accuracy of the three algorithms, we list the estimates in Table 5.2. From Table 5.2, we can see that the NUIC algorithm has the smallest deviation between the estimates and the true values. So we can conclude that NUIC algorithm has the fastest convergence speed and the best estimation accuracy among the three algorithms.

In this section, we introduced a novel information criterion for PSA and MSA and analyzed its landscape. Based on this criterion, a dual-purpose algorithm has been developed. Simulation results show that compared with other algorithms, the NUIC algorithm has lower computation complexity, higher estimation accuracy, and faster convergence speed.

5.5 Summary

In this chapter, several neural network-based dual-purpose PCA and MCA algorithms have been reviewed. Then, a UIC is introduced, and a dual-purpose principal and minor subspace gradient flow has been derived based on it. In this dual-purpose gradient flow, the weight matrix length is self-stabilizing. The energy function associated with the dual-purpose gradient flow exhibits a unique global minimum achieved if and only if its state matrices span the PS or MS of the autocorrelation matrix of a vector data stream. The other stationary points of this energy function are (unstable) saddle points. The dual-purpose gradient flow can efficiently track an orthonormal basis of the PS or MS. Simulations comparing the dual-purpose gradient flow with a number of existing dual-purpose algorithms have verified the feasibility and applicability of the dual-purpose algorithm. Finally, in order to further improve the performance of dual-purpose algorithms, another novel information criterion for PSA and MSA has been proposed, and its landscape has been analyzed. Based on this criterion, another dual-purpose algorithm has been developed. Simulation results show that compared with other algorithms, the NUIC algorithm has lower computation complexity, higher estimation accuracy, and faster convergence speed.

References

1. Feng, D. Z., Zheng, W. X., & Jia, Y. (2005). Neural network learning algorithms for tracking minor subspace in high-dimensional data stream. *IEEE Transactions on Neural Networks, 16* (3), 513–521.
2. Bannour, S., & Azimi-Sadjadi, R. (1995). Principal component extraction using recursive least squares learning. *IEEE Transactions on Neural Networks, 6*(2), 457–469.
3. Cichocki, A., Kasprzak, W., & Skarbek, W. (1996). Adaptive learning algorithm for principal component analysis with partial data. *Cybernetics Systems, 2*, 1014–1019.
4. Kung, S., Diamantaras, K., & Taur, J. (1994). Adaptive principal component extraction (APEX) and applications. *IEEE Transactions on Signal Processing, 42*(5), 1202–1217.
5. Möller, R., & Könies, A. (2004). Coupled principal component analysis. *IEEE Transactions on Neural Networks, 15*(1), 214–222.
6. Oja, E. (1982). A simplified neuron mode as a principal component analyzer. *Journal of Mathematics Biology, 15*(3), 167–273.
7. Ouyang, S., Bao, Z., & Liao, G. S. (2000). Robust recursive least squares learning algorithm for principal component analysis. *IEEE Transactions on Neural Networks, 11*(1), 215–221.
8. Sanger, T. D. (1989). Optimal unsupervised learning in a single-layer linear feedforward neural network. *Neural Networks, 2*(6), 459–473.

9. Xu, L. (1993). Least mean square error reconstruction principle for selforganizing neural-nets. *Neural Networks, 6*(5), 627–648.
10. Xu, L., Oja, E., & Suen, C. (1992). Modified Hebbian learning for curve and surface fitting. *Neural Networks, 5*(3), 441–457.
11. Oja, E. (1992). Principal component, minor component and linear neural networks. *Neural Networks, 5*(6), 927–935.
12. Feng, D. Z., Bao, Z., & Jiao, L. C. (1998). Total least mean squares algorithm. *IEEE Transactions on Signal Processing, 46*(6), 2122–2130.
13. Luo, F. L., & Unbehauen, R. (1997). A minor subspace analysis algorithm. *IEEE Transactions on Neural Networks, 8*(5), 1149–1155.
14. Cirrincione, G., Cirrincione, M., Herault, J., & Van Huffel, S. (2002). The MCA EXIN neuron for the minor component analysis. *IEEE Transactions on Neural Networks, 13*(1), 160–187.
15. Ouyang, S., Bao, Z., Liao, G. S., & Ching, P. C. (2001). Adaptive minor component extraction with modular structure. *IEEE Transactions on Signal Processing, 49*(9), 2127–2137.
16. Zhang, Q., & Leung, Y. W. (2000). A class of learning algorithms for principal component analysis and minor component analysis. *IEEE Transactions on Neural Networks, 11*(2), 529–533.
17. Möller, R. (2004). A self-stabilizing learning rule for minor component analysis. *International Journal of Neural System, 14*(1), 1–8.
18. Douglas, S. C., Kung, S. Y., & Amari, S. (2002). A self-stabilized minor subspace rule. *IEEE Signal Processing Letter, 5*(12), 1342–1352.
19. Oja, E. (1989). Neural networks, principal components, and subspaces. *International Journal of Neural Systems, 1*(1), 61–68.
20. Williams, R. J. (1985). *Feature discovery through error-correction learning*. Institute of Cognition Science, University of California, San Diego, Technical Report, 8501.
21. Baldi, P. (1989). Linear learning: Landscapes and algorithms. In D. S. Touretzky (Ed.), *Advances in Neural Information Processing Systems 1*. SanMateo, CA: Morgan Kaufmann.
22. Miao, Y. F., & Hua, Y. B. (1998). Fast subspace tracking and neural network learning by a novel information criterion. *IEEE Transactions on Signal Processing, 46*(7), 1967–1978.
23. Yang, B. (1995). Projection approximation subspace tracking. *IEEE Transactions on Signal Processing, 43*(1), 95–107.
24. Fu, Z., & Dowling, E. M. (1995). Conjugate gradient eigenstructure tracking for adaptive spectral estimation. *IEEE Transactions on Signal Processing, 43*(5), 1151–1160.
25. Mathew, G., Reddy, V. U., & Dasgupta, S. (1995). Adaptive estimation of eigensubspace. *IEEE Transactions on Signal Processing, 43*(2), 401–411.
26. Chen, T. (1997). Modified Oja's algorithms for principal and minor subspace extraction. *Neural Processing Letters, 5*(2), 105–110.
27. Chen, T., & Amari, S. (2001). Unified stabilization approach to principal and minor components extraction. *Neural Networks, 14*(10), 1377–1387.
28. Hasan, M. A. (2007). Self-normalizing dual systems for minor and principal component extraction. In *Proceedings of the ICASSP 2007 IEEE International Conference on Acoustic, Speech and Signal Processing* (Vol. 4, No. 4, pp. 885–888), April 15–20, 2007.
29. Peng, D. Z., Zhang, Y., & Xiang, Y. (2009). A unified learning algorithm to extract principal and minor components. *Digital Signal Processing, 19*(4), 640–649.
30. Chen, T., Amari, S. I., & Lin, Q. (1998). A unified algorithm for principal and minor component extraction. *Neural Networks, 11*(3), 365–369.
31. Chen, T., Amari, S. I., & Murata, N. (2001). Sequential extraction of minor components. *Neural Processing Letters, 13*(3), 195–201.
32. Jonathan, H. M., Uwe, H., & Iven, M. Y. M. (2005). A dual purpose principal and minor component flow. *Systems and Control Letters, 54*(8), 759–769.
33. Karhunen, J., & Joutsensalo, J. (1995). Generalizations of principal component analysis, optimization problems, and neural networks. *Neural Networks, 8*(4), 549–562.

34. Chatterjee, C., Kang, Z. J., & Poychowdhury, V. P. (2000). Algorithm for accelerated convergence of adaptive PCA. *IEEE Transactions on Neural Networks, 11*(2), 338–355.
35. Chen, T., Hua, Y., & Yan, W. (1998). Global convergence of Oja's subspace algorithm for principal component extraction. *IEEE Transactions on Neural Networks, 9*(1), 58–67.
36. Chauvin, Y. (1989). Principal component analysis by gradient descent on a constrained linear Hebbian cell. In *Proceedings of the Joint International Conference on Neural Networks* (pp. 373–380). San Diego, CA.
37. Magnus, J. R., & Neudecker, H. (1991). *Matrix differential calculus with applications in statistics and econometrics* (2nd ed.). New York: Wiley.
38. Kushner, H. J., & Clark, D. S. (1976). *Stochastic approximation methods for constrained and unconstrained systems*. New York: Springer.
39. Ljung, L. (1977). Analysis of recursive stochastic algorithms. *IEEE Transactions on Automatic Control, 22*(4), 551–575.
40. Kong, X. Y., Hu, C. H., & Han, C. Z. (2012). A dual purpose principal and minor subspace gradient flow. *IEEE Transactions on Signal Processing, 60*(1), 197–210.
41. LaSalle, J. P. (1976). *The stability of dynamical systems*. Philadelphia, PA: SIAM.
42. Kong, X. Y., Hu, C. H., & Han, C. Z. (2010). On the discrete time dynamics of a class of self-stabilizing MCA learning algorithms. *IEEE Transactions on Neural Networks, 21*(1), 175–181.
43. Ji, S. H., Xue, Y., & Carin, L. (2008). Bayesian compressive sensing. *IEEE Transactions on Signal Processing, 56*(6), 2346–2356.
44. Kang, Z. J., Chatterjee, C., & Roychowdhury, V. P. (2000). An adaptive quasi-Newton algorithm for eigensubspace estimation. *IEEE Transactions on Signal Processing, 48*(12), 3328–3333.

Chapter 6
Deterministic Discrete-Time System for the Analysis of Iterative Algorithms

6.1 Introduction

The convergence of neural network-based PCA or MCA learning algorithms is a difficult topic for direct study and analysis. Traditionally, based on the stochastic approximation theorem, the convergence of these algorithms is indirectly analyzed via corresponding DCT systems. The stochastic approximation theorem requires that some restrictive conditions must be satisfied. One important condition is that the learning rates of the algorithms must approach zero, which is not a reasonable requirement to be imposed in many practical applications. Clearly, the restrictive condition is difficult to be satisfied in many practical applications, where a constant learning rate is usually used due to computational roundoff issues and tracking requirements. Besides the DCT system, Lyapunov function method, differential equations method, etc., are also used to analyze the convergence of PCA algorithms. For example, in [1], a Lyapunov function was proposed for globally characterizing Oja's DCT model with a single neuron. Another single-neuron generalized version of Oja's DCT net was studied in [2] by explicitly solving the system of differential equations. The global behavior of a several-neuron Oja's DCT net was determined in [3] by explicitly solving the equations of the model, whereas [4] addressed a qualitative analysis of the generalized forms of this DCT network.

All these studies of DCT formulations are grounded on restrictive hypotheses so that the fundamental theorem of stochastic approximation can be applied. However, when some of these hypotheses cannot be satisfied, how to study the convergence of the original stochastic discrete formulation? In order to analyze the convergence of neural network-based PCA or MCA learning algorithms, several methods have been proposed, i.e., DCT, SDT, and DDT methods. The DCT method, first formalized by [5, 6], is based on a fundamental theorem of stochastic approximation theory. Thus, it is an approximation analysis method. The SDT method is a direct analysis method and it can analyze the temporal behavior of algorithm and derive

© Science Press, Beijing and Springer Nature Singapore Pte Ltd. 2017
X. Kong et al., *Principal Component Analysis Networks and Algorithms*,
DOI 10.1007/978-981-10-2915-8_6

the relation between the dynamic stability and learning rate [7]. The DDT method, as a bridge between DCT and SDT methods, transforming the original SDT system into a corresponding DDT system, and preserving the discrete-time nature of the original SDT systems, can shed some light on the convergence characteristics of SDT systems [8]. Recently, the convergence of many PCA or MCA algorithms has been widely studied via the DDT method [8–13].

The objective of this chapter is to study the DDT method, analyze the convergence of PCA or MCA algorithms via DDT method to obtain some sufficient conditions to guarantee the convergence, and analyze the stability of these algorithms. The remainder of this chapter is organized as follows. A review of performance analysis methods for neural network-based PCA/MCA algorithms is presented in Sect. 6.2. The main content, a DDT system of a novel MCA algorithm is introduced in Sect. 6.3. Furthermore, a DDT system of a unified PCA and MCA algorithm is introduced in Sect. 6.4, followed by the summary in Sect. 6.5.

6.2 Review of Performance Analysis Methods for Neural Network-Based PCA Algorithms

6.2.1 Deterministic Continuous-Time System Method

According to the stochastic approximation theory (see [5, 6]), if certain conditions are satisfied, its corresponding DCT systems can represent the SDT system effectively (i.e., their asymptotic paths are close with a large probability) and eventually the PCA/MCA solution tends with probability 1 to the uniformly asymptotically stable solution of the ODE. From a computational point of view, the most important conditions are the following:

1. $x(t)$ is zero-mean stationary and bounded with probability 1.
2. $\alpha(t)$ is a decreasing sequence of *positive* scalars.
3. $\Sigma_t \alpha(t) = \infty$.
4. $\Sigma_t \alpha^P(t) < \infty$ for some p.
5. $\lim_{t\to\infty} \sup\left[\frac{1}{\alpha(t)} - \frac{1}{\alpha(t-1)}\right] < \infty$.

For example, the sequence $\alpha(t) = \text{const} \cdot t^{-\gamma}$ satisfies Conditions 2–5 for $0 < \gamma \leq 1$. The fourth condition is less restrictive than the Robbins–Monro condition $\Sigma_t \alpha^2(t) < \infty$, which is satisfied, for example, only by $\alpha(t) = \text{const} \cdot t^{-\gamma}$ with $1/2 < \gamma \leq 1$.

For example, MCA EXIN algorithm can be written as follows:

$$\boldsymbol{w}(t+1) = \boldsymbol{w}(t) - \frac{\alpha(t)y(t)}{\|\boldsymbol{w}(t)\|_2^2}\left[\boldsymbol{x}(t) - \frac{y(t)\boldsymbol{w}(t)}{\|\boldsymbol{w}(t)\|_2^2}\right], \tag{6.1}$$

and its corresponding deterministic continuous-time (DCT) systems is

$$\frac{dw(t)}{dt} = -\frac{1}{\|w(t)\|_2^2}\left[R - \frac{w^{\mathrm{T}}(t)Rw(t)}{\|w(t)\|_2^2}\right]w(t) = -\frac{1}{\|w(t)\|_2^2}[R - r(w, R)I]w(t). \tag{6.2}$$

For the convergence proof using deterministic continuous-time system method, refer to the proof of Theorem 16 in [7] for details.

6.2.2 Stochastic Discrete-Time System Method

Using only the ODE approximation does not reveal some of the most important features of these algorithms [7]. For instance, it can be shown that the constancy of the weight modulus for OJAn, Luo, and MCA EXIN, which is the consequence of the use of the ODE, is not valid, except, as a very first approximation, in approaching the minor component [7]. The stochastic discrete-time system method has led to the very important problem of the sudden divergence [7]. In the following, we will analyze the performance of Luo MCA algorithm using the stochastic discrete-time system method.

In [14, 15], Luo proposed a MCA algorithm, which is

$$w(t+1) = w(t) - \alpha(t)\|w(t)\|_2^2\left[y(t)x(t) - \frac{y^2(t)}{\|w(t)\|_2^2}w(t)\right]. \tag{6.3}$$

Since (6.3) is the gradient flow of the RQ and using the property of orthogonality of RQ, it holds that

$$w^{\mathrm{T}}(t)\left\{y(t)x(t) - \frac{y^2(t)}{\|w(t)\|_2^2}w(t)\right\} = 0, \tag{6.4}$$

i.e., *the weight increment at each iteration is orthogonal to the weight direction.* The squared modulus of the weight vector at instant $t + 1$ is then given by

$$\|w(t+1)\|_2^2 = \|w(t)\|_2^2 + \frac{\alpha^2(t)}{4}\|w(t)\|_2^6\|x(t)\|_2^4\sin^2 2\vartheta_{xw}, \tag{6.5}$$

where ϑ_{xw} is the angle between the direction of $x(t)$ and $w(t)$. From (6.5), we can see that (1) Except for particular conditions, *the weight modulus always increases*, $\|w(t+1)\|_2^2 > \|w(t)\|_2^2$. These particular conditions, i.e., all data in exact particular directions, are too rare to be found in a noisy environment. (2) $\sin^2 2\vartheta_{xw}$ is a positive function with peaks within the interval $(-\pi, \pi]$. This is one of the possible interpretations of the oscillatory behavior of weight modulus.

The remaining part of this section is the convergence analysis of Dougla's MCA algorithm via the SDT Method. The purpose of this section is to analyze the temporal behavior of Dougla's MCA algorithm and the relation between the dynamic stability and learning rate, by using mainly the SDT system following the approach in [7].

Indeed, using only the ODE approximation does not reveal some of the most important features of MCA algorithms, and the ODE is only the very first approximation, in approaching the minor component. After the MC direction has been approached, how is the rule of the weight modulus?

From Dougla's MCA, it holds that

$$\begin{aligned}\|\boldsymbol{w}(t+1)\|^2 &= \boldsymbol{w}^{\mathrm{T}}(t+1)\boldsymbol{w}(t+1) = \{\boldsymbol{w}(t) - \alpha(t)[\|\boldsymbol{w}(t)\|^4 y(t)x(t) - y^2(t)\boldsymbol{w}(t)]\}^{\mathrm{T}} \cdot \{\boldsymbol{w}(t) - \alpha(t)[\|\boldsymbol{w}(t)\|^4 y(t)x(t) - y^2(t)\boldsymbol{w}(t)]\} \\ &= \|\boldsymbol{w}(t)\|^2 - 2\alpha(t)(\|\boldsymbol{w}(t)\|^4 y^2(t) - y^2(t)\|\boldsymbol{w}(t)\|^2) + \alpha^2(t)(\|\boldsymbol{w}(t)\|^8 y^2(t)\|x(t)\|^2 - 2\|\boldsymbol{w}(t)\|^4 y^4(t) + y^4(t)\|\boldsymbol{w}(t)\|^2) \\ &= \|\boldsymbol{w}(t)\|^2 + 2\alpha(t)y^2(t)\|\boldsymbol{w}(t)\|^2(1 - \|\boldsymbol{w}(t)\|^2) + O(\alpha^2(t)) \\ &\doteq \|\boldsymbol{w}(t)\|^2 + 2\alpha(t)y^2(t)\|\boldsymbol{w}(t)\|^2(1 - \|\boldsymbol{w}(t)\|^2).\end{aligned} \tag{6.6}$$

Hence, if the learning factor is small enough and the input vector is bounded, we can make such analysis as follows by neglecting the second-order terms of the $\alpha(t)$.

$$\frac{\|\boldsymbol{w}(t+1)\|^2}{\|\boldsymbol{w}(t)\|^2} \doteq 1 + 2\alpha(t)y^2(t)(1 - \|\boldsymbol{w}(t)\|^2) = \begin{cases} > 1 & \text{for} \quad \|\boldsymbol{w}(0)\|^2 < 1 \\ < 1 & \text{for} \quad \|\boldsymbol{w}(0)\|^2 < 1 \\ = 1 & \text{for} \quad \|\boldsymbol{w}(0)\|^2 = 1 \end{cases}. \tag{6.7}$$

This means that $\|\boldsymbol{w}(t+1)\|^2$ tends to one whether $\|\boldsymbol{w}(t)\|^2$ is equal to one or not, which is called the one-tending property (OTP), i.e., the weight modulus remains constant $\left(\|\boldsymbol{w}(t)\|^2 \rightarrow 1\right)$.

To use the stochastic discrete laws is a direct analytical method. In fact, the study of the stochastic discrete learning laws of the Douglas's algorithm is an analysis of their dynamics.

Define

$$r' = \frac{|\boldsymbol{w}^{\mathrm{T}}(t+1)\boldsymbol{x}(t)|^2}{\|\boldsymbol{w}(t+1)\|^2}, \quad r = \frac{|\boldsymbol{w}^{\mathrm{T}}(t)\boldsymbol{x}(t)|^2}{\|\boldsymbol{w}(t)\|^2},$$

$$\rho(\alpha) = \frac{r'}{r} \geq 1, \quad p = \|\boldsymbol{w}(t)\|^2, \quad u = y^2(t).$$

The two scalars r' and r represent, respectively, the squared perpendicular distance between the input $\boldsymbol{x}(t)$ and the data-fitting hyperplane whose normal is given by the weight and passes through the origin, after and before the weight increment. Recalling the definition of MC, we should have $r' \leq r$. If this inequality is not valid, this means that the learning law increases the estimation error due to the disturbances caused by noisy data. When this disturbance is too large, it will make $\boldsymbol{w}(t)$ deviate drastically from the normal learning, which may result in divergence or fluctuations (implying an increased learning time).

Theorem 6.1

$$\text{If } \alpha > \frac{2}{p\|\boldsymbol{x}(t)\|^2(p - 2\cos^2\theta_{xw})} \quad \wedge \quad p\|\boldsymbol{x}(t)\|^2(p - 2\cos^2\theta_{xw}) > 0,$$

then $r' > r$, which implies divergence.

Proof From Eq. (6.2), we have

$$\begin{aligned} \boldsymbol{w}^{\mathrm{T}}(t+1)\boldsymbol{x}(t) &= y(t) - \alpha[\|\boldsymbol{w}(t)\|^4 y(t)\|\boldsymbol{x}(t)\|^2 - y^3(t)] \\ &= y(t)(1 - \alpha[\|\boldsymbol{w}(t)\|^4\|\boldsymbol{x}(t)\|^2 - y^2(t)]) \end{aligned} \tag{6.8}$$

$$\begin{aligned} \|\boldsymbol{w}(t+1)\|^2 = \boldsymbol{w}^{\mathrm{T}}(t+1)\boldsymbol{w}(t+1) &= \|\boldsymbol{w}(t)\|^2 - 2\alpha(t)(\|\boldsymbol{w}(t)\|^4 y^2(t) - y^2(t)\|\boldsymbol{w}(t)\|^2) \\ &+ \alpha^2(t)(\|\boldsymbol{w}(t)\|^8 y^2(t)\|\boldsymbol{x}(t)\|^2 - 2\|\boldsymbol{w}(t)\|^4 y^4(t) + y^4(t)\|\boldsymbol{w}(t)\|^2). \end{aligned} \tag{6.9}$$

Therefore,

$$\begin{aligned} \rho(\alpha) = \frac{r'}{r} &= \frac{(\boldsymbol{w}^{\mathrm{T}}(t+1)\boldsymbol{x}(t))^2}{\|\boldsymbol{w}(t+1)\|^2}\frac{\|\boldsymbol{w}(t)\|^2}{(y(t))^2} = \frac{(1 - \alpha(t)[\|\boldsymbol{w}(t)\|^4\|\boldsymbol{x}(t)\|^2 - y^2(t)])^2}{1 - 2\alpha(t)y^2(t)(\|\boldsymbol{w}(t)\|^2 - 1) + \alpha^2 E} \\ &= \frac{(1 - \alpha q)^2}{1 - 2\alpha u(p-1) + \alpha^2 E}, \end{aligned} \tag{6.10}$$

where $q = \left(\|\boldsymbol{x}(t)\|^2 p^2 - u\right)$ and $E = \left(up^3\|\boldsymbol{x}(t)\|^2 - 2u^2 p + u^2\right)$.

Then, $\rho(\alpha) > 1$ (dynamic instability) if and only if

$$(1 - \alpha q)^2 > 1 - 2\alpha u(p-1) + \alpha^2\left(up^3\|\boldsymbol{x}(t)\|^2 - 2u^2 p + u^2\right). \tag{6.11}$$

Notice that $u/p = \|\boldsymbol{x}(t)\|^2\cos^2\theta_{zw}$.

From (6.11), it holds that

$$\begin{aligned} &\alpha^2\left[p^4\|\boldsymbol{x}(t)\|^4\sin^2\theta_{xw} - 2p^3\|\boldsymbol{x}(t)\|^4\cos^2\theta_{xw}\sin^2\theta_{xw}\right] \\ &\quad > 2\alpha\|\boldsymbol{x}(t)\|^2 p^2\sin^2\theta_{xw}. \end{aligned} \tag{6.12}$$

The dynamic instability condition is then

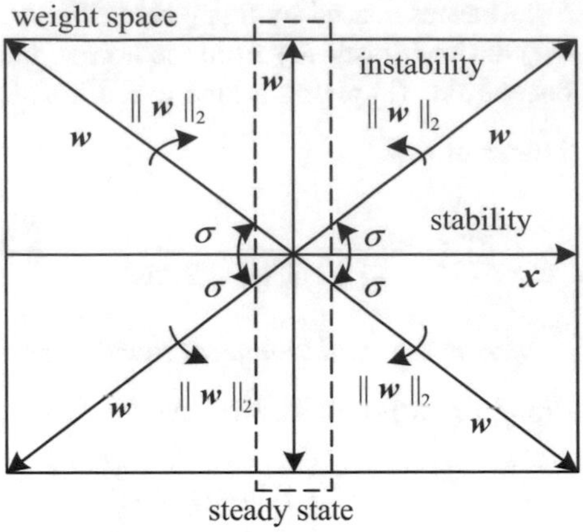

Fig. 6.1 The stability subspace of the weight space with regard to the input vector (two-dimensional space) of proposed algorithm

$$\alpha > \frac{2}{p\|\boldsymbol{x}(t)\|^2 (p - 2\cos^2 \theta_{xw})} \quad \wedge \quad p\|\boldsymbol{x}(t)\|^2 (p - 2\cos^2 \theta_{xw}) > 0. \tag{6.13}$$

The second condition implies the absence of the negative instability. It can be rewritten as

$$\cos^2 \theta_{xw} \le \frac{p}{2}. \tag{6.14}$$

In reality, the second condition is included in the first one. Considering the case $0 < \alpha_b \le \gamma < 1$, it holds that

$$\cos^2 \theta_{xw} \le \frac{p}{2} - \frac{1}{\gamma p \|\boldsymbol{x}(t)\|^2} = \Upsilon, \tag{6.15}$$

which is more restrictive than (6.14). Figure 6.1 shows this condition, where $\sigma = \arccos \sqrt{\Upsilon}$. From (6.15), we can see that the decrease of γ and p increases the domain of σ and then increases the stability. From Fig. 6.1, it is apparent that in the transient (in general low θ_{XW}), there are less fluctuations and this is beneficial to the stability.

This completes the proof.

6.2.3 Lyapunov Function Approach

Lyapunov function approach has also been applied in the convergence and stability analysis. For details, see references [7, 16, 17].

6.2.4 *Deterministic Discrete-Time System Method*

Traditionally, the convergence of neural network learning algorithms is analyzed via DCT systems based on a stochastic approximation theorem. However, there exist some restrictive conditions when using stochastic approximation theorem. One crucial condition is that the learning rate in the learning algorithm must converge to zero, which is not suitable in most practical applications because of the roundoff limitation and tracking requirements [8, 13]. In order to overcome the shortcomings of the DCT method, Zurifia proposed DDT method [8]. Different from the DCT method, the DDT method allows the learning rate to be a constant and can be used to indirectly analyze the dynamic behaviors of stochastic learning algorithms. Since the DDT method is more reasonable for studying the convergence of neural network algorithms than the traditional DCT method, it has been widely used to study many neural network algorithms [8, 10–13, 18–20].

6.3 DDT System of a Novel MCA Algorithm

In this section, we will analyze the convergence and stability of a class of self-stabilizing MCA algorithms via a DDT method. Some sufficient conditions are obtained to guarantee the convergence of these learning algorithms. Simulations are carried out to further illustrate the theoretical results achieved. It can be concluded that these self-stabilizing algorithms can efficiently extract the MCA, and they outperform some existing MCA methods.

In Sect. 6.3.1, a class of self-stabilizing learning algorithms is presented. In Sect. 6.3.2, the convergence and stability analysis of these algorithms via DDT method are given. In Sect. 6.3.3, computer simulation results on minor component extraction and some conclusions are presented.

6.3.1 *Self-stabilizing MCA Extraction Algorithms*

Consider a single linear neuron with the following input–output relation: $y(k) = \boldsymbol{W}^{\mathrm{T}}(k)\boldsymbol{X}(k), k = 0, 1, 2, \cdots$, where $y(k)$ is the neuron output, the input sequence $\{\boldsymbol{X}(k)|\boldsymbol{X}(k) \in \boldsymbol{R}^n(k = 0, 1, 2, \cdots)\}$ is a zero-mean stationary stochastic process, and $\boldsymbol{W}(k) \in \boldsymbol{R}^n(k = 0, 1, 2, \cdots)$ is the weight vector of the neuron. The target of MCA is to extract the minor component from the input data by updating the weight vector $\boldsymbol{W}(k)$ adaptively. Here, based on the OJA + algorithm [21], we add a penalty term $(1 - \|\boldsymbol{W}(t)\|^{2+\alpha})\boldsymbol{RW}$ to OJA + and present a class of MCA algorithms as follows:

$$\dot{\boldsymbol{W}} = -\|\boldsymbol{W}\|^{2+\alpha}\boldsymbol{R}\boldsymbol{W} + (\boldsymbol{W}^{\mathrm{T}}\boldsymbol{R}\boldsymbol{W} + 1 - \boldsymbol{W}^{\mathrm{T}}\boldsymbol{W})\boldsymbol{W}, \tag{6.16}$$

where $\boldsymbol{R} = E[\boldsymbol{X}(k)\boldsymbol{X}^{\mathrm{T}}(k)]$ is the correlation matrix of the input data and the integer $0 \leq \alpha \leq 2$. The parameter α can be real-valued. However, for the simplicity of theoretical analysis and practical computations, it would be convenient to choose α as an integer. Considering the needs in the proofs of latter theorems, the upper limit of α is 2. It is worth noting that Algorithm (6.16) coincides with the Chen rule for minor component analysis [22] in the case $\alpha = 0$. When $\alpha > 0$, these algorithms are very similar to the Chen algorithm and can be considered as modifications of the Chen algorithm. Therefore, for simplicity, we refer to all of them as Chen algorithms.

The stochastic discrete-time system of (6.16) can be written as follows:

$$\boldsymbol{W}(k+1) = \boldsymbol{W}(k) - \eta\left[\|\boldsymbol{W}(k)\|^{2+\alpha}y(k)\boldsymbol{X}(k) - (y^2(k) + 1 - \|\boldsymbol{W}(k)\|^2)\boldsymbol{W}(k)\right], \tag{6.17}$$

where $\eta(0 < \eta < 1)$ is the learning rate. From (6.17), it follows that

$$\begin{aligned}\|\boldsymbol{W}(k+1)\|^2 - \|\boldsymbol{W}(k)\|^2 &= -2\eta\|\boldsymbol{W}(k)\|^2\left[y^2(k)(\|\boldsymbol{W}(k)\|^{\alpha} - 1) + \left(\|\boldsymbol{W}(k)\|^2 - 1\right)\right] + O(\eta^2) \\ &\doteq -2\eta\|\boldsymbol{W}(k)\|^2(\|\boldsymbol{W}(k)\| - 1)Q\left(y^2(k), \|\boldsymbol{W}(k)\|\right),\end{aligned} \tag{6.18}$$

where $Q(y^2(k), \|\boldsymbol{W}(k)\|) = y^2(k)(\|\boldsymbol{W}(k)\|^{\alpha-1} + \|\boldsymbol{W}(k)\|^{\alpha-2} +, \cdots, \|\boldsymbol{W}(k)\| + 1) + (\|\boldsymbol{W}(k)\| + 1)$ is a positive efficient. For a relatively small constant learning rate, the second-order term is very small and can be omitted. Thus, from (6.18), we can claim that Algorithm (6.17) has self-stabilizing property [23].

6.3.2 *Convergence Analysis via DDT System*

From $y(k) = \boldsymbol{X}^{\mathrm{T}}(k)\boldsymbol{W}(k) = \boldsymbol{W}^{\mathrm{T}}(k)\boldsymbol{X}(k)$, by taking the conditional expectation $E\{\boldsymbol{W}(k+1)/\boldsymbol{W}(0), \boldsymbol{X}(i), i < k\}$ to (6.17) and identifying the conditional expectation as the next iterate, a DDT system can be obtained as

$$\begin{aligned}\boldsymbol{W}(k+1) &= \boldsymbol{W}(k) \\ &\quad - \eta\left[\|\boldsymbol{W}(k)\|^{2+\alpha}\boldsymbol{R}\boldsymbol{W}(k) - \left(\boldsymbol{W}^{\mathrm{T}}(k)\boldsymbol{R}\boldsymbol{W}(k) + 1 - \|\boldsymbol{W}(k)\|^2\right)\boldsymbol{W}(k)\right],\end{aligned} \tag{6.19}$$

where $\boldsymbol{R} = E[\boldsymbol{X}(k)\boldsymbol{X}^{\mathrm{T}}(k)]$ is the correlation matrix of the input data. Here, we analyze the dynamics of (6.19) subject to η being some smaller constant to interpret the convergence of Algorithm (6.17) indirectly.

For the convenience of analysis, we next give some preliminaries. Since $\boldsymbol{R}$ is a symmetric positive definite matrix, there exists an orthonormal basis of $\Re^n$ composed of the eigenvectors of $\boldsymbol{R}$. Obviously, the eigenvalues of the autocorrelation matrix $\boldsymbol{R}$ are nonnegative. Assume that $\lambda_1, \lambda_2, \cdots, \lambda_n$ are all eigenvalues of $\boldsymbol{R}$ ordered by $\lambda_1 \geq \lambda_2 \geq \cdots \geq \lambda_{n-1} > \lambda_n > 0$. Suppose that $\{\boldsymbol{V}_i | i = 1, 2, \cdots, n\}$ is an orthogonal basis of $\boldsymbol{R}^n$ such that each $\boldsymbol{V}_{\mathrm{i}}$ is unit eigenvector of $\boldsymbol{R}$ associated with the eigenvalue λ_i. Thus, for each $k \geq 0$, the weight vector $\boldsymbol{W}(k)$ can be represented as

$$\boldsymbol{W}(k) = \sum_{i=1}^{n} z_i(k)\boldsymbol{V}_i, \tag{6.20}$$

where $z_i(k)(i = 1, 2, \ldots, n)$ are some constants. From (6.19) and (6.20), it holds that

$$z_i(k+1) = \left[1 - \eta\lambda_i \|\boldsymbol{W}(k)\|^{2+\alpha} + \eta\left(\boldsymbol{W}^{\mathrm{T}}(k)\boldsymbol{R}\boldsymbol{W}(k) + 1 - \|\boldsymbol{W}(k)\|^2\right)\right] z_i(k) \tag{6.21}$$

$(i = 1, 2, \ldots, n)$, for all $k \geq 0$.

According to the properties of Rayleigh Quotient [7], it clearly holds that

$$\lambda_n \boldsymbol{W}^{\mathrm{T}}(k)\boldsymbol{W}(k) \leq \boldsymbol{W}^{\mathrm{T}}(k)\boldsymbol{R}\boldsymbol{W}(k) \leq \lambda_1 \boldsymbol{W}^{\mathrm{T}}(k)\boldsymbol{W}(k), \tag{6.22}$$

for all $\boldsymbol{W}(k) \neq 0$, and $k \geq 0$.

Next, we perform the convergence analysis of DDT system (6.19) via the following Theorems 6.2–6.6.

Theorem 6.2 *Suppose that* $\eta\lambda_1 < 0.125$ *and* $\eta < 0.25$. *If* $\boldsymbol{W}^{\mathrm{T}}(0)\boldsymbol{V}_n \neq 0$ *and* $\|\boldsymbol{W}(0)\| \leq 1$, *then it holds that* $\|\boldsymbol{W}(k)\| < (1 + \eta\lambda_1)$, *for all* $k \geq 0$.

Proof From (6.19) and (6.20), it follows that

$$\begin{aligned}
\|\boldsymbol{W}(k+1)\|^2 &= \sum_{i=1}^{n} z_i^2(k+1) \\
&= \sum_{i=1}^{n} [1 - \eta\lambda_i \|\boldsymbol{W}(k)\|^{2+\alpha} + \eta(\boldsymbol{W}^{\mathrm{T}}(k)\boldsymbol{R}\boldsymbol{W}(k) + 1 - \|\boldsymbol{W}(k)\|^2]^2 z_i^2(k) \\
&\leq \left[1 - \eta\left(\lambda_n \|\boldsymbol{W}(k)\|^{2+\alpha} - \lambda_1 \|\boldsymbol{W}(k)\|^2 + \|\boldsymbol{W}(k)\|^2 - 1\right)\right]^2 \sum_{i=1}^{n} z_i^2(k) \\
&\leq \left[1 - \eta\left(\lambda_n \|\boldsymbol{W}(k)\|^{2+\alpha} - \lambda_1 \|\boldsymbol{W}(k)\|^2 + \|\boldsymbol{W}(k)\|^2 - 1\right)\right]^2 \|\boldsymbol{W}(k)\|^2.
\end{aligned}$$

Thus, we have

$$\|\mathbf{W}(k+1)\|^2 \le [1+\eta(\lambda_1\|\mathbf{W}(k)\|^2+1-\|\mathbf{W}(k)\|^2)]^2\cdot\|\mathbf{W}(k)\|^2. \tag{6.23}$$

Define a differential function

$$f(s) = [1+\eta(\lambda_1 s+1-s)]^2 s, \tag{6.24}$$

over the interval [0,1]. It follows from (6.9) that

$$\dot{f}(s) = (1+\eta-\eta s(1-\lambda_1))(1+\eta-3\eta s(1-\lambda_1)),$$

for all $0<s<1$. Clearly,

$$\dot{f}(s)=0,\ if\quad s=(1+\eta)/(3\eta(1-\lambda_1))\quad or\quad s=(1+\eta)/\eta(1-\lambda_1)\ .$$

Denote

$$\theta = (1+\eta)/(3\eta(1-\lambda_1)).$$

Then,

$$\dot{f}(s)\begin{cases} >0, & \text{if}\quad 0<s<\theta \\ =0, & \text{if}\quad s=\theta \\ <0, & \text{if}\quad s>\theta. \end{cases} \tag{6.25}$$

By $\eta\lambda_1<0.125$ and $\eta<0.25$, clearly,

$$\theta = (1+\eta)/(3\eta(1-\lambda_1)) = (1/\eta+1)/(3(1-\lambda_1)) > 1. \tag{6.26}$$

From (6.25) and (6.26), it holds that

$$\dot{f}(s) > 0,$$

for all $0<s<1$. This means that $f(s)$ is monotonically increasing over the interval [0,1]. Then, we have

$$f(s) \le f(1) < (1+\eta\lambda_1)^2,$$

for all $0<s<1$.

Thus, $\|\mathbf{W}(k)\| < (1+\eta\lambda_1)$, for all $k\ge 0$.

This completes the proof.

Theorem 6.3 *Suppose that* $\eta\lambda_1<0.125$ *and* $\eta<0.25$. *If* $\mathbf{W}^{\mathrm{T}}(0)\mathbf{V}_n\neq 0$ *and* $\|\mathbf{W}(0)\|\le 1$, *then it holds that* $\|\mathbf{W}(k)\|>c$ *for all* $k\ge 0$, *where* $c=\min\{[1-\eta\lambda_1]\|\mathbf{W}(0)\|,[1-\eta\lambda_1(1+\eta\lambda_1)^4+\eta(1-(1+\eta\lambda_1)^2)]\}$.

Proof From Theorem 6.2, we have $\|\boldsymbol{W}(k)\| < (1+\eta\lambda_1)$ for all $k \geq 0$ under the conditions of Theorem 6.3. Next, two cases will be considered.

Case 1: $0 < \|\boldsymbol{W}(k)\| \leq 1$.

From (6.19) and (6.20), it follows that

$$\begin{aligned}
\|\boldsymbol{W}(k+1)\|^2 &= \sum_{i=1}^{n} [1 - \eta\lambda_i\|\boldsymbol{W}(k)\|^{2+\alpha} + \eta(\boldsymbol{W}^{\mathrm{T}}(k)\boldsymbol{R}\boldsymbol{W}(k) + 1 - \|\boldsymbol{W}(k)\|^2]^2 z_i^2(k) \\
&\geq \left[1 - \eta\left(\lambda_1\|\boldsymbol{W}(k)\|^{2+\alpha} - \lambda_n\|\boldsymbol{W}(k)\|^2\right) + \eta\left(1 - \|\boldsymbol{W}(k)\|^2\right)\right] \sum_{i=1}^{n} z_i^2(k) \\
&\geq \left[1 - \eta\left(\lambda_1\|\boldsymbol{W}(k)\|^{2+\alpha} - \lambda_n\|\boldsymbol{W}(k)\|^2\right)\right]^2 \|\boldsymbol{W}(k)\|^2 \\
&\geq \left[1 - \eta\lambda_1\|\boldsymbol{W}(k)\|^{2+\alpha}\right]^2 \|\boldsymbol{W}(k)\|^2 \\
&\geq [1 - \eta\lambda_1]^2 \|\boldsymbol{W}(k)\|^2.
\end{aligned}$$

Case 2: $1 < \|\boldsymbol{W}(k)\| < (1+\eta\lambda_1)$.

From (6.19) and (6.20), it follows that

$$\begin{aligned}
\|\boldsymbol{W}(k+1)\|^2 &\geq \left[1 - \eta\left(\lambda_1\|\boldsymbol{W}(k)\|^{2+\alpha} - \lambda_n\|\boldsymbol{W}(k)\|^2\right) + \eta\left(1 - \|\boldsymbol{W}(k)\|^2\right)\right]^2 \sum_{i=1}^{n} z_i^2(k) \\
&\geq \left[1 - \eta\lambda_1\|\boldsymbol{W}(k)\|^{2+\alpha} + \eta\left(1 - \|\boldsymbol{W}(k)\|^2\right)\right]^2 \|\boldsymbol{W}(k)\|^2 \\
&\geq \left[1 - \eta\lambda_1\|\boldsymbol{W}(k)\|^{2+\alpha} + \eta\left(1 - \|\boldsymbol{W}(k)\|^2\right)\right]^2 \\
&\geq \left[1 - \eta\lambda_1(1+\eta\lambda_1)^4 + \eta\left(1 - (1+\eta\lambda_1)^2\right)\right]^2.
\end{aligned}$$

Using the analysis of Cases 1 and 2, clearly,

$$\|\boldsymbol{W}(k)\| > c = \min\left\{[1 - \eta\lambda_1]\|\boldsymbol{W}(0)\|, \left[1 - \eta\lambda_1(1+\eta\lambda_1)^4 + \eta\left(1 - (1+\eta\lambda_1)^2\right)\right]\right\},$$

for all $k \geq 0$. From the conditions of Theorem 6.2, clearly, $c > 0$.

This completes the proof.

At this point, the boundness of DDT system (6.19) has been proven. Next, we will prove that under some mild conditions, $\lim_{k \to +\infty} \boldsymbol{W}(k) = \pm \boldsymbol{V}_n$, where $\boldsymbol{V}_n$ is the minor component. In order to analyze the convergence of DDT (6.19), we need to prove the following lemma first.

Lemma 6.1 *Suppose that* $\eta\lambda_1 < 0.125$ *and* $\eta < 0.25$. *If* $\boldsymbol{W}^{\mathrm{T}}(0)\boldsymbol{V}_n \neq 0$ *and* $\|\boldsymbol{W}(0)\| \leq 1$, *then it holds that*

$$1-\eta\lambda_i\|\boldsymbol{W}(k)\|^{2+\alpha}+\eta\left(\boldsymbol{W}^{\mathrm{T}}(k)\boldsymbol{R}\boldsymbol{W}(k)+1-\|\boldsymbol{W}(k)\|^{2}\right)>0.$$

Proof By Theorem 6.2, under the conditions of Lemma 6.1, it holds that $\|\boldsymbol{W}(k)\|<1+\eta\lambda_1$, for all $k\geq 0$. Next two cases will be considered.

Case 1: $0<\|\boldsymbol{W}(k)\|\leq 1$.

From (6.21) and (6.22), for each $i(1\leq i\leq n)$, we have

$$\begin{aligned}&1-\eta\lambda_i\|\boldsymbol{W}(k)\|^{2+\alpha}+\eta\left(\boldsymbol{W}^{\mathrm{T}}(k)\boldsymbol{R}\boldsymbol{W}(k)+1-\|\boldsymbol{W}(k)\|^{2}\right)\\&>1-\eta\lambda_1\|\boldsymbol{W}(k)\|^{2+\alpha}+\eta\lambda_n\|\boldsymbol{W}(k)\|^{2}\\&>1-\eta\lambda_1\|\boldsymbol{W}(k)\|^{2+\alpha}\\&>1-\eta\lambda_1\\&>0,\end{aligned}$$

for $k\geq 0$.

Case 2: $1<\|\boldsymbol{W}(k)\|<1+\eta\lambda_1$.

From (6.21) and (6.22), for each $i(1\leq i\leq n)$, we have

$$\begin{aligned}&1-\eta\lambda_i\|\boldsymbol{W}(k)\|^{2+\alpha}+\eta\left(\boldsymbol{W}^{\mathrm{T}}(k)\boldsymbol{R}\boldsymbol{W}(k)+1-\|\boldsymbol{W}(k)\|^{2}\right)\\&>1-\eta\lambda_1\|\boldsymbol{W}(k)\|^{2+\alpha}+\eta\lambda_n\|\boldsymbol{W}(k)\|^{2}-\eta\left(2\eta\lambda_1+\eta^2\lambda_1^2\right)\\&>1-\eta\lambda_1\|\boldsymbol{W}(k)\|^{2+\alpha}-0.25*\left(2\eta\lambda_1+\eta^2\lambda_1^2\right)\\&>1-\eta\lambda_1\|\boldsymbol{W}(k)\|^{2+\alpha}-\left(0.5*\eta\lambda_1+0.25*(\eta\lambda_1)^2\right)\\&>1-\eta\lambda_1\left((1+\eta\lambda_1)^4+0.5+0.25*\eta\lambda_1\right)\\&>0.\end{aligned}$$

This completes the proof.

Lemma 6.1 means that the projection of the weight vector $\boldsymbol{W}(k)$ on eigenvector $\boldsymbol{V}_i(i=1,2,\ldots,n)$, which is denoted as $z_i(k)=\boldsymbol{W}^T(k)\boldsymbol{V}_i(i=1,2,\ldots,n)$, does not change its sign in (6.21). From (6.20), we have $z_i(t)=\boldsymbol{W}^{\mathrm{T}}(t)\boldsymbol{V}_i$. Since $\boldsymbol{W}^{\mathrm{T}}(0)\boldsymbol{V}_n\neq 0$, we have $z_n(0)\neq 0$. It follows from (6.6) and Lemma 6.1 that $z_n(k)>0$ for all $k>0$ if $z_n(0)>0$; and $z_n(k)<0$ for all $k>0$ if $z_n(0)<0$. Without loss of generality, we assume that $z_n(0)>0$. Thus, $z_n(k)>0$ for all $k>0$.

From (6.20), for each $k\geq 0$, $\boldsymbol{W}(k)$ can be represented as

$$\boldsymbol{W}(k)=\sum_{i=1}^{n-1}z_i(k)\boldsymbol{V}_i+z_n(k)\boldsymbol{V}_n. \tag{6.27}$$

Clearly, the convergence of $\boldsymbol{W}(k)$ can be determined by the convergence of $z_i(k)$ $(i = 1,2,\ldots,n)$. Theorems 6.4 and 6.5 below provide the convergence of $z_i(k)$ $(i = 1,2,\ldots,n)$.

Theorem 6.4 *Suppose that* $\eta\lambda_1 < 0.125$ *and* $\eta < 0.25$. *If* $\boldsymbol{W}^{\mathrm{T}}(0)\boldsymbol{V}_n \neq 0$ *and* $\|\boldsymbol{W}(0)\| < 1$, *then* $\lim_{k\to\infty} z_i(k) = 0,\ (i = 1,2,\ldots,n-1)$.

Proof By Lemma 6.1, clearly,

$$1 - \eta\lambda_i\|\boldsymbol{W}(k)\|^{2+\alpha} + \eta\left(\boldsymbol{W}^{\mathrm{T}}(k)\boldsymbol{R}\boldsymbol{W}(k) + 1 - \|\boldsymbol{W}(k)\|^2\right) > 0, \quad (i = 1,2,\ldots,n) \tag{6.28}$$

for all $k \geq 0$. Using Theorems 6.2 and 6.3, it holds that $\|\boldsymbol{W}(k)\| > c$ and $\|\boldsymbol{W}(k)\| < (1+\eta\lambda_1)$ for all $k \geq 0$. Thus, it follows that for all $k \geq 0$

$$\begin{aligned}
&\left[\frac{1 - \eta\lambda_i\|\boldsymbol{W}(k)\|^{2+\alpha} + \eta\left(\boldsymbol{W}^{\mathrm{T}}(k)\boldsymbol{R}\boldsymbol{W}(k) + 1 - \|\boldsymbol{W}(k)\|^2\right)}{1 - \eta\lambda_n\|\boldsymbol{W}(k)\|^{2+\alpha} + \eta\left(\boldsymbol{W}^{\mathrm{T}}(k)\boldsymbol{R}\boldsymbol{W}(k) + 1 - \|\boldsymbol{W}(k)\|^2\right)}\right]^2 \\
&\overset{(1)}{=} \left[1 - \frac{\eta(\lambda_i - \lambda_n)\|\boldsymbol{W}(k)\|^{2+\alpha}}{1 - \eta\lambda_n\|\boldsymbol{W}(k)\|^{2+\alpha} + \eta\left(\boldsymbol{W}^{\mathrm{T}}(k)\boldsymbol{R}\boldsymbol{W}(k) + 1 - \|\boldsymbol{W}(k)\|^2\right)}\right]^2 \\
&\leq \left[1 - \frac{\eta(\lambda_i - \lambda_n)\|\boldsymbol{W}(k)\|^{2+\alpha}}{1 - \eta\lambda_n\|\boldsymbol{W}(k)\|^{2+\alpha} + \eta\left(\lambda_1\|\boldsymbol{W}(k)\|^2 + 1 - \|\boldsymbol{W}(k)\|^2\right)}\right]^2 \\
&= \left[1 - \frac{\eta(\lambda_i - \lambda_n)}{1/\|\boldsymbol{W}(k)\|^{2+\alpha} - \eta\lambda_n + \eta\left(\lambda_1\|\boldsymbol{W}(k)\|^{-\alpha} + 1/\|\boldsymbol{W}(k)\|^{2+\alpha} - \|\boldsymbol{W}(k)\|^{-\alpha}\right)}\right]^2 \\
&< \left[1 - \frac{\eta(\lambda_{n-1} - \lambda_n)}{1/c^{(2+\alpha)} - \eta\lambda_n + \eta[\lambda_1 c^{-\alpha} + 1/c^{(2+\alpha)} - (1+\eta\lambda_1)^{-\alpha}]}\right]^2, (i = 1,2,\ldots,n-1).
\end{aligned} \tag{6.29}$$

Denote

$$\theta = \left[1 - \frac{\eta(\lambda_{n-1} - \lambda_n)}{1/c^{(2+\alpha)} - \eta\lambda_n + \eta[\lambda_1 c^{-\alpha} + 1/c^{(2+\alpha)} - (1+\eta\lambda_1)^{-\alpha}]}\right]^2.$$

Clearly, θ is a constant and $0 < \theta < 1$. By $\boldsymbol{W}^{\mathrm{T}}(0)\boldsymbol{V}_n \neq 0$, clearly, $z_n(0) \neq 0$. Then, $z_n(k) \neq 0 (k > 0)$.

From (6.21), (6.28), and (6.29), it holds that

$$\left[\frac{z_i(k+1)}{z_n(k+1)}\right]^2 = \left[\frac{1-\eta\lambda_i\|\boldsymbol{W}(k)\|^{2+\alpha}+\eta\left(\boldsymbol{W}^{\mathrm{T}}(k)\boldsymbol{R}\boldsymbol{W}(k)+1-\|\boldsymbol{W}(k)\|^2\right)}{1-\eta\lambda_n\|\boldsymbol{W}(k)\|^{2+\alpha}+\eta\left(\boldsymbol{W}^{\mathrm{T}}(k)\boldsymbol{R}\boldsymbol{W}(k)+1-\|\boldsymbol{W}(k)\|^2\right)}\right]^2 \cdot \left[\frac{z_i(k)}{z_n(k)}\right]^2$$
$$\leq \theta \cdot \left[\frac{z_i(k)}{z_n(k)}\right]^2 \leq \theta^{k+1} \cdot \left[\frac{z_i(0)}{z_n(0)}\right]^2, (i=1,2,\ldots,n-1), \tag{6.30}$$

for all $k \geq 0$.

Thus, from $0<\theta<1\ (i=1,2,\ldots,n-1)$, we have

$$\lim_{k\to\infty}\frac{z_i(k)}{z_n(k)}=0, (i=1,2,\ldots,n-1).$$

By Theorems 6.2 and 6.3, $z_n(k)$ must be bounded. Then,

$$\lim_{k\to\infty} z_i(k)=0, (i=1,2,\ldots,n-1).$$

This completes the proof.

Theorem 6.5 *Suppose that* $\eta\lambda_1<0.125$ *and* $\eta<0.25$. *If* $\boldsymbol{W}^{\mathrm{T}}(0)\boldsymbol{V}_n \neq 0$ *and* $\|\boldsymbol{W}(0)\|<1$, *then it holds that* $\lim_{k\to\infty} z_n(k)=\pm 1$.

Proof Using Theorem 6.4, clearly, $\boldsymbol{W}(k)$ will converge to the direction of the minor component $\boldsymbol{V}_{\mathrm{n}}$, as $k\to\infty$. Suppose at time k_0, $\boldsymbol{W}(k)$ has converged to the direction of $\boldsymbol{V}_{\mathrm{n}}$, i.e., $\boldsymbol{W}(k_0)=z_n(k_0)\cdot\boldsymbol{V}_n$.

From (6.21), it holds that

$$\begin{aligned}
z_n(k+1) &= z_n(k)\left(1-\eta\lambda_n z_n^{(2+\alpha)}(k)+\eta(\lambda_n z_n^2(k)+1-z_n^2(k))\right)\\
&= z_n(k)\left(1+\eta[\lambda_n z_n^2(k)(1-z_n^{(\alpha)}(k))+1-z_n^2(k)]\right)\\
&= z_n(k)\left(1+\eta(1-z_n(k))(\lambda_n z_n^2(k)(z_n^{(\alpha-1)}(k)+z_n^{(\alpha-2)}(k)+\cdots+1)+(1+z_n(k)))\right)\\
&= z_n(k)\left(1+\eta(1-z_n(k))(\lambda_n(z_n^{(\alpha+1)}(k)+z_n^{(\alpha)}(k)+\cdots+z_n^2(k))+(1+z_n(k)))\right)\\
&= z_n(k)(1+\eta(1-z_n(k))Q(\lambda_n,z_n(k))),
\end{aligned} \tag{6.31}$$

where $Q(\lambda_n,z_n(k))=(\lambda_n(z_n^{(\alpha+1)}(k)+z_n^{(\alpha)}(k)+\cdots+z_n^2(k))+(1+z_n(k)))$ is a positive efficient, for all $k\geq k_0$.

From (6.31), it holds that

$$\begin{aligned}
z_n(k+1)-1 &= z_n(k)(1+\eta(1-z_n(k))Q(\lambda_n,z_n(k)))-1\\
&= [1-\eta z_n(k)Q(\lambda_n,z_n(k))](z_n(k)-1),
\end{aligned} \tag{6.32}$$

for $k>k_0$.

Since $z_n(k) < \|\boldsymbol{W}(\mathrm{k})\| \leq (1+\eta\lambda_1)$, we have

$$
\begin{aligned}
&1-\eta z_n(k)Q(\lambda_n, z_n(k)) \\
&\quad = 1-\eta z_n(k)(\lambda_n(z_n^{(\alpha+1)}(k)+z_n^{(\alpha)}(k)+\cdots+z_n^2(k))+(1+z_n(k))) \\
&\quad > 1-\eta(1+\eta\lambda_1)(\lambda_n((1+\eta\lambda_1)^{(\alpha+1)}+(1+\eta\lambda_1)^{(\alpha)}+\cdots+(1+\eta\lambda_1)^{(2)})+(1+(1+\eta\lambda_1))) \\
&\quad > 1-(1+\eta\lambda_1)(\eta\lambda_1((1+\eta\lambda_1)^{(\alpha+1)}+(1+\eta\lambda_1)^{(\alpha)}+\cdots+(1+\eta\lambda_1)^{(2)})+\eta(1+(1+\eta\lambda_1))) \\
&\quad > 1-(1+\eta\lambda_1)(\eta\lambda_1((1+\eta\lambda_1)^3+(1+\eta\lambda_1)^2)+\eta(1+(1+\eta\lambda_1))) \\
&\quad > 1-0.9980 \\
&\quad > 0,
\end{aligned}
\tag{6.33}
$$

for all $k \geq k_0$. Thus, denote $\delta = 1-\eta z_n(k)Q(\lambda_n, z_n(k))$, Clearly, it holds that $0<\delta<1$.

It follows from (6.32) and (6.33) that

$$|z_n(k+1)-1| \leq \delta|z_n(k)-1|,$$

for all $k > k_0$. Then, for $k > k_0$

$$|z_n(k+1)-1| \leq \delta^{k+1}|z_n(0)-1| \leq (k+1)\Pi \mathrm{e}^{-\theta(k+1)},$$

where $\theta = -\ln\delta$, $\Pi = |(1+\eta\lambda_1)-1|$.

Given any $\varepsilon > 0$, there exists a $K \geq 1$ such that

$$\frac{\Pi_2 K \mathrm{e}^{-\theta K}}{(1-\mathrm{e}^{-\theta})^2} \leq \varepsilon.$$

For any $k_1 > k_2 > k$, it follows from (6.21) that

$$
\begin{aligned}
|z_n(k_1)-z_n(k_2)| &= \left|\sum_{r=k_2}^{k_1-1}[z_n(r+1)-z_n(r)]\right| \leq \left|\sum_{r=k_2}^{k_1-1}\eta z_n(r)((1-z_n(r))Q(\lambda_n, z_n(r)))\right| \\
&\leq \sum_{r=k_2}^{k_1-1}|\eta z_n(r)((1-z_n(r))Q(\lambda_n, z_n(r)))| \leq \sum_{r=k_2}^{k_1-1}|\eta z_n(r)Q(\lambda_n, 1+\eta\lambda_1)(z_n(r)-1)| \\
&\leq \eta(1+\eta\lambda_1)Q(\lambda_n, 1+\eta\lambda_1)\sum_{r=k_2}^{k_1-1}|(z_n(r)-1)| \leq \Pi_2\sum_{r=k_2}^{k_1-1} r\mathrm{e}^{-\theta r} \\
&\leq \Pi_2\sum_{r=k}^{+\infty} r\mathrm{e}^{-\theta r} \leq \Pi_2 K\mathrm{e}^{-\theta K}\sum_{r=0}^{+\infty} r(\mathrm{e}^{-\theta})^{r-1} \leq \frac{\Pi_2 K\mathrm{e}^{-\theta K}}{(1-\mathrm{e}^{-\theta})^2} \\
&\leq \varepsilon.
\end{aligned}
$$

where $\Pi_2 = \eta(1+\eta\lambda_1)Q(\lambda_n, 1+\eta\lambda_1)(z_n(0)-1)$. This means that the sequence $\{z_n(k)\}$ is a Cauchy sequence. By the Cauchy convergence principle, there must exist a constant z^* such that $\lim_{x\to\infty} z_n(k) = z^*$.

From (6.27), we have $\lim_{k\to+\infty} \boldsymbol{W}(k) = z_n^* \cdot \boldsymbol{V}_n$. Since (6.17) has self-stabilizing property, it follows that $\lim_{x\to\infty} \boldsymbol{W}(k+1)/\boldsymbol{W}(k) = 1$. From (6.21), we have $1 = 1 - \eta[\lambda_n(z_n^*)^{2+\alpha} - (\lambda_n(z_n^*)^2 + 1 - (z_n^*)^2)]$, which means $z_n^* = \pm 1$.

This completes the proof.

Using (6.27), along with Theorems 6.4 and 6.5, we can draw the following conclusion:

Theorem 6.6 *Suppose that* $\eta\lambda_1 < 0.125$ *and* $\eta < 0.25$. *If* $\boldsymbol{W}^{\mathrm{T}}(0)\boldsymbol{V}_n \neq 0$ *and* $\|\boldsymbol{W}(0)\| < 1$, *then it holds that* $\lim_{k\to\infty} \boldsymbol{W}(k) = \pm\boldsymbol{V}_n$.

At this point, we have completed the proof of the convergence of DDT system (6.19). Next we will further study the stability of (6.19).

Theorem 6.7 *Suppose that* $\eta\lambda_1 < 0.125$ *and* $\eta < 0.25$. *Then the equilibrium points* $\boldsymbol{V}_n$ *and* $-\boldsymbol{V}_n$ *are locally asymptotical stable and other equilibrium points* (6.19) *are unstable.*

Proof Clearly, the set of all equilibrium points of (6.21) is $\{\boldsymbol{V}_1, \cdots, \boldsymbol{V}_n\} \cup \{-\boldsymbol{V}_1, \cdots, -\boldsymbol{V}_n\} \cup \{0\}$.

Denote

$$\begin{aligned} \boldsymbol{G}(\boldsymbol{W}) &= \boldsymbol{W}(k+1) \\ &= \boldsymbol{W}(k) - \eta\left[\|\boldsymbol{W}(k)\|^{2+\alpha}\boldsymbol{R}\boldsymbol{W}(k) - (\boldsymbol{W}^{\mathrm{T}}(k)\boldsymbol{R}\boldsymbol{W}(k) + 1 - \|\boldsymbol{W}(k)\|^2)\boldsymbol{W}(k)\right]. \end{aligned} \tag{6.34}$$

Then, we have

$$\begin{aligned} \frac{\partial \mathbf{G}}{\partial \mathbf{W}} = \mathbf{I} + \eta[&(\mathbf{W}^{\mathrm{T}}(k)\mathbf{R}\mathbf{W}(k) + 1 - \|\mathbf{W}(k)\|^2)\mathbf{I} - \|\mathbf{W}(k)\|^{2+\alpha}\mathbf{R} + 2\mathbf{R}\mathbf{W}(k)\mathbf{W}^{\mathrm{T}}(k) \\ &- 2\mathbf{W}(k)\mathbf{W}^{\mathrm{T}}(k) - (2+\alpha)\|\mathbf{W}(k)\|^{\alpha}\mathbf{R}\mathbf{W}(k)\mathbf{W}^{\mathrm{T}}(k)], \end{aligned} \tag{6.35}$$

where $\boldsymbol{I}$ is a unity matrix.

For the equilibrium point 0, it holds that

$$\left.\frac{\partial \mathbf{G}}{\partial \mathbf{W}}\right|_0 = \boldsymbol{I} + \eta\boldsymbol{I} = \boldsymbol{J}_0.$$

The eigenvalues of $\boldsymbol{J}_0$ are $\alpha_0^{(i)} = 1 + \eta > 1 \quad (i = 1, 2, \cdots, n)$. Thus, the equilibrium point is unstable.

For the equilibrium points $\pm \boldsymbol{V}_j (j = 1, 2, \cdots, n)$, it follows from (6.35) that

$$\left.\frac{\partial \boldsymbol{G}}{\partial \boldsymbol{W}}\right|_{\boldsymbol{V}_j} = \boldsymbol{I} + \eta \left[\lambda_j \boldsymbol{I} - \boldsymbol{R} - 2\boldsymbol{V}_j \boldsymbol{V}_j^{\mathrm{T}} - \alpha \lambda_j \boldsymbol{V}_j \boldsymbol{V}_j^{\mathrm{T}}\right] = \boldsymbol{J}_j. \tag{6.36}$$

After some simple manipulations, the eigenvalues of $\boldsymbol{J}_j$ are given by

$$\begin{cases} \alpha_j^{(i)} = 1 + \eta(\lambda_j - \lambda_i) & \text{if} \quad i \neq j. \\ \alpha_j^{(i)} = 1 - \eta(2 + \alpha\lambda_j) & \text{if} \quad i = j. \end{cases}$$

For any $j \neq n$, it holds that $\alpha_j^{(n)} = 1 + \eta(\lambda_j - \lambda_i) > 1$. Clearly, the equilibrium points $\pm \boldsymbol{V}_j (j \neq n)$ are unstable. For the equilibrium points $\pm \boldsymbol{V}_n$, from $\eta\lambda_n < \eta\lambda_1 < 0.125$, and $\eta < 0.25$, it holds that

$$\begin{cases} \alpha_n^{(i)} = 1 + \eta(\lambda_n - \lambda_i) < 1 & \text{if} \quad i \neq n. \\ \alpha_n^{(i)} = 1 - \eta(2 + \alpha\lambda_n) < 1 & \text{if} \quad i = n. \end{cases} \tag{6.37}$$

Thus, $\pm \boldsymbol{V}_n$ are asymptotical stable.

This completes the proof.

From (6.37), we can easily see that the only fixed points where the MCA condition is fulfilled are the attractors, and all others are repellers or saddle points. We conclude that the Algorithm (6.17) converges toward the minor eigenvector $\pm \boldsymbol{V}_n$ associated with the minor eigenvalue λ_n.

6.3.3 Computer Simulations

In this section, we provide simulation results to illustrate the convergence and stability of the MCA Algorithm (6.17) in a stochastic case. Since OJAm [17], Moller [23], and Peng [11] are self-stabilizing algorithms and have better convergence performance than some existing MCA algorithms, we compare performance of Algorithm (6.17) with these algorithms. In order to measure the convergence speed and accuracy of these algorithms, we compute the norm of $\boldsymbol{W}(k)$ and the direction cosine at the kth update. In the simulation, the input data sequence, which is generated by [17], $\boldsymbol{X}(k) = \boldsymbol{C}\, \boldsymbol{h}(\mathrm{k})$, where $\mathbf{C}$ = randn(5, 5)/5 and $\boldsymbol{h}(k) \in \boldsymbol{R}^{5\times 1}$, is Gaussian and randomly generated with zero-mean and unitary standard deviation. The above-mentioned four MCA algorithms are used to extract minor component from the input data sequence $\{x(k)\}$. The following learning curves show the convergence of $\boldsymbol{W}(k)$ and direction cosine(k) with the same initial norm for the weight vector and constant learning rate, respectively. All the learning curves below are obtained by averaging over 30 independent experiments. Figures 6.2 and 6.3 investigate the case $\|\boldsymbol{W}(0)\| = 1$, and Figs. 6.4 and 6.5 show the simulation results for higher-dimensional data (D = 12), using different learning rates and maximal eigenvalues, which satisfy the conditions of Theorem 6.6.

From Fig. 6.3, we can see that for all these MCA algorithms, direction cosine (k) converge to 1 at approximately the same speeds. However, from Fig. 6.2 we can see that the Moller and OJAm algorithms have approximately the same convergence for the weight vector length and there appear to be a residual deviation from unity for the weight vector length, and the norm of the weight vector in Peng algorithm has larger oscillations, and the norm of the weight vector in Algorithm (6.17) has a faster convergence, a better numerical stability and higher precision than other algorithms. From Figs. 6.4 and 6.5, it is obvious that even for higher-dimensional data, only if the conditions of Theorems 6.2–6.6 are satisfied, Algorithm (6.17) can satisfactorily extract the minor component of the input data stream.

In this section, dynamics of a class of algorithms are analyzed by the DDT method. It has been proved that if some mild conditions about the learning rate and the initial weight vector are satisfied, these algorithms will converge to the minor

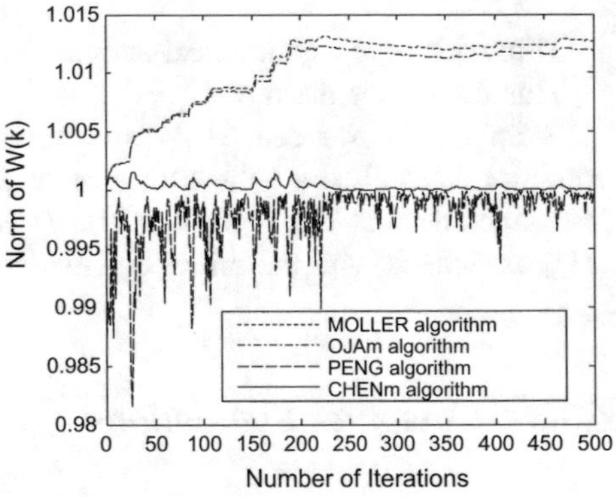

Fig. 6.2 Norm of $\boldsymbol{W}(k)$ ($\eta = 0.1$, $\|\boldsymbol{W}(0)\| = 1.0$, $\alpha = 2$)

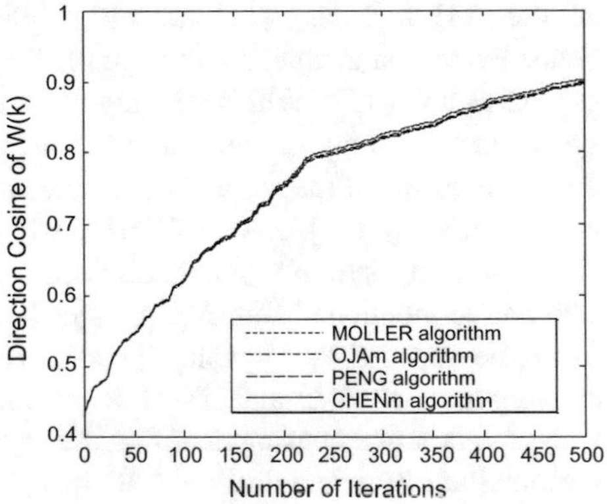

Fig. 6.3 Direction cosine of $\boldsymbol{W}(k)$ ($\eta = 0.1$, $\|\boldsymbol{W}(0)\| = 1.0$, $\alpha = 2$)

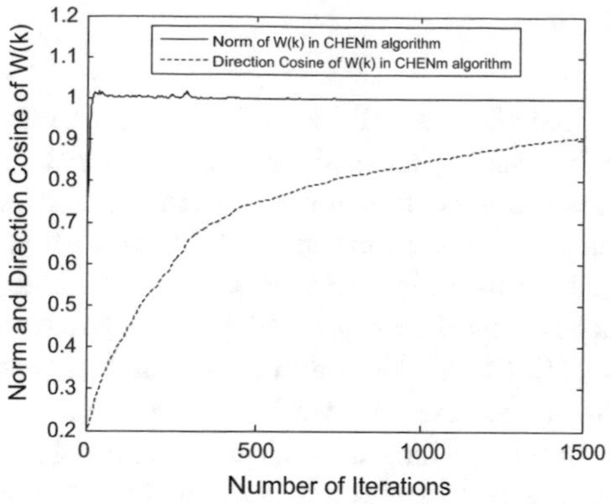

Fig. 6.4 $\eta = 0.1, \|\boldsymbol{W}(0)\| = 0.6$, $\lambda_1 = 1.1077$, $\alpha = 2$, and D = 12

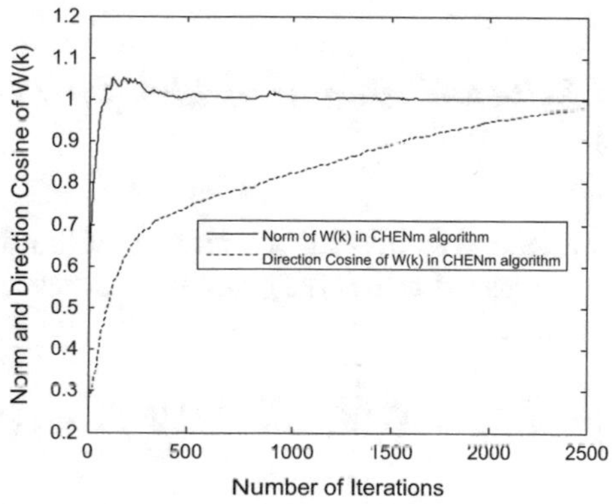

Fig. 6.5 $\eta = 0.01$, $\|\boldsymbol{W}(0)\| = 0.6$, $\lambda_1 = 12.1227$, $\alpha = 2$, and D = 12

component with unit norm. At the same time, stability analysis shows that the minor component is the asymptotical stable equilibrium point in these algorithms. Simulation results show that this class of self-stabilizing MCA algorithms outperforms some existing MCA algorithms.

6.4 DDT System of a Unified PCA and MCA Algorithm

In Sect. 6.3, the convergence of a MCA algorithm proposed by us is analyzed via DDT in details. However, in the above analysis, we made one assumption, i.e., the smallest eigenvalue of the correlation matrix of the input data is single. In this section, we will remove this assumption in the convergence analysis and analyze a unified PCA and MCA algorithm via the DDT method.

6.4.1 Introduction

Despite the large number of unified PCA and MCA algorithms proposed to date, there are few works that analyze these algorithms via the DDT method and derive the conditions to guarantee the convergence. Obviously, this is necessary from the point view of application. Among the unified PCA and MCA algorithms, Chen's algorithm [22] is regarded as a pioneering work. Other self-normalizing dual systems [24] or dual-purpose algorithms [19, 20] can be viewed as the generalizations of Chen's algorithm [22]. Chen's algorithm lays sound theoretical foundations for dual-purpose algorithms. However, no work has been done so far on the study of Chen's DDT system. In this section, the unified PCA and MCA algorithm proposed by Chen et al. [22] will be analyzed and some sufficient conditions to guarantee its convergence will be derived by the DDT method. These theoretical results will lay a solid foundation for the applications of this algorithm.

6.4.2 A Unified Self-stabilizing Algorithm for PCA and MCA

Chen et al. proposed a unified stabilizing learning algorithm for principal components and minor components extraction [22], and the stochastic discrete form of the algorithm can be written as

$$\boldsymbol{W}(k+1) = \boldsymbol{W}(k) \pm \eta\left[\|\boldsymbol{W}(k)\|^2 y(k)\boldsymbol{X}(k) - y^2(k)\boldsymbol{W}(k)\right] + \eta(1 - \|\boldsymbol{W}(k)\|^2)\boldsymbol{W}(k), \tag{6.38}$$

where η $(0<\eta<1)$ is the learning rate. Algorithm (6.38) can extract principal component if "+" is used. If the sign is simply altered, (6.38) can also serve as a minor component extractor. It is interesting that the only difference between the PCA algorithm and the MCA algorithm is the sign on the right hand of (6.38).

In order to derive some sufficient conditions to guarantee the convergence of Algorithm (6.38), next we analyze the dynamics of (6.38) via the DDT approach. The DDT system associated with (6.38) can be formulated as follows. Taking the conditional expectation $E\{\boldsymbol{W}(k+1)/\boldsymbol{W}(0), \boldsymbol{X}(i), i<k\}$ to (6.38) and identifying the conditional expectation as the next iterate, a DDT system can be obtained and given as

$$\begin{aligned}\boldsymbol{W}(k+1) = {} & \boldsymbol{W}(k) \pm \eta\left[\|\boldsymbol{W}(k)\|^2 \boldsymbol{R}\boldsymbol{W}(k) - \boldsymbol{W}^{\mathrm{T}}(k)\boldsymbol{R}\boldsymbol{W}(k)\boldsymbol{W}(k)\right] \\ & + \eta(1 - \|\boldsymbol{W}(k)\|^2)\boldsymbol{W}(k),\end{aligned} \tag{6.39}$$

where $R = E[X(k)X^{\mathrm{T}}(k)]$ is the correlation matrix of the input data. The main purpose of this section is to study the convergence of the weight vector $W(k)$ of (6.39) subject to the learning rate η being some constant.

6.4.3 Convergence Analysis

Since R is a symmetric positive definite matrix, there exists an orthonormal basis of $\Re^n$ composed of the eigenvectors of R. Let $\lambda_1, \lambda_2, \cdots, \lambda_n$ to be all the eigenvalues of R ordered by $\lambda_1 \geq \lambda_2 \geq \cdots \geq \lambda_{n-1} \geq \lambda_n > 0$. Denote by σ, the largest eigenvalue of R. Suppose that the multiplicity of σ is $m(1 \leq m \leq n)$. Then, $\sigma = \lambda_1 = \cdots = \lambda_m$. Suppose that $\{V_i | i = 1, 2, \cdots, n\}$ is an orthogonal basis of $\Re^n$ such that each V_i is a unitary eigenvector of R associated with the eigenvalue λ_i. Denote by V_σ the eigen-subspace of the largest eigenvalue σ, i.e., $V_\sigma = \mathrm{span}\{V_1, \cdots, V_m\}$. Denote by $V_\sigma^\perp$ the subspace which is perpendicular to V_σ. Clearly, $V_\sigma^\perp = \mathrm{span}\{V_{m+1}, \cdots, V_n\}$. Similarly, we can denote by V_τ the eigen-subspace of the smallest eigenvalue τ. Suppose that the multiplicity of τ is $p(1 \leq p \leq n-m)$. Then, $V_\tau = \mathrm{span}\{V_{n-p}, \cdots, V_n\}$ and $V_\tau^\perp = \mathrm{span}\{V_1, \cdots, V_{n-p-1}\}$.

Since the vector set $\{V_1, V_2, \cdots, V_n\}$ is an orthonormal basis of $\Re^n$, for each $k \geq 0$, $W(k)$ and $RW(k)$ can be represented, respectively, as

$$W(k) = \sum_{i=1}^{n} z_i(k) V_i, \quad RW(k) = \sum_{i=1}^{n} \lambda_i z_i(k) V_i, \tag{6.40}$$

where $z_i(k)(i = 1, 2, \ldots, n)$ are some constants.

From (6.39) and (6.40), it holds that

$$z_i(k+1) = [1 \pm \eta(\lambda_i \|W(k)\|^2 - W^{\mathrm{T}}(k) R W(k)) + \eta(1 - \|W(k)\|^2)] z_i(k), \tag{6.41}$$

$(i = 1, 2, \ldots, n)$, for all $k \geq 0$.

By letting $Q(R, W(k)) = \pm[\lambda_i \|W(k)\|^2 - W^{\mathrm{T}}(k) R W(k)]$,

(6.41) can be represented as

$$z_i(k+1) = [1 + \eta Q(R, W(k)) + \eta(1 - \|W(k)\|^2)] z_i(k), \tag{6.42}$$

$(i = 1, 2, \ldots, n)$, for all $k \geq 0$. According to the properties of the Rayleigh Quotient [7], it clearly holds that

$$\lambda_n W^{\mathrm{T}}(k) W(k) \leq W^{\mathrm{T}}(k) R W(k) \leq \lambda_1 W^{\mathrm{T}}(k) W(k), \tag{6.43}$$

for all $k \geq 0$. From (6.43), it holds that

$$Q_{\max} = (\lambda_1 - \lambda_n)\|\boldsymbol{W}(k)\|^2, \quad Q_{\min} = (\lambda_n - \lambda_1)\|\boldsymbol{W}(k)\|^2. \tag{6.44}$$

Next, we will analyze the convergence of DDT system (6.39) via the following Theorems 6.8–6.11.

Theorem 6.8 *Suppose that* $\eta \le 0.3$. *If* $\|\boldsymbol{W}(0)\| \le 1$ *and* $(\lambda_1 - \lambda_n) < 1$, *then it holds that* $\|\boldsymbol{W}(k)\| < (1 + \eta\lambda_1)$, *for all* $k \ge 0$.

Proof From (6.40)–(6.44), it follows that

$$\begin{aligned}
\|\boldsymbol{W}(k+1)\|^2 &= \sum_{i=1}^{n} z_i^2(k+1) = \sum_{i=1}^{n} [1 + \eta Q(\boldsymbol{R}, \boldsymbol{W}(k)) + \eta(1 - \|\boldsymbol{W}(k)\|^2)]^2 z_i^2(k) \\
&\le \sum_{i=1}^{n} \left[1 + \eta Q_{\max} + \eta(1 - \|\boldsymbol{W}(k)\|^2)\right]^2 z_i^2(k) \\
&\le \left[1 + \eta(\lambda_1 - \lambda_n)\|\boldsymbol{W}(k)\|^2 + \eta\left(1 - \|\boldsymbol{W}(k)\|^2\right)\right]^2 \sum_{i=1}^{n} z_i^2(k) \\
&\le [1 + \eta(\lambda_1 - \lambda_n)\|\boldsymbol{W}(k)\|^2 + \eta(1 - \|\boldsymbol{W}(k)\|^2)]^2 \|\boldsymbol{W}(k)\|^2.
\end{aligned}$$

Thus, it holds that $\|\boldsymbol{W}(k+1)\|^2 \le [1 + \eta(\lambda_1 - \lambda_n)\|\boldsymbol{W}(k)\|^2 + \eta(1 - \|\boldsymbol{W}(k)\|^2)]^2 \|\boldsymbol{W}(k)\|^2$.

Define a differential function $f(s) = [1 + \eta(\lambda_1 - \lambda_n - 1)s + \eta]^2 s$, over the interval [0, 1], where $s = \|\boldsymbol{W}(k)\|^2$ and $f(s) = \|\boldsymbol{W}(k+1)\|^2$. It follows that

$$\dot{f}(s) = (1 + \eta - \eta s(\lambda_n + 1 - \lambda_1))(1 + \eta - 3\eta s(\lambda_n + 1 - \lambda_1)), \tag{6.45}$$

for all $0 < s < 1$. Clearly,

$$\dot{f}(s) = 0, \quad \textit{if} \quad s = \frac{1+\eta}{3\eta(\lambda_n + 1 - \lambda_1)} \quad \textit{or} \quad s = \frac{1+\eta}{\eta(\lambda_n + 1 - \lambda_1)}.$$

Denote $\theta = (1+\eta)/(3\eta(\lambda_n + 1 - \lambda_1))$. Then, we have

$$\dot{f}(s)\begin{cases} > 0, & \text{if } 0 < s < \theta \\ = 0, & \text{if } s = \theta \\ < 0, & \text{if } s > \theta. \end{cases} \tag{6.46}$$

By $\eta \le 0.3$, clearly,

$$\theta = (1+\eta)/(3\eta(\lambda_n + 1 - \lambda_1)) = (1 + 1/\eta)/(3[1 - (\lambda_1 - \lambda_n)]) > 1. \tag{6.47}$$

From (6.46) and (6.47), it holds that $\dot{f}(s) > 0$ for all $0 < s < 1$. This means that $f(s)$ is monotonically increasing over the interval [0,1]. Then, for all $0 < s < 1$, it follows that

$$f(s) \le f(1) = [1+\eta(\lambda_1 - \lambda_n)]^2 < (1+\eta\lambda_1)^2.$$

Thus, we have $\|\boldsymbol{W}(k)\| < (1+\eta\lambda_1)$ for all $k \ge 0$.

This completes the proof.

Theorem 6.8 shows that there exists an upper bound for $\|\boldsymbol{W}(k)\|$ in the DDT system (6.39), for all $k \ge 0$.

Theorem 6.9 *Suppose that* $\eta \le 0.3$. *If* $\|\boldsymbol{W}(0)\| \le 1$, *then it holds that* $\|\boldsymbol{W}(k)\| > c$ *for all* $k \ge 0$, *where* $c = \min\left\{[1-\eta\lambda_1]\|\boldsymbol{W}(0)\|, [1-\eta\lambda_1(1+\eta\lambda_1)^2 - \eta(2\eta\lambda_1 + \eta^2\lambda_1^2)]\right\}$.

Proof From Theorem 6.8, we have $\|\boldsymbol{W}(k)\| < (1+\eta\lambda_1)$ for all $k \ge 0$ under the conditions of Theorem 6.9. Next, two cases will be considered.

Case 1: $0 < \|\boldsymbol{W}(k)\| \le 1$.

From (6.40)–(6.44), it follows that

$$\begin{aligned}
\|\boldsymbol{W}(k+1)\|^2 &\ge \sum_{i=1}^{n}\left[1+\eta Q_{\min} + \eta(1-\|\boldsymbol{W}(k)\|^2)\right]^2 z_i^2(k) \\
&\ge \left[1+\eta(\lambda_n - \lambda_1)\|\boldsymbol{W}(k)\|^2 + \eta\left(1-\|\boldsymbol{W}(k)\|^2\right)\right]^2 \sum_{i=1}^{n} z_i^2(k) \\
&\ge \left[1+\eta(\lambda_n - \lambda_1)\|\boldsymbol{W}(k)\|^2\right]^2 \|\boldsymbol{W}(k)\|^2 \ge \left[1-\eta\lambda_1\|\boldsymbol{W}(k)\|^2\right]^2 \|\boldsymbol{W}(k)\|^2 \\
&\ge [1-\eta\lambda_1]^2 \|\boldsymbol{W}(k)\|^2.
\end{aligned}$$

Case 2: $1 < \|\boldsymbol{W}(k)\| \le (1+\eta\lambda_1)$.

From (6.40)–(6.44), it follows that

$$\begin{aligned}
\|\boldsymbol{W}(k+1)\|^2 &\ge \sum_{i=1}^{n}[1+\eta Q_{\min} + \eta(1-\|\boldsymbol{W}(k)\|^2)]^2 z_i^2(k) \\
&= [1+\eta(\lambda_n - \lambda_1)\|\boldsymbol{W}(k)\|^2 + \eta(1-\|\boldsymbol{W}(k)\|^2)]^2 \sum_{i=1}^{n} z_i^2(k) \\
&\ge \left[1-\eta\lambda_1\|\boldsymbol{W}(k)\|^2 + \eta\left(-2\eta\lambda_1 - \eta^2\lambda_1^2\right)\right]^2 \|\boldsymbol{W}(k)\|^2 \\
&\ge \left[1-\eta\lambda_1(1+\eta\lambda_1)^2 - \eta\left(2\eta\lambda_1 + \eta^2\lambda_1^2\right)\right]^2.
\end{aligned}$$

From the above analysis, clearly,

$$\|\boldsymbol{W}(k)\| > c = \min\{[1-\eta\lambda_1]\|\boldsymbol{W}(0)\|, [1-\eta\lambda_1(1+\eta\lambda_1)^2 - \eta(2\eta\lambda_1 + \eta^2\lambda_1^2)]\},$$

for all $k \ge 0$. From the conditions of Theorem 6.2, clearly, it holds that $c > 0$.

This completes the proof.

At this point, the boundness of DDT system (6.39) has been proved. Next, we will prove that under some mild conditions, $\lim_{k\to+\infty} \boldsymbol{W}(k) = \sum_{i=1}^{m} z_i^* \boldsymbol{V}_i \in \boldsymbol{V}$ for PCA and $\lim_{k\to+\infty} \boldsymbol{W}(k) = \sum_{i=n-p}^{n} z_i^* \boldsymbol{V}_i \in \boldsymbol{V}_\tau$ for MCA.

In order to analyze the convergence of DDT (6.39), we need to prove some preliminary results.

From (6.40), for each $k \geq 0$, $\boldsymbol{W}(k)$ can be represented as

$$\begin{cases} \boldsymbol{W}(k) = \sum\limits_{i=1}^{m} z_i(k)\boldsymbol{V}_i + \sum\limits_{j=m+1}^{n} z_j(k)\boldsymbol{V}_j & \textit{for} \quad \text{PCA} \\ \boldsymbol{W}(k) = \sum\limits_{i=1}^{n-p} z_i(k)\boldsymbol{V}_i + \sum\limits_{j=n-p+1}^{n} z_j(k)\boldsymbol{V}_j & \textit{for} \quad \text{MCA.} \end{cases}$$

Clearly, the convergence of $\boldsymbol{W}(k)$ can be determined by the convergence of $z_i(k)$ $(i = 1,2,\ldots,n)$. The following Lemmas 6.2–6.4 provide the convergence of $z_i(k)$ $(i = 1,2,\ldots,n)$ for PCA, and Lemmas 6.5–6.7 provide the convergence of $z_i(k)$ $(i = 1,2,\ldots,n)$ for MCA.

In the following Lemmas 6.2–6.4, we will prove that all $z_i(k)$ $(i = 2,3,\ldots,n)$ will converge to zero under some mild conditions.

Lemma 6.2 Suppose that $\eta \leq 0.3$. If $\boldsymbol{W}(0) \notin \boldsymbol{V}_\sigma^\perp$ and $\|\boldsymbol{W}(0)\| \leq 1$, then for PCA algorithm of (6.39) there exist constants $\theta_1 > 0$ and $\Pi_1 \geq 0$ such that $\sum_{j=m+1}^{n} z_j^2(k) \leq \prod_1 \cdot \mathrm{e}^{-\theta_1 k}$ for all $k \geq 0$, where $\theta_1 = -\ln\beta > 0$ and $\beta = [1 - \eta(\sigma - \lambda_{m+1})/(1/c^2 + \eta(\sigma - \tau) + \eta(1/c^2 - 1))]^2$. Clearly, β is a constant and $0 < \beta < 1$.

Proof Since $\boldsymbol{W}(0) \notin \boldsymbol{V}_\sigma^\perp$, there must exist some $i(1 \leq i \leq m)$ such that $z_i(0) \neq 0$. Without loss of generality, assume $z_1(0) \neq 0$. For PCA, it follows from (6.41) that

$$\begin{aligned} z_i(k+1) = [1 + \eta(\sigma\|\mathbf{W}(k)\|^2 - \mathbf{W}^{\mathrm{T}}(k)\mathbf{R}\mathbf{W}(k)) \\ + \eta(1 - \|\mathbf{W}(k)\|^2)]z_i(k), \quad (1 \leq i \leq m) \end{aligned} \tag{6.48}$$

and

$$\begin{aligned} z_j(k+1) = [1 + \eta(\lambda_j\|\mathbf{W}(k)\|^2 - \mathbf{W}^{\mathrm{T}}(k)\mathbf{R}\mathbf{W}(k)) \\ + \eta(1 - \|\mathbf{W}(k)\|^2)]z_j(k), \quad m+1 \leq j \leq n \end{aligned} \tag{6.49}$$

for $k \geq 0$.

Using Theorem 6.9, it holds that $\|\boldsymbol{W}(k)\| > c$ for all $k \geq 0$. Then, from (6.48) and (6.49), for each $j(m+1 \leq j \leq n)$, we have

$$
\begin{aligned}
\left[\frac{z_j(k+1)}{z_1(k+1)}\right]^2 &= \left[\frac{1+\eta(\lambda_j\|\boldsymbol{W}(k)\|^2-(\boldsymbol{W}^{\mathrm{T}}(k)\boldsymbol{R}\boldsymbol{W}(k))+\eta(1-\|\boldsymbol{W}(k)\|^2)}{1+\eta(\sigma\|\boldsymbol{W}(k)\|^2-(\boldsymbol{W}^{\mathrm{T}}(k)\boldsymbol{R}\boldsymbol{W}(k))+\eta(1-\|\boldsymbol{W}(k)\|^2)}\right]^2\cdot\left[\frac{z_j(k)}{z_1(k)}\right]^2 \\
&= \left[1-\frac{\eta(\sigma-\lambda_j)\|\boldsymbol{W}(k)\|^2}{1+\eta(\sigma\|\boldsymbol{W}(k)\|^2-(\boldsymbol{W}^{\mathrm{T}}(k)\boldsymbol{R}\boldsymbol{W}(k))+\eta(1-\|\boldsymbol{W}(k)\|^2)}\right]^2\cdot\left[\frac{z_j(k)}{z_1(k)}\right]^2 \\
&= \left[1-\frac{\eta(\sigma-\lambda_j)}{1/\|\boldsymbol{W}(k)\|^2+\eta(\sigma-\tau)+\eta(1/\|\boldsymbol{W}(k)\|^2-1)}\right]^2\cdot\left[\frac{z_j(k)}{z_1(k)}\right]^2 \\
&\le \left[1-\frac{\eta(\sigma-\lambda_{m+1})}{1/c^2+\eta(\sigma-\tau)+\eta(1/c^2-1)}\right]^2\cdot\left[\frac{z_j(k)}{z_1(k)}\right]^2 \\
&= \beta\frac{z_j^2(k)}{z_1^2(k)} \le \beta^{k+1}\frac{z_j^2(0)}{z_1^2(0)} = \frac{z_j^2(0)}{z_1^2(0)}\mathrm{e}^{-\theta_1(k+1)},
\end{aligned}
\tag{6.50}
$$

for all $k \geq 0$, where $\theta_1 = -\ln\beta > 0$. Since $\|\boldsymbol{W}(k)\| < (1+\eta\lambda_1)$, $z_1(k)$ must be bounded, i.e., there exists a constant $d > 0$ such that $z_1^2(k) \leq d$ for all $k \geq 0$. Then,

$$
\sum_{j=m+1}^{n} z_j^2(k) = \sum_{j=m+1}^{n}\left[\frac{z_j(k)}{z_1(k)}\right]^2\cdot z_1^2(k) \le \prod_1 \mathrm{e}^{-\theta_1 k},
$$

for $k \geq 0$ where $\prod_1 = d\sum_{j=m+1}^{n}\left[\frac{z_j(0)}{z_1(0)}\right]^2 \geq 0$.

This completes the proof.

Based on the Lemma, we have Lemma 6.3.

Lemma 6.3 *Suppose that* $\eta\lambda_1 < 0.25$ *and* $\eta \leq 0.3$. *Then for PCA algorithm of* (6.39) *there exist constants* $\theta_2 > 0$ *and* $\prod_2 > 0$ *such that*

$$
\begin{aligned}
&\left|1-(1-\sigma)\|\boldsymbol{W}(k+1)\|^2-\boldsymbol{W}^{\mathrm{T}}(k+1)\boldsymbol{R}\boldsymbol{W}(k+1)\right| \le (k+1) \\
&\cdot\prod_2\cdot[\mathrm{e}^{-\theta_2(k+1)}+\max\{\mathrm{e}^{-\theta_2 k},\mathrm{e}^{-\theta_1 k}\}],
\end{aligned}
$$

for all $k \geq 0$.

Proof For PCA, it follows from (6.41) that

$$
\begin{aligned}
\|\boldsymbol{W}(k+1)\|^2 &= \sum_{i=1}^{n}[1+\eta(\lambda_i\|\boldsymbol{W}(k)\|^2-\boldsymbol{W}^{\mathrm{T}}(k)\boldsymbol{R}\boldsymbol{W}(k))+\eta(1-\|\boldsymbol{W}(k)\|^2)]^2 z_i^2(k) \\
&= \sum_{i=1}^{n}[1+\eta(\sigma\|\boldsymbol{W}(k)\|^2-\boldsymbol{W}^{\mathrm{T}}(k)\boldsymbol{R}\boldsymbol{W}(k))+\eta(1-\|\boldsymbol{W}(k)\|^2)]^2 z_i^2(k) \\
&\quad+\sum_{i=m+1}^{n}[1+\eta(\lambda_i\|\boldsymbol{W}(k)\|^2-\boldsymbol{W}^{\mathrm{T}}(k)\boldsymbol{R}\boldsymbol{W}(k))+\eta(1-\|\boldsymbol{W}(k)\|^2)]^2 z_i^2(k) \\
&\quad-\sum_{i=m+1}^{n}[1+\eta(\sigma\|\boldsymbol{W}(k)\|^2-\boldsymbol{W}^{\mathrm{T}}(k)\boldsymbol{R}\boldsymbol{W}(k))+\eta(1-\|\boldsymbol{W}(k)\|^2)]^2 z_i^2(k) \\
&= [1+\eta(\sigma\|\boldsymbol{W}(k)\|^2-\boldsymbol{W}^{\mathrm{T}}(k)\boldsymbol{R}\boldsymbol{W}(k))+\eta(1-\|\boldsymbol{W}(k)\|^2)]^2\|\boldsymbol{W}(k)\|^2+H(k),
\end{aligned}
\tag{6.51}
$$

for any $k \geq 0$,
where $H(k)=\sum_{i=m+1}^{n}\left[(2+\eta(\lambda_i+\sigma)\|\boldsymbol{W}(k)\|^2+2\eta(1-\|\boldsymbol{W}(k)\|^2-\boldsymbol{W}^{\mathrm{T}}(k)\boldsymbol{R}\boldsymbol{W}(k)))\cdot\eta(\lambda_i-\sigma)\|\boldsymbol{W}(k)\|^2\cdot z_i^2(k)\right]$.

Clearly,

$$
\begin{aligned}
\boldsymbol{W}^{\mathrm{T}}(k+1)\boldsymbol{R}\boldsymbol{W}(k+1) &= \sum_{i=1}^{n}\lambda_i[1+\eta(\lambda_i\|\boldsymbol{W}(k)\|^2-\boldsymbol{W}^{\mathrm{T}}(k)\boldsymbol{R}\boldsymbol{W}(k))+\eta(1-\|\boldsymbol{W}(k)\|^2)]^2 z_i^2(k) \\
&= \sum_{i=1}^{n}\lambda_i[1+\eta(\sigma\|\boldsymbol{W}(k)\|^2-\boldsymbol{W}^{\mathrm{T}}(k)\boldsymbol{R}\boldsymbol{W}(k))+\eta(1-\|\boldsymbol{W}(k)\|^2)]^2 z_i^2(k) \\
&\quad+\sum_{i=m+1}^{n}\lambda_i[1+\eta(\lambda_i\|\boldsymbol{W}(k)\|^2-\boldsymbol{W}^{\mathrm{T}}(k)\boldsymbol{R}\boldsymbol{W}(k))+\eta(1-\|\boldsymbol{W}(k)\|^2)]^2 z_i^2(k) \\
&\quad-\sum_{i=m+1}^{n}\lambda_i[1+\eta(\sigma\|\boldsymbol{W}(k)\|^2-\boldsymbol{W}^{\mathrm{T}}(k)\boldsymbol{R}\boldsymbol{W}(k))+\eta(1-\|\boldsymbol{W}(k)\|^2)]^2 z_i^2(k) \\
&= [1+\eta(\sigma\|\boldsymbol{W}(k)\|^2-\boldsymbol{W}^{\mathrm{T}}(k)\boldsymbol{R}\boldsymbol{W}(k))+\eta(1-\|\boldsymbol{W}(k)\|^2)]^2\boldsymbol{W}^{\mathrm{T}}(k)\boldsymbol{R}\boldsymbol{W}(k)+H'(k),
\end{aligned}
\tag{6.52}
$$

for any $k \geq 0$,
where $H'(k)=\sum_{i=m+1}^{n}\Big[(2+\eta(\lambda_i+\sigma)\|\boldsymbol{W}(k)\|^2+2\eta(1-\|\boldsymbol{W}(k)\|^2-\boldsymbol{W}^{\mathrm{T}}(k)\boldsymbol{R}\boldsymbol{W}(k)))\cdot\eta(\lambda_i-\sigma)\|\boldsymbol{W}(k)\|^2\cdot$ $\lambda_i z_i^2(k)\Big]$. Then, it follows from (6.51) and (6.52) that

$$
\begin{aligned}
&1-(1-\sigma)\|\boldsymbol{W}(k+1)\|^2-\boldsymbol{W}^{\mathrm{T}}(k+1)\boldsymbol{R}\boldsymbol{W}(k+1) \\
&\quad=(1-(1-\sigma)\|\boldsymbol{W}(k)\|^2-\boldsymbol{W}^{\mathrm{T}}(k)\boldsymbol{R}\boldsymbol{W}(k))\{1-[2\eta+\eta^2(1-(1-\sigma)\|\boldsymbol{W}(k)\|^2 \\
&\qquad-\boldsymbol{W}^{\mathrm{T}}(k)\boldsymbol{R}\boldsymbol{W}(k))]((1-\sigma)\|\boldsymbol{W}(k)\|^2+\boldsymbol{W}^{\mathrm{T}}(k)\boldsymbol{R}\boldsymbol{W}(k))\}-(1-\sigma)H(k)-H'(k)
\end{aligned}
$$

for all $k \geq 0$.

Denote

$$V(k) = \left|1 - (1-\sigma)\|\boldsymbol{W}(k)\|^2 - \boldsymbol{W}^{\mathrm{T}}(k)\boldsymbol{R}\boldsymbol{W}(k)\right|,$$

for any $k \geq 0$. Clearly,

$$\begin{aligned} V(k+1) \leq V(k)\big|\{1 - [2\eta + \eta^2(1-(1-\sigma)\|\mathbf{W}(k)\|^2 - \mathbf{W}^{\mathrm{T}}(k)\mathbf{R}\mathbf{W}(k))]((1-\sigma)\|\mathbf{W}(k)\|^2 \\ + \mathbf{W}^{\mathrm{T}}(k)\mathbf{R}\mathbf{W}(k))\}\big| + |(1-\sigma)H(k) + H'(k)|. \end{aligned}$$

Denote

$$\delta = \left|\{1 - [2\eta + \eta^2(1-(1-\sigma)\|\boldsymbol{W}(k)\|^2 - \boldsymbol{W}^{\mathrm{T}}(k)\boldsymbol{R}\boldsymbol{W}(k))]((1-\sigma)\|\boldsymbol{W}(k)\|^2 + \boldsymbol{W}^{\mathrm{T}}(k)\boldsymbol{R}\boldsymbol{W}(k))\}\right|.$$

From Theorem 6.8, $\eta\lambda_1 < 0.25$, $\eta \leq 0.3$ and (6.43), it holds that

$$\begin{aligned} &[2\eta + \eta^2(1-(1-\sigma)\|\boldsymbol{W}(k)\|^2 - \boldsymbol{W}^{\mathrm{T}}(k)\boldsymbol{R}\boldsymbol{W}(k))]((1-\sigma)\|\boldsymbol{W}(k)\|^2 + \boldsymbol{W}^{\mathrm{T}}(k)\boldsymbol{R}\boldsymbol{W}(k)) \\ &\quad < [2\eta + \eta^2(1-(1-\sigma)\|\boldsymbol{W}(k)\|^2 - \boldsymbol{W}^{\mathrm{T}}(k)\boldsymbol{R}\boldsymbol{W}(k))] \\ &\quad < [2\eta + \eta^2(1-(1-\sigma)\|\boldsymbol{W}(k)\|^2 - \lambda_n\|\boldsymbol{W}(k)\|^2)] \\ &\quad < 2\eta + \eta^2(1+\sigma\|\boldsymbol{W}(k)\|^2) \leq 2\eta + \eta(\eta + \eta\lambda_1(1+\eta\lambda_1)^2) \leq 0.8071, \end{aligned}$$

Clearly, $0 < \delta < 1$. Then,

$$V(k+1) \leq \delta V(k) + |(1-\sigma)H(k) + H'(k)|, k \geq 0.$$

Since

$$\begin{aligned} |(1-\sigma)H(k) + H'(k)| &\leq (2 + 2\eta\sigma\|\boldsymbol{W}(k)\|^2 + 2\eta)(\eta\sigma\|\boldsymbol{W}(k)\|^2)\sum_{i=m+1}^{n} z_i^2(k)[(1-\sigma) + \lambda_i] \\ &\leq (2 + 2\eta\sigma\|\boldsymbol{W}(k)\|^2 + 2\eta)(\eta\sigma\|\boldsymbol{W}(k)\|^2)\sum_{i=m+1}^{n} z_i^2(k) \leq \phi\prod_1 \mathrm{e}^{-\theta_1 k}, \end{aligned}$$

for any $k \geq 0$, where $\phi = (2 + 2\eta\sigma(1+\eta\lambda_1)^2 + 2\eta)\cdot\eta\sigma(1+\eta\lambda_1)^2$, then

$$\begin{aligned} V(k+1) &\leq \delta^{k+1}V(0) + \phi\prod\nolimits_1\sum_{r=0}^{k}(\delta\mathrm{e}^{\theta_1})^r\mathrm{e}^{-\theta_1 k} \\ &\leq \delta^{k+1}V(0) + (k+1)\phi\prod\nolimits_1\max\{\delta^k, \mathrm{e}^{-\theta_1 k}\} \\ &\leq (k+1)\prod\nolimits_2\left[\mathrm{e}^{-\theta_2(k+1)} + \max\{\mathrm{e}^{-\theta_2 k}, \mathrm{e}^{-\theta_1 k}\}\right], \end{aligned}$$

where $\theta_2 = -\ln\delta > 0$ and $\prod_2 = \max\{\left|1-(1-\sigma)\|\boldsymbol{W}(0)\|^2 - \boldsymbol{W}^{\mathrm{T}}(0)\boldsymbol{R}\boldsymbol{W}(0)\right|, \phi\prod_1\} > 0$.

This completes the proof.

Based on Lemmas 6.2 and 6.3, we have Lemma 6.4.

Lemma 6.4 *For PCA algorithm of* (6.39), *suppose there exist constants* $\theta > 0$ *and* $\prod > 0$ *such that*

$$\eta\left|(1-(1-\sigma)\|\boldsymbol{W}(k+1)\|^2 - \boldsymbol{W}^{\mathrm{T}}(k+1)\boldsymbol{R}\boldsymbol{W}(k+1))z_i(k+1)\right| \leq (k+1)\prod \mathrm{e}^{-\theta(k+1)}, \quad (i=1,\ldots,m)$$

for all $k \geq 0$. *Then,* $\lim_{k\to\infty} z_i(k) = z_i^*, (i=1,\ldots,m)$, *where* $z_i^*, (i=1,\ldots,m)$ *are constants.*

Proof Given any $\varepsilon > 0$, there exists a $K \geq 1$ such that

$$\frac{\prod K \mathrm{e}^{-\theta K}}{(1-\mathrm{e}^{-\theta})^2} \leq \varepsilon.$$

For any $k_1 > k_2 > K$, it follows that

$$\begin{aligned}
|z_i(k_1) - z_i(k_2)| &= \left|\sum_{r=k_2}^{k_1-1}[z_i(r+1) - z_i(r)]\right| \leq \eta \sum_{r=k_2}^{k_1-1}\left|(\sigma\|\boldsymbol{W}(r)\|^2 - \boldsymbol{W}(r)^{\mathrm{T}}\boldsymbol{R}\boldsymbol{W}(r) + 1 - \|\boldsymbol{W}(r)\|^2)z_i(r)\right| \\
&= \eta\sum_{r=k_2}^{k_1-1}\left|(1-(1-\sigma)\|\boldsymbol{W}(r)\|^2 - \boldsymbol{W}(r)^{\mathrm{T}}\boldsymbol{R}\boldsymbol{W}(r))z_i(r)\right| \\
&\leq \prod\sum_{r=k_2}^{k_1-1} r\mathrm{e}^{-\theta r} \leq \prod\sum_{r=K}^{+\infty} r\mathrm{e}^{-\theta r} \leq \prod K\mathrm{e}^{-\theta K}\sum_{r=0}^{+\infty} r(\mathrm{e}^{-\theta})^{r-1} \\
&\leq \frac{\prod K\mathrm{e}^{-\theta K}}{(1-\mathrm{e}^{-\theta})^2} \leq \varepsilon, (i=1,\ldots,m).
\end{aligned}$$

This means that the sequence $\{\ z_i(k)\ \}$ is a Cauchy sequence. By the Cauchy convergence principle, there must exist a constant $z^*(i=1,\ldots,m)$ such that $\lim_{k\to+\infty} z_i(k) = z^*, (i=1,\ldots,m)$.

This completes the proof.

Using the above theorems and lemmas, the convergence of DDT system (6.39) for PCA can be proved as in Theorem 6.10 next.

Theorem 6.10 *Suppose that* $\eta\lambda_1 < 0.25$ *and* $\eta \leq 0.3$. *If* $\boldsymbol{W}(0) \notin \boldsymbol{V}_\sigma^\perp$ *and* $\|\boldsymbol{W}(0)\| \leq 1$, *then the weight vector of* (6.39) *for PCA will converge to a unitary eigenvector associated with the largest eigenvalue of the correlation matrix.*

Proof By *Lemma* 6.2, there exist constants $\theta_1 > 0$ and $\Pi_1 \geq 0$ such that $\sum_{j=m+1}^{n} z_j^2(k) \leq \prod_1 \cdot \mathrm{e}^{-\theta_1 k}$, for all $k \geq 0$. By *Lemma* 6.3, there exist constants $\theta_2 > 0$ and $\prod_2 > 0$ such that

$$\left|(1-(1-\sigma)\|\boldsymbol{W}(k+1)\|^2-\boldsymbol{W}^{\mathrm{T}}(k+1)\boldsymbol{R}\boldsymbol{W}(k+1))\right| \le (k+1)\cdot \prod\nolimits_2 \cdot\left[\mathrm{e}^{-\theta_2(k+1)}+\max\{\mathrm{e}^{-\theta_2 k},\mathrm{e}^{-\theta_1 k}\}\right],$$

for all $k \ge 0$. Obviously, there exist constants $\theta > 0$ and $\prod > 0$ such that

$$\eta\left|(1-(1-\sigma)\|\boldsymbol{W}(k+1)\|^2-\boldsymbol{W}^{\mathrm{T}}(k+1)\boldsymbol{R}\boldsymbol{W}(k+1))z_i(k+1)\right| \le (k+1)\prod \mathrm{e}^{-\theta(k+1)},\quad (i=1,\ldots,m)$$

for $k \ge 0$. Using *Lemmas* 6.4 and 6.2, it follows that

$$\begin{cases}\lim\limits_{k\to+\infty} z_i(k)=z_i^*,\,(i=1,\ldots,m)\\ \lim\limits_{k\to+\infty} z_i(k)=0,\,(i=m+1,\ldots,n).\end{cases}$$

Then, $\lim\limits_{k\to+\infty} \boldsymbol{W}(k)=\sum_{i=1}^m z_i^*\boldsymbol{V}_i \in \boldsymbol{V}_\sigma$. It can be easily seen that $\lim\limits_{k\to+\infty}\|\boldsymbol{W}(k)\|^2=\sum_{i=1}^m (z_i^*)^2=1$.

This completes the proof.

After proving the convergence of DDT system (6.39) for PCA, we can also prove the convergence of DDT system (6.39) for MCA using similar method. In order to prove the convergence of the weight vector of (6.39) for MCA, we can use the following Lemmas 6.5–6.7 and Theorem 6.11, the proofs of which are similar to those of Lemmas 6.2–6.4 and Theorem 6.10. Here, only these lemmas and theorem will be given and their proofs are omitted.

Lemma 6.5 *Suppose that* $\eta \le 0.3$ *.If* $\boldsymbol{W}(0) \notin \boldsymbol{V}_\tau^\perp$ *and* $\|\boldsymbol{W}(0)\| \le 1$*, then for MCA algorithm of* (6.39) *there exist constants* $\theta_1' > 0$ *and* $\Pi_1' \ge 0$ *such that* $\sum_{j=1}^{n-p} z_j^2(k) \le \prod_1' \cdot \mathrm{e}^{-\theta_1 k}$*, for all* $k \ge 0$*, where* $\theta_1' = -\ln\beta' > 0$*, and* $\beta' = [1-\eta(\lambda_{n-p-1}-\tau)/(1/c^2-\eta(\tau-\sigma)+\eta(1/c^2-1))]^2$*. Clearly,* β' *is a constant and* $0 < \beta' < 1$.

Proof For MCA, it follows from (6.41) that

$$\begin{aligned}\|\boldsymbol{W}(k+1)\|^2 &= \sum_{i=1}^n [1-\eta(\lambda_i\|\boldsymbol{W}(k)\|^2-\boldsymbol{W}^{\mathrm{T}}(k)\boldsymbol{R}\boldsymbol{W}(k))+\eta(1-\|\boldsymbol{W}(k)\|^2)]^2 z_i^2(k)\\ &= [1-\eta(\tau\|\boldsymbol{W}(k)\|^2-\boldsymbol{W}^{\mathrm{T}}(k)\boldsymbol{R}\boldsymbol{W}(k))+\eta(1-\|\boldsymbol{W}(k)\|^2)]^2\|\boldsymbol{W}(k)\|^2+\bar{H}(k),\end{aligned} \tag{6.53}$$

for any $k \ge 0$ where

$$\begin{aligned}\bar{H}(k) = \sum_{i=1}^{n-p}\Big[&(2-\eta(\lambda_i+\tau)\|\boldsymbol{W}(k)\|^2+2\eta(1\\ &-\|\boldsymbol{W}(k)\|^2+\boldsymbol{W}^{\mathrm{T}}(k)\boldsymbol{R}\boldsymbol{W}(k)))\cdot\eta(\tau-\lambda_i)\|\boldsymbol{W}(k)\|^2\cdot z_i^2(k)\Big].\end{aligned}$$

and,

$$\begin{aligned} \boldsymbol{W}^{\mathrm{T}}(k+1)\boldsymbol{R}\boldsymbol{W}(k+1) &= \sum_{i=1}^{n} \lambda_i[1-\eta(\lambda_i\|\boldsymbol{W}(k)\|^2 - \boldsymbol{W}^{\mathrm{T}}(k)\boldsymbol{R}\boldsymbol{W}(k)) + \eta(1-\|\boldsymbol{W}(k)\|^2)]^2 z_i^2(k) \\ &= [1-\eta(\tau\|\boldsymbol{W}(k)\|^2 - \boldsymbol{W}^{\mathrm{T}}(k)\boldsymbol{R}\boldsymbol{W}(k)) + \eta(1-\|\boldsymbol{W}(k)\|^2)]^2 \boldsymbol{W}^{\mathrm{T}}(k)\boldsymbol{R}\boldsymbol{W}(k) + \overline{H}'(k), \end{aligned} \tag{6.54}$$

for any $k \geq 0$ where

$$\bar{H}'(k) = \sum_{i=1}^{n-p} \left[(2-\eta(\lambda_i+\tau)\|\boldsymbol{W}(k)\|^2 + 2\eta(1-\|\boldsymbol{W}(k)\|^2 + \boldsymbol{W}^{\mathrm{T}}(k)\boldsymbol{R}\boldsymbol{W}(k)))\cdot\eta(\tau-\lambda_i)\|\boldsymbol{W}(k)\|^2\cdot\lambda_i z_i^2(k)\right].$$

Then, it follows from (6.53) and (6.54) that

$$\begin{aligned} &1-(1+\tau)\|\boldsymbol{W}(k+1)\|^2 + \boldsymbol{W}^{\mathrm{T}}(k+1)\boldsymbol{R}\boldsymbol{W}(k+1) \\ &\quad = (1-(1+\tau)\|\boldsymbol{W}(k)\|^2 + \boldsymbol{W}^{\mathrm{T}}(k)\boldsymbol{R}\boldsymbol{W}(k))\{1+[-2\eta+\eta^2(1-(1+\tau)\|\boldsymbol{W}(k)\|^2 \\ &\qquad + \boldsymbol{W}^{\mathrm{T}}(k)\boldsymbol{R}\boldsymbol{W}(k))][\boldsymbol{W}^{\mathrm{T}}(k)\boldsymbol{R}\boldsymbol{W}(k) - (1+\tau)\|\boldsymbol{W}(k)\|^2]\} - (1+\tau)\bar{H}(k) + \bar{H}'(k) \end{aligned}$$

for all $k \geq 0$.

Denote

$$\bar{V}(k) = \left|1-(1+\tau)\|\boldsymbol{W}(k)\|^2 + \boldsymbol{W}^{\mathrm{T}}(k)\boldsymbol{R}\boldsymbol{W}(k)\right|,$$

for any $k \geq 0$. Clearly,

$$\begin{aligned} \bar{V}(k+1) \leq \bar{V}(k)\big|\{1-[2\eta-\eta^2(1-(1+\tau)\|\mathbf{W}(k)\|^2 + \mathbf{W}^T(k)\mathbf{R}\mathbf{W}(k))][\mathbf{W}^T(k)\mathbf{R}\mathbf{W}(k) \\ -(1+\tau)\|\mathbf{W}(k)\|^2]\}\big| + |H'(k) - (1+\tau)H(k)|. \end{aligned}$$

Denote

$$\delta' = \left|\{1-[2\eta-\eta^2(1-(1+\tau)\|\boldsymbol{W}(k)\|^2 + \boldsymbol{W}^{\mathrm{T}}(k)\boldsymbol{R}\boldsymbol{W}(k))][\boldsymbol{W}^{\mathrm{T}}(k)\boldsymbol{R}\boldsymbol{W}(k) - (1+\tau)\|\boldsymbol{W}(k)\|^2]\}\right|.$$

From Theorem 6.8, $\eta\lambda_1 < 0.25$, $\eta \leq 0.3$, and (6.43), it holds that

$$\begin{aligned} &[2\eta-\eta^2(1-(1+\tau)\|\boldsymbol{W}(k)\|^2 + \boldsymbol{W}^{\mathrm{T}}(k)\boldsymbol{R}\boldsymbol{W}(k))][\sigma\|\boldsymbol{W}(k)\|^2 - (1+\tau)\|\boldsymbol{W}(k)\|^2] \\ &\quad = [2\eta-\eta^2(1-(1+\tau)\|\boldsymbol{W}(k)\|^2 + \boldsymbol{W}^{\mathrm{T}}(k)\boldsymbol{R}\boldsymbol{W}(k))][(\sigma-(1+\tau))\|\boldsymbol{W}(k)\|^2] \\ &\quad < [2\eta-\eta^2(1-\|\boldsymbol{W}(k)\|^2)][\sigma\|\boldsymbol{W}(k)\|^2] < (\eta\lambda_1)[2-\eta(1-\|\boldsymbol{W}(k)\|^2)][\|\boldsymbol{W}(k)\|^2] \\ &\quad \leq 0.25 * [2-0.3+0.3*(1+0.25)^2](1+0.25)^2 \\ &\quad = 0.8471. \end{aligned}$$

Clearly, $0 < \delta' < 1$. Then,

$$\bar{V}(k+1) \leq \delta \bar{V}(k) + |H'(k) - (1+\tau)H(k)|, k \geq 0.$$

Since

$$\begin{aligned}
&|H'(k) - (1+\tau)H(k)| \\
&\leq \left| (2 - 2\eta\tau\|\boldsymbol{W}(k)\|^2 + 2\eta(1+\sigma\|\boldsymbol{W}(k)\|^2))(\eta\sigma\|\boldsymbol{W}(k)\|^2)\sum_{i=1}^{n-p} z_i^2(k)[\lambda_i - (1+\tau)] \right| \\
&\leq (2 + 2\eta(1+\sigma\|\boldsymbol{W}(k)\|^2)) \cdot (\eta\sigma\|\boldsymbol{W}(k)\|^2) \cdot |\sigma - (1+\tau)| \cdot \sum_{i=1}^{n-p} z_i^2(k) \\
&\leq \phi' \prod\nolimits_1' \mathrm{e}^{-\theta_1' k},
\end{aligned}$$

for any $k \geq 0$, where $\phi' = (2 + 2\eta(1+\sigma(1+\eta\lambda_1)^2)) \cdot (\eta\sigma(1+\eta\lambda_1)^2) \cdot |\sigma - (1+\tau)|$, we have

$$\begin{aligned}
\bar{V}(k+1) &\leq \delta'^{k+1}\bar{V}(0) + \phi' \prod_1' \sum_{r=0}^{k} (\delta' \mathrm{e}^{\theta_1'})^r \mathrm{e}^{-\theta_1' k} \\
&\leq \delta'^{k+1}\bar{V}(0) + (k+1)\phi' \prod_1' \max\{\delta'^k, \mathrm{e}^{-\theta_1' k}\} \\
&\leq (k+1) \prod_2' \left[\mathrm{e}^{-\theta_2'(k+1)} + \max\{\mathrm{e}^{-\theta_2' k}, \mathrm{e}^{-\theta_1' k}\} \right],
\end{aligned}$$

where $\theta_2' = -\ln\delta' > 0$ and $\prod_2' = \max\left\{ |1 - (1+\tau)\|\boldsymbol{W}(0)\|^2 + \boldsymbol{W}^{\mathrm{T}}(0)\boldsymbol{R}\boldsymbol{W}(0)|, \phi' \prod_1' \right\} > 0$.

This completes the proof.

Lemma 6.6 *Suppose that* $\eta\lambda_1 < 0.25$ *and* $\eta \leq 0.3$. *Then for MCA algorithm of* (6.39) *there exist constants* $\theta_2' > 0$ *and* $\prod_2' > 0$ *such that*

$$\begin{aligned}
&|1 - (1+\tau)\|\boldsymbol{W}(k+1)\|^2 + \boldsymbol{W}^{\mathrm{T}}(k+1)\boldsymbol{R}\boldsymbol{W}(k+1)| \leq (k+1) \cdot \prod\nolimits_2' \\
&\quad \cdot [\mathrm{e}^{-\theta_2'(k+1)} + \max\{\mathrm{e}^{-\theta_2' k}, \mathrm{e}^{-\theta_1' k}\}],
\end{aligned}$$

for all $k \geq 0$.

For the proof of this lemma, refer to Lemma 6.3.

Lemma 6.7 *For MCA algorithm of* (6.39), *suppose there exists constants* $\theta' > 0$ *and* $\prod' > 0$ *such that*

$$\eta\left|(1-(1+\tau)\|\boldsymbol{W}(k+1)\|^2+\boldsymbol{W}^{\mathrm{T}}(k+1)\boldsymbol{R}\boldsymbol{W}(k+1))z_i(k+1)\right| \leq (k+1)\prod{}' \mathrm{e}^{-\theta'(k+1)},$$
$$(i=n-p+1,\ldots,n)$$

for $k \geq 0$. *Then,* $\lim_{k\to\infty} z_i(k) = z_i^*, (i=n-p+1,\ldots,n)$, *where* $z_i^*, (i=n-p+1,\ldots,n)$ *are constants.*

For the proof of this lemma, refer to Lemma 6.4.

Theorem 6.11 *Suppose that* $\eta\lambda_1 < 0.25$ *and* $\eta \leq 0.3$. *If* $\boldsymbol{W}(0) \notin \boldsymbol{V}_\tau^\perp$ *and* $\|\boldsymbol{W}(0)\| \leq 1$, *then the weight vector of* (6.39) *for MCA will converge to a unitary eigenvector associated with the smallest eigenvalue of the correlation matrix.*

From Lemmas 6.5–6.7, clearly Theorem 6.11 holds.

At this point, we have completed the proof of the convergence of DDT system (6.39). From Theorems 6.8 and 6.9, we can see that the weight norm of PCA algorithm and MCA algorithm of DDT system (6.39) have the same bounds, and from Theorems 6.8–6.11, it is obvious that the sufficient conditions to guarantee the convergence of the two algorithms are also same, which is in favored in practical applications.

6.4.4 Computer Simulations

In this section, we provide simulation results to illustrate the performance of Chen's algorithm. This experiment mainly shows the convergence of Algorithm (6.39) under the condition of Theorems 6.10 and 6.11.

In this simulation, we randomly generate a 12×12 correlation matrix and its eigenvalues are $\lambda_1 = 0.2733$, $\lambda_2 = 0.2116$, $\lambda_3 = 0.1543$, …and $\lambda_{12} = 0.0001$. The initial weight vector is Gaussian and randomly generated with zero-mean and unitary standard deviation, and its norm is less than 1. In the following experiments, the learning rate for PCA is $\eta = 0.05$ and the learning rate for MCA is $\eta = 0.20$, which satisfies the condition of $\eta\lambda_1 \leq 0.25$ and $\eta \leq 0.3$. Figure 6.6 shows that the convergence of the component $z_i(k)$ of $\boldsymbol{W}(k)$ in (6.39) for PCA where $z_i(k) = \boldsymbol{W}^{\mathrm{T}}(k)\boldsymbol{V}_i$ is the coordinate of $\boldsymbol{W}(k)$ in the direction of the eigenvector $\boldsymbol{V}_i (i = 1,2,3,4,\ldots,12)$. In the simulation result, $z_i(k)(i = 2,3,4,\ldots,12)$ converges to zero and $z_1(k)$ converges to a constant 1, as $k \to \infty$, which is consistent with the convergence results in Theorem 6.10. Figure 6.7 shows the convergence of the component $z_i(k)$ of $\boldsymbol{W}(k)$ in (6.39) for MCA. In the simulation result, $z_i(k)(i = 1,2,3,\ldots,11)$ converges to zero and $z_{12}(k)$ converges to a constant 1, as $k \to \infty$, which is consistent with the convergence results in Theorem 6.11.

From the simulation results shown in Figs. 6.6 and 6.7, we can see that on conditions of $\eta\lambda_1 \leq 0.25$, $\eta \leq 0.3$, and $\|\boldsymbol{W}(0)\| \leq 1$, Algorithm (6.39) for PCA converge to the direction of the principal component of the correlation matrix. And if we simply switch the sign in the same learning rule, Algorithm (6.39) for MCA also converge to the direction of minor component of the correlation matrix. Besides, further simulations with high dimensions, e.g., 16, 20, and 30, also show

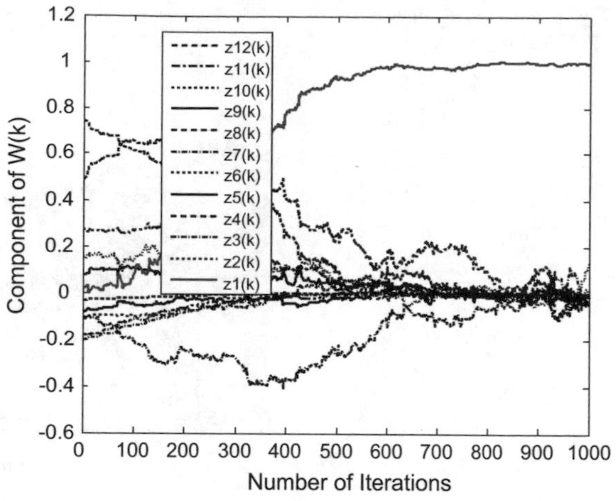

Fig. 6.6 Convergence of $\boldsymbol{W}(k)$ for PCA

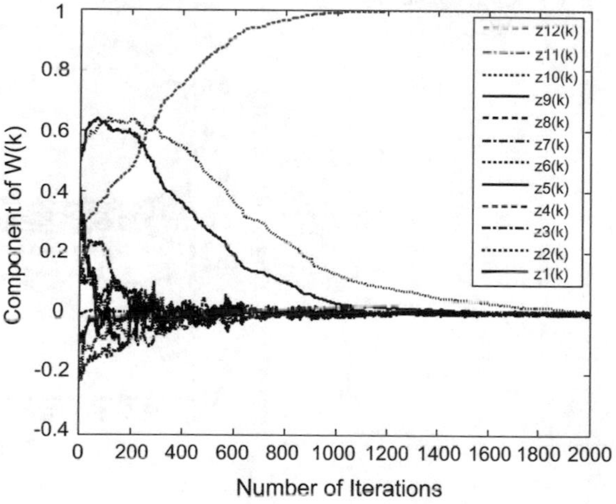

Fig. 6.7 Convergence of $\boldsymbol{W}(k)$ for MCA

that Algorithm (6.39) has satisfactory convergence under the conditions of Theorems 6.10 and 6.11. Figures 6.8 and 6.9 show the simulation results of Chen's PCA and MCA algorithm with dimension 20, respectively, where the learning rate for PCA is $\eta = 0.05$ and the learning rate for MCA is $\eta = 0.20$, which satisfy the condition of $\eta\lambda_1 \leq 0.25$ and $\eta \leq 0.3$.

In this section, dynamics of a unified self-stability learning algorithm for principal and minor components extraction are analyzed by the DDT method. The learning rate is assumed to be constant and thus not required to approach zero as required by the DCT method. Some sufficient conditions to guarantee the convergence are derived.

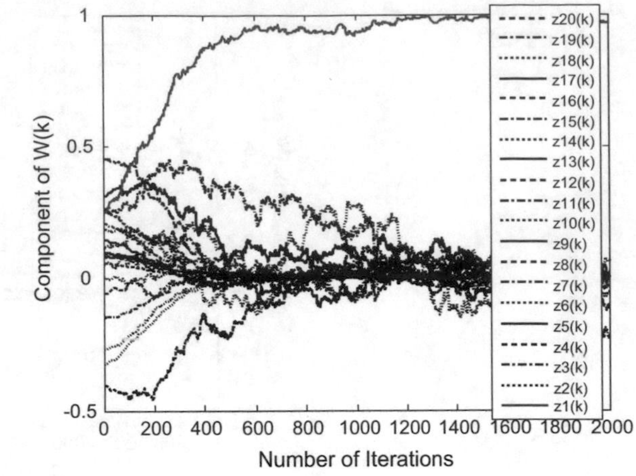

Fig. 6.8 Convergence of $\boldsymbol{W}(k)$ for PCA

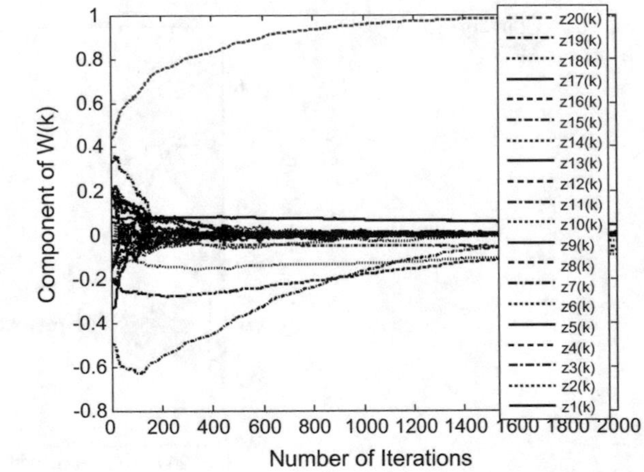

Fig. 6.9 Convergence of $\boldsymbol{W}(k)$ for MCA

6.5 Summary

In this chapter, we have analyzed the DDT systems of neural network principal/minor component analysis algorithms in details. First, we have reviewed several convergence or stability performance analysis methods for neural network-based PCA/MCA algorithms. Then, a DDT system of a novel MCA algorithm proposed by us has been analyzed. Finally, we have removed the assumption that the smallest eigenvalue of the correlation matrix of the input data is single, and a DDT system of a unified PCA and MCA algorithm has been analyzed.

References

1. Zhang, Q., & Leung, Y. W. (1995). Energy function for one-unit Oja algorithm. *IEEE Transactions on Neural Networks, 6*(9), 1291–1293.
2. Zhang, Q., & Bao, Z. (1995). Dynamical system for computing the eigenvectors with the largest eigenvalue of a positive definite matrix. *IEEE Transactions on Neural Networks, 6*(5), 790–791.
3. Yan, W. Y., Helmke, U., & Moore, J. B. (1994). Global analysis of Oja's flow for neural networks. *IEEE Transactions on Neural Networks, 5*(9), 674–683.
4. Plumbley, M. D. (1995). Lyapunov functions for convergence of principal component algorithms. *Neural Networks, 8*(1), 11–23.
5. Kushner, H. J., & Clark, D. S. (1976). *Stochastic approximation methods for constrained and unconstrained systems*. New York: Springer-Verlag.
6. Ljung, L. (1977). Analysis of recursive stochastic algorithms. *IEEE Transactions on Automatic Control, 22*(4), 551–575.
7. Cirrincione, G., Cirrincione, M., Herault, J., & Van Huffel, S. (2002). The MCA EXIN neuron for the minor component analysis. *IEEE Transactions on Neural Networks, 13*(1), 160–187.
8. Zufiria, P. J. (2002). On the discrete-time dynamics of the basic Hebbian neural network node. *IEEE Transactions on Neural Networks, 13*(6), 1342–1352.
9. Zhang, Q. (2003). On the discrete-time dynamics of a PCA learning algorithm. *Neurocomputing, 55*(10), 761–769.
10. PengD, Z., Zhang, Y., & Xiang, Y. (2008). On the discrete time dynamics of a self-stabilizing MCA learning algorithm. *Mathematical and Computer Modeling, 47*(9–10), 903–916.
11. Peng, D. Z., Zhang, Y., Lv, J. C., & Xiang, Y. (2008). A stable MCA learning algorithm. *Computers and Mathematics with Applications, 56*(4), 847–860.
12. Ye, M., Fan, X., & Li, X. (2006). A class of self-stabilizing MCA learning algorithms. *IEEE Transactions on Neural Networks, 17*(6), 1634–1638.
13. Zhang, Y., Ye, M., Lv, J. C., & Tan, K. K. (2005). Convergence analysis of a deterministic discrete time system of Oja's PCA learning algorithm. *IEEE Transactions on Neural Networks, 16*(6), 1318–1328.
14. Luo, F. L., & Unbehauen, R. (1997). A minor subspace analysis algorithm. *IEEE Transactions on Neural Networks, 8*(5), 1149–1155.
15. Luo F. L., & Unbehauen R. (1997). A generalized learning algorithm of minor component. In *Proceeding of International Conference on Acoustic, Speech, Signal Processing*, *4*(4), 3229–3232
16. Feng, D. Z., Bao, Z., & Jiao, L. C. (1998). Total least mean squares algorithm. *IEEE Transactions on Signal Processing, 46*(6), 2122–2130.
17. Feng, D. Z., Zheng, W. X., & Jia, Y. (2005). Neural network Learning algorithms for tracking minor subspace in high-dimensional data stream. *IEEE Transactions on Neural Network, 16* (3), 513–521.
18. Kong, X. Y., Hu, C. H., & Han, C. Z. (2010). On the discrete-time dynamics of a class of self-stabilizing MCA extraction algorithms. *IEEE Transactions on Neural Networks, 21*(1), 175–181.
19. Peng, D. Z., Zhang, Y., & Xiang, Y. (2009). A unified learning algorithm to extract principal and minor components. *Digital Signal Processing, 19*(4), 640–649.
20. Kong, X. Y., An, Q. S., Ma, H. G., Han, C. Z., & Zhang, Q. (2012). Convergence analysis of deterministic discrete time system of a unified self-stabilizing algorithm for PCA and MCA. *Neural Networks, 36*(8), 64–72.
21. Oja, E. (1992). Principal component, minor component and linear neural networks. *Neural Networks, 5*(6), 927–935.
22. Chen, T., & Amari, S. (2001). Unified stabilization approach to principal and minor components extraction algorithms. *Neural Networks, 14*(10), 1377–1387.

23. Möller, R. (2004). A self-stabilizing learning rule for minor component analysis. *International Journal of Neural Systems, 14*(1), 1–8.
24. Hasan M. A. (2007). Self-normalizing dual systems for minor and principal component extraction. In *Proceeding of International Conference on Acoustic, Speech, and Signal Processing,4*, 885–888.
25. Reddy, V. U., Egardt, B., & Kailath, T. (1982). Least squares type algorithm for adaptive implementation of Pisarenko's harmonic retrieval method. *IEEE Transactions on Acoustic, Speech, and Signal Processing, 30*(3), 399–405.
26. Xu, L., Oja, E., & Suen, C. (1992). Modified Hebbian learning for curve and surface fitting. *Neural Networks, 5*(3), 441–457.
27. Ouyang, S., Bao, Z., Liao, G. S., & Ching, P. C. (2001). Adaptive minor component extraction with modular structure. *IEEE Transactions on Signal Processing, 49*(9), 2127–2137.
28. Zhang, Q., & Leung, Y. W. (2000). A class of learning algorithms for principal component analysis and minor component analysis. *IEEE Transactions on Neural Networks, 11*(2), 529–533.
29. Douglas, S. C., Kung, S. Y., & Amari, S. (1998). A self-stabilized minor subspace rule. *IEEE Signal Processing Letter, 5*(12), 328–330.
30. LaSalle, J. P. (1976). *The stability of dynamical systems*. Philadelphia, PA: SIAM.
31. Magnus, J. R., & Neudecker, H. (1991). *Matrix differential calculus with applications in statistics and econometrics* (2nd ed.). New York: Wiley.
32. Kong, X. Y., Han, C. Z., & Wei, R. X. (2006). Modified gradient algorithm for total least squares filtering. *Neurocomputing, 70*(1), 568–576.

Chapter 7
Generalized Principal Component Analysis

7.1 Introduction

Recently, as a powerful feature extraction technique, generalized eigen decomposition (GED) has been attracting great attention and been widely used in many fields, e.g., spectral estimation [1], blind source separation [2], digital mobile communications [3], and antenna array processing [4, 5]. The GED problem is to find a vector $\boldsymbol{v}$ and a scalar λ such that

$$\boldsymbol{R}_y \boldsymbol{v} = \lambda \boldsymbol{R}_x \boldsymbol{v}, \tag{7.1}$$

where $\boldsymbol{R}_x$ and $\boldsymbol{R}_y$ are $n \times n$ symmetric positive definite matrices. The positive scalar λ and the associated vector $\boldsymbol{v}$ are called the generalized eigenvalue and generalized eigenvector, respectively, of the matrix pencil $(\boldsymbol{R}_y, \boldsymbol{R}_x)$. According to the matrix theory, this matrix pencil has n positive generalized eigenvalue, $\lambda_1, \lambda_2, \ldots, \lambda_n$, and associated $\boldsymbol{R}_x$-orthonormal generalized eigenvectors, $\boldsymbol{v}_1, \boldsymbol{v}_2, \ldots, \boldsymbol{v}_n$, i.e.,

$$\boldsymbol{R}_y \boldsymbol{v}_i = \lambda_i \boldsymbol{R}_x \boldsymbol{v}_i, \tag{7.2}$$

$$\boldsymbol{v}_i^{\mathrm{T}} \boldsymbol{R}_x \boldsymbol{v}_j = \delta_{ij} \quad i,j \in \{1, 2, \ldots, n\}, \tag{7.3}$$

where $(\cdot)^{\mathrm{T}}$ stands for the transpose of a vector or a matrix and δ_{ij} is the Kronecker delta function.

In order to solve the GED problem, some algebraic algorithms have been proposed for given $\boldsymbol{R}_y$ and $\boldsymbol{R}_x$ [6, 7, 8–11]. Using equivalence transformations, Moler [11] proposed a *QZ* algorithm, and Kaufman [9] proposed an LZ algorithm for solving it iteratively. However, their methods do not exploit the structure in $\boldsymbol{R}_y$ and $\boldsymbol{R}_x$. In the case of symmetric $\boldsymbol{R}_y$ and symmetric positive definite $\boldsymbol{R}_x$, several efficient approaches were proposed. By using the Cholesky factorization of $\boldsymbol{R}_x$, this problem can be reduced to the standard eigenvalue problem as reported by Martin in [11]. Bunse-Gerstner [6] proposed an approach using congruence transformations for the

© Science Press, Beijing and Springer Nature Singapore Pte Ltd. 2017

X. Kong et al., *Principal Component Analysis Networks and Algorithms*,
DOI 10.1007/978-981-10-2915-8_7

simultaneous diagonalization of $\boldsymbol{R}_y$ and $\boldsymbol{R}_x$. Shougen [7] reported an algorithm that makes use of Cholesky, QR, and SVD when $\boldsymbol{R}_y$ is also positive definite. This algorithm is stable, faster than the QZ algorithm, and is superior to that of [10]. Auchmuty [8] proposed and analyzed certain cost functions that are minimized at the eigenvectors associated with some specific eigenvalues. He also developed two iterative algorithms for numerically minimizing these functions.

The approaches mentioned above are for the case where $\boldsymbol{R}_y$ and $\boldsymbol{R}_x$ are fixed matrices. In adaptive signal processing applications, however, $\boldsymbol{R}_y$ and $\boldsymbol{R}_x$ correspond to asymptotic covariance matrices and need to be estimated. On the other hand, these algebraic algorithms are computationally intensive and belong to batch processing algorithms [12]. They are inefficient or infeasible in many real applications. Therefore, it is valuable to develop adaptive algorithms for the GED. Neural networks can be used to solve this problem, which possess many obvious advantages. For example, neural network algorithms have lower computational complexity than algebraic algorithms. Besides, neural network methods are suitable for the tracking of nonstationary distributions and high-dimensional data, since they can avoid the computation of the covariance matrix [13].

In recent years, some adaptive or neural network methods have been proposed to solve the GED problem. In [14], the generalized symmetric eigenvalue problem, where the underlying matrix pencil consists of symmetric positive definite matrices, was recasted into an unconstrained minimization problem by constructing an appropriate cost function. Then, it is extended to the case of multiple eigenvectors using an inflation technique. Based on this asymptotic formulation, a quasi-Newton-based adaptive algorithm for estimating the required generalized eigenvectors in the data case was derived. The resulting algorithm is modular and parallel, and it is globally convergent with probability one. Mao et al. proposed a two-step PCA approach to solve the GED in [15]. They used the Rubner–Tavan model [16, 17] for the PCA steps. Thus, the convergence of their method depends only on the convergence of the PCA algorithms. In [18], Chatterjee et al. proposed a gradient algorithm by building a two-layer linear heteroassociative network. However, this algorithm has low convergence speed and difficulty for selecting an appropriate step size [18]. Xu et al. [19] developed an online and local algorithm for the GED. The rule for extracting the first generalized eigenvector is similar to the LDA algorithm in [18]. But, they used a lateral inhibition network similar to the APEX algorithm for PCA [20] for extracting the minor components. Most of the above-mentioned algorithms are based on gradient methods and they involve the selection of right step sizes to ensure convergence. In general, the step sizes have an upper bound that is a function of the eigenvalues of the input data as shown in [18]. This fact makes it very hard on many occasions to choose a proper step size. If we use better optimization procedures, the computational complexity is also a key issue.

In order to resolve these issues, Rao et al. developed a RLS-like, not true RLS, algorithm for the GED, which is more computationally feasible and converges faster than gradient algorithm [21]. The true RLS-based adaptive algorithm was proposed by Yang et al. in [22]. Although the RLS algorithm can make use of the data to estimate the covariance matrices and the generalized eigenvectors, it is still computationally very costly. Another approach for the GED problem is a so-called reduced-rank generalized eigenvector extraction (R-GEVE) algorithm [23], which searches for the generalized eigenvectors in a subspace spanned by the previous estimate and the current input. However, the R-GEVE algorithm has a limitation in fast tracking because the dimension of the subspace where the eigenvectors are searched is much smaller than that of the signal space [24]. In order to obtain fast adaptive algorithm, Tanaka proposed a power method-based fast generalized eigenvector extraction (PM-FGE), which reformulated the GED as a classical eigenvector problem [24]. Inspired by the learning rule [25], Yang et al. presented an unconstrained cost criterion, whose global minimum is exactly the generalized eigenvectors, and derived a fast adaptive algorithm by gradient method [12, 26].

The above-mentioned algorithms are very efficient in computing the principal generalized eigenvectors, which are the eigenvector associated with the largest generalized eigenvalue of a matrix pencil. However, the minor generalized eigenvectors are also needed, since minor generalized eigenvectors also play vital roles in many signal processing applications [27–31]. In [32], Ye et al. proposed an adaptive algorithm to extract the minor generalized eigenvector by using a single-layer linear forward neural network. In [33], Nguyen et al. derived a fast and stable algorithm for the GED problem by extending the Möller algorithm [34]. Up to now, there are very few adaptive algorithms for the minor generalized eigenvector extraction [33, 35]. In this chapter, we will develop several adaptive algorithms for minor generalized eigenvector extraction. These algorithms are self-stabilizing, and they have faster convergence speed and better estimation accuracy compared with some existing algorithms. Their convergence will be analyzed via DDT method or Lyapunov function approach.

In this chapter, we will review and discuss the existing generalized principal or minor component analysis algorithms. Two minor generalized eigenvector extraction algorithms proposed by us will be analyzed in detail. The remainder of this chapter is organized as follows. An overview of the existing generalized principal or minor component analysis algorithms is presented in Sect. 7.2. A minor generalized eigenvector extraction algorithm and its convergence analysis via the DDT method are discussed in Sect. 7.3. An information criterion for generalized minor component and its extension to extract multiple generalized minor components and their algorithms and performance analyses by Lyapunov function approach are presented in Sect. 7.4, followed by summary in Sect. 7.5.

7.2 Review of Generalized Feature Extraction Algorithm

7.2.1 Mathew's Quasi-Newton Algorithm for Generalized Symmetric Eigenvalue Problem

In [14], the problem Mathew addressed can be formulated as follows. Given the time series $y(n)$ and $\hat{\boldsymbol{R}}_x$, develop an adaptive algorithm for estimating the first d

$(d < N)$ $\boldsymbol{R}_x$-orthogonal eigenvectors of the pencil $(\boldsymbol{R}_y, \boldsymbol{R}_x)$ using $(\boldsymbol{R}_y(n), \hat{\boldsymbol{R}}_x)$ as an estimate of $(\boldsymbol{R}_y, \boldsymbol{R}_x)$.

Consider the following problem [14]:

$$\min_{\boldsymbol{a}} \boldsymbol{a}^{\mathrm{T}} \boldsymbol{R}_y \boldsymbol{a} \text{ subject to } \boldsymbol{a}^{\mathrm{T}} \boldsymbol{R}_x \boldsymbol{a} = 1. \tag{7.4}$$

Using the penalty function method, the constrained minimization problem (7.4) can be converted into an equivalent unconstrained minimization of the following cost function:

$$J(\boldsymbol{a}, \mu) = \frac{\boldsymbol{a}^{\mathrm{T}} \boldsymbol{R}_y \boldsymbol{a}}{2} + \frac{\mu(\boldsymbol{a}^{\mathrm{T}} \boldsymbol{R}_x \boldsymbol{a} - 1)^2}{4}, \tag{7.5}$$

where μ a positive scalar. Mathew proved that $\boldsymbol{a}^*$ is a global minimizer of $J(\boldsymbol{a}, \mu)$ if and only if $\boldsymbol{a}^*$ is the minimum eigenvector of $(\boldsymbol{R}_y, \boldsymbol{R}_x)$ associated with eigenvalue $\lambda_{\min} = \mu(1 - \boldsymbol{a}^{*\mathrm{T}} \boldsymbol{R}_x \boldsymbol{a}^*)$.

Let $\boldsymbol{a}_i^*, i = 1, 2, \ldots, k - 1$ (with $2 \le k \le D$) be the $\boldsymbol{R}_x$-orthogonal eigenvectors of $(\boldsymbol{R}_y, \boldsymbol{R}_x)$ associated with the eigenvalues $\lambda_i = \mu(1 - \boldsymbol{a}_i^{*\mathrm{T}} \boldsymbol{R}_x \boldsymbol{a}_i^*)$. To obtain the next $\boldsymbol{R}_x$-orthogonal eigenvector $\boldsymbol{a}_k^*$, consider the following cost function:

$$J_k(\boldsymbol{a}_k, \mu, \alpha) = \frac{\boldsymbol{a}_k^{\mathrm{T}} \boldsymbol{R}_{yk} \boldsymbol{a}_k}{2} + \frac{\mu}{4} (\boldsymbol{a}_k^{\mathrm{T}} \boldsymbol{R}_x \boldsymbol{a}_k - 1)^2, \tag{7.6}$$

where $\boldsymbol{a}_k \in \Re^N$, α is a positive scalar, and

$$\boldsymbol{R}_{yk} = \boldsymbol{R}_y + \alpha \sum_{i=1}^{k-1} (\boldsymbol{R}_x \boldsymbol{a}_i^*)(\boldsymbol{R}_x \boldsymbol{a}_i^*)^{\mathrm{T}} = \boldsymbol{R}_{yk-1} + \alpha (\boldsymbol{R}_x \boldsymbol{a}_{k-1}^*)(\boldsymbol{R}_x \boldsymbol{a}_{k-1}^*)^{\mathrm{T}}, \quad k \ge 2, \tag{7.7}$$

with $\boldsymbol{R}_{y1} = \boldsymbol{R}_y$. Equation (7.7) represents the inflation step. Mathew has shown in [14] that the minimizer of $J_k(\boldsymbol{a}_k, \mu, \alpha)$ is the kth $\boldsymbol{R}_x$-orthogonal eigenvector of $(\boldsymbol{R}_y, \boldsymbol{R}_x)$ associated with eigenvalue $\lambda_k = \mu(1 - \boldsymbol{a}_k^{*T} \boldsymbol{R}_x \boldsymbol{a}_k^*)$.

Let $\boldsymbol{a}_k(n), k = 1, 2, \ldots, D$ be the estimates of these vectors at the nth adaptation instant. The Newton-based algorithm updating $\boldsymbol{a}_k(n-1)$ to $\boldsymbol{a}_k(n)$ is of the form

$$\boldsymbol{a}_k(n) = \boldsymbol{a}_k(n-1) - \boldsymbol{H}_k^{-1}(n-1) \boldsymbol{g}_k(n-1), \tag{7.8}$$

where $\boldsymbol{H}_k(n-1)$ and $\boldsymbol{g}_k(n-1)$ are the Hessian matrix and gradient vector, respectively, of $J_k(\boldsymbol{a}_k, \mu, \alpha)$ evaluated at $\boldsymbol{a}_k = \boldsymbol{a}_k(n-1)$. By computing $\boldsymbol{g}_k(n-1)$ and $\boldsymbol{H}_k(n-1)$, and approximating the Hessian to reduce the computation and guarantee positive definiteness, quasi-Newton adaptive algorithm can be obtained, which is summarized as follows [14]:

$$\boldsymbol{a}_k(n) = l_k(n-1) \boldsymbol{R}_{y_k}^{-1}(n) \boldsymbol{R}_x \boldsymbol{a}_k(n-1), \quad k = 1, 2, \ldots, D, \tag{7.9}$$

$$l_k(n-1) = \mu \frac{1 + \boldsymbol{a}_k^{\mathrm{T}}(n-1)\boldsymbol{R}_x \boldsymbol{a}_k(n-1)}{1 + 2\mu[\boldsymbol{R}_x \boldsymbol{a}_k(n-1)]^{\mathrm{T}} \boldsymbol{R}_{\boldsymbol{y}_k}^{-1}(n)[\boldsymbol{R}_x \boldsymbol{a}_k(n-1)]}, \quad k = 1, 2, \ldots, D, \tag{7.10}$$

$$\boldsymbol{R}_{\boldsymbol{y}_k}^{-1}(n) = \boldsymbol{R}_{\boldsymbol{y}_{k-1}}^{-1}(n) - \frac{\boldsymbol{R}_{\boldsymbol{y}_{k-1}}^{-1}(n)[\boldsymbol{R}_x \boldsymbol{a}_{k-1}(n-1)][\boldsymbol{R}_x \boldsymbol{a}_{k-1}(n-1)]^{\mathrm{T}} \boldsymbol{R}_{\boldsymbol{y}_{k-1}}^{-1}(n)}{1/\alpha + [\boldsymbol{R}_x \boldsymbol{a}_{k-1}(n-1)]^{\mathrm{T}} \boldsymbol{R}_{\boldsymbol{y}_{k-1}}^{-1}(n)[\boldsymbol{R}_x \boldsymbol{a}_{k-1}(n-1)]}, \quad k \geq 2, \tag{7.11}$$

$$\boldsymbol{R}_{\boldsymbol{y}_1}^{-1}(n) = \boldsymbol{R}_{\boldsymbol{y}}^{-1}(n) = \frac{n}{n-1}\left[\boldsymbol{R}_{\boldsymbol{y}}^{-1}(n-1) - \frac{\boldsymbol{R}_{\boldsymbol{y}}^{-1}(n-1)\boldsymbol{y}(n)\boldsymbol{y}^{\mathrm{T}}(n)\boldsymbol{R}_{\boldsymbol{y}}^{-1}(n-1)}{n - 1 + \boldsymbol{y}^{\mathrm{T}}(n)\boldsymbol{R}_{\boldsymbol{y}}^{-1}(n-1)\boldsymbol{y}(n)}\right], \quad n \geq 2, \tag{7.12}$$

where $\boldsymbol{y}(n) = [y(n), y(n-1), \ldots, y(n-N+1)]^{\mathrm{T}}$ is the data vector at the nth instant.

It is worth noting that the above algorithm is modular and parallel, and it is globally convergent with probability one. Simulation results demonstrate that the performance of this algorithm is almost identical to that of the rank-one updating algorithm of Karasalo. Furthermore, it does not suffer from the error buildup problems when a large number of updates are performed.

7.2.2 Self-organizing Algorithms for Generalized Eigen Decomposition

In [18], Chatterjee et al. proposed a gradient algorithm based on linear discriminant analysis (LDA). They proposed an online algorithm for extracting the first generalized eigenvector and then used deflation procedure for estimating the minor components. The algorithm is summarized as follows:

$$\boldsymbol{W}(k+1) = \boldsymbol{W}(k) + \eta_k(\boldsymbol{A}(k)\boldsymbol{W}(k) - \boldsymbol{B}(k)\boldsymbol{W}(k)UT[\boldsymbol{W}^{\mathrm{T}}(k)\boldsymbol{A}(k)\boldsymbol{W}(k)]), \tag{7.13}$$

$$\boldsymbol{A}(k) = \boldsymbol{A}(k-1) + \gamma_k(\boldsymbol{x}(k)\boldsymbol{x}^{\mathrm{T}}(k) - \boldsymbol{A}(k-1)), \tag{7.14}$$

$$\boldsymbol{B}(k) = \boldsymbol{B}(k-1) + \gamma_k(\boldsymbol{y}(k)\boldsymbol{y}^{\mathrm{T}}(k) - \boldsymbol{B}(k-1)). \tag{7.15}$$

where $\boldsymbol{W}(k) \in \Re^{n \times p}(p \leq n)$, $\boldsymbol{A}_0$ and $\boldsymbol{B}_0$ are symmetric, $\{\eta_k\}$ and $\{\gamma_k\}$ are sequences of scalar gains, respectively.

The main drawback of this method is that the algorithm is based on simple gradient techniques and this makes convergence dependent on the step sizes that are difficult to be set a priori.

7.2.3 Fast RLS-like Algorithm for Generalized Eigen Decomposition

In [21], the GED equation was rewritten as

$$\boldsymbol{R}_1\boldsymbol{w} = \frac{\boldsymbol{w}^T\boldsymbol{R}_1\boldsymbol{w}}{\boldsymbol{w}^T\boldsymbol{R}_2\boldsymbol{w}}\boldsymbol{R}_2\boldsymbol{w}. \tag{7.16}$$

If $\boldsymbol{R}_2 = \boldsymbol{I}$, then (7.16) will be reduced to the Rayleigh quotient and the generalized eigenvalue problem will be degencrated to PCA. Premultiplying (7.16) by $\boldsymbol{R}_2^{-1}$ and rearranging the terms, it holds that

$$\boldsymbol{w} = \frac{\boldsymbol{w}^T\boldsymbol{R}_2\boldsymbol{w}}{\boldsymbol{w}^T\boldsymbol{R}_1\boldsymbol{w}}\boldsymbol{R}_2^{-1}\boldsymbol{R}_1\boldsymbol{w}. \tag{7.17}$$

Equation (7.17) is the basis of Rao's iterative algorithm. Let the weight vector $\boldsymbol{w}(n-1)$ at iteration $(n-1)$ be the estimate of the principal generalized eigenvector. Then, the estimate of the new weight vector at iteration n according to (7.17) is

$$\boldsymbol{w}(n) = \frac{\boldsymbol{w}^T(n-1)\boldsymbol{R}_2(n)\boldsymbol{w}(n-1)}{\boldsymbol{w}^T(n-1)\boldsymbol{R}_1(n)\boldsymbol{w}(n-1)}\boldsymbol{R}_2^{-1}(n)\boldsymbol{R}_1(n)\boldsymbol{w}(n-1). \tag{7.18}$$

It can be observed that (7.18) tracks the GED equation at every time step, and this is a fixed-point update. The fixed-point algorithms are known to be faster compared to the gradient algorithms. However, many fixed-point algorithms work in batch mode, which means that the weight update is done over a window of time [21]. This can be a potential drawback of the fixed-point methods. In the following, the fixed-point update in (7.18) is transformed into a form that can be implemented online.

By using Sherman–Morrison–Woodbury matrix-inversion lemma, it holds that

$$\boldsymbol{R}_2^{-1}(n) = \boldsymbol{R}_2^{-1}(n-1) - \frac{\boldsymbol{R}_2^{-1}(n-1)\boldsymbol{x}_2(n)\boldsymbol{x}_2^{\mathrm{T}}(n)\boldsymbol{R}_2^{-1}(n-1)}{1+\boldsymbol{x}_2^{\mathrm{T}}(n)\boldsymbol{R}_2^{-1}(n-1)\boldsymbol{x}_2(n)}. \tag{7.19}$$

If $\boldsymbol{w}$ is the weight vector of a single-layer feedforward network, then we can define $y_1(n) = \boldsymbol{w}^{\mathrm{T}}(n-1)\boldsymbol{x}_1(n)$ and $y_2(n) = \boldsymbol{w}^{\mathrm{T}}(n-1)\boldsymbol{x}_2(n)$ as the outputs of the network for signals $\boldsymbol{x}_1(n)$ and $\boldsymbol{x}_2(n)$, respectively. With this, it follows easily that $\boldsymbol{w}^{\mathrm{T}}(n-1)\boldsymbol{R}_1(n)\boldsymbol{w}(n-1) = (1/n)\sum_{i=1}^{n} y_1^2(i)$ and $\boldsymbol{w}^{\mathrm{T}}(n-1)\boldsymbol{R}_2(n)\boldsymbol{w}(n-1) = (1/n)\sum_{i=1}^{n} y_2^2(i)$. This is true for the stationary cases when sample variance estimators can be used instead of the expectation operators. However, for nonstationary signals, a simple forgetting factor can be included with a trivial change in the update equation [18].

With these simplifications, the modified update equation for the stationary case can be written as

$$w(n)=\frac{\sum_{i=1}^{n} y_2^2(i)}{\sum_{i=1}^{n} y_1^2(i)} R_2^{-1}(n)\frac{1}{n}\sum_{i=1}^{n} x_1(i)y_1(i), \tag{7.20}$$

where $R_2^{-1}(n)$ is estimated using (7.19). The fixed-point GED algorithm for principal GED vector can be summarized as follows:

(a) Initialize $w(0) \in \Re^{n\times 1}$ as a random vector.
(b) Initialize $P(0) \in \Re^{n\times 1}$ as a vector with small random values.
(c) Fill $Q(0) \in \Re^{n\times n}$ with small random values.
(d) Initialize $C_1(0), C_2(0)$ as zeroes for $j > 0$.
(e) Compute $y_1(j) = w^{\mathrm{T}}(j-1)x_1(j)$ and $y_2(j) = w^{\mathrm{T}}(j-1)x_2(j)$.
(f) Update P by $P(j) = \left[1-\frac{1}{j}\right]P(j-1) + \left[\frac{1}{j}\right]x_1(j)y_1(j)$.
(g) Update Q by $Q(j) = Q(j-1) - \frac{Q(j-1)x_2(j)x_2^{\mathrm{T}}(j)Q(j-1)}{1+x_2^{\mathrm{T}}(j)Q(j-1)x_2(j)}$.
(h) Update C_1, C_2 by $C_i(j) = \left[1-\frac{1}{j}\right]C_i(j-1) + \left[\frac{1}{j}\right]y_i^2(j), \quad i = 1,2$.
(i) Update the weight vector by $w(j) = \frac{C_2(j)}{C_1(j)}Q(j)P(j)$.
(j) Normalize the weight vector
(k) Go back to step (e) and repeat until convergence is reached.

The above algorithm extracts the principal generalized eigenvector. For the minor components, it can resort to the deflation technique. For detail, see [21].

The convergence of the fixed-point GED algorithm is exponential, whereas the convergence of the online gradient methods is linear. Gradient algorithms are dependent on step sizes, which result in nonrobust performance. In contrast, a fixed-point algorithm does not require a step size for the updates. Like the gradient methods, the above fixed-point algorithm has an online implementation that is computationally feasible. The computational complexity is $O(N^2)$ where N is the dimensionality of the input signals.

7.2.4 Generalized Eigenvector Extraction Algorithm Based on RLS Method

In [22], in order to derive efficient online adaptive algorithms, Yang formulated a novel unconstrained quadratic cost function for the GED problem. By applying appropriate projection approximation [36], the cost function is modified to be fit for the RLS learning rule. First, a parallel iterative algorithm for estimating the basis for r-dimensional dominant generalized eigen subspace is derived. Then, starting from the parallel algorithm for one vector case ($r = 1$), a sequential algorithm for explicitly estimating the first dominant generalized eigenvectors by using a deflation method is developed. Furthermore, the algorithm for the generalized eigen subspace is extended to estimate the first principal generalized eigenvectors in parallel.

Consider the following scalar function:

$$J(\boldsymbol{W}) = E\left\|\boldsymbol{R}_x^{-1}\boldsymbol{y} - \boldsymbol{W}\boldsymbol{W}^H\boldsymbol{y}\right\|_{\boldsymbol{R}_x}^2. \tag{7.21}$$

It has been shown that the proposed cost function $\boldsymbol{J}(\boldsymbol{W})$ has a global minimum at which the columns of $\boldsymbol{W}$ span r-dimensional dominant generalized eigen subspace of $(\boldsymbol{R}_y, \boldsymbol{R}_x)$ and no other local minima [22]. This implies that one can search the global minimizer of $\boldsymbol{J}(W)$ by iterative methods [22].

Consider the following exponentially weighted sum instead of the expectation (7.21):

$$J(\boldsymbol{W}[k]) = \sum_{j=1}^{k} \beta^{k-j}\left\|\boldsymbol{R}_x^{-1}\boldsymbol{y}[j] - \boldsymbol{W}[k]\boldsymbol{W}^H[k]\boldsymbol{y}[j]\right\|_{\boldsymbol{R}_x}^2, \tag{7.22}$$

where the forgetting factor β is between 0 and 1. If $\beta = 1$, all the samples are given the same weights, and no forgetting of old data takes place. Choosing $\beta<1$ is especially useful in tracking nonstationary changes.

Considering the projection approximation technique [36], the computation can be simplified by approximating $\boldsymbol{W}^H[k]\boldsymbol{y}[j]$ in (7.22) with $\boldsymbol{z}[j] = \boldsymbol{W}^H[j-1]\boldsymbol{y}[j]$. These vectors can be easily computed because the estimated weight matrices $\boldsymbol{W}[j-1]$ for the previous iteration steps $j = 1, 2, \ldots, n$ are already known at step n. This approximation yields the modified least squares-type criterion

$$J'(\boldsymbol{W}[k]) = \sum_{j=1}^{k} \beta^{k-j}\left\|\boldsymbol{R}_x^{-1}\boldsymbol{y}[j] - \boldsymbol{W}[k]\boldsymbol{z}[j]\right\|_{\boldsymbol{R}_x}^2. \tag{7.23}$$

Applying the RLS technique to minimize the modified criterion $J'(\boldsymbol{W}[k])$, the following recursive algorithm to solve the basis for r-dimensional dominant generalized eigen subspace can be derived. The parallel RLS-based adaptive algorithm for principal generalized eigen subspace can be written as

$$\boldsymbol{z}(k) = \boldsymbol{W}^H(k-1)\boldsymbol{y}(k) \tag{7.24}$$

$$\boldsymbol{h}(k) = \boldsymbol{P}(k-1)\boldsymbol{z}(k) \tag{7.25}$$

$$\boldsymbol{g}(k) = \boldsymbol{h}(k)/(\beta + \boldsymbol{z}^H(k)\boldsymbol{h}(k)) \tag{7.26}$$

$$\boldsymbol{P}(k) = \mathrm{Tri}\left(\frac{1}{\beta}\left(\boldsymbol{P}(k-1) - \boldsymbol{g}(k)\boldsymbol{h}^H(k)\right)\right) \tag{7.27}$$

$$\boldsymbol{e}(k) = \boldsymbol{Q}(k)\boldsymbol{y}(k) - \boldsymbol{W}(k-1)\boldsymbol{z}(k) \tag{7.28}$$

$$\boldsymbol{W}(k) = \boldsymbol{W}(k-1) + \boldsymbol{e}(k)\boldsymbol{g}^H(k) \tag{7.29}$$

$$\boldsymbol{Q}(k) = \frac{1}{\mu}\left(\boldsymbol{I} - \frac{\boldsymbol{Q}(k-1)\boldsymbol{x}[k]\boldsymbol{x}^H[k]}{\mu + \boldsymbol{x}^H[k]\boldsymbol{Q}(k-1)\boldsymbol{x}[k]}\right)\boldsymbol{Q}(k-1) \tag{7.30}$$

where the notation Tri(·) means that only the upper triangular part of the argument is computed and its transpose is duplicated to the lower triangular part, thus making the matrix $\boldsymbol{P}[k]$ symmetric. The total computational complexity of the algorithm is $O(N^2) + 3Nr + O(r^2)$ per update.

Furthermore, the sequential adaptive algorithm for the first r principal generalized eigenvectors can be summarized as [22]

$$\boldsymbol{y}_1(k) = \boldsymbol{y}(k) \tag{7.31}$$

$$\boldsymbol{Q}_x(k) = \mathrm{Tri}\left(\frac{1}{\mu}\left(\boldsymbol{I} - \frac{\boldsymbol{Q}_x(k-1)\boldsymbol{x}(k)\boldsymbol{x}^H(k)}{\mu + \boldsymbol{x}^H(k)\boldsymbol{Q}_x(k-1)\boldsymbol{x}(k)}\right)\boldsymbol{Q}_x(k-1)\right) \tag{7.32}$$

$$\begin{gathered} \text{for i} = 1, 2, \ldots, \text{r} \quad \text{do} \\ z_i(k) = \boldsymbol{c}_i^H(k-1)\boldsymbol{Q}(k)\boldsymbol{y}_i(k) \end{gathered} \tag{7.33}$$

$$d_i(k) = \beta d_{i-1}(k) + |z_i(k)|^2 \tag{7.34}$$

$$\boldsymbol{s}_i(k) = \boldsymbol{y}_i(k) - \boldsymbol{c}_i(k-1)z_i(k) \tag{7.35}$$

$$\boldsymbol{c}_i(k) = \boldsymbol{c}_i(k-1) + \boldsymbol{s}_i(k)(z_i^H(k)/d_i(k)) \tag{7.36}$$

$$\boldsymbol{w}_i(k) = \boldsymbol{Q}(k)\boldsymbol{c}_i(k) \tag{7.37}$$

$$\boldsymbol{y}_{i+1}(k) = \boldsymbol{y}_i(k) - \boldsymbol{c}_i(k)z_i(k) \tag{7.38}$$

The above algorithm requires $4Nr + N^2r + O(N^2) + O(r)$ operations per update. In contrast to the parallel algorithm, this method enables an explicit computation of the generalized eigenvectors. On the other hand, the deflation technique causes a slightly increased computational complexity [22].

Although the sequential version of the algorithm can give r principal generalized eigenvectors, the minor generalized eigenvectors may converge slowly because of the estimation error propagation. In order to overcome these difficulties, Yang et al. extended the parallel algorithm to find the principal generalized eigenvectors in parallel. The modified parallel algorithm for r principal generalized eigenvectors can be summarized as follows:

$$z(k) = W^{H}(k-1)y(k) \tag{7.39}$$

$$h(k) = P(k-1)z(k) \tag{7.40}$$

$$g(k) = h(k)/(\beta + z^{H}(k)h(k)) \tag{7.41}$$

$$P(k) = \mathrm{Tri}\left(\frac{1}{\beta}(P(k-1) - g(k)h^{H}(k))\right) \tag{7.42}$$

$$Q_{x}(k) = \mathrm{Tri}\left(\frac{1}{\mu}\left(I - \frac{Q_{x}(k-1)x(k)x^{H}(k)}{\mu + x^{H}(k)Q_{x}(k-1)x(k)}\right)Q_{x}(k-1)\right) \tag{7.43}$$

$$e(k) = Q(k)y(k) - W(k-1)z(k) \tag{7.44}$$

$$W(k) = W(k-1) + e(k)g^{H}(k) \tag{7.45}$$

$$R_{y}(k) = \eta R_{y}(k-1) + y(k)y^{H}(k) \tag{7.46}$$

$$W(k) = R_{y}\text{-orthogonalize}(W(k)) \tag{7.47}$$

Finally, the convergence properties of the above algorithms were analyzed in [22]. Obviously, although the RLS algorithm makes use of the data to estimate the covariance matrices, it is still computationally very costly.

7.2.5 *Fast Adaptive Algorithm for the Generalized Symmetric Eigenvalue Problem*

In [23], Samir Attallah et al. proposed a new adaptive algorithm for the generalized symmetric eigenvalue problem, which can extract the principal and minor generalized eigenvectors, as well as their associated subspaces, at a low computational cost. It is based, essentially, on the idea of reduced rank. That is, using some appropriate transformation, the dimension of the problem is first reduced. The algorithm has the following advantages: (1) lower computational complexity; (2) faster convergence for large dimensional matrix and large number of principal (or minor) generalized eigenvectors to be extracted; and (3) it can estimate both principal and minor generalized eigenvectors, as well as their associated subspaces, which are spanned by the principal and minor generalized eigenvectors. The idea of reduced rank appeared for the first time in the context of generalized subspace estimation [23].

Let $\boldsymbol{x}_1(t)$ and $\boldsymbol{x}_2(t)$ be two sequences of $n \times 1$ random vectors with covariance matrices $\boldsymbol{C}_1 = E[\boldsymbol{x}_1(t)\boldsymbol{x}_1^H(t)]$ and $\boldsymbol{C}_2 = E[\boldsymbol{x}_2(t)\boldsymbol{x}_2^H(t)]$, respectively. The following lemma holds [23].

Lemma 7.1 *Assume* $p<n$ *and* $(\boldsymbol{C}_1, \boldsymbol{C}_2)$ *are* $n \times n$ *positive definite covariance matrices. Let* $\boldsymbol{L} = \mathrm{diag}(l_1, \ldots, l_p)$ *with* $l_1 > l_2 \ldots > l_p > 0$ *be a diagonal matrix and* $\boldsymbol{W}$ *a* $\boldsymbol{C}_2$*- orthogonal matrix satisfying* $\boldsymbol{W}^H\boldsymbol{C}_2\boldsymbol{W} = \boldsymbol{I}$. *Then, the function* $J(\boldsymbol{W}) = \mathrm{tr}(\boldsymbol{L}\boldsymbol{W}^H\boldsymbol{C}_1\boldsymbol{W})$ *reaches its maximum (resp. minimum) when the column vectors of* $\boldsymbol{W}$ *correspond to the p principal (resp. minor) generalized eigenvectors of* $(\boldsymbol{C}_1, \boldsymbol{C}_2)$.

Thus, the principal (minor) generalized eigen decomposition (GEV) of $(\boldsymbol{C}_1,\boldsymbol{C}_2)$ can be obtained. It should be noted that by choosing $\boldsymbol{L} = \boldsymbol{I}$, the estimated is the generalized signal (resp. minor) subspace instead of the principal (resp. minor) GEV.

Following the same idea as the approach in [37], Samir Attallah et al. proposed to reduce the cost of the previous optimization problem by limiting the search of $\boldsymbol{W}(k)$ to a reduced-rank subspace, which is chosen to be the range space of $\boldsymbol{W}(k-1)$ plus one or two additional search directions. Here, they proposed to search for $\boldsymbol{W}(k)$ in the subspace spanned by $\boldsymbol{V}(t) \triangleq [\boldsymbol{W}(k-1)\boldsymbol{x}_1(t)]$ so that one can have

$$\boldsymbol{W}(k) = \boldsymbol{V}(t)\boldsymbol{U}(t), \tag{7.48}$$

where $\boldsymbol{U}(t)$ is a $(p+1) \times p$ matrix. By doing so, one ends up with a similar GEV problem but with dimensionality-reduced matrices, i.e., one now seeks a matrix $\boldsymbol{U}(t)$ to minimize (resp. maximize), under the constraint $\boldsymbol{U}^H\bar{\boldsymbol{C}}_2\boldsymbol{U} = \boldsymbol{I}$, the cost function

$$\bar{J}(\boldsymbol{U}) = \mathrm{tr}(\boldsymbol{L}\boldsymbol{U}^H\bar{\boldsymbol{C}}_1\boldsymbol{U}), \tag{7.49}$$

where

$$\bar{\boldsymbol{C}}_i \triangleq \boldsymbol{V}^H\boldsymbol{C}_i\boldsymbol{V}, \quad i = 1, 2. \tag{7.50}$$

In other words, $\boldsymbol{U}$ is the matrix of the least (resp. principal) GEV of $(\bar{\boldsymbol{C}}_1, \bar{\boldsymbol{C}}_2)$ and can be computed using a brute force method in $O(p^3)$. The R-GEVE algorithm can be summarized as follows: At time instant t

(a) Compute $\bar{\boldsymbol{C}}_1(t)$ and $\bar{\boldsymbol{C}}_2(t)$.
(b) Compute $\boldsymbol{U}(t)$ of the p minor (resp. principal) GEV of $[\bar{\boldsymbol{C}}_1(t), \bar{\boldsymbol{C}}_2(t)]$.
(c) Compute $\boldsymbol{W}(t)$ using (7.48).

The reduced-rank generalized eigenvalue extraction (R-GEVE) algorithm is given as follows:

Initialization:

$\boldsymbol{W}(0)$, $\boldsymbol{C}_i(0)$, $\boldsymbol{C}_{yi}(0) = \boldsymbol{W}^H(0)\boldsymbol{C}_i(0)\boldsymbol{W}(0), i = 1, 2,$ and $0<\beta<1$.

For t = 1 onward and i = 1, 2:

$$\boldsymbol{y}_i(t) = \boldsymbol{W}^H(t-1)\boldsymbol{x}_i(t) \tag{7.51}$$

$$z_i(t) = \boldsymbol{C}_i(t-1)\boldsymbol{x}_1(t) \tag{7.52}$$

$$\bar{\boldsymbol{y}}_i(t) = \boldsymbol{W}^H(t-1)z_i(t) \tag{7.53}$$

$$\tilde{\boldsymbol{y}}_i(t) = \beta\bar{\boldsymbol{y}}_i(t) + (\boldsymbol{x}_i^H(t)\boldsymbol{x}_1(t))\boldsymbol{y}_i(t) \tag{7.54}$$

$$c_i(t) = \beta\boldsymbol{x}_1^H(t)z_i(t) + \left|\boldsymbol{x}_i^H(t)\boldsymbol{x}_1(t)\right|^2 \tag{7.55}$$

$$\boldsymbol{C}_{yi}(t-1) = \boldsymbol{U}^H(t-1)\boldsymbol{C}_i(t-1)\boldsymbol{U}(t-1) \tag{7.56}$$

$$\tilde{\boldsymbol{C}}_i(t) = \beta\boldsymbol{C}_{yi}(t-1) + \boldsymbol{y}_i(t)\boldsymbol{y}_i^H(t) \tag{7.57}$$

$$\bar{\boldsymbol{C}}_i(t) = \begin{bmatrix} \tilde{\boldsymbol{C}}_i(t) & \tilde{\boldsymbol{y}}_i(t) \\ \tilde{\boldsymbol{y}}_i^H(t) & c_i(t) \end{bmatrix} \tag{7.58}$$

$$\boldsymbol{C}_i(t) = \beta\boldsymbol{C}_i(t-1) + \boldsymbol{x}_i(t)\boldsymbol{x}_i^H(t) \tag{7.59}$$

$$\boldsymbol{U}(t) \leftarrow \text{GEV}(\bar{\boldsymbol{C}}_1(t), \bar{\boldsymbol{C}}_2(t)) \tag{7.60}$$

$$\boldsymbol{V}(t) = [\,\boldsymbol{W}(t-1) \quad \boldsymbol{x}_1(t)\,] \tag{7.61}$$

$$\boldsymbol{W}(t) = \boldsymbol{V}(t)\boldsymbol{U}(t) \tag{7.62}$$

R-GEVE is also suitable for the extraction of both principal and minor generalized eigenvectors, as well as their associated subspaces, at a low computational complexity.

7.2.6 *Fast Generalized Eigenvector Tracking Based on the Power Method*

The power method is a classical way to find the subspace (principal subspace) spanned by eigenvectors associated with the largest eigenvalues of a Hermitian matrix. The power method-based algorithm can track generalized eigenvectors quite fast. An iterative algorithm based on a power method can be shown as follows:

$$\boldsymbol{R}_y(k) = \alpha\boldsymbol{R}_y(k-1) + \boldsymbol{y}(k)\boldsymbol{y}^H(k) \tag{7.63}$$

$$\boldsymbol{R}_x(k) = \beta\boldsymbol{R}_x(k-1) + \boldsymbol{x}(k)\boldsymbol{x}^H(k) \tag{7.64}$$

$$\boldsymbol{K}(k) = \boldsymbol{R}_x^{-1/2}(k) \tag{7.65}$$

$$\boldsymbol{C}(k) = \boldsymbol{K}(k)\boldsymbol{R}_y(k)\boldsymbol{K}^H(k) \tag{7.66}$$

$$\tilde{\boldsymbol{W}}(k) = \boldsymbol{C}(k)\overline{\boldsymbol{W}}(k-1) \tag{7.67}$$

$$\tilde{\boldsymbol{W}}(k) = \boldsymbol{Q}(k)\boldsymbol{R}(k),\, \overline{\boldsymbol{W}}(k) = \text{First}\, r\, \text{columns of}\ \boldsymbol{Q}(k) \tag{7.68}$$

$$\boldsymbol{W}(k) = \boldsymbol{K}^H(k)\overline{\boldsymbol{W}}(k), \tag{7.69}$$

in which the first two equations are used to update $\boldsymbol{R}_y$ and $\boldsymbol{R}_x$, and the first three equations are devoted to obtaining $\boldsymbol{R}_x^{-1/2}(k)\boldsymbol{R}_y(k)\boldsymbol{R}_x^{-H/2}(k)$, followed by the rest standard steps in the power method. The fifth equation in this algorithm is the QR decomposition. It can obtain the principal generalized eigenvectors, but not the eigen subspace. Also, it can be replaced by either the Gram–Schmidt orthonormalization or the Householder transformation. It can be seen that this algorithm obtains the generalized principal eigenvector of the matrix pencil ($\boldsymbol{R}_y$, $\boldsymbol{R}_x$) by extracting the principal eigenvector of $\boldsymbol{R}_x^{-1/2}(k)\boldsymbol{R}_y(k)\boldsymbol{R}_x^{-H/2}(k)$.

In order to avoid the computation of a matrix square root and its inverse, $\boldsymbol{K}(k) = \boldsymbol{R}_x^{-1/2}(k)$ can be obtained as follows:

$$\boldsymbol{K}(k) = (1/\sqrt{\beta})\left(\boldsymbol{I} + r(k)\tilde{\boldsymbol{x}}(k)\tilde{\boldsymbol{x}}^H(k)\right)\boldsymbol{K}(k-1), \tag{7.70}$$

in which

$$\tilde{\boldsymbol{x}}(k) = (1/\sqrt{\beta})\boldsymbol{K}(k-1)\boldsymbol{x}(k), \tag{7.71}$$

$$r(k) = \frac{1}{\|\tilde{\boldsymbol{x}}(k)\|^2}\left(\frac{1}{\sqrt{1 + \|\tilde{\boldsymbol{x}}(k)\|^2}} - 1\right). \tag{7.72}$$

Thus, $\boldsymbol{C}(k)$ in (7.66) can be written as

$$\begin{aligned}\boldsymbol{C}(k) &= \boldsymbol{K}(k)\boldsymbol{R}_y(k)\boldsymbol{K}^H(k) = \boldsymbol{K}(k)\left(\alpha\boldsymbol{R}_y(k-1) + \boldsymbol{y}(k)\boldsymbol{y}^H(k)\right)\boldsymbol{K}^H(k)\\ &= \frac{1}{\beta}\left(\alpha\boldsymbol{C}(k-1) + \tilde{\boldsymbol{y}}(k)\tilde{\boldsymbol{y}}^H(k) + \delta(k)\tilde{\boldsymbol{x}}(k)\tilde{\boldsymbol{x}}^H(k) + \gamma(k)\tilde{\boldsymbol{x}}(k)\boldsymbol{h}^H(k) + \left(\gamma(k)\tilde{\boldsymbol{x}}(k)\boldsymbol{h}^H(k)\right)^H\right).\end{aligned} \tag{7.73}$$

Fast generalized eigenvector tracking based on the power method proposed by Toshihisa Tanaka can be summarized as follows [24]:

$$\tilde{\boldsymbol{y}}(k) = \boldsymbol{K}(k-1)\boldsymbol{y}(k) \tag{7.74}$$

$$\tilde{\boldsymbol{x}}(k) = (1/\sqrt{\beta})\boldsymbol{K}(k-1)\boldsymbol{x}(k) \tag{7.75}$$

$$\bar{\boldsymbol{x}}(k) = (1/\sqrt{\beta})\boldsymbol{K}^H(k-1)\tilde{\boldsymbol{x}}(k) \tag{7.76}$$

$$z(k) = \boldsymbol{C}(k-1)\tilde{\boldsymbol{y}}(k) \tag{7.77}$$

$$r(k) = (1/\|\tilde{\boldsymbol{n}}(k)\|^2)\left(1/\sqrt{1+\|\tilde{\boldsymbol{n}}(k)\|^2} - 1\right) \tag{7.78}$$

$$e(k) = \tilde{\boldsymbol{y}}^H(k)\tilde{\boldsymbol{x}}(k) \tag{7.79}$$

$$\delta(k) = r(k)^2(\alpha\tilde{\boldsymbol{x}}^H(k)z(k) + |e(k)|^2) \tag{7.80}$$

$$\boldsymbol{h}(k) = \alpha z(k) + e(k)\tilde{\boldsymbol{y}}(k) \tag{7.81}$$

$$\begin{aligned}\boldsymbol{C}(k) = (1/\beta)(\alpha\boldsymbol{C}(k-1) + \tilde{\boldsymbol{y}}(k)\tilde{\boldsymbol{y}}^H(k) + \delta(k)\tilde{\boldsymbol{x}}(k)\tilde{\boldsymbol{x}}^H(k)\\ + \gamma(k)\tilde{\boldsymbol{x}}(k)\boldsymbol{h}^H(k) + (\gamma(k)\tilde{\boldsymbol{x}}(k)\boldsymbol{h}^H(k))^H)\end{aligned} \tag{7.82}$$

$$\tilde{\boldsymbol{W}}(k) = \boldsymbol{C}(k)\overline{\boldsymbol{W}}(k-1) \tag{7.83}$$

$$\tilde{\boldsymbol{W}}(k) = \boldsymbol{Q}(k)\boldsymbol{R}(k), \overline{\boldsymbol{W}}(k) = \text{First r columns of } \boldsymbol{Q}(k) \tag{7.84}$$

$$\boldsymbol{K}(k) = (1/\beta)\boldsymbol{K}(k-1) + r(k)\tilde{\boldsymbol{x}}(k)\tilde{\boldsymbol{x}}^H(k) \tag{7.85}$$

$$\boldsymbol{W}(k) = \boldsymbol{K}^H(k)\overline{\boldsymbol{W}}(k). \tag{7.86}$$

In summary, the above algorithm has complexity $O(rN^2)$, but the tracking speed achieves almost the same as the direct SVD computation.

7.2.7 Generalized Eigenvector Extraction Algorithm Based on Newton Method

In order to deal with the adaptive signal receiver problem in multicarrier DS-CDMA system, Yang et al. proposed an adaptive algorithm based on Newton method [38] as follows:

$$\boldsymbol{P}(k+1) = \frac{1}{\mu}\boldsymbol{P}(k)\left(\boldsymbol{I} - \frac{\boldsymbol{x}(k+1)\boldsymbol{x}^H(k+1)\boldsymbol{P}(k)}{\mu + \boldsymbol{x}^H(k+1)\boldsymbol{P}(k)\boldsymbol{x}(k+1)}\right) \tag{7.87}$$

$$c(k+1) = \boldsymbol{w}^H(k)\boldsymbol{y}(k+1) \tag{7.88}$$

$$\boldsymbol{r}(k+1) = \beta \boldsymbol{r}(k+1) + (1-\beta)\boldsymbol{y}(k+1)c^*(k+1) \tag{7.89}$$

$$d(k+1) = \beta d(k+1) + (1-\beta)c(k+1)c^*(k+1) \tag{7.90}$$

$$\tilde{\boldsymbol{w}}(k+1) = \boldsymbol{r}(k+1)/d(k+1) \tag{7.91}$$

$$\boldsymbol{w}(k+1) = \frac{2\boldsymbol{P}(k+1)\tilde{\boldsymbol{w}}(k+1)}{1+\tilde{\boldsymbol{w}}^H(k+1)\boldsymbol{P}(k+1)\tilde{\boldsymbol{w}}(k+1)} \tag{7.92}$$

which is used to extract the maximum generalized eigenvector. In this algorithm, the initial condition can be simply determined as $\boldsymbol{P}(0) = \eta_1 \boldsymbol{I}$, $\boldsymbol{w}(0) = \boldsymbol{r}(0) = \eta_2[1, 0, \ldots, 0]^{\mathrm{T}}$ and $d(0) = \eta_3$,in which η_i (i = 1, 2, 3) is a suitable positive number. Since this algorithm has the subtraction operation in the updating of $\boldsymbol{P}(k)$, $\boldsymbol{P}(k)$ may lose definiteness in the computation procedure, while $\boldsymbol{P}(k)$ is theoretically Hermitian positive definite. One efficient and robust method to solve this problem is to use the *QR* decomposition to resolve the matrix root $\boldsymbol{P}^{1/2}(k)$ of $\boldsymbol{P}(k)$ and then obtain $\boldsymbol{P}(k)$ by $\boldsymbol{P}(k) = \boldsymbol{P}^{1/2}(k)\boldsymbol{P}^{H/2}(k)$ to keep the definite nature of $\boldsymbol{P}(k)$.

However, the above algorithm can only extract the generalized principal eigenvector. Later, Yang et al. also proposed another similar algorithm [26]

$$\boldsymbol{P}(k) = \frac{1}{\alpha}\left(\boldsymbol{P}(k-1) - \frac{\boldsymbol{P}(k-1)z(k)z^{\mathrm{T}}(k)\boldsymbol{P}(k-1)}{\alpha + z^{\mathrm{T}}(k)\boldsymbol{P}(k-1)z(k)}\right) \tag{7.93}$$

$$\boldsymbol{c}(k) = \boldsymbol{W}^{\mathrm{T}}(k-1)\boldsymbol{x}(k) \tag{7.94}$$

$$\boldsymbol{g}(k) = \frac{\boldsymbol{Q}(k-1)\boldsymbol{c}(k-1)}{\beta + \boldsymbol{c}^{\mathrm{T}}(k)\boldsymbol{Q}(k-1)\boldsymbol{c}(k)} \tag{7.95}$$

$$\boldsymbol{Q}(k) = (1/\beta)(\boldsymbol{Q}(k-1) - \boldsymbol{g}(k)\boldsymbol{c}^{\mathrm{T}}(k)\boldsymbol{Q}(k-1)) \tag{7.96}$$

$$\hat{\boldsymbol{d}}(k) = \boldsymbol{Q}(k)\boldsymbol{c}(k) \tag{7.97}$$

$$\hat{\boldsymbol{c}}(k) = \boldsymbol{W}^{\mathrm{T}}(k-1)\boldsymbol{c}(k) \tag{7.98}$$

$$\boldsymbol{W}(k) = \boldsymbol{W}(k-1) + \boldsymbol{P}(k)\boldsymbol{x}(k)\hat{\boldsymbol{d}}^{\mathrm{T}}(k) - \hat{\boldsymbol{c}}(k)\boldsymbol{g}^{\mathrm{T}}(k) \tag{7.99}$$

which is used to extract multiple generalized eigenvectors, whose corresponding initial conditions of variables are set to $\boldsymbol{P}(0) = \delta_1\boldsymbol{I}_m$, $\boldsymbol{Q}(0) = \delta_2\boldsymbol{I}_{\mathrm{n}}$, and $\boldsymbol{W}(0) = \delta_3[\boldsymbol{e}_1, \ldots, \boldsymbol{e}_n]$, in which δ_i (i = 1, 2, 3) are suitable positive numbers, $\boldsymbol{e}_i$ is the ith column vector with its ith element being 1, and the rest are zeros.

7.2.8 Online Algorithms for Extracting Minor Generalized Eigenvector

The previous adaptive algorithms all focus on extracting principal generalized eigenvectors. However, in many practical applications, e.g., dimensionality reduction and signal processing, extracting the minor generalized eigenvectors adaptively is also needed. The eigenvector associated with the smallest generalized eigenvalue of a matrix pencil is called minor generalized eigenvector. In [32], several approaches that led to a few algorithms for extracting minor generalized eigenvectors were proposed and discussed.

(1) Algorithm for Extracting the First Minor LDA Transform

First, an adaptive algorithm was derived by using a single-layer linear forward neural network from the viewpoint of linear discriminant analysis. Here, the minor generalized eigenvector extracting algorithm is discussed as follows. Consider a two-layer linear forward network with weight matrix $\boldsymbol{W}$ from the input-to-hidden layer and the weight matrix $\boldsymbol{V}$ from the hidden-to-output layer.

Based on the criterion,

$$J(\boldsymbol{w},\boldsymbol{v}) = E\left[\left\|d - \boldsymbol{v}\boldsymbol{w}^{\mathrm{T}}\boldsymbol{x}\right\|^{2}\right] + \mu(\boldsymbol{w}^{\mathrm{T}}\boldsymbol{S}_{b}\boldsymbol{w} - 1), \tag{7.100}$$

and using gradient ascent method, the iterative algorithm for computing the first principal generalized eigenvector can be derived:

$$\boldsymbol{w}(k+1) = \boldsymbol{w}(k) + \eta(\boldsymbol{S}_{m}\boldsymbol{w}(k) - \boldsymbol{S}_{b}\boldsymbol{w}(k)\boldsymbol{w}^{\mathrm{T}}(k)\boldsymbol{S}_{m}\boldsymbol{w}(k)), \tag{7.101}$$

where η is the learning rate. It should be noted that $\boldsymbol{S}_m = E[\boldsymbol{x}\boldsymbol{x}^{\mathrm{T}}]$, $\boldsymbol{S}_b = \boldsymbol{M}\boldsymbol{M}^{\mathrm{T}}$, where $M = E[\boldsymbol{x}\boldsymbol{d}^{\mathrm{T}}]$. When (7.101) converges, the weight vector $\boldsymbol{w}(k) \to \boldsymbol{w}$, where $\boldsymbol{w}$ is the first principal generalized eigenvector of matrix pencil $(\boldsymbol{S}_m, \boldsymbol{S}_b)$. The first minor generalized eigenvector $(\boldsymbol{S}_m, \boldsymbol{S}_b)$ can be obtained as follows:

$$\tilde{\boldsymbol{w}} = \frac{\boldsymbol{w}}{\sqrt{\boldsymbol{w}^{\mathrm{T}}\boldsymbol{S}_{m}\boldsymbol{w}}}. \tag{7.102}$$

(2) Algorithm for Extracting Multiple Minor LDA Transforms

Assume that there are p hidden neurons and the $(j-1)$th neurons have already been trained whose input weights $\boldsymbol{W} = [\boldsymbol{w}_1, \ldots, \boldsymbol{w}_{j-1}]$ converge to the *(j-1)* principal generalized eigenvectors. By using the objective function,

$$J_j(\boldsymbol{w}_j,\boldsymbol{v}_j) = E\left[\left\|d - \boldsymbol{v}_j\boldsymbol{w}_j^{\mathrm{T}}\boldsymbol{x}\right\|^{2}\right] + \sum_{i=1}^{j-1}\alpha_i \boldsymbol{w}_i^{\mathrm{T}}\boldsymbol{S}_b\boldsymbol{w}_i + \mu(\boldsymbol{w}_j^{\mathrm{T}}\boldsymbol{S}_b\boldsymbol{w}_j - 1), \tag{7.103}$$

and the gradient ascent method, an adaptive algorithm for $\boldsymbol{w}_j$ is

$$\begin{aligned}\boldsymbol{w}_j(k+1) = \boldsymbol{w}_j(k) + \eta(\boldsymbol{S}_m\boldsymbol{w}_j(k) - \boldsymbol{S}_b\boldsymbol{w}_j(k)\boldsymbol{w}_j^{\mathrm{T}}(k)\boldsymbol{S}_m\boldsymbol{w}_j(k) \\ - \boldsymbol{S}_b\sum_{i=1}^{j-1}\boldsymbol{w}_i(k)\boldsymbol{w}_i^{\mathrm{T}}(k)\boldsymbol{S}_m\boldsymbol{w}_i(k)).\end{aligned} \tag{7.104}$$

For matrix form, the above algorithm can be rewritten as

$$\boldsymbol{W}(k+1) = \boldsymbol{W}(k) + \eta(\boldsymbol{S}_m\boldsymbol{W}(k) - \boldsymbol{S}_b\boldsymbol{W}(k)UT[\boldsymbol{W}^{\mathrm{T}}(k)\boldsymbol{S}_m\boldsymbol{W}(k)]), \tag{7.105}$$

where $\boldsymbol{W}(k) = [\boldsymbol{w}_1(k), \ldots, \boldsymbol{w}_j](j \leq m)$ and UT[] sets all elements below the diagonal to zero, i.e., makes the matrix upper triangular. Since the above algorithms compute the principal generalized eigenvectors of matrix pencil $(\boldsymbol{S}_m, \boldsymbol{S}_b)$, for a weight vector $\boldsymbol{w}_i$ at convergence, $\tilde{\boldsymbol{w}}_i$ is the desired minor generalized eigenvector by using the following computation:

$$\tilde{\boldsymbol{w}}_i = \frac{\boldsymbol{w}_i}{\sqrt{\boldsymbol{w}_i^{\mathrm{T}}\boldsymbol{S}_m\boldsymbol{w}_i}}. \tag{7.106}$$

In [32], other objective functions were also proposed and the corresponding algorithms for extracting the first principal generalized eigenvector and multiple principal generalized eigenvector were derived. For detail, see [32].

(3) Extensions to the Generalized Eigen Decomposition

The algorithms (7.101), (7.104), (7.105), etc., can be easily generalized to obtain online algorithms for extracting minor generalized eigenvectors of matrix pencil ($\boldsymbol{A}$, $\boldsymbol{B}$) if matrices $\boldsymbol{A}$, $\boldsymbol{B}$ are considered as $\boldsymbol{S}_b$, $\boldsymbol{S}_m$ respectively, where $\boldsymbol{A}$, $\boldsymbol{B}$ are assumed to be positive definite and symmetric. In real situation, matrices $\boldsymbol{A}$ and $\boldsymbol{B}$ cannot be obtained directly. Two sequences of random matrices $\{\boldsymbol{A}_k \in \Re^{N\times N}\}$ and $\{\boldsymbol{B}_k \in \Re^{N\times N}\}$ with $\lim_{k\to\infty} E[\boldsymbol{A}_k] = \boldsymbol{A}$ and $\lim_{k\to\infty} E[\boldsymbol{B}_k] = \boldsymbol{B}$ are known. Thus, the algorithms (7.101) and (7.105) can be extended to generalized eigen decomposition problem as

$$\boldsymbol{w}(k+1) = \boldsymbol{w}(k) + \eta(\boldsymbol{B}_k\boldsymbol{w}(k) - \boldsymbol{A}_k\boldsymbol{w}(k)\boldsymbol{w}^{\mathrm{T}}(k)\boldsymbol{B}_k\boldsymbol{w}(k)), \tag{7.107}$$

and

$$\boldsymbol{W}(k+1) = \boldsymbol{W}(k) + \eta(\boldsymbol{B}_k\boldsymbol{W}(k) - \boldsymbol{A}_k\boldsymbol{W}(k)UT[\boldsymbol{W}^{\mathrm{T}}(k)\boldsymbol{B}_k\boldsymbol{W}(k)]), \tag{7.108}$$

where sequences $\{\boldsymbol{A}_k\}$ and $\{\boldsymbol{B}_k\}$ are obtained through

$$\boldsymbol{A}_k = \boldsymbol{A}_{k-1} + \gamma_k(\boldsymbol{x}_k\boldsymbol{x}_k^{\mathrm{T}} - \boldsymbol{A}_{k-1}), \tag{7.109}$$

and

$$\boldsymbol{B}_k = \boldsymbol{B}_{k-1} + \gamma_k(\boldsymbol{y}_k\boldsymbol{y}_k^{\mathrm{T}} - \boldsymbol{B}_{k-1}), \tag{7.110}$$

and $\{\gamma_k\}$ is a constant sequence. When the weight vector $\boldsymbol{w}_i$ converges, the desired minor generalized eigenvector $\tilde{\boldsymbol{w}}_i$ can be computed as follows:

$$\tilde{\boldsymbol{w}}_i = \frac{\boldsymbol{w}_i}{\sqrt{\boldsymbol{w}_i^{\mathrm{T}}\boldsymbol{B}\boldsymbol{w}_i}}. \tag{7.111}$$

The stability and convergence of these algorithms were analyzed theoretically, and simulations validated the efficiency and effectiveness of these algorithms.

7.3 A Novel Minor Generalized Eigenvector Extraction Algorithm

In this section, we will introduce an adaptive algorithm to extract the minor generalized eigenvector of a matrix pencil. In this algorithm, although there is no normalization step, the norm of the weight vector is self-stabilizing, i.e., moving toward unit length at each learning step.

In [39], Cirrincione studied some neural network learning algorithms and divided them into two classes. In the rules of the first class, the norm of the weight vector is undermined. When a numerical procedure is applied, these algorithms are plagued by "divergence or sudden divergence" of the weight vector length. However, the algorithms in the second class are self-stabilizing, which means that the weight vector length converges toward a fixed value independent of the input vector. Since all algorithms lacking self-stabilization are prone to fluctuations and divergence in weight vector norm, self-stability is an indispensable property for neural network learning algorithms. However, the algorithms mentioned above are only suitable for extracting the eigenvector of a matrix, not a matrix pencil. Up to now, almost all analyses of self-stabilizing property are in allusion to eigenvector extraction algorithms of a matrix, not generalized eigenvector extractions of a matrix pencil. In this section, the concept of self-stabilizing property will be extended to the generalized algorithms and the self-stabilizing property of our algorithm will be also analyzed.

Nowadays, the convergence behavior of many neural network algorithms [40–43] has been analyzed via the DDT method. However, the algorithms analyzed above are all eigenvector extraction algorithms. Up to now, there have been few works which analyze generalized eigenvector extraction algorithms via the DDT method. In this section, the convergence of our generalized eigenvector extraction algorithm will be analyzed via the DDT method and some sufficient conditions to guarantee its convergence will be derived.

7.3.1 Algorithm Description

Definition 7.1 Given an $n \times 1$ vector $\boldsymbol{w}$ and an $n \times n$ matrix $\boldsymbol{R}$, define the R-norm of vector $\boldsymbol{w}$ as $\|\boldsymbol{w}\|_{\boldsymbol{R}} = \sqrt{\boldsymbol{w}^{\mathrm{T}}\boldsymbol{R}\boldsymbol{w}}$.

Suppose that the two vector sequences $\boldsymbol{x}_k$ and $\boldsymbol{y}_k$, $k = 1, 2, \ldots$ are two stationary stochastic processes with zero means and autocorrelation matrices $\boldsymbol{R}_x = E[\boldsymbol{x}_k \boldsymbol{x}_k^{\mathrm{T}}]$ and $\boldsymbol{R}_y = E[\boldsymbol{y}_k \boldsymbol{y}_k^{\mathrm{T}}]$, respectively. Then, we propose the following algorithm to estimate the generalized eigenvector associated with the smallest generalized eigenvalue of matrix pencil $(\boldsymbol{R}_y, \boldsymbol{R}_x)$:

$$\begin{aligned} \boldsymbol{w}(k+1) = \boldsymbol{w}(k) + \eta \big[& \boldsymbol{w}^{\mathrm{T}}(k)\boldsymbol{R}_y \boldsymbol{w}(k)\boldsymbol{w}(k) + \boldsymbol{w}(k) \\ & - \boldsymbol{w}^{\mathrm{T}}(k)\boldsymbol{R}_x \boldsymbol{w}(k)(\boldsymbol{I} + (\boldsymbol{R}_x)^{-1}\boldsymbol{R}_y)\boldsymbol{w}(k) \big], \end{aligned} \tag{7.112}$$

where η is the learning rate and $\boldsymbol{w}_k$ is the estimate of the minor generalized eigenvector at time k.

In some practical signal processing applications, the autocorrelation $\boldsymbol{R}_x$ and $\boldsymbol{R}_y$ may be unknown and have to be estimated online from their respective sample data sequences. Therefore, it is very necessary to develop adaptive algorithms, especially in nonstationary signal environments. In this section, we will change the above algorithms into adaptive algorithms by exponentially weighted method. The sample autocorrelation matrices are estimated by the following equations:

$$\hat{\boldsymbol{R}}_y(k+1) = \beta \hat{\boldsymbol{R}}_y(k) + \boldsymbol{y}^{\mathrm{T}}(k+1)\boldsymbol{y}(k+1), \tag{7.113}$$

$$\hat{\boldsymbol{R}}_x(k+1) = \alpha \hat{\boldsymbol{R}}_x(k) + \boldsymbol{x}^{\mathrm{T}}(k+1)\boldsymbol{x}(k+1), \tag{7.114}$$

where $0 < \alpha, \beta \leq 1$ are the forgetting factors. As to how to choose a proper forgetting factor, see [23] for detail.

By using the matrix-inversion lemma, we can write the time update for the inverse matrix $\boldsymbol{Q}(k) = \boldsymbol{R}_x^{-1}(k)$ as

$$\boldsymbol{Q}(k+1) = \frac{1}{\alpha}\left[\boldsymbol{Q}(k) - \frac{\boldsymbol{Q}(k)\boldsymbol{x}(k+1)\boldsymbol{x}^{\mathrm{T}}(k+1)\boldsymbol{Q}(k)}{\alpha + \boldsymbol{x}^{\mathrm{T}}(k+1)\boldsymbol{Q}(k)\boldsymbol{x}(k+1)}\right]. \tag{7.115}$$

Table 7.1 Adaptive version of the proposed algorithm

Initialization: Set $k = 0$, and set initial estimates $\hat{\boldsymbol{R}}_y(0), \hat{\boldsymbol{R}}_x(0), \boldsymbol{Q}_x(0)$ and randomly generate a weight vector $\boldsymbol{w}(0)$
Iteration: Step 1. Set $k = k+1$ and update $\hat{\boldsymbol{R}}_y(k), \hat{\boldsymbol{R}}_x(k), \boldsymbol{Q}_x(k)$ by (7.113)–(7.115)   Step 2. Update $\boldsymbol{w}(k)$ by the following algorithm: $\boldsymbol{w}(k+1) = \boldsymbol{w}(k) + \eta[\boldsymbol{w}^{\mathrm{T}}(k)\hat{\boldsymbol{R}}_y \boldsymbol{w}(k)\boldsymbol{w}(k) + \boldsymbol{w}(k) - \boldsymbol{w}^{\mathrm{T}}(k)\hat{\boldsymbol{R}}_x \boldsymbol{w}(k)(\boldsymbol{I} + \boldsymbol{Q}_x\hat{\boldsymbol{R}}_y)\boldsymbol{w}(k)]$ (7.116)

By replacing $\boldsymbol{R}_y, \boldsymbol{R}_x, \boldsymbol{R}_x^{-1}$ in the above algorithm by $\hat{\boldsymbol{R}}_y(k+1), \hat{\boldsymbol{R}}_x(k+1), \boldsymbol{Q}(k+1)$, respectively, we can obtain the following adaptive algorithm as in Table 7.1.

7.3.2 Self-stabilizing Analysis

Next, we will study the self-stability of (7.112).

Theorem 7.1 *If the learning rate* η *is small enough and the input vector is bounded, then the* R_x*-norm of the weight vector in the algorithm* (7.112) *for extracting the minor generalized eigenvector will converge to one.*

Proof Since the learning rate η is small enough and the input vector is bounded, we have

$$\begin{aligned}
\|\boldsymbol{w}(k+1)\|_{\boldsymbol{R}_x}^2 &= \boldsymbol{w}^{\mathrm{T}}(k+1)\boldsymbol{R}_x\boldsymbol{w}(k+1)\\
&= \left[\boldsymbol{w}(k)+\eta(\boldsymbol{w}^{\mathrm{T}}(k)\boldsymbol{R}_y\boldsymbol{w}(k)\boldsymbol{w}(k)+\boldsymbol{w}(k) - \boldsymbol{w}(k)^{\mathrm{T}}\boldsymbol{R}_x\boldsymbol{w}(k)\times(\boldsymbol{I}+\boldsymbol{R}_x^{-1}\boldsymbol{R}_y)\boldsymbol{w}(k))\right]^{\mathrm{T}}\\
&\quad \boldsymbol{R}_x\left[\boldsymbol{w}(k)+\eta(\boldsymbol{w}^{\mathrm{T}}(k)\boldsymbol{R}_y\boldsymbol{w}(k)\boldsymbol{w}(k)+\boldsymbol{w}(k) - \boldsymbol{w}(k)^{\mathrm{T}}\boldsymbol{R}_x\boldsymbol{w}(k)(\boldsymbol{I}+\boldsymbol{R}_x^{-1}\boldsymbol{R}_y)\boldsymbol{w}(k))\right]\\
&= \boldsymbol{w}(k)^{\mathrm{T}}\boldsymbol{R}_x\boldsymbol{w}(k)+2\eta\boldsymbol{w}(k)^{\mathrm{T}}\boldsymbol{R}_x\boldsymbol{w}(k)\left[1-\boldsymbol{w}(k)^{\mathrm{T}}\boldsymbol{R}_x\boldsymbol{w}(k)\right]\\
&\quad +\eta^2\Big[\left(\boldsymbol{w}(k)^{\mathrm{T}}\boldsymbol{R}_x\boldsymbol{w}(k)\right)^3-2\left(\boldsymbol{w}(k)^{\mathrm{T}}\boldsymbol{R}_x\boldsymbol{w}(k)\right)^2+\boldsymbol{w}(k)^{\mathrm{T}}\boldsymbol{R}_x\boldsymbol{w}(k)\\
&\quad -\boldsymbol{w}^{\mathrm{T}}(k)\boldsymbol{R}_y\boldsymbol{w}(k)\left(\boldsymbol{w}(k)^{\mathrm{T}}\boldsymbol{R}_x\boldsymbol{w}(k)\right)^2+2\boldsymbol{w}^{\mathrm{T}}(k)\boldsymbol{R}_y\boldsymbol{w}(k)\boldsymbol{w}(k)^{\mathrm{T}}\boldsymbol{R}_x\boldsymbol{w}(k)\\
&\quad -\boldsymbol{w}(k)^{\mathrm{T}}\boldsymbol{R}_x\boldsymbol{w}(k)\left(\boldsymbol{w}^{\mathrm{T}}(k)\boldsymbol{R}_y\boldsymbol{w}(k)\right)^2-\boldsymbol{w}^{\mathrm{T}}(k)\boldsymbol{R}_y\boldsymbol{w}(k)\Big]\\
&= \boldsymbol{w}(k)^{\mathrm{T}}\boldsymbol{R}_x\boldsymbol{w}(k)+2\eta\boldsymbol{w}(k)^{\mathrm{T}}\boldsymbol{R}_x\boldsymbol{w}(k)\left[1-\boldsymbol{w}(k)^{\mathrm{T}}\boldsymbol{R}_x\boldsymbol{w}(k)\right]+o(\eta)\\
&\approx \boldsymbol{w}(t)^{\mathrm{T}}\boldsymbol{R}_x\boldsymbol{w}(t)\left\{1+2\eta\left[1-\boldsymbol{w}(t)^{\mathrm{T}}\boldsymbol{R}_x\boldsymbol{w}(t)\right]\right\}.
\end{aligned} \tag{7.117}$$

Note that the second-order terms associated with the learning rate in the above equation are neglected. Then, it holds that

$$\begin{aligned}
\frac{\|\boldsymbol{w}(k+1)\|_{\boldsymbol{R}_x}^2}{\|\boldsymbol{w}(k)\|_{\boldsymbol{R}_x}^2} &\approx \frac{\boldsymbol{w}(k)^{\mathrm{T}}\boldsymbol{R}_x\boldsymbol{w}(k)\left\{1+2\eta\left[1-\boldsymbol{w}(k)^{\mathrm{T}}\boldsymbol{R}_x\boldsymbol{w}(k)\right]\right\}}{\boldsymbol{w}(k)^{\mathrm{T}}\boldsymbol{R}_x\boldsymbol{w}(k)}\\
&= 1+2\eta\left[1-\boldsymbol{w}(k)^{\mathrm{T}}\boldsymbol{R}_x\boldsymbol{w}(k)\right]\\
&= \begin{cases} >1 & \text{if} \quad \|\boldsymbol{w}(t)\|_{\boldsymbol{R}_x}^2<1\\ =1 & \text{if} \quad \|\boldsymbol{w}(t)\|_{\boldsymbol{R}_x}^2=1\\ <1 & \text{if} \quad \|\boldsymbol{w}(t)\|_{\boldsymbol{R}_x}^2>1. \end{cases}
\end{aligned} \tag{7.118}$$

It can be easily seen that the norm of weight vector $\|w(t)\|^2_{R_x}$ in algorithm (7.112) will tend to one whether the norm of the initial weight vector is equal to one or not. We denote this property as one-tending property. In other words, the algorithm has self-stabilizing property.

This completes the proof.

7.3.3 Convergence Analysis

In this section, we will present convergence analysis of algorithm (7.112) by the DDT method. By taking the conditional expectation $E\{w(k+1)/w(0), x(i), i<k\}$ to (7.116) and identifying the conditional expectation as the next iterate, the DDT system of (7.116) can be obtained, which is the same as (7.112). That is, (7.112) is also the DDT system of its adaptive algorithm (7.116).

Since the autocorrelation matrices $\boldsymbol{R}_x$ and $\boldsymbol{R}_y$ are two symmetric nonnegative definite matrices, their generalized eigenvalues are nonnegative and their generalized eigenvectors compose an $\boldsymbol{R}_x$-orthonormal basis of $\Re^n$. For convenience, arrange the generalized eigenvectors in such a way that the associated eigenvalues are in a nonascending order, i.e., $\lambda_1 > \lambda_2 > \cdots > \lambda_n \geq 0$. Then, the weight vector $\boldsymbol{w}(k)$ can be represented as

$$\boldsymbol{w}(k) = \sum_{i=1}^{n} z_i(k)\boldsymbol{v}_i, \tag{7.119}$$

where $z_i(k) = \boldsymbol{v}_i^{\mathrm{T}}\boldsymbol{R}_x\boldsymbol{w}(k)(i = 1, 2, \ldots, n)$ are some constants. By substituting (7.119) into (7.112), we have

$$z_i(k+1) = \left[1 + \eta(1 + \boldsymbol{w}^{\mathrm{T}}(k)\boldsymbol{R}_y\boldsymbol{w}(k) - (\lambda_i + 1)\boldsymbol{w}^{\mathrm{T}}(k)\boldsymbol{R}_x\boldsymbol{w}(k))\right]z_i(k), \tag{7.120}$$

for all $k \geq 0$.

According to the properties of the Rayleigh quotient [41], it holds that

$$0 \leq \lambda_n < \frac{\boldsymbol{w}^{\mathrm{T}}(k)\boldsymbol{R}_y\boldsymbol{w}(k)}{\boldsymbol{w}^{\mathrm{T}}(k)\boldsymbol{R}_x\boldsymbol{w}(k)} < \lambda_1, \tag{7.121}$$

for each $k \geq 0$.

Next, we will discuss some conditions under which the weight vector will converge to the minor generalized eigenvector of the matrix pencil $(\boldsymbol{R}_y, \boldsymbol{R}_x)$. Before this, the following theorems are needed.

Theorem 7.2 *Suppose that* $\eta \le 0.3$ *and* $\eta\lambda_1 \le 0.35$. *If the initial vector satisfies* $\boldsymbol{w}^{\mathrm{T}}(0)\boldsymbol{R}_x\boldsymbol{v}_n \ne 0$, *then it holds that* $\|\boldsymbol{w}(k)\|_{\boldsymbol{R}_x} < 1+\eta\lambda_1$ *for all* $k \ge 0$.

Proof From (7.119) and (7.120), it follows that

$$\begin{aligned}\|\boldsymbol{w}(k+1)\|_{\boldsymbol{R}_x}^2 &= \sum_{i=1}^{n} z_i^2(k+1)\\ &= \sum_{i=1}^{n}\left[1+\eta(1+\boldsymbol{w}^{\mathrm{T}}(k)\boldsymbol{R}_y\boldsymbol{w}(k) - (\lambda_i+1)\boldsymbol{w}^{\mathrm{T}}(k)\boldsymbol{R}_x\boldsymbol{w}(k)\right]^2 z_i^2(k)\\ &< \left[1+\eta(1+\lambda_1\boldsymbol{w}^{\mathrm{T}}(k)\boldsymbol{R}_x\boldsymbol{w}(k) - (\lambda_i+1)\boldsymbol{w}^{\mathrm{T}}(k)\boldsymbol{R}_x\boldsymbol{w}(k)\right]^2 \sum_{i=1}^{n} z_i^2(k)\\ &< \left[1+\eta(1+(\lambda_1-1))\boldsymbol{w}^{\mathrm{T}}(k)\boldsymbol{R}_x\boldsymbol{w}(k)\right]^2\|\boldsymbol{w}(k)\|_{\boldsymbol{R}_x}^2.\end{aligned} \tag{7.122}$$

Then, we have

$$\|\boldsymbol{w}(k+1)\|_{\boldsymbol{R}_x}^2 < \left[1+\eta(1+(\lambda_1-1))\|\boldsymbol{w}(k)\|_{\boldsymbol{R}_x}^2\right]^2\|\boldsymbol{w}(k)\|_{\boldsymbol{R}_x}^2. \tag{7.123}$$

Over the interval $[0, 1]$, define a differential function

$$f(s) = (1+\eta(1+(\lambda_1-1)s))^2 s, \tag{7.124}$$

where $s = \|\boldsymbol{w}(k)\|_{\boldsymbol{R}_x}^2$ and $f(s) = \|\boldsymbol{w}(k+1)\|_{\boldsymbol{R}_x}^2$. Then, the gradient of $f(s)$ with respect to s is given by

$$\dot{f}(s) = (1+\eta+\eta(\lambda_1-1)s)(1+\eta+3\eta(\lambda_1-1)s). \tag{7.125}$$

Clearly, the roots of the function $\dot{f}(s) = 0$ are

$$s_1 = -\frac{1+\eta}{\eta(\lambda_1-1)},\quad s_2 = -\frac{1+\eta}{3\eta(\lambda_1-1)}. \tag{7.126}$$

Next, two cases are considered.
Case 1: $\lambda_1 > 1$
Since $\eta > 0, \lambda_1 > 1$, we have $s_1 < s_2 < 0$. That is, for all $0 < s < 1$,we have

$$\dot{f}(s) > 0. \tag{7.127}$$

Case 2: $\lambda_1 < 1$
From $\eta < 0.3$, we have $s_1 > s_2 > 1$. Then, it follows that for all $0 < s < 1$

$$\dot{f}(s) > 0. \tag{7.128}$$

Equations (7.127) and (7.128) mean that over the interval $[0, 1]$, $f(s)$ is a monotonically increasing function. Then, for all $0 < s < 1$, we have

$$f(s) \leq f(1) = (1 + \eta\lambda_1)^2. \tag{7.129}$$

So for all $k \geq 0$, we have

$$\|\boldsymbol{w}(k)\|_{\boldsymbol{R}_x} < 1 + \eta\lambda_1. \tag{7.130}$$

This completes the proof.

Theorem 7.3 Suppose that $\eta \leq 0.3$ and $\eta\lambda_1 \leq 0.35$. If the initial vector satisfies $\boldsymbol{w}^{\mathrm{T}}(0)\boldsymbol{R}_x\boldsymbol{v}_n \neq 0$, then it holds that $\|\boldsymbol{w}(k)\|_{\boldsymbol{R}_x} > c$ for all $k \geq 0$, where c is a constant and equals $(1 + \eta\lambda_1)\left[1 - \eta(1 + \eta\lambda_1)^2\right]$.

Proof From (7.119) and (7.120), it follows that

$$\begin{aligned}
\|\boldsymbol{w}(k+1)\|_{\boldsymbol{R}_x}^2 &= \sum_{i=1}^{n} z_i^2(k+1) \\
&= \sum_{i=1}^{n} \left[1 + \eta(1 + \boldsymbol{w}^{\mathrm{T}}(k)\boldsymbol{R}_y\boldsymbol{w}(k) - (\lambda_i + 1)\boldsymbol{w}^{\mathrm{T}}(k)\boldsymbol{R}_x\boldsymbol{w}(k)\right]^2 z_i^2(k) \\
&> \left[1 - \eta\boldsymbol{w}^{\mathrm{T}}(k)\boldsymbol{R}_x\boldsymbol{w}(k)\right]^2 \sum_{i=1}^{n} z_i^2(k).
\end{aligned} \tag{7.131}$$

Next, two cases are considered to complete the proof.

Case 1: $1 + \boldsymbol{w}^{\mathrm{T}}(k)\boldsymbol{R}_y\boldsymbol{w}(k) - (\lambda_i + 1)\boldsymbol{w}^{\mathrm{T}}(k)\boldsymbol{R}_x\boldsymbol{w}(k) > 0$

From Theorem 7.2, we have $\|\boldsymbol{w}(k)\|_{\boldsymbol{R}_x} < 1 + \eta\lambda_1$ for all $k \geq 0$. By using Theorem 7.2, we have

$$\begin{aligned}
&1 + \eta\left[(1 + \boldsymbol{w}^{\mathrm{T}}(k)\boldsymbol{R}_y\boldsymbol{w}(k) - (\lambda_i + 1)\boldsymbol{w}^{\mathrm{T}}(k)\boldsymbol{R}_x\boldsymbol{w}(k)\right] \\
&> 1 + \eta - \eta(1 + \lambda_1)\boldsymbol{w}^{\mathrm{T}}(k)\boldsymbol{R}_x\boldsymbol{w}(k) \\
&\geq 1 - \eta(\eta\lambda_1)(2 + \eta\lambda_1) - \eta\lambda_1(1 + \eta\lambda_1)^2 \\
&\geq 1 - 0.3 \times 0.35 \times (2 + 0.35) - 0.35 \times (1 + 0.35)^2 \\
&= 0.1154 > 0
\end{aligned} \tag{7.132}$$

By using (7.131) and (7.132), we can obtain that

$$\|\boldsymbol{w}(k+1)\|_{\boldsymbol{R}_x}^2 \geq \left[1+\eta-\eta(1+\lambda_1)(1+\eta\lambda_1)^2\right]^2(1+\eta\lambda_1)^2 \tag{7.133}$$

Case 2: $1+\boldsymbol{w}^{\mathrm{T}}(k)\boldsymbol{R}_y\boldsymbol{w}(k)-(\lambda_i+1)\boldsymbol{w}^{\mathrm{T}}(k)\boldsymbol{R}_x\boldsymbol{w}(k)<0$
In this case, we can obtain the following in equation

$$(\lambda_i+1)\boldsymbol{w}^{\mathrm{T}}(k)\boldsymbol{R}_x\boldsymbol{w}(k)-1>\boldsymbol{w}^{\mathrm{T}}(k)\boldsymbol{R}_y\boldsymbol{w}(k)>0 \tag{7.134}$$

From Theorem 7.2, we have

$$\begin{aligned}
&1-\eta\left[(\lambda_i+1)\boldsymbol{w}^{\mathrm{T}}(k)\boldsymbol{R}_x\boldsymbol{w}(k)-1-\boldsymbol{w}^{\mathrm{T}}(k)\boldsymbol{R}_y\boldsymbol{w}(k)\right]\\
&>1+\eta-\eta(\lambda_1+1)\boldsymbol{w}^{\mathrm{T}}(k)\boldsymbol{R}_x\boldsymbol{w}(k)\\
&\geq 1-\eta(\eta\lambda_1)(2+\eta\lambda_1)-\eta\lambda_1(1+\eta\lambda_1)^2\\
&\geq 1-0.3\times 0.35\times(2+0.35)-0.35\times(1+0.35)^2\\
&=0.1154>0
\end{aligned} \tag{7.135}$$

By using (7.131) and (7.135), we have

$$\|\boldsymbol{w}(k+1)\|_{\boldsymbol{R}_x}^2 \geq \left[1+\eta-\eta(1+\lambda_1)(1+\eta\lambda_1)^2\right]^2(1+\eta\lambda_1)^2 \tag{7.136}$$

Denote $c=\left[1+\eta-\eta(1+\lambda_1)(1+\eta\lambda_1)^2\right](1+\eta\lambda_1)$, and then, by using $\eta\lambda_1\leq 0.35$ and $\eta<0.3$, we can obtain that

$$\begin{aligned}
c&=\left[1+\eta-\eta(1+\lambda_1)(1+\eta\lambda_1)^2\right](1+\eta\lambda_1)\\
&>(1+0.35)\left[1+0.3-(0.3+0.35)(1+0.35)^2\right]\\
&=0.1558>0
\end{aligned} \tag{7.137}$$

Clearly, we have

$$\|\boldsymbol{w}(k+1)\|_{\boldsymbol{R}_x}\geq c \tag{7.138}$$

for all $k\geq 0$.

This completes the proof.

At this point, the boundness of the DDT system (7.112) of our algorithm has been proved. Next, we will prove that under some mild conditions, it holds that $\lim_{k\to\infty}\boldsymbol{w}(k)=a\boldsymbol{v}_n$, where a is a constant. In order to analyze the convergence of the DDT system (7.112), the following preliminary results are needed.

Lemma 7.2 *Suppose that* $\eta \leq 0.3$ *and* $\eta\lambda_1 \leq 0.35$. *If the initial vector satisfies* $\boldsymbol{w}^{\mathrm{T}}(0)\boldsymbol{R}_x\boldsymbol{v}_n \neq 0$, *then it holds that* $1+\eta(1+\boldsymbol{w}^{\mathrm{T}}(k)\boldsymbol{R}_y\boldsymbol{w}(k) - (\lambda_i+1)\boldsymbol{w}^{\mathrm{T}}(k)\boldsymbol{R}_x\boldsymbol{w}(k)) > 0$ *for all* $k \geq 0$, *where* $i = 1,2,\ldots,n$.

Proof From Theorem 7.1 and $\eta\lambda_1 \leq 0.2$, we have

$$\begin{aligned} & 1+\eta(1+\boldsymbol{w}^{\mathrm{T}}(k)\boldsymbol{R}_y\boldsymbol{w}(k) - (\lambda_i+1)\boldsymbol{w}^{\mathrm{T}}(k)\boldsymbol{R}_x\boldsymbol{w}(k)) \\ & > 1 - \eta\boldsymbol{w}^{\mathrm{T}}(k)\boldsymbol{R}_x\boldsymbol{w}(k) \\ & \geq 1 - \eta(1+\eta\lambda_1)^2 \\ & \geq 1 - 0.3 \times (1+0.35)^2 \\ & = 0.4532 > 0. \end{aligned} \tag{7.139}$$

This completes the proof.

Lemma 7.2 shows that $z_i(k) = \boldsymbol{v}_i^{\mathrm{T}}\boldsymbol{R}_x\boldsymbol{w}(k)$ $(i = 1,2,\ldots,n)$, which is the $\boldsymbol{R}_x$-orthogonal projection of the weight vector $\boldsymbol{w}(k)$ onto the generalized eigenvector $\boldsymbol{v}_i$ and does not change its sign in (7.120). Since $\boldsymbol{w}^{\mathrm{T}}(0)\boldsymbol{R}_x\boldsymbol{v}_n \neq 0$, we have

$$z_n(k) = \boldsymbol{v}_n^{\mathrm{T}}\boldsymbol{R}_x\boldsymbol{w}(k) \neq 0. \tag{7.140}$$

From (7.119), for all $k \geq 0$, $\boldsymbol{w}(k)$ can be represented as

$$\boldsymbol{w}(k) = \sum_{i=1}^{n-1} z_i(k)\boldsymbol{v}_i + z_n(k)\boldsymbol{v}_n. \tag{7.141}$$

Clearly, the convergence of $\boldsymbol{w}(k)$ can bc determined by the convergence of $z_i(k)(i = 1,2,\ldots,n)$. Lemmas 7.3 and 7.4 will provide the convergence analysis of $z_i(k)(i = 1,2,\ldots,n)$.

Lemma 7.3 *Suppose that* $\eta \leq 0.3$ *and* $\eta\lambda_1 \leq 0.35$. *If the initial vector satisfies* $\boldsymbol{w}^{\mathrm{T}}(0)\boldsymbol{R}_x\boldsymbol{v}_n \neq 0$, *then it holds that* $\lim_{k\to\infty} z_i(k) = 0,\ (i = 1,2,\ldots,n-1)$.

Proof From Lemma 7.2, we have

$$1+\eta\boldsymbol{w}^{\mathrm{T}}(k)\boldsymbol{R}_y\boldsymbol{w}(k) - \eta\lambda_i(\boldsymbol{w}^{\mathrm{T}}(k)\boldsymbol{R}_x\boldsymbol{w}(k))^2 > 0, \tag{7.142}$$

for all $k \geq 0$, where $i = 1,2,\ldots,n$. Furthermore, from Theorem 7.1 and Theorem 7.2, it holds that

$$c < \|\boldsymbol{w}(k)\|_{\boldsymbol{R}_x} < 1+\eta\lambda_1, \tag{7.143}$$

for all $k \geq 0$. Then, we have

$$\begin{aligned}
&\left[\frac{z_i(k+1)}{z_n(k+1)}\right]^2 \\
&= \left[\frac{1+\eta(1+\boldsymbol{w}^{\mathrm{T}}(k)\boldsymbol{R}_y\boldsymbol{w}(k)-(\lambda_i+1)\boldsymbol{w}^{\mathrm{T}}(k)\boldsymbol{R}_x\boldsymbol{w}(k))}{1+\eta(1+\boldsymbol{w}^{\mathrm{T}}(k)\boldsymbol{R}_y\boldsymbol{w}(k)-(\lambda_n+1)\boldsymbol{w}^{\mathrm{T}}(k)\boldsymbol{R}_x\boldsymbol{w}(k))}\right]^2\frac{z_i^2(k)}{z_n^2(k)} \\
&= \left[1-\frac{\eta(\lambda_i-\lambda_n)\boldsymbol{w}^{\mathrm{T}}(k)\boldsymbol{R}_x\boldsymbol{w}(k)}{1+\eta(1+\boldsymbol{w}^{\mathrm{T}}(k)\boldsymbol{R}_y\boldsymbol{w}(k)-(\lambda_n+1)\boldsymbol{w}^{\mathrm{T}}(k)\boldsymbol{R}_x\boldsymbol{w}(k))}\right]^2\frac{z_i^2(k)}{z_n^2(k)} \\
&\le \left[1-\frac{\eta(\lambda_i-\lambda_n)\boldsymbol{w}^{\mathrm{T}}(k)\boldsymbol{R}_x\boldsymbol{w}(k)}{1+\eta(1+(\lambda_1-\lambda_n-1)\boldsymbol{w}^{\mathrm{T}}(k)\boldsymbol{R}_x\boldsymbol{w}(k))}\right]^2\frac{z_i^2(k)}{z_n^2(k)},
\end{aligned} \tag{7.144}$$

for all $k \ge 0$, where $i = 1, 2, \ldots, n-1$.

Let $\delta_k = \frac{\eta(\lambda_i-\lambda_n)\boldsymbol{w}^{\mathrm{T}}(k)\boldsymbol{R}_x\boldsymbol{w}(k)}{1+\eta(1+(\lambda_1-\lambda_n-1)\boldsymbol{w}^{\mathrm{T}}(k)\boldsymbol{R}_x\boldsymbol{w}(k))}$. Then, we will prove that $0 < \delta_k < 2$.

First, we will prove that $\delta_k > 0$. From $\eta \le 0.3$ and $\eta\lambda_1 \le 0.2$, we have

$$\begin{aligned}
&1+\eta(1+(\lambda_1-\lambda_n-1)\boldsymbol{w}^{\mathrm{T}}(k)\boldsymbol{R}_x\boldsymbol{w}(k)) \\
&> 1-\eta\boldsymbol{w}^{\mathrm{T}}(k)\boldsymbol{R}_x\boldsymbol{w}(k) \\
&> 1-\eta(1+\eta\lambda_1)^2 \\
&\ge 1-0.3\times(1+0.35)^2 \\
&= 0.4532 > 0.
\end{aligned} \tag{7.145}$$

From $\lambda_i > \lambda_n$ and $\boldsymbol{w}^{\mathrm{T}}(k)\boldsymbol{R}_x\boldsymbol{w}(k) > 0$, we have $\delta_k > 0$.

Next, we will prove that $\delta_k < 2$.

$$\begin{aligned}
\delta_k &= \frac{\eta(\lambda_i-\lambda_n)\boldsymbol{w}^{\mathrm{T}}(k)\boldsymbol{R}_x\boldsymbol{w}(k)}{1+\eta(1+(\lambda_1-\lambda_n-1)\boldsymbol{w}^{\mathrm{T}}(k)\boldsymbol{R}_x\boldsymbol{w}(k))} \\
&< \frac{\eta\lambda_1\boldsymbol{w}^{\mathrm{T}}(k)\boldsymbol{R}_x\boldsymbol{w}(k)}{1+\eta(1-\boldsymbol{w}^{\mathrm{T}}(k)\boldsymbol{R}_x\boldsymbol{w}(k))} \\
&< \frac{\eta\lambda_1(1+\eta\lambda_1)^2}{1-\eta(1+\eta\lambda_1)^2} \\
&\le \frac{0.35\times(1+0.35)^2}{1-0.3\times(1+0.35)^2} \\
&= 1.4073 < 2.
\end{aligned} \tag{7.146}$$

Denote $\theta_k = (1-\delta_k)^2$. Then, $0 < \theta_k < 1$. From (7.144), we have

$$\left[\frac{z_i(k+1)}{z_n(k+1)}\right]^2 \le \theta_k\frac{z_i^2(k)}{z_n^2(k)} \le \cdots \le \prod_{j=0}^{k}\theta_j\frac{z_i^2(0)}{z_n^2(0)}, \tag{7.147}$$

for all $k \ge 0$, where $i = 1, 2, \ldots, n-1$.

Denote $\theta = \max(\theta_0, \theta_1, \cdots \theta_k, \cdots)$. Clearly, $0 < \theta < 1$. Then, it holds that

$$\left[\frac{z_i(k+1)}{z_n(k+1)}\right]^2 \leq \theta^{k+1} \frac{z_i^2(0)}{z_n^2(0)}. \tag{7.148}$$

Since $0 < \theta < 1$, then it follows that

$$\lim_{k \to \infty} \frac{z_i(k)}{z_n(k)} = 0 \quad i = 1, 2, \ldots, n-1. \tag{7.149}$$

From Theorem 7.1 and Theorem 7.2, we can see that $z_n(k)$ must be bounded. Then, we have

$$\lim_{k \to \infty} z_i(k) = 0, \quad i = 1, 2, \ldots, n-1. \tag{7.150}$$

This completes the proof.

Lemma 7.4 *Suppose that* $\eta \leq 0.3$ *and* $\eta\lambda_1 \leq 0.35$. *If the initial vector satisfies* $\boldsymbol{w}^{\mathrm{T}}(0)\boldsymbol{R}_x\boldsymbol{v}_n \neq 0$, *then it holds that* $\lim_{k \to \infty} z_n(k) = a$, *where a is a constant.*

Proof From Lemma 7.3, we know that $w(k)$ will converge to the direction of the minor generalized eigenvector $\boldsymbol{v}_n$, as $k \to \infty$. Suppose that $\boldsymbol{w}(k)$ has converged to the direction of $\boldsymbol{v}_n$ at time k_0, i.e., $\boldsymbol{w}(k_0) = z_n(k_0)\boldsymbol{v}_n$.

From (7.120), we have

$$\begin{aligned} z_n(k+1) &= \left[1 + \eta\left[1 + \lambda_n z_n^2(k) - (\lambda_n + 1)z_n^2(k)\right]\right] z_n(k) \\ &= \left[1 + \eta(1 - z_n^2(k))\right] z_n(k). \end{aligned} \tag{7.151}$$

From (7.151), it holds that

$$\begin{aligned} z_n(k+1) - 1 &= \left[1 + \eta(1 - z_n^2(k))\right] z_n(k) - 1 \\ &= [1 - \eta[z_n(k) - 1][z_n(k) + 1]] z_n(k) - 1 \\ &= [z_n(k) - 1][1 - \eta z_n(k)[z_n(k) + 1]], \end{aligned} \tag{7.152}$$

for all $k \geq k_0$. Then, from $\eta \leq 0.3$ and $\eta\lambda_1 \leq 0.35$, we have

$$\begin{aligned} 1 - \eta z_n(k)[z_n(k) + 1] &\geq 1 - \eta(1 + \eta\lambda_1)(1 + \eta\lambda_1 + 1) \\ &\geq 1 - 0.3 \times (1 + 0.35)(1 + 0.35 + 1) \\ &= 0.0482 > 0, \end{aligned} \tag{7.153}$$

for all $k \geq k_0$.

Denote $\beta = |1 - \eta z_n(k)[z_n(k) + 1]|$. Clearly, $0 < \beta < 1$. From (7.152) and (7.153), it follows that for all $k \ge k_0$

$$|z_n(k+1) - 1| \le \beta |z_n(k) - 1| \le \cdots \le \beta^{k+1} |z_n(0) - 1|. \tag{7.154}$$

Denote $\alpha = -\ln \beta$ and $\Pi_1 = |z_n(0) - 1|$. Substituting them into (7.154), then we have

$$|z_n(k+1) - 1| \le (k+1)\Pi_1 \mathrm{e}^{-\alpha(k+1)}. \tag{7.155}$$

Given any $\varepsilon > 0$, there exists a constant $K > 1$ such that

$$\frac{\Pi_2 K \mathrm{e}^{-\alpha K}}{(1 - \mathrm{e}^{-\alpha})^2} \le \varepsilon. \tag{7.156}$$

For any $k_1 > k_2 > K$, it follows from (7.120) and (7.155) that

$$\begin{aligned} |z_n(k_1) - z_n(k_2)| &= \left| \sum_{r=k_2}^{k_1-1} [z_n(r+1) - z_n(r)] \right| \\ &\le \left| \sum_{r=k_2}^{k_1-1} [\eta z_n(r)(1 - z_n^2(r))] \right| \le \sum_{r=k_2}^{k_1-1} \eta |z_n(r)||z_n(r) + 1||z_n(r) - 1| \\ &\le \eta(1 + \eta\lambda_1)(2 + \eta\lambda_1) \sum_{r=k_2}^{k_1} |z_n(r) - 1| \le \Pi_2 \sum_{r=k_2}^{k_1} r\mathrm{e}^{-\alpha r} \\ &\le \Pi_2 \sum_{r=K}^{+\infty} r\mathrm{e}^{-\alpha r} \le \Pi_2 K \mathrm{e}^{-\alpha K} \sum_{r=0}^{+\infty} r(\mathrm{e}^{-\alpha})^{r-1} \le \frac{\Pi_2 K \mathrm{e}^{-\alpha K}}{(1 - \mathrm{e}^{-\alpha})^2} \le \varepsilon, \end{aligned} \tag{7.157}$$

where $\Pi_2 = \eta(1 + \eta\lambda_1)(2 + \eta\lambda_1)|z_n(0) - 1|$.

This implies that the sequence $\{z_n(k)\}$ is a Cauchy sequence. By using the Cauchy convergence principle, there must exist a constant a such that $\lim_{k\to\infty} z_n(k) = a$.

This completes the proof.

Theorem 7.4 *Suppose that* $\eta \le 0.3$ *and* $\eta\lambda_1 \le 0.35$. *If the initial vector satisfies* $\boldsymbol{w}^{\mathrm{T}}(0)\boldsymbol{R}_x\boldsymbol{v}_n \ne 0$, *then it holds that* $\lim_{k\to\infty} \boldsymbol{w}(k) = a\boldsymbol{v}_n$.

Proof From (7.141) and Lemmas 7.3–7.4, we have

$$\begin{aligned}
&\lim_{k\to\infty} \boldsymbol{w}(k) \\
&= \lim_{k\to\infty}\left(\sum_{i=1}^{n-1} z_i(k)\boldsymbol{v}_i + z_n(k)\boldsymbol{v}_n\right) \\
&= \lim_{k\to\infty}\left(\sum_{i=1}^{n-1} z_i(k)\boldsymbol{v}_i\right) + \lim_{k\to\infty}(z_n(k)\boldsymbol{v}_n) \\
&= a\boldsymbol{v}_n.
\end{aligned} \tag{7.158}$$

This completes the proof.

At this point, we have finished the proof of the convergence analysis of algorithm (7.51). Next, we will provide some remarks on the convergence conditions obtained in Theorem 7.4.

Remark 7.1 From the conditions $\eta \leq 0.3$ and $\eta\lambda_1 \leq 0.35$ in Theorem 7.4, we can see that the selection of the learning rate is related to the largest generalized eigenvalue. In many signal processing fields, although the largest generalized eigenvalue is unknown, its upper can be estimated based on the problem-specific knowledge [42]. So it is very easy to choose an appropriate learning rate to satisfy the two conditions $\eta \leq 0.3$ and $\eta\lambda_1 \leq 0.35$. Another required condition is that the initial weight vector must satisfy $\boldsymbol{w}^{\mathrm{T}}(0)\boldsymbol{R}_x\boldsymbol{v}_n \neq 0$. In practical applications, when the initial weight vector is randomly generated, the probability for the above condition to be satisfied will be one. So the conditions in Theorem 7.4 are reasonable and easy to be satisfied in practical applications.

Remark 7.2 It is worth to note that by simply switching some signs in (7.112), the proposed algorithm can become a principal generalized eigenvector extraction algorithm. Similar to Eq. (7.112), the principal generalized eigenvector extraction algorithm also has self-stabilizing property. The analysis about the self-stabilizing property and the convergence is very similar to that for Eq. (7.112), so we omit the proofs here.

7.3.4 Computer Simulations

In this section, we will provide simulation results to illustrate the performance of our algorithm for extracting the minor generalized eigenvectors. The first simulation is designed to extract minor generalized eigenvectors from two random vector processes and compare our algorithm with the other algorithm. The second simulation mainly shows the convergence of our algorithm under the conditions in Theorem 7.4. The simulation results in Figs. 7.1, 7.2, 7.3, 7.4, and 7.6 are obtained by averaging over 100 independent experiments.

In order to evaluate the convergence speed and the estimation accuracy of our algorithm, we compute the $\boldsymbol{R}_x$-norm of the weight vector at the kth update

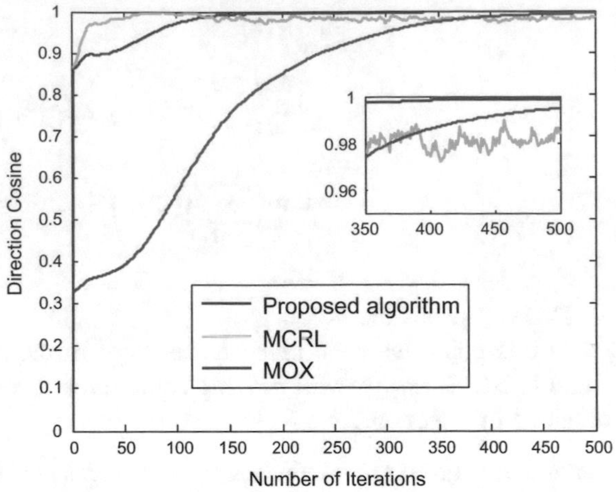

Fig. 7.1 DC curves of the minor generalized eigenvector estimated by using three algorithms

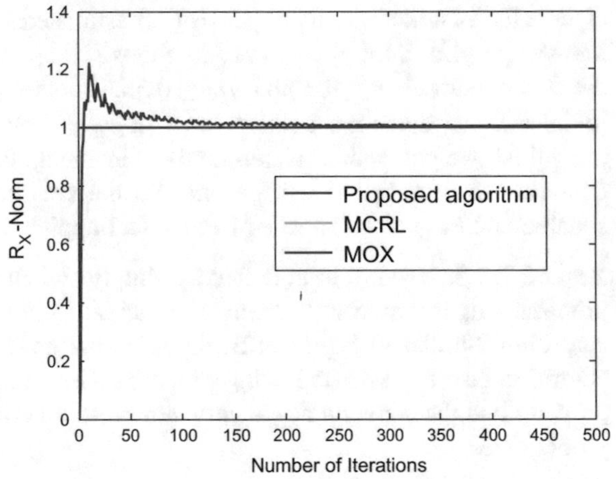

Fig. 7.2 $\boldsymbol{R}_x$-norm curves of the minor generalized eigenvector estimated by using three algorithms

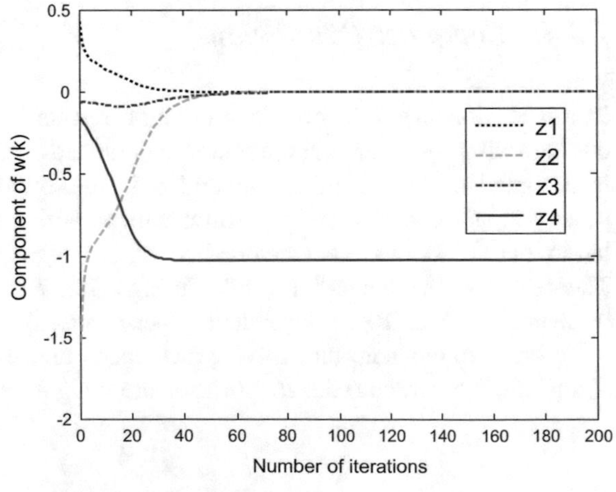

Fig. 7.3 Convergence of $\boldsymbol{w}(k)$

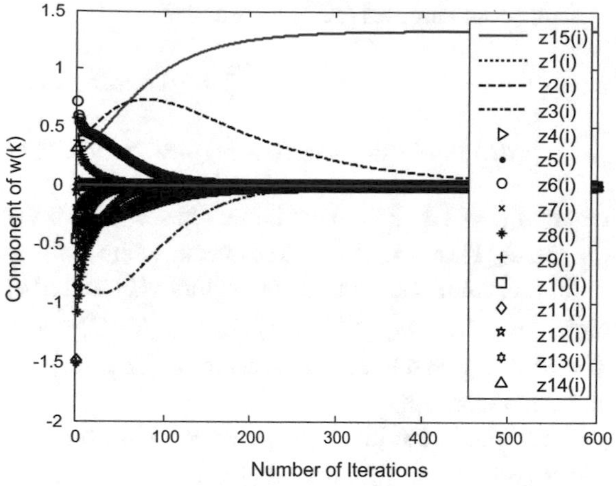

Fig. 7.4 Convergence of $\boldsymbol{w}(k)$ with high-dimensional matrices

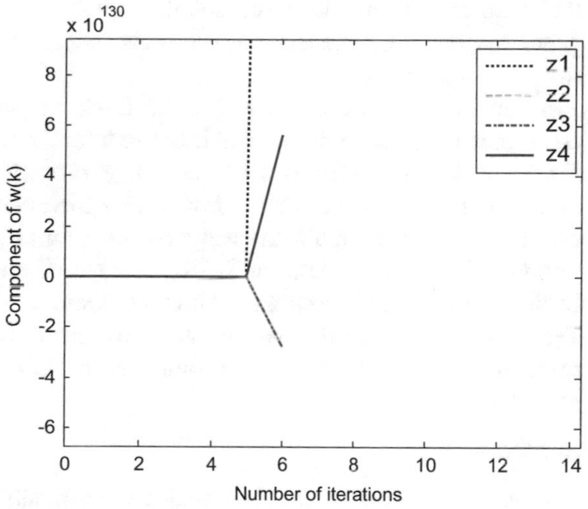

Fig. 7.5 Divergence of $\boldsymbol{w}(k)$

$$\rho(\boldsymbol{w}_k) = \|\boldsymbol{w}(k)\|_{\boldsymbol{R}_x} = \sqrt{\boldsymbol{w}^{\mathrm{T}}(k)\boldsymbol{R}_x\boldsymbol{w}(k)}, \tag{7.159}$$

and the direction cosine(k) (DC(k)).

Simulation 1

In this simulation, we compare our algorithm with the modified coupled learning rule (MCLR) [31] and the modified Oja–Xu algorithm (MOX) [33], which are two well-known minor generalized eigenvector extraction algorithms proposed in recent years.

The input samples are generated by

$$y(k) = \sqrt{2}\sin(0.62\pi k + \theta_1) + n_1(k), \tag{7.160}$$

$$x(k) = \sqrt{2}\sin(0.46\pi k + \theta_2) + \sqrt{2}\sin(0.74\pi k + \theta_3) + n_2(k), \tag{7.161}$$

where $\theta_i(i = 1,2,3)$ are the initial phases, which follow uniform distributions over $[0, 2\pi]$, $n_1(k)$ and $n_2(k)$ are zero-mean white noises with variances $\sigma_1^2 = \sigma_2^2 = 0.1$.

In this simulation, the input vectors $y(k)$ and $x(k)$ are arranged in blocks of size eight ($n = 8$), i.e., $\boldsymbol{y}(k) = [y(k), \ldots, y(k+7)]^{\mathrm{T}}$ and $\boldsymbol{x}(k) = [x(k), \ldots, x(k+7)]^{\mathrm{T}}$. The minor generalized eigenvectors of $(\boldsymbol{R}_y, \boldsymbol{R}_x)$ are extracted via MCLR, MOX, and our algorithm.

The initial conditions in the three algorithms are as follows. For MCLR, $\gamma = 0.998$ and $\lambda(0) = 100$. For all algorithms, $\alpha = \beta = 0.998$, $\hat{\boldsymbol{R}}_y(0) = \hat{\boldsymbol{R}}_x(0) = \boldsymbol{0}$, and $\boldsymbol{Q}_x(0) = \delta\boldsymbol{I}$, where δ is a small positive number and $\boldsymbol{0}$ is an $n \times n$ zero matrix. For the three algorithms, the same initial weight vector is used, which is randomly generated, and the same learning rate is also used. The simulation results are shown in Figs. 7.1 and 7.2.

From Fig. 7.1, we can see that the direction cosine curve of our algorithm converges to 1, which means that it can extract the minor generalized eigenvector. From the last 150 iterations of the whole procedure, we can observe that our algorithm has the least deviation from one, which means our algorithm has the best estimation accuracy among these algorithms. From Fig. 7.2, we can see that the $\boldsymbol{R}_x$-norms of the weight vector in MCRL and MOX are equal to 1 due to a normalization step in the algorithms. Although there is no normalization step in our algorithm, the $\boldsymbol{R}_x$-norm curve in our algorithm also converges to one. Since the normalization step is omitted, the computational complexity is also reduced for our algorithm.

Simulation 2

Consider the following two nonnegative symmetric definite matrices, which are randomly generated. The generalized eigenvalues of the matrix pencil $(\boldsymbol{R}_y, \boldsymbol{R}_x)$ are $\lambda_1 = 1.8688, \lambda_2 = 1.1209, \lambda_3 = 1.0324, \lambda_4 = 0.7464$.

$$\boldsymbol{R}_y = \begin{bmatrix} 0.8313 & 0.0876 & -0.0510 & 0.1362 \\ 0.0876 & 0.8895 & 0.0790 & -0.0351 \\ -0.0510 & 0.0790 & 0.9123 & -0.0032 \\ 0.1362 & -0.0351 & -0.0032 & 0.7159 \end{bmatrix} \tag{7.162}$$

$$\boldsymbol{R}_x = \begin{bmatrix} 0.7307 & -0.1700 & -0.0721 & 0.0129 \\ -0.1700 & 0.6426 & 0.0734 & -0.0011 \\ -0.0721 & 0.0734 & 0.8202 & -0.0102 \\ 0.0129 & -0.0011 & -0.0102 & 0.7931 \end{bmatrix} \tag{7.163}$$

The initial weight vector is randomly generated, and the learning rate is $\eta = 0.1$, which satisfies the condition that $\eta \leq 0.3$ and $\eta\lambda_1 \leq 0.35$. Figure 7.3 shows that the convergence of the component $z_i(k)$ of $\boldsymbol{w}(k)$, where $z_i(k) = \boldsymbol{w}^{\mathrm{T}}(k)\boldsymbol{R}_x\boldsymbol{v}_i$ is the coordinate of $\boldsymbol{w}(k)$ in the direction of the generalized eigenvector $\boldsymbol{v}_i(i = 1, 2, 3, 4)$. In the simulation result, $z_i(k)(i = 1, 2, 3)$ converges to zero and $z_4(k)$ converges to a constant, which is consistent with the results in Theorem 7.4.

Next, we will provide a simulation to show that the proposed algorithm can deal with high-dimensional matrix pencil. We randomly generate two 15×15 matrices, and their largest generalized eigenvalue is $\lambda_1 = 9.428$. The other initial conditions are the same as in the above simulation. The simulation results are shown in Fig. 7.4. From this figure, we can see the same results as in the above simulations. So we can say that our algorithm can deal with high-dimensional matrix pencil.

The last simulation is an example, in which the learning rate does not satisfy the conditions and the norm of $\boldsymbol{w}(k)$ diverges. We still use the two matrices (7.162) and (7.163), and the learning rate is $\eta = 0.5$. When the initial vector is $\boldsymbol{w} = [0.1765, -0.7914, 1.3320, 2.3299]^{\mathrm{T}}$, we can obtain the following simulation result, as shown in Fig. 7.5. From this figure, we can see that the divergence may occur when the learning rate does not satisfy the conditions in Theorem 7.4.

In Sect. 7.3, we have provided an algorithm for extracting the minor generalized eigenvector of a matrix pencil, which has the self-stabilizing property. The convergence of the algorithm has been analyzed by the DDT method, and some sufficient conditions have also been obtained to guarantee its convergence. Simulation experiments show that compared with other algorithms, this algorithm has faster convergence speed and better estimation accuracy and can deal with high-dimensional matrix pencil.

7.4 Novel Multiple GMC Extraction Algorithm

In Sect. 7.3, we propose an adaptive GMC algorithm. However, this algorithm can only extract one minor generalized eigenvector of a matrix pencil. In some applications, estimating multiple GMCs or the subspace spanned by these GMCs is of interest. The methods for solving the multiple eigenvector extraction have the "inflation" procedure and parallel algorithms. The purpose of this part is to develop an information criterion and an algorithm for multiple GMC extraction.

7.4.1 An Inflation Algorithm for Multiple GMC Extraction

Here, multiple GMC extraction algorithm is to estimate the first r GMCs of the matrix pencil $(\boldsymbol{R}_y, \boldsymbol{R}_x)$. Through some modification, the information criterion in [44] can be extended into a GMC extraction information criterion (GIC) as follows:

$$\begin{aligned} \boldsymbol{w}^{*} &= \arg\max J_{\mathrm{GIC}}(\boldsymbol{w}) \\ J_{\mathrm{GIC}}(\boldsymbol{w}) &= \frac{1}{2}\mathrm{tr}\left[\ln\left(\boldsymbol{w}^{\mathrm{T}}\boldsymbol{R}_x\boldsymbol{w}\right) - \boldsymbol{w}^{\mathrm{T}}\boldsymbol{R}_y\boldsymbol{w}\right] \end{aligned} \tag{7.164}$$

where $\boldsymbol{w} \in \mathbb{R}^{n\times 1}$ is a vector. GIC can only estimate the first GMC of the matrix pencil $(\boldsymbol{R}_y, \boldsymbol{R}_x)$.

In the domain $\Omega = \{\boldsymbol{w}|0 < \boldsymbol{w}^{\mathrm{T}}\boldsymbol{R}_x\boldsymbol{w} < \infty\}$, $\boldsymbol{w}^{\mathrm{T}}\boldsymbol{R}_x\boldsymbol{w}$ is a positive scalar. Thus, the gradient of $J_{\mathrm{GIC}}(\boldsymbol{w})$ with respect to $\boldsymbol{w}$ exists and is given by

$$\boldsymbol{g} = \nabla J_{\mathrm{GIC}}(\boldsymbol{w}) = \boldsymbol{R}_x\boldsymbol{w}\left(\boldsymbol{w}^{\mathrm{T}}\boldsymbol{R}_x\boldsymbol{w}\right)^{-1} - \boldsymbol{R}_y\boldsymbol{w} \tag{7.165}$$

After some discretization operations, we can obtain the following nonlinear stochastic learning rule:

$$\boldsymbol{w}(k+1) = \boldsymbol{w}(k) + \eta\left[\boldsymbol{R}_x\boldsymbol{w}(k)\left(\boldsymbol{w}^{\mathrm{T}}(k)\boldsymbol{R}_x\boldsymbol{w}(k)\right)^{-1} - \boldsymbol{R}_y\boldsymbol{w}(k)\right] \tag{7.166}$$

where η is the learning rate and satisfies $0 < \eta < 1$. Generally, gradient method cannot lead to a fast algorithm, so we use the quasi-Newton method to derive a fast algorithm. By using (7.165), we can obtain the Hessian matrix of $J_{\mathrm{GIC}}(\boldsymbol{w})$:

$$\boldsymbol{H}_1 = \boldsymbol{R}_x\left(\boldsymbol{w}^{\mathrm{T}}\boldsymbol{R}_x\boldsymbol{w}\right)^{-1} - 2\left(\boldsymbol{w}^{\mathrm{T}}\boldsymbol{R}_x\boldsymbol{w}\right)^{-2}\boldsymbol{R}_x\boldsymbol{w}\boldsymbol{w}^{\mathrm{T}}\boldsymbol{R}_x - \boldsymbol{R}_y \tag{7.167}$$

In order to use the Quasi-Newton method, we use some approximation in the Hessian matrix $\boldsymbol{H}_1$ and define another matrix, which can be written as

$$\boldsymbol{H}_2 = 2\left(\boldsymbol{w}^{\mathrm{T}}\boldsymbol{R}_x\boldsymbol{w}\right)^{-2}\boldsymbol{R}_x\boldsymbol{w}\boldsymbol{w}^{\mathrm{T}}\boldsymbol{R}_x + \boldsymbol{R}_y \tag{7.168}$$

Obviously, we can obtain that $\boldsymbol{H}_1 \approx -\boldsymbol{H}_2$. Then, the inverse of $\boldsymbol{H}_2$ is given by

$$\boldsymbol{H}_2^{-1} = \boldsymbol{R}_y^{-1} - \frac{2\boldsymbol{R}_y^{-1}\boldsymbol{R}_x\boldsymbol{w}\boldsymbol{w}^{\mathrm{T}}\boldsymbol{R}_x\boldsymbol{R}_y^{-1}}{(\boldsymbol{w}^{\mathrm{T}}\boldsymbol{R}_x\boldsymbol{w})^2 + 2\boldsymbol{w}^{\mathrm{T}}\boldsymbol{R}_x\boldsymbol{R}_y^{-1}\boldsymbol{R}_x\boldsymbol{w}} \tag{7.169}$$

Following the procedures of the quasi-Newton method, we can obtain the following fast learning rule for updating the state vector of the neural network:

$$\begin{aligned} \boldsymbol{w}(k+1) &= \boldsymbol{w}(k) + \eta\boldsymbol{H}_2^{-1} * \boldsymbol{g}|_{\boldsymbol{w}=\boldsymbol{w}(k)} \\ &= (1-\eta)\boldsymbol{w}(k) + \eta\frac{3(\boldsymbol{w}^{\mathrm{T}}(k)\boldsymbol{R}_x\boldsymbol{w}(k))\boldsymbol{R}_y^{-1}\boldsymbol{R}_x\boldsymbol{w}(k)}{(\boldsymbol{w}^{\mathrm{T}}(k)\boldsymbol{R}_x\boldsymbol{w}(k))^2 + 2\boldsymbol{w}^{\mathrm{T}}(k)\boldsymbol{R}_x\boldsymbol{R}_y^{-1}\boldsymbol{R}_x\boldsymbol{w}(k)} \end{aligned} \tag{7.170}$$

Considering the autocorrelation matrices ($\boldsymbol{R}_x$ and $\boldsymbol{R}_y$) are often unknown in prior, we use the exponentially weighted sample correlation matrices $\hat{\boldsymbol{R}}_x(k)$ and $\hat{\boldsymbol{R}}_y(k)$ instead of $\boldsymbol{R}_x$ and $\boldsymbol{R}_y$, respectively. The recursive update equation can be written as

$$\hat{\boldsymbol{R}}_x(k+1) = \alpha_1 \hat{\boldsymbol{R}}_x(k) + \boldsymbol{x}(k+1)\boldsymbol{x}^{\mathrm{T}}(k+1) \tag{7.171}$$

$$\hat{\boldsymbol{R}}_y(k+1) = \alpha_2 \hat{\boldsymbol{R}}_y(k) + \boldsymbol{y}(k+1)\boldsymbol{y}^{\mathrm{T}}(k+1) \tag{7.172}$$

By using the matrix-inversion lemma and Eq. (7.172), we can also obtain the updating equation for $\boldsymbol{Q}_y(k) = \hat{\boldsymbol{R}}_y^{-1}(k)$:

$$\boldsymbol{Q}_y(k+1) = \frac{1}{\alpha_2}\left[\boldsymbol{Q}_y(k) - \frac{\boldsymbol{Q}_y(k)\boldsymbol{y}(k+1)\boldsymbol{y}^{\mathrm{T}}(k+1)\boldsymbol{Q}_y(k)}{\alpha_2 + \boldsymbol{y}^{\mathrm{T}}(k+1)\boldsymbol{Q}_y(k)\boldsymbol{y}(k+1)}\right] \tag{7.173}$$

where α_1, α_2 denote the forgetting factors and satisfy $0 < \alpha_1, \alpha_2 \leq 1$. The rules for choosing the values of α_1, α_2 can be found in [45], and the interested readers may refer to it for detail.

The above learning rule has fast convergence speed, but it can only extract one GMC. By using the inflation technique, the above learning rule can accomplish the extraction of multiple GMCs, whose detail implementation procedures are shown in Table 7.2.

In the above table, λ_n is the largest generalized eigenvalue of the matrix pencil $(\boldsymbol{R}_y, \boldsymbol{R}_x)$. In many practical applications, although the value of λ_n is unknown, its upper bound can be estimated based on the problem-specific knowledge [46]. After the iteration procedure in Table 7.2, we can obtain vector sequences $\boldsymbol{w}_i (i = 1, 2, \ldots, r)$, which can justly compose the first r GMCs of the matrix pencil $(\boldsymbol{R}_y, \boldsymbol{R}_x)$.

The above inflation procedure is computationally expensive when extracting higher-order components, and it belongs to sequential methods and this method usually produces a long processing delay since the different components are extracted one after another, but also it needs more memory to store the input samples since they are repeatedly used. Different from the sequential methods, the parallel methods can extract the components simultaneously and overcome the drawbacks of the sequential methods. In the following, we will develop a parallel multiple GMC extraction algorithm.

Table 7.2 The iteration procedure for multiple GMC extraction

Multiple GMC extraction by inflation method
Initialization: Set $i = 1$ and $\boldsymbol{R}_1 = \boldsymbol{R}_y$ Iteration: For $i = 1, 2, \ldots, r$, Step 1. Calculate the first FMC of the matrix pencil $(\boldsymbol{R}_i, \boldsymbol{R}_x)$ by using Eq. (7.170), and denote it as $\boldsymbol{w}_i$ Step 2. Set $i = i + 1$, and update the matrix $\boldsymbol{R}_i$ by the following equation: $\boldsymbol{R}_i = \boldsymbol{R}_{i-1} + \tau \boldsymbol{R}_x \boldsymbol{w}_{i-1} \boldsymbol{w}_{i-1}^{\mathrm{T}} \boldsymbol{R}_x / (\boldsymbol{w}_{i-1}^{\mathrm{T}} \boldsymbol{R}_x \boldsymbol{w}_{i-1})$ (7.174)
where τ is some constant and satisfies $\tau > \lambda_n$

7.4.2 A Weighted Information Criterion and Corresponding Multiple GMC Extraction

(1) A weighted information criterion for multiple GMC extraction

In this section, Eq. (7.99) is modified into a multiple GMC extraction information by using the weighted matrix method. Here, we denote this information criterion as WGIC, which can be expressed as

$$\begin{aligned} \boldsymbol{W}^* &= \arg\max J_{\text{WGIC}}(\boldsymbol{W}) \\ J_{\text{WGIC}}(\boldsymbol{W}) &= \frac{1}{2}\text{tr}\left[\ln\left(\boldsymbol{W}^{\text{T}}\boldsymbol{R}_x\boldsymbol{W}\boldsymbol{A}\right) - \boldsymbol{W}^{\text{T}}\boldsymbol{R}_y\boldsymbol{W}\right] \end{aligned} \tag{7.175}$$

where $\boldsymbol{W} \in \mathbb{R}^{n\times r}$ is the state matrix of the neural networks and $\boldsymbol{A} = \text{diag}(a_1, a_2, \ldots, a_r)$ is a diagonal matrix and its diagonal elements satisfy $a_1 > a_2 > \cdots > a_r$. The landscape of the WGIC is described by the following two theorems.

(2) Landscape of Generalized Information Criterion

Theorem 7.5 *In the domain* $\Omega = \left\{\boldsymbol{W}|0 < \boldsymbol{W}^{\text{T}}\boldsymbol{R}_x\boldsymbol{W} < \infty\right\}$, $\boldsymbol{W}$ *is a stationary point of* $J_{\text{WGIC}}(\boldsymbol{W})$ *if and only if* $\boldsymbol{W} = \boldsymbol{L}_r\boldsymbol{\Lambda}_r^{-\frac{1}{2}}\boldsymbol{Q}'$, *where* $\boldsymbol{\Lambda}_r$ *is a* $r \times r$ *diagonal matrix, whose diagonal elements are any r distinct generalized eigenvalues of the matrix pencil* $(\boldsymbol{R}_y, \boldsymbol{R}_x)$, $\boldsymbol{L}_r$ *is an* $n \times r$ *matrix and is composed of the corresponding generalized eigenvectors, and* $\boldsymbol{Q}'$ *is a permutation matrix.*

Proof Since $\boldsymbol{W}^{\text{T}}\boldsymbol{R}_x\boldsymbol{W}$ is positive define in the domain Ω, it is invertible. By using (7.175), we can calculate the first-order differential of $J_{\text{WGIC}}(\boldsymbol{W})$

$$\begin{aligned} \text{d}J_{\text{WGIC}}(\text{W}) &= \text{d}\left\{\text{tr}\left[log(\boldsymbol{W}^{\text{T}}\boldsymbol{R}_x\boldsymbol{W}\boldsymbol{A})\right] - \text{tr}(\boldsymbol{W}^{\text{T}}\boldsymbol{R}_y\boldsymbol{W})\right\} \\ &= \text{tr}\left[(\boldsymbol{W}^{\text{T}}\boldsymbol{R}_x\boldsymbol{W}\boldsymbol{A})^{-1}\boldsymbol{A}\boldsymbol{W}^{\text{T}}\boldsymbol{R}_x\text{d}\boldsymbol{W}\right] - \text{tr}(\boldsymbol{W}^{\text{T}}\boldsymbol{R}_y\text{d}\boldsymbol{W}) \end{aligned} \tag{7.176}$$

Then, we can obtain the gradient of $J_{\text{WGIC}}(\boldsymbol{W})$ with respect to $\boldsymbol{W}$:

$$\nabla J_{\text{WGIC}}(\boldsymbol{W}) = \boldsymbol{R}_x\boldsymbol{W}\left(\boldsymbol{A}\boldsymbol{W}^{\text{T}}\boldsymbol{R}_x\boldsymbol{W}\boldsymbol{A}^{-1}\right)^{-1} - \boldsymbol{R}_y\boldsymbol{W} \tag{7.177}$$

If $\boldsymbol{W} = \boldsymbol{L}_r\boldsymbol{\Lambda}^{-\frac{1}{2}}\boldsymbol{Q}'$, it is easy to show that $\nabla J_{\text{WGIC}}(\boldsymbol{W}) = \boldsymbol{0}$. Conversely, by definition, the stationary point of $J_{\text{WGIC}}(\boldsymbol{W})$ satisfies $\nabla J_{\text{WGIC}}(\boldsymbol{W}) = \boldsymbol{0}$. Then, we have

$$\boldsymbol{R}_x\boldsymbol{W}\left(\boldsymbol{A}\boldsymbol{W}^{\text{T}}\boldsymbol{R}_x\boldsymbol{W}\boldsymbol{A}^{-1}\right)^{-1} = \boldsymbol{R}_y\boldsymbol{W} \tag{7.178}$$

Premultiplying both sides of (7.178) by $\boldsymbol{W}^{\mathrm{T}}$, we can obtain

$$\boldsymbol{W}^{\mathrm{T}}\boldsymbol{R}_x\boldsymbol{W}\left(\boldsymbol{A}\boldsymbol{W}^{\mathrm{T}}\boldsymbol{R}_x\boldsymbol{W}\boldsymbol{A}^{-1}\right)^{-1} = \boldsymbol{W}^{\mathrm{T}}\boldsymbol{R}_y\boldsymbol{W} \tag{7.179}$$

Let $\boldsymbol{W}^{\mathrm{T}}\boldsymbol{R}_x\boldsymbol{W} = \boldsymbol{Q}_1\boldsymbol{\Lambda}_p\boldsymbol{Q}_1^{\mathrm{T}}$ and $\boldsymbol{A}\boldsymbol{W}^{\mathrm{T}}\boldsymbol{R}_x\boldsymbol{W}\boldsymbol{A}^{-1} = \boldsymbol{Q}_2\boldsymbol{\Lambda}_p\boldsymbol{Q}_2^{\mathrm{T}}$ be the EVD of the two matrices, respectively, where $\boldsymbol{Q}_1$ and $\boldsymbol{Q}_2$ are two orthonormal matrices. According to the matrix theory, $\boldsymbol{W}^{\mathrm{T}}\boldsymbol{R}_x\boldsymbol{W}$ has the same eigenvalues as matrix $\boldsymbol{A}\boldsymbol{W}^{\mathrm{T}}\boldsymbol{R}_x\boldsymbol{W}\boldsymbol{A}^{-1}$. Then, we can obtain

$$\boldsymbol{Q}_2\boldsymbol{\Lambda}_p\boldsymbol{Q}_2^{\mathrm{T}} = \boldsymbol{A}\boldsymbol{Q}_1\boldsymbol{\Lambda}_p\boldsymbol{Q}_1^{\mathrm{T}}\boldsymbol{A}^{-1} \tag{7.180}$$

From the above equation, we have $\boldsymbol{\Lambda}_p\boldsymbol{Q}_2^{\mathrm{T}}\boldsymbol{A}\boldsymbol{Q}_1 = \boldsymbol{Q}_2^{\mathrm{T}}\boldsymbol{A}\boldsymbol{Q}_1\boldsymbol{\Lambda}_p$, which implies that $\boldsymbol{Q}_2^{\mathrm{T}}\boldsymbol{A}\boldsymbol{Q}_1$ must be a diagonal matrix since $\boldsymbol{\Lambda}_p$ has different diagonal elements. Since $\boldsymbol{A}$ is also a diagonal matrix and $\boldsymbol{Q}_1, \boldsymbol{Q}_2$ are two orthonormal matrices, we have $\boldsymbol{Q}_1 = \boldsymbol{Q}_2 = \boldsymbol{I}$. By using this fact, we can obtain that $\boldsymbol{W}^{\mathrm{T}}\boldsymbol{R}_x\boldsymbol{W} = \boldsymbol{\Lambda}_p$. Substituting it into Eq. (7.179) yields

$$\boldsymbol{W}^{\mathrm{T}}\boldsymbol{R}_y\boldsymbol{W} - \boldsymbol{I}_r \tag{7.181}$$

Substituting generalized eigen decomposition of $\boldsymbol{R}_y$ into the above formula, we have $(\boldsymbol{Q}')^{\mathrm{T}}\boldsymbol{Q}' = \boldsymbol{I}_r$, where $\boldsymbol{Q}' = \boldsymbol{\Lambda}^{\frac{1}{2}}\boldsymbol{V}^{\mathrm{T}}\boldsymbol{R}_x\boldsymbol{W}$, which means that the columns of $\boldsymbol{Q}'$ are orthonormal at any point of $J_{\mathrm{WGIC}}(\boldsymbol{W})$. Hence, we can obtain that $\boldsymbol{W} = \boldsymbol{L}_r\boldsymbol{\Lambda}^{\frac{1}{2}}\boldsymbol{Q}'$ is a stationary point of $J_{\mathrm{WGIC}}(\boldsymbol{W})$.

Theorem 7.5 establishes the property of all the stationary point of $J_{\mathrm{WGIC}}(\boldsymbol{W})$. The following theorem will distinguish the global maximum point set attained by $\boldsymbol{W}$ composing the desired GMCs from any other stationary point, which are saddle points.

This completes the proof of Theorem 7.5.

Theorem 7.6 *In the domain* Ω, $J_{\mathrm{WGIC}}(\boldsymbol{W})$ *has a global maximum that is attained if and only if* $\boldsymbol{W} = \boldsymbol{L}_1\boldsymbol{\Lambda}_1^{-\frac{1}{2}}$, *where* $\boldsymbol{\Lambda}_1 = \mathrm{diag}([\lambda_1, \lambda_2, \ldots, \lambda_r])$ *is a diagonal matrix and its diagonal elements are composed by the first r generalized eigenvalues of the matrix pencil* $(\boldsymbol{R}_y, \boldsymbol{R}_x)$, $\boldsymbol{L}_1 = [\boldsymbol{v}_1, \boldsymbol{v}_2, \ldots, \boldsymbol{v}_r]$ *is an* $n \times r$ *matrix and is composed of the corresponding generalized eigenvectors. All the other stationary points are saddle points of* $J_{\mathrm{WGIC}}(\boldsymbol{W})$. *At this global maximum, we have*

$$J_{\mathrm{WGIC}}(\boldsymbol{W}) = (1/2)\left(\sum_{i=1}^{r} a_i/\lambda_i - r\right) \tag{7.182}$$

Proof From (7.176), it is very easy to obtain the second order of $J_{\mathrm{WGIC}}(\boldsymbol{W})$:

$$
\begin{aligned}
\mathrm{d}^2 J_{\mathrm{WGIC}}(W) &= \mathrm{d}\Big\{\mathrm{tr}\Big[(\boldsymbol{W}^{\mathrm{T}}\boldsymbol{R}_x\boldsymbol{W}\boldsymbol{A})^{-1}\boldsymbol{A}\boldsymbol{W}^{\mathrm{T}}\boldsymbol{R}_x\mathrm{d}\boldsymbol{W}\Big] - \mathrm{tr}(\boldsymbol{W}^{\mathrm{T}}\boldsymbol{R}_y\mathrm{d}\boldsymbol{W})\Big\} \\
&= -\mathrm{tr}\big[\mathrm{d}(\boldsymbol{W}^{\mathrm{T}})\boldsymbol{R}_y\mathrm{d}\boldsymbol{W}\big] + \mathrm{tr}\Big[\big(\boldsymbol{A}^{-1}\boldsymbol{W}^{\mathrm{T}}\boldsymbol{R}_x\boldsymbol{W}\boldsymbol{A}\big)^{-1}\mathrm{d}(\boldsymbol{W}^{\mathrm{T}})\boldsymbol{R}_x\mathrm{d}\boldsymbol{W}\Big] \\
&\quad - \mathrm{tr}\Big[(\boldsymbol{A}^{-1}\boldsymbol{W}^{\mathrm{T}}\boldsymbol{R}_x\boldsymbol{W}\boldsymbol{A})^{-1}\mathrm{d}(\boldsymbol{A}^{-1}\boldsymbol{W}^{\mathrm{T}}\boldsymbol{R}_x\boldsymbol{W}\boldsymbol{A})(\boldsymbol{A}^{-1}\boldsymbol{W}^{\mathrm{T}}\boldsymbol{R}_x\boldsymbol{W}\boldsymbol{A})^{-1}\boldsymbol{W}^{\mathrm{T}}\boldsymbol{R}_x\mathrm{d}\boldsymbol{W}\Big] \\
&= -\mathrm{tr}\big[\mathrm{d}(\boldsymbol{W}^{\mathrm{T}})\boldsymbol{R}_y\mathrm{d}\boldsymbol{W}\big] + \mathrm{tr}\Big[\big(\boldsymbol{A}^{-1}\boldsymbol{W}^{\mathrm{T}}\boldsymbol{R}_x\boldsymbol{W}\boldsymbol{A}\big)^{-1}\mathrm{d}(\boldsymbol{W}^{\mathrm{T}})\boldsymbol{R}_x\mathrm{d}\boldsymbol{W}\Big] \\
&\quad - \mathrm{tr}\Big[(\boldsymbol{A}^{-1}\boldsymbol{W}^{\mathrm{T}}\boldsymbol{R}_x\boldsymbol{W}\boldsymbol{A})^{-1}\big(\boldsymbol{A}^{-1}\mathrm{d}(\boldsymbol{W}^{\mathrm{T}})\boldsymbol{R}_x\boldsymbol{W}\boldsymbol{A} + \boldsymbol{A}^{-1}\boldsymbol{W}^{\mathrm{T}}\boldsymbol{R}_x\mathrm{d}\boldsymbol{W}\boldsymbol{A}\big) \\
&\quad \times (\boldsymbol{A}^{-1}\boldsymbol{W}^{\mathrm{T}}\boldsymbol{R}_x\boldsymbol{W}\boldsymbol{A})^{-1}\boldsymbol{W}^{\mathrm{T}}\boldsymbol{R}_x\mathrm{d}\boldsymbol{W}\Big] \\
&= -\mathrm{tr}\big[\mathrm{d}(\boldsymbol{W}^{\mathrm{T}})\boldsymbol{R}_y\mathrm{d}\boldsymbol{W}\big] + \mathrm{tr}\Big[\big(\boldsymbol{A}^{-1}\boldsymbol{W}^{\mathrm{T}}\boldsymbol{R}_x\boldsymbol{W}\boldsymbol{A}\big)^{-1}\mathrm{d}(\boldsymbol{W}^{\mathrm{T}})\boldsymbol{R}_x\mathrm{d}\boldsymbol{W}\Big] \\
&\quad - \mathrm{tr}\Big[(\boldsymbol{W}^{\mathrm{T}}\boldsymbol{R}_x\boldsymbol{W}\boldsymbol{A})^{-1}\mathrm{d}(\boldsymbol{W}^{\mathrm{T}})\boldsymbol{R}_x\boldsymbol{W}(\boldsymbol{A}^{-1}\boldsymbol{W}^{\mathrm{T}}\boldsymbol{R}_x\boldsymbol{W})^{-1}\boldsymbol{W}^{\mathrm{T}}\boldsymbol{R}_x\mathrm{d}\boldsymbol{W}\Big] \\
&\quad - \mathrm{tr}\Big[(\boldsymbol{W}^{\mathrm{T}}\boldsymbol{R}_x\boldsymbol{W}\boldsymbol{A})^{-1}\boldsymbol{W}^{\mathrm{T}}\boldsymbol{R}_x\mathrm{d}(\boldsymbol{W})(\boldsymbol{A}^{-1}\boldsymbol{W}^{\mathrm{T}}\boldsymbol{R}_x\boldsymbol{W})^{-1}\boldsymbol{W}^{\mathrm{T}}\boldsymbol{R}_x\mathrm{d}\boldsymbol{W}\Big]
\end{aligned}
\tag{7.183}
$$

Let $\boldsymbol{H}J_{\mathrm{WGIC}}(\boldsymbol{W})$ be the Hessian matrix of $J_{\mathrm{WGIC}}(\boldsymbol{W})$ with respect to the nr-dimensional vector $\mathrm{vec}(\boldsymbol{W}) = [\boldsymbol{w}_1^{\mathrm{T}}, \boldsymbol{w}_2^{\mathrm{T}}, \ldots, \boldsymbol{w}_r^{\mathrm{T}}]$, which is defined by

$$
\boldsymbol{H}J_{\mathrm{WGIC}}(\boldsymbol{W}) = \frac{\partial}{\partial(\mathrm{vec}(\boldsymbol{W}))^{\mathrm{T}}}\left(\frac{\partial J_{\mathrm{WGIC}}(\boldsymbol{W})}{\partial(\mathrm{vec}(\boldsymbol{W}))^{\mathrm{T}}}\right)^{\mathrm{T}}
\tag{7.184}
$$

By using (7.183), we have

$$
\begin{aligned}
\boldsymbol{H}J_{\mathrm{WGIC}}(\boldsymbol{W}) &= -\boldsymbol{I}_r \otimes \boldsymbol{R}_y + \frac{1}{2}\Big[\big(\boldsymbol{A}\boldsymbol{W}^{\mathrm{T}}\boldsymbol{R}_x\boldsymbol{W}\boldsymbol{A}^{-1}\big)^{-1}\Big] \otimes \boldsymbol{R}_x + \frac{1}{2}\Big[\big(\boldsymbol{A}^{-1}\boldsymbol{W}^{\mathrm{T}}\boldsymbol{R}_x\boldsymbol{W}\boldsymbol{A}\big)^{-1}\Big] \otimes \boldsymbol{R}_x \\
&\quad - \frac{1}{2}\Big[(\boldsymbol{A}\boldsymbol{W}^{\mathrm{T}}\boldsymbol{R}_x\boldsymbol{W})^{-1}\Big] \otimes \Big[\boldsymbol{R}_x\boldsymbol{W}(\boldsymbol{A}^{-1}\boldsymbol{W}^{\mathrm{T}}\boldsymbol{R}_x\boldsymbol{W})^{-1}\boldsymbol{W}^{\mathrm{T}}\boldsymbol{R}_x\Big] \\
&\quad - \frac{1}{2}\Big[(\boldsymbol{W}^{\mathrm{T}}\boldsymbol{R}_x\boldsymbol{W}\boldsymbol{A})^{-1}\Big] \otimes \Big[\boldsymbol{R}_x\boldsymbol{W}(\boldsymbol{W}^{\mathrm{T}}\boldsymbol{R}_x\boldsymbol{W}\boldsymbol{A}^{-1})^{-1}\boldsymbol{W}^{\mathrm{T}}\boldsymbol{R}_x\Big] \\
&\quad - \frac{1}{2}\boldsymbol{K}_{rn}\Big\{\Big[\boldsymbol{R}_x\boldsymbol{W}(\boldsymbol{A}\boldsymbol{W}^{\mathrm{T}}\boldsymbol{R}_x\boldsymbol{W})^{-1}\Big] \otimes \Big[(\boldsymbol{A}^{-1}\boldsymbol{W}^{\mathrm{T}}\boldsymbol{R}_x\boldsymbol{W})^{-1}\boldsymbol{W}^{\mathrm{T}}\boldsymbol{R}_x\Big] \\
&\quad + \Big[\boldsymbol{R}_x\boldsymbol{W}(\boldsymbol{W}^{\mathrm{T}}\boldsymbol{R}_x\boldsymbol{W}\boldsymbol{A}^{-1})^{-1}\Big] \otimes \Big[(\boldsymbol{W}^{\mathrm{T}}\boldsymbol{R}_x\boldsymbol{W}\boldsymbol{A})^{-1}\boldsymbol{W}^{\mathrm{T}}\boldsymbol{R}_x\Big]\Big\}
\end{aligned}
\tag{7.185}
$$

where $\otimes$ means the Kronecker product and $\boldsymbol{K}_{rn}$ is the $rn \times rn$ commutation matrix such that $\boldsymbol{K}_{rn}\mathrm{vec}(\boldsymbol{M}) = vec(\boldsymbol{M}^{\mathrm{T}})$ for any matrix $\boldsymbol{M}$. The commutation matrix $\boldsymbol{K}_{rn}$ has some good properties, such as for any matrices $\boldsymbol{M}_1 \in R^{m\times n}$ and $\boldsymbol{M}_2 \in R^{p\times q}$, there is

$$\boldsymbol{K}_{pm}(\boldsymbol{M}_1 \otimes \boldsymbol{M}_2) = (\boldsymbol{M}_2 \otimes \boldsymbol{M}_1)\boldsymbol{K}_{qn} \tag{7.186}$$

This property will be used in the following derivations. Substituting Eq. (7.178) into (7.185) and then calculating $\boldsymbol{H}J_{\text{WGIC}}(\boldsymbol{W})$ at the stationary point $\boldsymbol{W} = \boldsymbol{L}_1\boldsymbol{\Lambda}_1^{-\frac{1}{2}}$, we obtain

$$\begin{aligned}
\boldsymbol{H}^* &= \boldsymbol{H}J_{\text{WGIC}}(\boldsymbol{W})\big|_{\boldsymbol{W}=\boldsymbol{L}_1\boldsymbol{\Lambda}_1^{-\frac{1}{2}}} \\
&= -\boldsymbol{I}_r \otimes \boldsymbol{R}_y + \boldsymbol{\Lambda}_1 \otimes \boldsymbol{R}_x - \left[\boldsymbol{\Lambda}_1\boldsymbol{A}^{-1}\right] \otimes \left[\boldsymbol{R}_x\boldsymbol{L}_1\boldsymbol{A}\boldsymbol{L}_1^{\mathrm{T}}\boldsymbol{R}_x\right] \\
&\quad -\frac{1}{2}\boldsymbol{K}_{rn}[(\boldsymbol{R}_x\boldsymbol{L}_1) \otimes \boldsymbol{I}_r]\left[\left(\boldsymbol{\Lambda}_1^{\frac{1}{2}}\boldsymbol{A}^{-1}\right) \otimes \left(\boldsymbol{A}\boldsymbol{\Lambda}_1^{\frac{1}{2}}\right) + \left(\boldsymbol{\Lambda}_1^{\frac{1}{2}}\boldsymbol{A}\right) \otimes \left(\boldsymbol{A}^{-1}\boldsymbol{\Lambda}_1^{\frac{1}{2}}\right)\right]\left[\boldsymbol{I}_r \otimes (\boldsymbol{L}_1^{\mathrm{T}}\boldsymbol{R}_x)\right]
\end{aligned} \tag{7.187}$$

Applying (7.186) to the third term on the right-hand side of (7.187) yields

$$\boldsymbol{K}_{rn}[(\boldsymbol{R}_x\boldsymbol{L}_1) \otimes \boldsymbol{I}_r] = [\boldsymbol{I}_r \otimes (\boldsymbol{R}_x\boldsymbol{L}_1)]\boldsymbol{K}_{rr} \tag{7.188}$$

Substituting the generalized eigen decomposition of $\boldsymbol{R}_y = \boldsymbol{R}_x\boldsymbol{L}_1\boldsymbol{\Lambda}_1\boldsymbol{L}_1^{\mathrm{T}}\boldsymbol{R}_x + \boldsymbol{R}_x\boldsymbol{L}_n\boldsymbol{\Lambda}_n\boldsymbol{L}_n^{\mathrm{T}}\boldsymbol{R}_x$ and the fact $\boldsymbol{R}_x = \boldsymbol{R}_x\boldsymbol{L}_1\boldsymbol{L}_1^{\mathrm{T}}\boldsymbol{R}_x + \boldsymbol{R}_x\boldsymbol{L}_n\boldsymbol{L}_n^{\mathrm{T}}\boldsymbol{R}_x$ into (7.187), after some proper manipulations, we get

$$\begin{aligned}
\boldsymbol{H}^* &= -[\boldsymbol{I}_r \otimes (\boldsymbol{R}_x\boldsymbol{L}_1)][\boldsymbol{I}_r \otimes \boldsymbol{\Lambda}_1]\left[\boldsymbol{I}_r \otimes (\boldsymbol{L}_1^{\mathrm{T}}\boldsymbol{R}_x)\right] \\
&\quad - [\boldsymbol{I}_r \otimes (\boldsymbol{R}_x\boldsymbol{L}_n)][\boldsymbol{I}_r \otimes \boldsymbol{\Lambda}_n]\left[\boldsymbol{I}_r \otimes (\boldsymbol{L}_n^{\mathrm{T}}\boldsymbol{R}_x)\right] \\
&\quad + [\boldsymbol{I}_r \otimes (\boldsymbol{R}_x\boldsymbol{L}_1)][\boldsymbol{\Lambda}_1 \otimes \boldsymbol{I}_r]\left[\boldsymbol{I}_r \otimes (\boldsymbol{L}_1^{\mathrm{T}}\boldsymbol{R}_x)\right] \\
&\quad + [\boldsymbol{I}_r \otimes (\boldsymbol{R}_x\boldsymbol{L}_n)][\boldsymbol{\Lambda}_1 \otimes \boldsymbol{I}_r]\left[\boldsymbol{I}_r \otimes (\boldsymbol{L}_n^{\mathrm{T}}\boldsymbol{R}_x)\right] \\
&\quad - [\boldsymbol{I}_r \otimes (\boldsymbol{R}_x\boldsymbol{L}_1)]\left[(\boldsymbol{\Lambda}_1\boldsymbol{A}^{-1}) \otimes \boldsymbol{A}\right]\left[\boldsymbol{I}_r \otimes (\boldsymbol{L}_1^{\mathrm{T}}\boldsymbol{R}_x)\right] \\
&\quad - \frac{1}{2}[\boldsymbol{I}_r \otimes (\boldsymbol{R}_x\boldsymbol{L}_1)]\left[\boldsymbol{I}_r \otimes (\boldsymbol{L}_1^{\mathrm{T}}\boldsymbol{R}_x)\right] \\
&\quad \times \boldsymbol{K}_{rr}\left[\left(\boldsymbol{\Lambda}_1^{\frac{1}{2}}\boldsymbol{A}^{-1}\right) \otimes \left(\boldsymbol{A}\boldsymbol{\Lambda}_1^{\frac{1}{2}}\right) + \left(\boldsymbol{\Lambda}_1^{\frac{1}{2}}\boldsymbol{A}\right) \otimes \left(\boldsymbol{A}^{-1}\boldsymbol{\Lambda}_1^{\frac{1}{2}}\right)\right]
\end{aligned} \tag{7.189}$$

Denote

$$\begin{aligned}
\boldsymbol{D}_1 &= -[\boldsymbol{I}_r \otimes \boldsymbol{\Lambda}_1] + [\boldsymbol{\Lambda}_1 \otimes \boldsymbol{I}_r] - \left[(\boldsymbol{\Lambda}_1\boldsymbol{A}^{-1}) \otimes \boldsymbol{A}\right] \\
&\quad -\frac{1}{2}\boldsymbol{K}_{rr}\left[\left(\boldsymbol{\Lambda}_1^{\frac{1}{2}}\boldsymbol{A}^{-1}\right) \otimes \left(\boldsymbol{A}\boldsymbol{\Lambda}_1^{\frac{1}{2}}\right) + \left(\boldsymbol{\Lambda}_1^{\frac{1}{2}}\boldsymbol{A}\right) \otimes \left(\boldsymbol{A}^{-1}\boldsymbol{\Lambda}_1^{\frac{1}{2}}\right)\right]
\end{aligned} \tag{7.190}$$

Then, it has the following EVD

$$\boldsymbol{D}_1 = \boldsymbol{U}\boldsymbol{S}\boldsymbol{U}^{\mathrm{T}} \tag{7.191}$$

where $\boldsymbol{S} = \mathrm{diag}(s_1, s_2, \ldots, s_{r^2})$ is a diagonal matrix whose diagonal elements are composed by the r^2 eigenvalues of $\boldsymbol{D}_1$ and $\boldsymbol{V}$ is an $r^2 \times r^2$ orthogonal matrix consisting of corresponding eigenvectors.

Substituting (7.191) into (7.189), we have

$$\begin{aligned}\boldsymbol{H}^* &= [\boldsymbol{I}_r \otimes (\boldsymbol{R}_x\boldsymbol{L}_1)](\boldsymbol{U}\boldsymbol{S}\boldsymbol{U}^{\mathrm{T}})\left[\boldsymbol{I}_r \otimes (\boldsymbol{L}_1^{\mathrm{T}}\boldsymbol{R}_x)\right] + [\boldsymbol{I}_r \otimes (\boldsymbol{R}_x\boldsymbol{L}_n)]\{\boldsymbol{D}_2\}\left[\boldsymbol{I}_r \otimes (\boldsymbol{L}_n^{\mathrm{T}}\boldsymbol{R}_x)\right] \\ &= \begin{bmatrix} \boldsymbol{U}^{\mathrm{T}}(\boldsymbol{I}_r \otimes (\boldsymbol{L}_1^{\mathrm{T}}\boldsymbol{R}_x)) \\ \boldsymbol{I}_r \otimes (\boldsymbol{L}_n^{\mathrm{T}}\boldsymbol{R}_x) \end{bmatrix}^{\mathrm{T}} \begin{bmatrix} \boldsymbol{S} & \\ & \boldsymbol{D}_2 \end{bmatrix} \begin{bmatrix} \boldsymbol{I}_r \otimes (\boldsymbol{L}_1^{\mathrm{T}}\boldsymbol{R}_x) \\ \boldsymbol{I}_r \otimes (\boldsymbol{L}_n^{\mathrm{T}}\boldsymbol{R}_x) \end{bmatrix}\end{aligned} \tag{7.192}$$

where

$$\boldsymbol{D}_2 = [\boldsymbol{\Lambda}_1 \otimes \boldsymbol{I}_r] - [\boldsymbol{I}_r \otimes \boldsymbol{\Lambda}_n] \tag{7.193}$$

It is obvious that $\boldsymbol{D}_2$ is a diagonal matrix. Equation (7.189) is actually the EVD of $\boldsymbol{H}^*$ whose eigenvalues are also the diagonal elements of both $\boldsymbol{S}$ and $\boldsymbol{D}_2$.

It is obvious that all the diagonal elements of $\boldsymbol{D}_2$ must be negative as long as $\max\{\lambda_1, \lambda_2, \ldots, \lambda_{n-r}\} \leq \min\{\lambda_{n-r+1}, \lambda_{n-r+2}, \ldots, \lambda_n\}$. So we only need to determine whether the eigenvalues of $\boldsymbol{G}$ are negative or not. In order to illustrate this problem, we follow the proof procedures in [47]. Let us consider one case for $r = 2$, then, we have

$$\boldsymbol{D}_1 = \begin{bmatrix} -2\lambda_1 & & & \\ & -\lambda_2 + \lambda_1 - \frac{a_2}{a_1}\lambda_1 & -\frac{a_1\sqrt{\lambda_1\lambda_2}}{2a_2} - \frac{a_2\sqrt{\lambda_1\lambda_2}}{2a_1} & \\ & -\frac{a_1\sqrt{\lambda_1\lambda_2}}{2a_2} - \frac{a_2\sqrt{\lambda_1\lambda_2}}{2a_1} & -\lambda_1 + \lambda_2 - \frac{a_1}{a_2}\lambda_2 & \\ & & & -2\lambda_2 \end{bmatrix} \tag{7.194}$$

Without loss of generality, let $a_1 = 1$ and $d = \lambda_1/\lambda_2$, then we can obtain the four eigenvalues of the matrix $\boldsymbol{G}$: They are given by $s_1 = -2\lambda_1$, $s_4 = -2\lambda_2$, $s_2 = -(\mathrm{b} - \sqrt{\mathrm{b}^2 + 4\mathrm{c}})/2$, and $s_3 = -(\mathrm{b} + \sqrt{\mathrm{b}^2 + 4\mathrm{c}})/2$, where $b = a_2\lambda_1 - \lambda_2/a_2$ and

$$c = \lambda_1\lambda_2\left[(-1 + d - a_2 d)(-1 + 1/d - 1/a_2 d) - (1/2a_2 + a_2/2)^2\right] \tag{7.195}$$

Clearly, we only need to calculate the conditions where the eigenvalue s_3 is negative, i.e., we need to solve the inequality $-\mathrm{b} + \sqrt{\mathrm{b}^2 + 4\mathrm{c}} < 0$, where $c < 0$. After some manipulations, we have

$$a_2^4 + 4(1 - d)a_2^3 + [4d - 10 + 4/d]a_2^2 - 4(1/d - 1)a_2 + 1 > 0 \tag{7.196}$$

The above equation provides the relationship between a_2 and d. Solving (7.196) will give a required positive real root. For illustrative purpose, Fig. 7.6 shows the

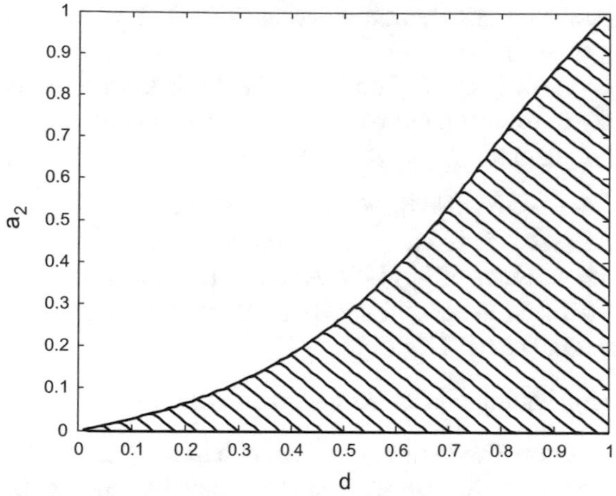

Fig. 7.6 Curve of the relation of a_2 to d

feasible domain of a_2 versus d, which is highlighted by the shadow. From Fig. 7.6, we can see that there always exists the properly chosen a_2 so that (7.196) holds. It is easy to verify the above conclusions hold for $r > 2$. So we can conclude that all eigenvalues of $\boldsymbol{H}^*$ are negative as long as the following inequality holds:

$$\max\{\lambda_1, \lambda_2, \ldots, \lambda_{n-r}\} \le \min\{\lambda_{n-r+1}, \lambda_{n-r+2}, \ldots, \lambda_n\} \tag{7.197}$$

Since $\boldsymbol{H}^*$ is nonpositive definite if and only if $\min\{\lambda_1, \lambda_2, \ldots, \lambda_{n-r}\} \ge \max\{\lambda_{n-r+1}, \lambda_{n-r+2}, \ldots, \lambda_n\}$. So at the stationary point $\boldsymbol{W} = \boldsymbol{L}_n \boldsymbol{\Lambda}^{-\frac{1}{2}}$, $J_{\text{WGIC}}(\boldsymbol{W})$ has the local maximum. Since $J_{\text{WGIC}}(\boldsymbol{W})$ is unbounded as $\boldsymbol{W}$ tends to infinity, this is also the global minimum. Except $\boldsymbol{W} = \boldsymbol{L}_n \boldsymbol{\Lambda}^{-\frac{1}{2}}$, all other stationary points result in $\boldsymbol{H}^*$ being indefinite (having both positive and negative eigenvalues) and, thus, are saddle points.

At the stationary point $\boldsymbol{W} = \boldsymbol{L}_n \boldsymbol{\Lambda}^{-\frac{1}{2}} \boldsymbol{Q}$, $J_{\text{GIC}}(\boldsymbol{W})$ attains the global maximum value

$$J_{\text{WGIC}}(\boldsymbol{W}) = (1/2)\left(\sum_{i=1}^{r} a_i/\lambda_i - r\right) \tag{7.198}$$

This completes the proof of Theorem 7.6.

Comparing Eq. (7.164) with Eq. (7.175), we can see that the differences of the two information criteria are the weighted matrix and the dimension of the neural network state matrix. If the state matrix in Eq. (7.175) is reduced to a vector and the weighted matrix is set as $\boldsymbol{A} = 1$, then Eq. (7.175) will be equal to Eq. (7.164). Since we have accomplished landscape analysis of (7.175) through Theorem 7.5 and Theorem 7.6, it will be a repeated work to analyze the landscape of Eq. (7.164). Furthermore, if we set $\boldsymbol{A} = \boldsymbol{I}_r$ and $\boldsymbol{R}_y = \boldsymbol{I}_n$, the information criterion $J_{\text{WGIC}}(\boldsymbol{W})$ is reduced to the AMEX criterion in [44], which is proposed for MCA. Therefore, the proposed information criterion (7.175) is a universal criterion for MCA and GED problems.

From [44], we know that the AMEX criterion has a symmetrical construction. By introducing the weighted matrix $\boldsymbol{A}$, we proposed an asymmetrical criterion. Let us denote $\boldsymbol{R}_{xy} = \boldsymbol{R}_y^{-1}\boldsymbol{R}_x\boldsymbol{W}$ and $\boldsymbol{R}_{xw} = \boldsymbol{W}^{\mathrm{T}}\boldsymbol{R}_x\boldsymbol{W}$, and substitute them into Eq. (7.178). Then, we have $\boldsymbol{W} = \boldsymbol{R}_{xy}(\boldsymbol{A}\boldsymbol{R}_{xw}\boldsymbol{A}^{-1})^{-1}$. The function of the matrix $\boldsymbol{A}\boldsymbol{R}_{xw}\boldsymbol{A}^{-1}$ is to carry out an implicit Gram–Schmidt orthonormalization (GSO) on the columns of $\boldsymbol{R}_{xy}$ [48]. As a result of the GSO operation, the columns of the state matrix in the neural network will exactly correspond to the different GMCs of the matrix pencil $(\boldsymbol{R}_y, \boldsymbol{R}_x)$.

(3) Multiple GMC extraction algorithm

From Theorem 7.6, we can obtain that $J_{\mathrm{WGIC}}(\boldsymbol{W})$ has a global maximum and no local ones. So the iterative algorithms like the gradient ascent search algorithm can be used for finding the global maximum point of $J_{\mathrm{WGIC}}(\boldsymbol{W})$. Given the gradient of $J_{\mathrm{WGIC}}(\boldsymbol{W})$ with respect to $\boldsymbol{W}$ in (7.176), we can get the following gradient ascent algorithm:

$$\boldsymbol{W}(k+1) = \boldsymbol{W}(k) + \eta\left[\boldsymbol{R}_x\boldsymbol{W}(k)\left(\boldsymbol{A}\boldsymbol{W}^{\mathrm{T}}(k)\boldsymbol{R}_x\boldsymbol{W}(k)\boldsymbol{A}^{-1}\right)^{-1} - \boldsymbol{R}_y\boldsymbol{W}(k)\right] \tag{7.199}$$

If the matrices $\boldsymbol{R}_y$ and $\boldsymbol{R}_x$ are unknown, then Eqs. (7.171)–(7.173) can be used to estimate them. It is well known that the matrix inverse can cause an adaptive learning rate and improve the properties of the neural network algorithms. From this point, we rewrite (7.199) as

$$\boldsymbol{W}(k+1) = \boldsymbol{W}(k) + \eta\left[\boldsymbol{R}_y^{-1}\boldsymbol{R}_x\boldsymbol{W}(k)\left(\boldsymbol{A}\boldsymbol{W}^{\mathrm{T}}(k)\boldsymbol{R}_x\boldsymbol{W}(k)\boldsymbol{A}^{-1}\right)^{-1} - \boldsymbol{W}(k)\right] \tag{7.200}$$

It is obvious that Eq. (7.200) has the same equilibrium point as Eq. (7.199). Although the only difference between (7.200) and (7.199) is the location of the matrix $\boldsymbol{R}_y$, this modification actually changes the learning rate into an adjustable value by using $\eta\boldsymbol{R}_y^{-1}$. This modification can improve the performance of the original gradient algorithm [44].

It should be noted that the proposed WGIC algorithm can extract the multiple GMCs of the matrix pencil $(\boldsymbol{R}_y, \boldsymbol{R}_x)$ in parallel, not a basis of the generalized minor space. This algorithm is also suitable for the cases where the generalized minor space is only needed, since the GMCs can also be regarded as a special basis of the generalized minor space; however, this solution is not the best way. Since the weighted matrix $\boldsymbol{A}$ is to implement the GSO operation on the state matrix of the neural network. So if we set $\boldsymbol{A} = \boldsymbol{I}$ in (7.199), then it will become a generalized minor space tracking algorithm. Here, it is worth noting that algorithm (7.200) has the self-stabilizing property, proof of which can refer to Sect. 7.3.2.

In the following, we study the global convergence property of algorithm (7.199) by the Lyapunov function approach.

Under the conditions that $\boldsymbol{x}$ and $\boldsymbol{y}$ are two zero-mean stationary process and the learning rate η is small enough, the discrete-time difference Eq. (7.177) can be approximated by the following continuous-time ordinary differential equation (ODE):

$$\frac{\mathrm{d}\boldsymbol{W}(t)}{\mathrm{d}t} = \boldsymbol{R}_x\boldsymbol{W}\left(\boldsymbol{A}\boldsymbol{W}^{\mathrm{T}}\boldsymbol{R}_x\boldsymbol{W}\boldsymbol{A}^{-1}\right)^{-1} - \boldsymbol{R}_y\boldsymbol{W} \tag{7.201}$$

where $t = \eta k$. By analyzing the global convergence properties of (7.201), we will establish the conditions for the global convergence of the batch algorithm (7.199). In particular, we will answer the following questions:

1. Can the dynamical system be able to globally converge to the GMCs?
2. What is the domain of attraction around the stationary point attained at the GMCs or equivalently, what is the initial condition to ensure the global convergence?

In order to answer the above two questions, we define a function as follows:

$$L(\boldsymbol{W}) = \frac{1}{2}\mathrm{tr}(\boldsymbol{W}^{\mathrm{T}}\boldsymbol{R}_y\boldsymbol{W}) - \frac{1}{2}\mathrm{tr}\left[\ln(\boldsymbol{W}^{\mathrm{T}}\boldsymbol{R}_x\boldsymbol{W}\boldsymbol{A})\right] \tag{7.202}$$

Then, we have that $L(\boldsymbol{W})$ is a bounded function in the region $\Omega = \left\{\boldsymbol{W}|0 < \boldsymbol{W}^{\mathrm{T}}\boldsymbol{R}_x\boldsymbol{W} < \infty\right\}$. According to the Lyapunov function approach, we need to probe the first-order derivative of $L(\boldsymbol{W})$ is nonpositive. By the chain rule of differential matrix [49], we have

$$\begin{aligned}\frac{\mathrm{d}L(\boldsymbol{W})}{\mathrm{d}t} &= \mathrm{tr}\left[\boldsymbol{W}^{\mathrm{T}}\boldsymbol{R}_y\frac{\mathrm{d}\boldsymbol{W}}{\mathrm{d}t} - (\boldsymbol{A}^{-1}\boldsymbol{W}^{\mathrm{T}}\boldsymbol{R}_x\boldsymbol{W}\boldsymbol{A})^{-1}\boldsymbol{W}^{\mathrm{T}}\boldsymbol{R}_x\frac{\mathrm{d}\boldsymbol{W}}{\mathrm{d}t}\right] \\ &= -\mathrm{tr}\left(\frac{\mathrm{d}\boldsymbol{W}^{\mathrm{T}}}{\mathrm{d}t}\frac{\mathrm{d}\boldsymbol{W}}{\mathrm{d}t}\right)\end{aligned} \tag{7.203}$$

Note that the dependence on t in the above formula has been dropped for convenience. From (7.203), it is easy to see that in the domain of attraction $\boldsymbol{W} \in \Omega - \left\{\boldsymbol{W}|\boldsymbol{W} = \boldsymbol{L}_n\boldsymbol{\Lambda}^{-\frac{1}{2}}\boldsymbol{P}\right\}$, we have $\mathrm{d}\boldsymbol{W}/dt \neq 0$ and $\mathrm{d}L(\boldsymbol{W})/\mathrm{d}t < 0$, then $\mathrm{d}L(\boldsymbol{W})/\mathrm{d}t = 0$ if and only if $\mathrm{d}\boldsymbol{W}/\mathrm{d}t = 0$ for $\boldsymbol{W} = \boldsymbol{L}_n\boldsymbol{\Lambda}^{-\frac{1}{2}}\boldsymbol{P}$. So $L(\boldsymbol{W})$ will strictly and monotonically decrease from any initial value of $\boldsymbol{W}$ in Ω. That is to say, $L(\boldsymbol{W})$ is a Lyapunov function for the domain Ω and $\boldsymbol{W}(t)$ will globally asymptotically converge to the needed GMCs from any $\boldsymbol{W}(0) \in \Omega$.

7.4.3 Simulations and Application Experiments

In this section, we provide several experiments to demonstrate the behavior and applicability of the proposed algorithms. The first experiment mainly shows the capability of the WGIC algorithm to extract the multiple minor components, and the second provides one example of practical applications.

(1) The ability to extract multiple GMCs

In this experiment, we provide simulation results to show the performance of the proposed algorithm for multiple GMC extraction of a randomly generated matrix pencil, which are given by

$$\boldsymbol{R}_y = \begin{bmatrix} 0.5444 & -0.0596 & 0.1235 & 0.0197 & -0.0611 & -0.1309 & -0.0055 \\ -0.0596 & 0.3892 & -0.0583 & -0.1300 & 0.0984 & 0.0138 & 0.1919 \\ 0.1235 & -0.0583 & 0.5093 & 0.0570 & 0.0394 & 0.0582 & -0.0140 \\ 0.0197 & -0.1300 & 0.0570 & 0.3229 & -0.0350 & 0.2035 & -0.1035 \\ -0.0611 & 0.0984 & 0.0394 & -0.0350 & 0.4960 & -0.0191 & -0.1087 \\ -0.1309 & 0.0138 & 0.0582 & 0.2035 & -0.0191 & 0.3148 & 0.0212 \\ -0.0055 & 0.1919 & -0.0140 & -0.1035 & -0.1087 & 0.0212 & 0.2819 \end{bmatrix} \tag{7.204}$$

and

$$\boldsymbol{R}_x = \begin{bmatrix} 0.3979 & 0.0633 & 0.0294 & 0.0805 & -0.0199 & -0.1911 & -0.1447 \\ 0.0633 & 0.5778 & -0.0858 & -0.0067 & 0.0971 & 0.0691 & -0.0046 \\ 0.0294 & -0.0858 & 0.5221 & -0.0413 & 0.1534 & 0.1488 & -0.1716 \\ 0.0805 & -0.0067 & -0.0413 & 0.6070 & 0.0896 & -0.1212 & 0.0724 \\ -0.0199 & 0.0971 & 0.1534 & 0.0896 & 0.5633 & 0.1213 & 0.0399 \\ -0.1911 & 0.0691 & 0.1488 & -0.1212 & 0.1213 & 0.3171 & 0.1417 \\ -0.1447 & -0.0046 & -0.1716 & 0.0724 & 0.0399 & 0.1417 & 0.5541 \end{bmatrix} \tag{7.205}$$

By using MATLAB toolboxes, we can obtain the generalized eigenvalues of matrix pencil $(\boldsymbol{R}_y, \boldsymbol{R}_x)$ are $\lambda_1 = 0.1021$, $\lambda_2 = 0.1612$, $\lambda_3 = 0.6464$, $\lambda_4 = 0.8352$, $\lambda_5 = 1.3525$, $\lambda_6 = 2.1360$, $\lambda_7 = 5.2276$. Then, we use two different algorithms to extract the first three GMCs of this matrix pencil, that is to say: $r = 3$. The two algorithms are

Algorithm 1: sequential extracting algorithm based on the deflation technique given by Table 7.1.

Algorithm 2: gradient algorithm based on the weighted matrix given by (7.200).

The initial parameters for the two algorithms are set as follows: for Algorithm 1, $\tau = 100$ and $\eta = 0.2$, and for Algorithm 2: $\boldsymbol{A} = \mathrm{diag}([3, 2, 1])$ and $\eta = 0.2$. By these settings, the two algorithms have the same learning rate. The initial state vector or matrix is randomly generated. The simulation results are shown in Fig. 7.7.

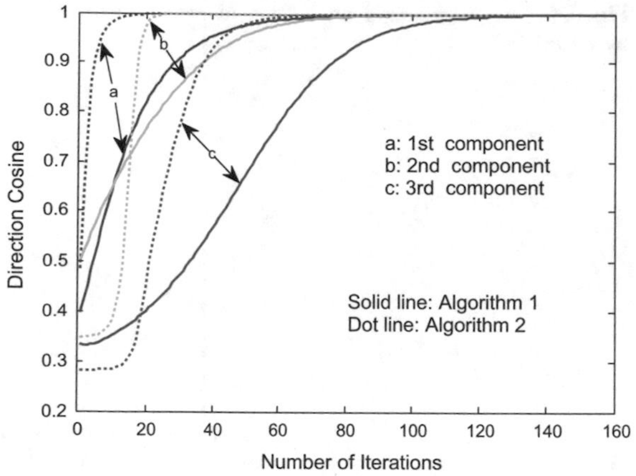

Fig. 7.7 Direction cosine curves of the two algorithms for extracting the first three GMC

From Fig. 7.7, we can see that after about 140 iteration steps, all the direction cosine curves have converged to one, which means that both the Algorithm 1 and the Algorithm 2 have the ability to extract the multiple GMCs of a matrix pencil. Notice that the Algorithm 1 is a sequential extraction algorithm, so the next component starts after the previous one has converged. That is to say, the actual time instant for the start of the second GMC is $k = 160$ instead of $k = 0$ and that of the third GMC is $k = 320$ instead of $k = 0$. The starting points of the last two GMCs have been moved to the base point in this figure in order to save space. It is shown in Fig. 7.7 that when using the Algorithm 2, the averaged convergence time for each GMC is 80 iterations. By using this fact, we can obtain that the actual total iterations for the convergence of three GMCs are about 240 iterations. However, the total iteration by the Algorithm 1 is about 130, which is cheaper than that of the Algorithm 2. Since derivation between the total iterations of the two algorithms will increase as the extracted number of the GMCs, the Algorithm 2 is very suitable for real applications, where multiple GMCs need fast parallel extraction.

(2) Practical applications

In this section, we provide one example of practical application to show the effectiveness of our algorithm. An important application of the minor generalized eigenvector is to solve data classification problem. In [50], Mangasarian and Wild proposed a new approach called generalized eigenvalue proximal support vector machine (GEPSVM) based on GED and provided an interesting example to illustrate the effectiveness of GEPSVM. In this section, we use our algorithm to classify two data sets.

The input data vectors are generated through the following method. Firstly, randomly take 15 points from the line $y_1 = x_1 + 2$ and the line $y_2 = -2x_2 + 5$, respectively, and then add Gaussian noises to the 30 points; finally, we can obtain

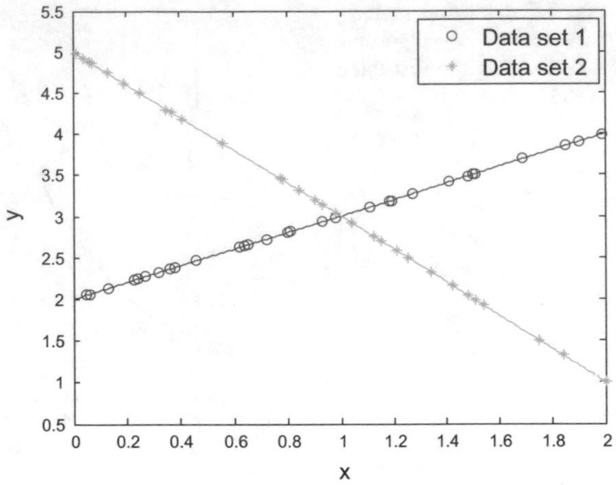

Fig. 7.8 The planes obtained by GEPSVM

two data sets, which are close to one of two intersecting "cross-planes" in $\mathbb{R}^2$. The data points are shown in Fig. 7.8. The problem of data classification is to find two planes such that each plane is as close as possible to one of the data sets and as far as possible from the other data set. After some simplified calculation, the data classification problem by GEPSVM method can be changed into a problem of calculating the minor generalized eigenvector of two data matrices.

Figure 7.8 gives the two planes obtained by our proposed algorithms. From Fig. 7.8, we can see that training set correctness is 100 percent, which means the proposed algorithm has a satisfactory performance for solving data classification problems.

7.5 Summary

In this chapter, the generalized eigen decomposition problem has been briefly reviewed. Several well-known algorithms, e.g., generalized eigenvector extraction algorithm based on Newton or quasi-Newton method, fast generalized eigenvector tracking based on the power method, and generalized eigenvector extraction algorithm based on RLS method, have been analyzed. Then, a minor generalized eigenvector extraction algorithm proposed by us has been introduced, and its convergence analysis has been performed via the DDT method. Finally, an information criterion for GMCA is proposed, and a fast GMCA algorithm is derived by using quasi-Newton method. This information criterion is extended into a weighted one through the weighed matrix method so as to extract multiple generalized minor components. A gradient algorithm is also derived based on this weighted information criterion, and its convergence is analyzed by Lyapunov function approach.

References

1. Huanqun, C., Sarkar, T. K., Dianat, S. A., & Brule, J. D. (1986). Adaptive spectral estimation by the conjugate gradient method. *IEEE Transactions on Acoustics, Speech, and Signal Processing, 34*(2), 272–284.
2. Chang, C., Ding, Z., Yau, S. F., & Chan, F. H. Y. (2000). A matrix pencil approach to blind separation of colored nonstationary signals. *IEEE Transactions on Signal Processing, 48*(3), 900–907.
3. Comon, P., & Golub, C. H. (1990). Tracking a few extreme singular values and vectors in signal processing. *Proceedings of the IEEE, 78*(8), 1327–1343.
4. Choi, S., Choi, J., Im, H. J., & Choi, B. (2002). A novel adaptive beamforming algorithm for antenna array CDMA systems with strong interferers. *IEEE Transactions on Vehicular Technology, 51*(5), 808–816.
5. Morgan, D. R. (2003). Downlink adaptive array algorithms for cellular mobile communications. *IEEE Transactions on Communications, 51*(3), 476–488.
6. Bunse-Gerstner A. (1984). An algorithm for the symmetric generalized eigenvalue problem. *Linear Algebra and its Applications, 58*(ARR), 43–68.
7. Shougen, W., & Shuqin, Z. (1991). An algorithm for Ax = λBx with symmetric and positive-definite A and B. *SIAM Journal on Matrix Analysis and Applications, 12*, 654–660.
8. Auchmuty, G. (1991). Globally and rapidly convergent algorithms for symmetric eigenproblems. *SIAM Journal of Matrix Analysis and Applications, 12*(4), 690–706.
9. Kaufman, L. (1974). The LZ algorithm to solve the generalized eigenvalue problem. *SIAM Journal of Numerical Analysis, 11*(5), 997–1024.
10. Martin R. S., &Wilkinson J. H. (1968). Reduction of the symmetric eigenproblem Ax = λBx and related problems to standard form. *Numerical Mathematics, 11*, 99–110.
11. Moler, C. B., & Stewart, G. W. (1973). An algorithm for generalized matrix eigenvalue problems. *SIAM Journal of Numerical Analysis, 10*(2), 241–256.
12. Yang, J., Hu, H., & Xi, H. (2013). Weighted non-linear criterion-based adaptive generalized eigendecomposition. *IET Signal Processing, 7*(4), 285–295.
13. Kong, X. Y., Hu, C. H., & Han, C. Z. (2012). A dual purpose principal and minor subspace gradient flow. *IEEE Transactions on Signal Processing, 60*(1), 197–210.
14. Mathew, G., & Reddy, V. U. (1996). Aquasi-Newton adaptive algorithm for generalized symmetric eigenvalue problem. *IEEE Transactions on Signal Processing, 44*(10), 2413–2422.
15. Mao, J., & Jain, A. K. (1995). Artificial neural networks for feature extraction and multivariate data projection. *IEEE Transactions on Neural Networks, 6*(2), 296–317.
16. Rubner, J., & Tavan, P. (1989). A self-organizing network for principal component analysis. *Europhysics Letters, 10*(7), 693–698.
17. Rubner, J., & Tavan, P. (1990). Development of feature detectors by self organization. *Biology Cybernetics, 62*(62), 193–199.
18. Chatterjee, C., Roychowdhury, V. P., Ramos, J., & Zoltowski, M. D. (1997). Self-organizing algorithms for generalized eigen-decomposition. *IEEE Transactions on Neural Networks, 8* (6), 1518–1530.
19. Xu, D., Principe, J. C., & Wu, H. C. (1998). Generalized eigendecomposition with an on-line local algorithm. *IEEE Signal Processing Letters, 5*(11), 298–301.
20. Diamantaras K.I., & Kung S.Y. (1996). *Principal component neural networks, theory and applications*. New York: Wiley.
21. Rao, Y. N., Principe, J. C., Wong, T. F., & Abdi, H. (2004). Fast RLS-like algorithm for generalized eigendecomposition and its applications. *Journal of VLSI Signal Processing, 37* (2–3), 333–344.
22. Yang, J., Xi, H., Yang, F., & Yu, Z. (2006). RLS-based adaptive algorithms for generalized eigen -decomposition. *IEEE Transactions on Signal Processing, 54*(4), 1177–1188.
23. Attallah, S., & Abed-Meraim, K. (2008). A fast adaptive algorithm for the generalized symmetric eigenvalue problem. *IEEE Signal Processing Letters, 15*, 797–800.

24. Tanaka, T. (2009). Fast generalized eigenvector tracking based on the power method. *IEEE Signal Processing Letters, 16*(11), 969–972.
25. Xu, L. (1993). Least mean square error reconstruction principle for self-organizing neural-nets. *Neural Networks, 6*(5), 627–648.
26. Yang, J., Zhao, Y., & Xi, H. (2011). Weighted rule based adaptive algorithm for simultaneously extracting generalized eigenvectors. *IEEE Transactions on Neural Networks, 22*(5), 800–806.
27. Martinez, A. M., & Zhu, M. (2005). Where are linear feature extraction methods applicable. *IEEE Transactions on Pattern Analysis and Machine Intelligence, 27*(12), 1934–1944.
28. Mangasarian, O. L., & Wild, E. W. (2006). Multisurface proximal support vector machine classification via generalized eigenvalues. *IEEE Transactions on Pattern Analysis and Machine Intelligence, 28*(1), 69–74.
29. Bello, L., Cruz, W. L., & Raydan, M. (2010). Residual algorithm for large-scale positive definite generalized eigenvalue problems. *Computation Optimization and Applications, 46*(2), 217–227.
30. Stone, J. V. (2002). Blind deconvolution using temporal predictability. *Neurocomputing, 49* (1–4), 79–86.
31. Shahbazpanahi, S., Gershman, A. B., Luo, Z., & Wong, K. M. (2003). Robust adaptive beamforming for general-rank signal models. *IEEE Transactions on Signal Processing, 51*(9), 2257–2269.
32. Ye M., Liu Y., Wu H., & Liu Q. (2008). A few online algorithms for extracting minor generalized eigenvectors. In *International Joint Conference on Neural Networks* (pp. 1714–1720).
33. Nguyen, T. D., & Yamada, I. (2013). Adaptive normalized quasi-Newton algorithms for extraction of generalized eigen-pairs and their convergence analysis. *IEEE Transactions on Signal Processing, 61*(61), 1404–1418.
34. Möller, R., & Könies, A. (2004). Coupled principal component analysis. *IEEE Transactions on Neural Networks, 15*(1), 214–222.
35. Nguyen, T. D., Takahashi, N., & Yamada, I. (2013). An adaptive extraction of generalized eigensubspace by using exact nested orthogonal complement structure. *Multidimensional System and Signal Processing, 24*(3), 457–483.
36. Yang, B. (1995). Projection approximation subspace tracking. *IEEE Transactions on Signal Processing, 43*(1), 95–107.
37. Davila, C. E. (2000). Efficient, high performance, subspace tracking for time-domain data. *IEEE Transactions on Signal Processing, 48*(12), 3307–3315.
38. Yang, J., Yang, F., & Xi, H. S. (2007). Robust adaptive modified Newton algorithm for generalized eigendecomposition and its application. *EURASIP Journal on Advances in Signal Processing, 2007*(2), 1–10.
39. Cirrincione, G., Cirrincione, M., Hérault, J., & Van Huffel, S. (2002). The MCA EXIN neuron for the minor component analysis. *IEEE Transactions on Neural Networks, 13*(1), 160–187.
40. Peng, D., & Zhang, Y. (2006). Convergence analysis of a deterministic discrete time system of feng's MCA learning algorithm. *IEEE Transactions on Signal Processing, 54*(9), 3626–3632.
41. Peng, D., Zhang, Y., & Luo, W. (2007). Convergence analysis of a simple minor component analysis algorithm. *Neural Networks, 20*(7), 842–850.
42. Peng, D., Zhang, Y., & Yong, X. (2008). On the discrete time dynamics of a self-stabilizing MCA learning algorithm. *Mathematical and Computer Modeling, 47*(9–10), 903–916.
43. Lv, J., Zhang, Y., & Tan, K. K. (2006). Convergence analysis of Xu's LMSER learning algorithm via deterministic discrete time system method. *Neurocomputing, 70*(1), 362–372.
44. Ouyang, S., Bao, Z., Liao, G., & Ching, P. C. (2001). Adaptive minor component extraction with modular structure. *IEEE Transactions on Signal Processing, 49*(9), 2127–2137.
45. Miao, Y., & Hua, Y. (1998). Fast subspace tracking and neural network learning by a novel information criterion. *IEEE Transactions on Signal Processing, 46*(7), 1967–1979.

46. Kong, X. Y., An, Q. S., Ma, H. G., Han, C. Z., & Zhang, Q. (2012). Convergence analysis of deterministic discrete time system of a unified self-stabilizing algorithm for PCA and MCA. *Neural Networks, 36*(8), 64–72.
47. Ouyang, S., & Bao, Z. (2001). Fast principal component extraction by a weighted information criterion. *IEEE Transactions on Signal Processing, 50*(8), 1994–2002.
48. Oja, E. (1992). Principal components, minor components and linear neural networks. *Neural Networks, 5*(6), 927–935.
49. Magnus, J. R., & Neudecker, H. (1991). *Matrix differential calculus with applications in statistics and econometrics*. New York: Wiley.
50. Mangasarian, O. L., & Wild, E. W. (2006). Multisurface proximal support vector machine classification via generalized eigenvalues. *IEEE Transactions on Pattern Analysis and Machine Intelligence, 28*(1), 69–74.
51. Zufiria, P. J. (2002). On the discrete-time dynamics of the basic Hebbian neural-network node. *IEEE Transactions on Neural Networks, 13*(6), 1342–1352.
52. Parlett, B. N. (1998). *The symmetric eigenvalue problem*. Philadelphia: SIAM.
53. Zhang, Y., Ye, M., Lv, J., & Tan, K. K. (2005). Convergence analysis of a deterministic discrete time system of Oja's PCA learning algorithm. *IEEE Transactions on Neural Networks, 16*(6), 1318–1328.
54. Moller, R. (2004). A self-stabilizing learning rule for minor component analysis. *International Journal of Neural Systems, 14*(1), 1–8.

Chapter 8
Coupled Principal Component Analysis

8.1 Introduction

Among neural network-based PCA or MCA algorithms, most previously reviewed do not consider eigenvalue estimates in the update equations of the weights, except an attempt to control the learning rate based on the eigenvalue estimates [1]. In [2], Moller provided a framework for a special class of learning rules where eigenvectors and eigenvalues are simultaneously estimated in coupled update equations, and has proved that coupled learning algorithms are solutions for the speed stability problem that plagues most noncoupled learning algorithms. The convergence speed of a system depends on the eigenvalues of its Jacobian, which vary with the eigenvalues of the covariance matrix in noncoupled PCA/MCA algorithms [2]. Moller showed that, in noncoupled PCA algorithms, the eigen motion in all directions mainly depends on the principal eigenvalue of the covariance matrix [2]. Numerical stability and fast convergence of algorithms can only be achieved by guessing this eigenvalue in advance [2]. In particular for chains of principal component analyzers which simultaneously estimate the first few principal eigenvectors [3], choosing the right learning rates for all stages may be difficult. The problem is even more severe for MCA algorithms. MCA algorithms exhibit a wide range of convergence speeds in different eigen directions, since the eigenvalues of the Jacobian cover approximately the same range as the eigenvalues of the covariance matrix. Using small enough learning rates to still guarantee the stability of the numerical procedure, noncoupled MCA algorithms may converge very slowly [2].

In [2], Moller derived a coupled learning rule by applying Newton's method to a common information criterion. A Newton descent yields learning rules with approximately equal convergence speeds in all eigen directions of the system. Moreover, all eigenvalues of the Jacobian of such a system are approximately. Thus, the dependence on the eigenvalues of the covariance matrix can be eliminated [2]. Moller showed that with respect to averaged differential equations, this

© Science Press, Beijing and Springer Nature Singapore Pte Ltd. 2017

X. Kong et al., *Principal Component Analysis Networks and Algorithms*,
DOI 10.1007/978-981-10-2915-8_8

approach solves the speed stability problem for both PCA and MCA rules. However, these differential equations can only be turned into the aforementioned online rules for the PCA but not for the MCA case, leaving the more severe MCA stability problem still unresolved [2]. Interestingly, unlike most existing adaptive algorithms, the coupled learning rule for the HEP effectively utilizes the latest estimate of the eigenvalue to update the estimate of the eigenvector [4]. Numerical examples in [2] showed that this algorithm achieves fast and stable convergence for both low-dimensional data and high-dimensional data. Unfortunately, there has been no report about any explicit convergence analysis for the coupled learning rule. Thus, the condition for the convergence to the desired eigen pair is not clear; e.g., the region within which the initial estimate of the eigen pair must be chosen to guarantee the convergence to the desired eigen pair has not yet been known [4].

Recently, Tuan Duong Nguyen et al. proposed novel algorithms in [4] for given explicit knowledge of the matrix pencil ($\boldsymbol{R}_y$, $\boldsymbol{R}_x$). These algorithms for estimating the generalized eigen pair associated with the largest/smallest generalized eigenvalue are designed (i) based on a new characterization of the generalized eigen pair as a stationary point of a certain function and (ii) by combining a normalization step and quasi-Newton step at each update. Moreover, the rigorous convergence analysis of the algorithms was established by the DDT approach. For adaptive implementation of the algorithms, Tuan Duong Nguyen et al. proposed to use the exponentially weighted sample covariance matrices and the Sherman–Morrison–Woodbury matrix-inversion lemma.

The aim of this chapter was to develop some coupled PCA or coupled generalized PCA algorithms. First, on the basis of a special information criterion in [5], we propose a coupled dynamical system by modifying Newton's method in this chapter. Based on the coupled system and some approximation, we derive two CMCA algorithms and two CPCA algorithms; thus, two unified coupled algorithms are obtained [6]. Then, we propose a coupled generalized system in this chapter, which is obtained by using the Newton's method and a novel generalized information criterion. Based on this coupled generalized system, we obtain two coupled algorithms with normalization steps for minor/principal generalized eigen pair extraction. The technique of multiple generalized eigen pair extraction is also introduced in this chapter. The convergence of algorithms is justified by DDT system.

In this chapter, we will review and discuss the existing coupled PCA or coupled generalized PCA algorithms. Two coupled algorithms proposed by us will be analyzed in detail. The remainder of this chapter is organized as follows. An overview of the existing coupled PCA or coupled generalized PCA algorithms is presented in Sect. 8.2. An unified and coupled self-stabilizing algorithm for minor and principal eigen pair extraction algorithms are discussed in Sect. 8.3. An adaptive generalized eigen pair extraction algorithms and their convergence analysis via DDT method are presented in Sect. 8.4, followed by summary in Sect. 8.5.

8.2 Review of Coupled Principal Component Analysis

8.2.1 Moller's Coupled PCA Algorithm

Learning rules for principal component analysis are often derived by optimizing some information criterion, e.g., by maximizing the variance of the projected data or by minimizing the reconstruction error [2, 7]. In [2], Moller proposed the following information criterion as the starting point of his analysis

$$p = \boldsymbol{w}^{\mathrm{T}}\boldsymbol{C}\boldsymbol{w}\lambda^{-1} - \boldsymbol{w}^{\mathrm{T}}\boldsymbol{w} + \ln\lambda. \tag{8.1}$$

where $\boldsymbol{w}$ denotes an n-dimensional weight vector, i.e., the estimate of the eigenvector, λ is the eigenvalue estimate, and $\boldsymbol{C} = E\{\boldsymbol{x}\boldsymbol{x}^{\mathrm{T}}\}$ is the $n \times n$ covariance matrix of the data. From (8.1), by using the gradient method and the Newton descent, Moller derived a coupled system of differential equations for the PCA case

$$\dot{\boldsymbol{w}} = \boldsymbol{C}\boldsymbol{w}\lambda^{-1} - \boldsymbol{w}\boldsymbol{w}^{\mathrm{T}}\boldsymbol{C}\boldsymbol{w}\lambda^{-1} - \frac{1}{2}\boldsymbol{w}\left(1 - \boldsymbol{w}^{\mathrm{T}}\boldsymbol{w}\right), \tag{8.2}$$

$$\dot{\lambda} = \boldsymbol{w}^{\mathrm{T}}\boldsymbol{C}\boldsymbol{w} - \boldsymbol{w}^{\mathrm{T}}\boldsymbol{w}\lambda, \tag{8.3}$$

and another for MCA case

$$\dot{\boldsymbol{w}} = \boldsymbol{C}^{-1}\boldsymbol{w}\lambda + \boldsymbol{w}\boldsymbol{w}^{\mathrm{T}}\boldsymbol{C}\boldsymbol{w}\lambda^{-1} - \frac{1}{2}\boldsymbol{w}\left(1 + 3\boldsymbol{w}^{\mathrm{T}}\boldsymbol{w}\right), \tag{8.4}$$

$$\dot{\lambda} = \boldsymbol{w}^{\mathrm{T}}\boldsymbol{C}\boldsymbol{w} - \boldsymbol{w}^{\mathrm{T}}\boldsymbol{w}\lambda. \tag{8.5}$$

For the stability of the above algorithms, see [2]. It has been shown that for the above coupled PCA system, if we assume $\lambda_j \ll \lambda_1$, the system converges with approximately equal speeds in all its eigen directions, and this speed is widely independent of the eigenvalues λ_j of the covariance matrix. And for the above coupled MCA system, if we assume $\lambda_1 \ll \lambda_j$, then the convergence speed is again about equal in all eigen directions and independent of the eigenvalues of $\boldsymbol{C}$.

By informally approximating $\boldsymbol{C} \approx \boldsymbol{x}\boldsymbol{x}^{\mathrm{T}}$, the averaged differential equations of (8.2) and (8.3) can be turned into an online learning rule:

$$\dot{\boldsymbol{w}} = \gamma\left[y\lambda^{-1}(\boldsymbol{x} - \boldsymbol{w}y) - \frac{1}{2}\boldsymbol{w}\left(1 - \boldsymbol{w}^{\mathrm{T}}\boldsymbol{w}\right)\right], \tag{8.6}$$

$$\dot{\lambda} = \gamma(y^2 - \boldsymbol{w}^{\mathrm{T}}\boldsymbol{w}\lambda). \tag{8.7}$$

According to the stochastic approximation theory, the resulting stochastic differential equation has the same convergence goal as the deterministic averaged equation if certain conditions are fulfilled, the most important of which is that a learning rate decreases to zero over time. The online rules (8.6) and (8.7) can be understood as a learning rule for the weight vector $\boldsymbol{w}$ of a linear neuron which computes its output y from the scalar product of weight vector and input vector $\boldsymbol{y} = \boldsymbol{w}^{\mathrm{T}}\boldsymbol{x}$.

In [2], the analysis of the temporal derivative of the (squared) weight vector length in (8.6) has shown that the weight vector length may in general be fluctuating. By further approximating $\boldsymbol{w}^{\mathrm{T}}\boldsymbol{w} \approx 1$ (which is fulfilled in the vicinity of the stationary points) in the averaged systems (8.2) and (8.3), the following system can be derived

$$\dot{\boldsymbol{w}} = \boldsymbol{C}\boldsymbol{w}\lambda^{-1} - \boldsymbol{w}\boldsymbol{w}^{\mathrm{T}}\boldsymbol{C}\boldsymbol{w}\lambda^{-1}, \tag{8.8}$$

$$\dot{\lambda} = \boldsymbol{w}^{\mathrm{T}}\boldsymbol{C}\boldsymbol{w} - \lambda. \tag{8.9}$$

This learning rule system is known as ALA [1]. The eigenvalues of the system's Jacobian are still approximately equal and widely independent of the eigenvalues of the covariance matrix. The corresponding online system is given by

$$\dot{\boldsymbol{w}} = \gamma y\lambda^{-1}(\boldsymbol{x} - \boldsymbol{w}y), \tag{8.10}$$

$$\dot{\lambda} = \gamma(y^2 - \lambda). \tag{8.11}$$

It is obvious that ALA can be interpreted as an instance of Oja's PCA rule.

From (8.4) and (8.5), it has been shown that having a Jacobian with eigenvalues that are equal and widely independent of the eigenvalues of the covariance matrix appears to be a solution for the speed stability problem. However, when attempting to turn this system into an online rule, a problem is encountered when replacing the inverse covariance matrix $\boldsymbol{C}^{-1}$ by a quantity including the input vector $\boldsymbol{x}$. An averaged equation linearly depending on C takes the form $\dot{\boldsymbol{w}} = f(\boldsymbol{C}, \boldsymbol{w}) = f(E\{\boldsymbol{x}\boldsymbol{x}^{\mathrm{T}}\}, \boldsymbol{w}) = E\{f(\boldsymbol{x}\boldsymbol{x}^{\mathrm{T}}, \boldsymbol{w})\}$. In an online rule, the expectation of the gradient is approximated by slowly following $\dot{\boldsymbol{w}} = \gamma f(\boldsymbol{x}\boldsymbol{x}^{\mathrm{T}}, \boldsymbol{w})$ for subsequent observations of $\boldsymbol{x}$. This transition is obviously not possible if the equation contains $\boldsymbol{C}^{-1}$. Thus, there is no online version for the MCA systems (8.4) and (8.5). Despite using the ALA-style normalization, the convergence speed in different eigen directions still depends on the entire range of eigenvalues of the covariance matrix. So the speed stability problem still exists.

8.2.2 Nguyen's Coupled Generalized Eigen pairs Extraction Algorithm

In [8], Nguyen proposed a generalized principal component analysis algorithm and its differential equation form is given as:

$$\dot{\boldsymbol{w}} = \boldsymbol{R}_x^{-1}\boldsymbol{R}_y\boldsymbol{w} - \boldsymbol{w}^H\boldsymbol{R}_y\boldsymbol{w}\boldsymbol{w}. \tag{8.12}$$

Let $W = [w_1, w_2, \ldots, w_N]$, in which $w_1, w_2, \ldots, w_N$ are the generalized eigenvectors of matrix pencil (R_y, R_x). The Jacobian in the stationary point is given as:

$$J(w_1) = \frac{\partial \dot{w}}{\partial w^{\mathrm{T}}}|_{w=w_1} = R_x^{-1}R_y - \lambda_1 I - 2\lambda_1 w_1 w_1^H R_x. \tag{8.13}$$

Solving for the eigenvectors of J can be simplified to the solving for the eigenvector of its similar diagonally matrix $J^* = P^{-1}JP$, since J^* and J have the same eigenvectors and eigenvalues, and the eigenvalues of diagonal matrix J^* are easy to be obtained. Considering $W^H R_x W = I$, let $P = W$. Then we have $P^{-1} = W^H R_x$. Thus, it holds that

$$\begin{aligned} J^*(w_1) &= W^H R_x (R_x^{-1} R_y - \lambda_1 I - 2\lambda_1 w_1 w_1^H R_x) W \\ &= \Lambda - \lambda_1 I - 2\lambda_1 W^H R_x w_1 (W^H R_x w_1)^H \frac{\Delta y}{\Delta x}. \end{aligned} \tag{8.14}$$

Since $W^H R_x w_1 = e_1 = [1, 0, \ldots, 0]^{\mathrm{T}}$, (8.14) will be reduced to

$$J^*(w_1) = \Lambda - \lambda_1 I - 2\lambda_1 e_1 e_1^H. \tag{8.15}$$

The eigenvalues α determined from $\det(J^* - \alpha I) = 0$ are given as:

$$\alpha_1 = -2\lambda_1, \; \alpha_j = \lambda_j - \lambda_1, \; j = 2, \ldots, N. \tag{8.16}$$

Since the stability requires $\alpha < 0$ and thus $\lambda_1 \gg \lambda_j$, $j = 2, 3, \ldots, n$, it can be seen that only principal eigenvector–eigenvalue pairs are stable stationary points, and all other stationary points are saddles or repellers, which can still testify that (8.12) is a generalized PCA algorithm. In the practical signal processing applications, it always holds that $\lambda_1 \gg \lambda_j$, $j = 2, 3, \ldots, n$. Thus, $\alpha_j \approx -\lambda_1$, i.e., the eigen motion in all directions in algorithm (8.12) depends on the principal eigenvalue of the covariance matrix. Thus, this algorithm has the speed stability problem.

In [4], an adaptive normalized quasi-Newton algorithm for generalized eigen pair extraction was proposed and its convergence analysis was conducted. This algorithm is a coupled generalized eigen pair extraction algorithm, which can be interpreted as natural combinations of the normalization step and quasi-Newton steps for finding the stationary points of the function

$$\xi(w, \lambda) = w^H R_y w \lambda^{-1} - w^H R_x w + \ln \lambda, \tag{8.17}$$

which is a generalization of the information criterion introduced in [2] for the HEP. The stationary point of ξ is defined as a zero of

$$\begin{pmatrix} \frac{\partial \xi}{\partial w} \\ \frac{\partial \xi}{\partial \lambda} \end{pmatrix} = \begin{pmatrix} 2R_y w \lambda^{-1} - 2R_x w \\ -w^H R_y w \lambda^{-2} + \lambda^{-1} \end{pmatrix}. \tag{8.18}$$

Hence, from

$$\begin{cases} \boldsymbol{R}_y\bar{\boldsymbol{w}} = \bar{\lambda}\boldsymbol{R}_x\bar{\boldsymbol{w}} \\ \bar{\boldsymbol{w}}^H\boldsymbol{R}_y\bar{\boldsymbol{w}} = \bar{\lambda} \end{cases}, \tag{8.19}$$

it can be seen that $(\bar{\boldsymbol{w}}, \bar{\lambda}) \in C^N \times \Re$ is a stationary point of ξ, which implies that a stationary point $(\bar{\boldsymbol{w}}, \bar{\lambda})$ of ξ is a generalized eigen pair of the matrix pencil $(\boldsymbol{R}_y, \boldsymbol{R}_x)$. To avoid these computational difficulties encountered in Newton's method, Nguyen proposed to use approximations of $\boldsymbol{H}^{-1}(\boldsymbol{w}, \lambda)$ in the vicinity of two stationary points of their special interest

$$\boldsymbol{H}^{-1}(\boldsymbol{w}, \lambda) \approx \widetilde{\boldsymbol{H}}_P^{-1}(\boldsymbol{w}, \lambda) = \frac{1}{2}\begin{pmatrix} \frac{1}{2}\boldsymbol{w}\boldsymbol{w}^H - \boldsymbol{R}_x^{-1} & -\boldsymbol{w}\lambda \\ -\boldsymbol{w}^H\lambda & 0 \end{pmatrix}, \tag{8.20}$$

for $(\boldsymbol{w}, \lambda) \approx (\boldsymbol{v}_N, \lambda_N)$, and

$$\boldsymbol{H}^{-1}(\boldsymbol{w}, \lambda) \approx \widetilde{\boldsymbol{H}}_M^{-1}(\boldsymbol{w}, \lambda) = \frac{1}{2}\begin{pmatrix} \boldsymbol{R}_y^{-1}\lambda - \frac{3}{2}\boldsymbol{w}\boldsymbol{w}^H & -\boldsymbol{w}\lambda \\ -\boldsymbol{w}^H\lambda & 0 \end{pmatrix}, \tag{8.21}$$

for $(\boldsymbol{w}, \lambda) \approx (\boldsymbol{v}_1, \lambda_1)$. By applying Newton's strategy for finding the stationary point of ξ using the gradient (8.18) and the approximations (8.20) and (8.21), a learning rule for estimating the generalized eigen pair associated with the largest generalized eigenvalue was obtained as:

$$\begin{aligned} \boldsymbol{w}(k+1) = \boldsymbol{w}(k) + \eta_1\{&\boldsymbol{R}_x^{-1}\boldsymbol{R}_y\boldsymbol{w}(k)\lambda^{-1}(k) \\ &-\boldsymbol{w}^H(k)\boldsymbol{R}_y\boldsymbol{w}(k)\boldsymbol{w}(k)\lambda^{-1}(k) - \frac{1}{2}\boldsymbol{w}(k)[1 - \boldsymbol{w}^H(k)\boldsymbol{R}_x\boldsymbol{w}(k)]\Big\}, \end{aligned} \tag{8.22}$$

$$\lambda(k+1) = \lambda(k) + \gamma_1\left[\boldsymbol{w}^H(k+1)\boldsymbol{R}_y\boldsymbol{w}(k+1) - \boldsymbol{w}^H(k+1)\boldsymbol{R}_x\boldsymbol{w}(k+1)\lambda(k)\right], \tag{8.23}$$

and a learning rule for estimating the generalized eigen pair associated with the smallest generalized eigenvalue was obtained as:

$$\begin{aligned} \boldsymbol{w}(k+1) = \boldsymbol{w}(k) + \eta_2\Big\{&\boldsymbol{R}_y^{-1}\boldsymbol{R}_x\boldsymbol{w}(k)\lambda(k) + \boldsymbol{w}^H(k)\boldsymbol{R}_y\boldsymbol{w}(k)\boldsymbol{w}(k)\lambda^{-1}(k) \\ &-\frac{1}{2}\boldsymbol{w}(k)\left[1 + 3\boldsymbol{w}^H(k)\boldsymbol{R}_x\boldsymbol{w}(k)\right]\Big\}, \end{aligned} \tag{8.24}$$

$$\lambda(k+1) = \lambda(k) + \gamma_2\left[\boldsymbol{w}^H(k+1)\boldsymbol{R}_y\boldsymbol{w}(k+1) - \boldsymbol{w}^H(k+1)\boldsymbol{R}_x\boldsymbol{w}(k+1)\lambda(k)\right], \tag{8.25}$$

where $\eta_1, \gamma_1, \eta_2, \gamma_2 > 0$ are the step sizes, and $[\boldsymbol{w}(k), \lambda(k)]$ is the estimate at time k of the generalized eigen pair associated with the largest/smallest generalized eigenvalue. By introducing the normalization step in the above learning rules at each update, using the exponentially weighted sample covariance matrices $\widehat{\boldsymbol{R}}_y$ and

$\widehat{\boldsymbol{R}}_x$ which are updated recursively, and using the Sherman–Morrison–Woodbury matrix-inversion lemma, Nguyen's coupled generalized eigen pair extraction algorithm was obtained as follows.

Adaptive coupled generalized PCA algorithm:

$$\begin{aligned}\tilde{\boldsymbol{w}}(k) = \boldsymbol{w}(k-1) + \frac{\eta_1}{\lambda(k-1)}\Big(&\boldsymbol{Q}_x(k)\widehat{\boldsymbol{R}}_y(k)\boldsymbol{w}(k-1) \\ &-\boldsymbol{w}^H(k-1)\widehat{\boldsymbol{R}}_y(k)\boldsymbol{w}(k-1)\boldsymbol{w}(k-1)\Big),\end{aligned} \tag{8.26}$$

$$\boldsymbol{w}(k) = \frac{\tilde{\boldsymbol{w}}(k)}{\|\tilde{\boldsymbol{w}}(k)\|_{\widehat{\boldsymbol{R}}_x(k)}}, \tag{8.27}$$

$$\lambda(k) = (1-\gamma_1)\lambda(k-1) + \gamma_1\boldsymbol{w}^H(k)\widehat{\boldsymbol{R}}_y(k)\boldsymbol{w}(k), \tag{8.28}$$

and adaptive coupled generalized MCA algorithm:

$$\begin{aligned}\tilde{\boldsymbol{w}}(k) = \boldsymbol{w}(k-1) + \eta_2\Big(&\boldsymbol{Q}_y(k)\widehat{\boldsymbol{R}}_x(k)\boldsymbol{w}(k-1)\lambda(k-1) \\ &+\boldsymbol{w}^H(k-1)\widehat{\boldsymbol{R}}_y(k)\boldsymbol{w}(k-1)\boldsymbol{w}(k-1)\lambda^{-1}(k-1) - 2\boldsymbol{w}(k-1)\Big),\end{aligned} \tag{8.29}$$

$$\boldsymbol{w}(k) = \frac{\tilde{\boldsymbol{w}}(k)}{\|\tilde{\boldsymbol{w}}(k)\|_{\widehat{\boldsymbol{R}}_x(k)}}, \tag{8.30}$$

$$\lambda(k) = (1-\gamma_2)\lambda(k-1) + \gamma_2\boldsymbol{w}^H(k)\widehat{\boldsymbol{R}}_y(k)\boldsymbol{w}(k), \tag{8.31}$$

where $\|\boldsymbol{u}\|_{\boldsymbol{R}_x} = \sqrt{\boldsymbol{u}^H\boldsymbol{R}_x\boldsymbol{u}}$ is defined as the $\boldsymbol{R}_x$-norm, $\boldsymbol{Q}_x = \boldsymbol{R}_x^{-1}$, $\boldsymbol{Q}_y = \boldsymbol{R}_y^{-1}$, which are updated recursively as follows:

$$\widehat{\boldsymbol{R}}_y(k+1) = \beta\widehat{\boldsymbol{R}}_y(k) + \boldsymbol{y}(k+1)\boldsymbol{y}^H(k+1), \tag{8.32}$$

$$\widehat{\boldsymbol{R}}_x(k+1) = \alpha\widehat{\boldsymbol{R}}_x(k) + \boldsymbol{x}(k+1)\boldsymbol{x}^H(k+1), \tag{8.33}$$

$$\boldsymbol{Q}_x(k+1) = \frac{1}{\alpha}\left(\boldsymbol{Q}_x(k) - \frac{\boldsymbol{Q}_x(k)\boldsymbol{x}(k+1)\boldsymbol{x}^H(k+1)\boldsymbol{Q}_x(k)}{\alpha + \boldsymbol{x}^H(k+1)\boldsymbol{Q}_x(k)\boldsymbol{x}(k+1)}\right), \tag{8.34}$$

$$\boldsymbol{Q}_y(k+1) = \frac{1}{\beta}\left(\boldsymbol{Q}_y(k) - \frac{\boldsymbol{Q}_y(k)\boldsymbol{y}(k+1)\boldsymbol{y}^H(k+1)\boldsymbol{Q}_y(k)}{\beta + \boldsymbol{y}^H(k+1)\boldsymbol{Q}_y(k)\boldsymbol{y}(k+1)}\right), \tag{8.35}$$

where $\alpha, \beta \in (0, 1)$ are the forgetting factors.

Different from the analysis of Möller's coupled algorithm, the convergence analysis of Nguyen algorithm in [4] was not conducted via the eigenvalue of Jacobian matrix. Nguyen established rigorous analysis of the DDT systems showing that, for a step size within a certain range, the algorithm converges to the orthogonal projection of the initial estimate onto the generalized eigen-subspace associated with the largest/smallest generalized eigenvalue.

Next, we analyze the convergence of Nguyen's algorithm via the eigenvalues of Jacobian matrix.

By ignoring the normalization step (8.27), the differential equation form of GPCA algorithms (8.26) and (8.28) can be written as:

$$\dot{\boldsymbol{w}} = \lambda^{-1}\left(\boldsymbol{R}_x^{-1}\boldsymbol{R}_y\boldsymbol{w} - \boldsymbol{w}^H\boldsymbol{R}_y\boldsymbol{w}\boldsymbol{w}\right), \tag{8.36}$$

$$\dot{\lambda} = \boldsymbol{w}^H\boldsymbol{R}_y\boldsymbol{w} - \lambda. \tag{8.37}$$

The Jacobian matrix at the stationary point $(\boldsymbol{w}_1, \lambda_1)$ is given as:

$$\begin{aligned} \boldsymbol{J}(\boldsymbol{w}_1, \lambda_1) &= \left.\begin{pmatrix} \frac{\partial \dot{\boldsymbol{w}}}{\partial \boldsymbol{w}^{\mathrm{T}}} & \frac{\partial \dot{\boldsymbol{w}}}{\partial \lambda} \\ \frac{\partial \dot{\lambda}}{\partial \boldsymbol{w}^{\mathrm{T}}} & \frac{\partial \dot{\lambda}}{\partial \lambda} \end{pmatrix}\right|_{(\boldsymbol{w}_1, \lambda_1)}, \\ &= \begin{pmatrix} \lambda_1^{-1}\boldsymbol{R}_x^{-1}\boldsymbol{R}_y - \boldsymbol{I} - 2\boldsymbol{w}_1\boldsymbol{w}_1^H\boldsymbol{R}_x & \boldsymbol{0} \\ 2\lambda_1\boldsymbol{w}_1^H\boldsymbol{R}_x & -1 \end{pmatrix}. \end{aligned} \tag{8.38}$$

Let

$$\boldsymbol{P} = \begin{pmatrix} \boldsymbol{W} & \boldsymbol{0} \\ \boldsymbol{0}^{\mathrm{T}} & 1 \end{pmatrix}. \tag{8.39}$$

Then, it can be easily seen that

$$\boldsymbol{P}^{-1} = \begin{pmatrix} \boldsymbol{W}^H\boldsymbol{R}_x & \boldsymbol{0} \\ \boldsymbol{0}^{\mathrm{T}} & 1 \end{pmatrix}. \tag{8.40}$$

Solving for the eigenvectors of $\boldsymbol{J}$ can then be simplified to the solving for the eigenvector of its similar diagonally matrix $\boldsymbol{J}^* = \boldsymbol{P}^{-1}\boldsymbol{J}\boldsymbol{P}$. Then it holds that

$$\boldsymbol{J}^*(\boldsymbol{w}_1, \lambda_1) = \begin{pmatrix} \lambda_1^{-1}\boldsymbol{\Lambda} - \boldsymbol{I} - 2\boldsymbol{e}_1\boldsymbol{e}_1^H & \boldsymbol{0} \\ 2\lambda_1\boldsymbol{e}_1^H & -1 \end{pmatrix}. \tag{8.41}$$

The eigenvalues α determined from $\det(\boldsymbol{J}^* - \alpha\boldsymbol{I}) = 0$ are

$$\alpha_1 = -2,\ \alpha_{N+1} = -1,\ \alpha_j = \frac{\lambda_j}{\lambda_1} - 1,\ j = 2, \ldots, N. \tag{8.42}$$

Since the stability requires $\alpha < 0$ and thus $\lambda_j < \lambda_1$, $j = 2, 3, \ldots, n$, it can be seen that only principal eigenvector–eigenvalue pairs are stable stationary points, and all other stationary points are saddles or repellers. If we further assume that $\lambda_1 \gg \lambda_j$, then $\alpha_j \approx -1, j = 2, 3, \ldots, n$. That is to say, the eigen motion in all directions in the algorithm do not depend on the generalized eigenvalue of the covariance matrix of input signal. Thus, this algorithm does not have the speed stability problem. Similar analysis can be applied to the GMCA algorithms (8.29) and (8.31).

8.2.3 Coupled Singular Value Decomposition of a Cross-Covariance Matrix

In [9], a coupled online learning rule for the singular value decomposition (SVD) of a cross-covariance matrix was derived. In coupled SVD rules, the singular value is estimated alongside the singular vectors, and the effective learning rates for the singular vector rules are influenced by the singular value estimates [9]. In addition, a first-order approximation of Gram–Schmidt orthonormalization as decorrelation method for the estimation of multiple singular vectors and singular values was used. It has been shown that the coupled learning rules converge faster than Hebbian learning rules and that the first-order approximation of Gram–Schmidt orthonormalization produces more precise estimates and better orthonormality than the standard deflation method [9].

The neural network and its learning algorithm for the singular value decomposition of a cross-covariance matrix will be discussed in Chap. 9, in which the coupled online learning rules for the SVD of a cross-covariance matrix will be analyzed in detail.

8.3 Unified and Coupled Algorithm for Minor and Principal Eigen Pair Extraction

Coupled algorithm can mitigate the speed stability problem which exists in most noncoupled algorithms. Though unified algorithm and coupled algorithm have these advantages over single purpose algorithm and noncoupled algorithm, respectively, there are only few of unified algorithms, and coupled algorithms have been proposed. Moreover, to the best of the authors' knowledge, there are no both unified and coupled algorithms which have been proposed. In this chapter, based on a novel information criterion, we propose two self-stabilizing algorithms which are both unified and coupled. In the derivation of our algorithms, it is easier to obtain the results compared with traditional methods, because there is no need to calculate the inverse Hessian matrix. Experiment results show that the proposed algorithms perform better than existing coupled algorithms and unified algorithms.

8.3.1 Couple Dynamical System

The derivation of neural network learning rules often starts with an information criterion, e.g., by maximization of the variance of the projected data or by minimization of the reconstruction error [7]. However, as stated in [10], the freedom of choosing an information criterion is greater if Newton's method is applied because the criterion just has to have stationary points in the desired solutions. Thus in [2], Moller proposed a special criterion. Based on this criterion and by using Newton's method, Moller derived some CPCA learning rules and a CMCA learning rule. Based on another criterion, Hou [5] derived the same CPCA and CMCA learning rules as that of Moller's, and Appendix 2 of [5] showed that it is easier and clearer to approximate the inverse of the Hessian.

To start the analysis, we use the same information criterion as Hou's, which is

$$p = \boldsymbol{w}^{\mathrm{T}}\boldsymbol{C}\boldsymbol{w} - \boldsymbol{w}^{\mathrm{T}}\boldsymbol{w}\lambda + \lambda \tag{8.43}$$

where $\boldsymbol{C} = E\{\boldsymbol{x}\boldsymbol{x}^{\mathrm{T}}\} \in \Re^{n\times n}$ is the covariance matrix of the n-dimensional input data sequence $\boldsymbol{x}$, $\boldsymbol{w} \in \Re^{n\times 1}$ and $\lambda \in \Re$ denotes the estimation of eigenvector (weight vector) and eigenvalue of $\boldsymbol{C}$, respectively.

It is found that

$$\frac{\partial p}{\partial \boldsymbol{w}} = 2\boldsymbol{C}\boldsymbol{w} - 2\lambda\boldsymbol{w} \tag{8.44}$$

$$\frac{\partial p}{\partial \lambda} = -\boldsymbol{w}^{\mathrm{T}}\boldsymbol{w} + 1. \tag{8.45}$$

Thus, the stationary points $(\bar{\boldsymbol{w}}, \bar{\lambda})$ of (8.43) are defined by

$$\left.\frac{\partial p}{\partial \boldsymbol{w}}\right|_{(\bar{\boldsymbol{w}},\bar{\lambda})} = \boldsymbol{0}, \left.\frac{\partial p}{\partial \lambda}\right|_{(\bar{\boldsymbol{w}},\bar{\lambda})} = 0. \tag{8.46}$$

Then, we can obtain

$$\boldsymbol{C}\bar{\boldsymbol{w}} = \bar{\lambda}\bar{\boldsymbol{w}}, \tag{8.47}$$

$$\bar{\boldsymbol{w}}^{\mathrm{T}}\bar{\boldsymbol{w}} = 1 \tag{8.48}$$

from which we can also conclude that $\bar{\boldsymbol{w}}^{\mathrm{T}}\boldsymbol{C}\bar{\boldsymbol{w}} = \bar{\lambda}$. Thus, the criterion (8.43) fulfills the aforementioned requirement: The stationary points include all associated eigenvectors and eigenvalues of $\boldsymbol{C}$. The Hessian of the criterion is given as:

$$\boldsymbol{H}(\boldsymbol{w},\lambda)=\begin{pmatrix}\frac{\partial^2 p}{\partial \boldsymbol{w}^2} & \frac{\partial^2 p}{\partial \boldsymbol{w}\partial\lambda}\\ \frac{\partial^2 p}{\partial\lambda\partial \boldsymbol{w}} & \frac{\partial^2 p}{\partial\lambda^2}\end{pmatrix}=2\begin{pmatrix}\boldsymbol{C}-\lambda\boldsymbol{I} & -\boldsymbol{w}\\ -\boldsymbol{w}^{\mathrm{T}} & 0\end{pmatrix}. \tag{8.49}$$

Based on the Newton's method, the equation used by Moller and Hou to derive the differential equations can be written as:

$$\begin{pmatrix}\dot{\boldsymbol{w}}\\ \dot{\lambda}\end{pmatrix}=-\boldsymbol{H}^{-1}(\boldsymbol{w},\lambda)\begin{pmatrix}\frac{\partial p}{\partial \boldsymbol{w}}\\ \frac{\partial p}{\partial\lambda}\end{pmatrix}. \tag{8.50}$$

Based on different information criteria, both Moller and Hou tried to find the inverse of their Hessian $\boldsymbol{H}^{-1}(\boldsymbol{w},\lambda)$. Although the inverse Hessian of Moller and Hou is different, they finally obtained the same CPCA and CMCA rules [5]. Here we propose to derive the differential equation with another technical, which is

$$\boldsymbol{H}(\boldsymbol{w},\lambda)\begin{pmatrix}\dot{\boldsymbol{w}}\\ \dot{\lambda}\end{pmatrix}=-\begin{pmatrix}\frac{\partial p}{\partial \boldsymbol{w}}\\ \frac{\partial p}{\partial\lambda}\end{pmatrix}. \tag{8.51}$$

In this case, there is no need to calculate the inverse Hessian. Substituting (8.44), (8.45), and (8.49) into (8.51), it yields

$$2\begin{pmatrix}\boldsymbol{C}-\lambda\boldsymbol{I} & -\boldsymbol{w}\\ -\boldsymbol{w}^{\mathrm{T}} & 0\end{pmatrix}\begin{pmatrix}\dot{\boldsymbol{w}}\\ \dot{\lambda}\end{pmatrix}=-\begin{pmatrix}2\boldsymbol{C}\boldsymbol{w}-2\lambda\boldsymbol{w}\\ -\boldsymbol{w}^{\mathrm{T}}\boldsymbol{w}+1\end{pmatrix}. \tag{8.52}$$

Then we can get

$$(\boldsymbol{C}-\lambda\boldsymbol{I})\dot{\boldsymbol{w}}-\boldsymbol{w}\dot{\lambda}=-(\boldsymbol{C}-\lambda\boldsymbol{I})\boldsymbol{w} \tag{8.53}$$

$$2\boldsymbol{w}^{\mathrm{T}}\dot{\boldsymbol{w}}=\boldsymbol{w}^{\mathrm{T}}\boldsymbol{w}-1. \tag{8.54}$$

In the vicinity of the stationary point $(\boldsymbol{w}_1,\lambda_1)$, by approximating $\boldsymbol{w}\approx\boldsymbol{w}_1,\ \lambda\approx\lambda_1\ll\lambda_j\ (2\le j\le n)$, and after some manipulations (see Appendix A in [6]), we get a coupled dynamical system as

$$\dot{\boldsymbol{w}}=\frac{\boldsymbol{C}^{-1}\boldsymbol{w}(\boldsymbol{w}^{\mathrm{T}}\boldsymbol{w}+1)}{2\boldsymbol{w}^{\mathrm{T}}\boldsymbol{C}^{-1}\boldsymbol{w}}-\boldsymbol{w} \tag{8.55}$$

$$\dot{\lambda}=\frac{\boldsymbol{w}^{\mathrm{T}}\boldsymbol{w}+1}{2}\left(\frac{1}{\boldsymbol{w}^{\mathrm{T}}\boldsymbol{C}^{-1}\boldsymbol{w}}-\lambda\right). \tag{8.56}$$

8.3.2 *The Unified and Coupled Learning Algorithms*

8.3.2.1 Coupled MCA Algorithms

The differential equations can be turned into the online form by informally approximating $\boldsymbol{C} = \boldsymbol{x}(k)\boldsymbol{x}^{\mathrm{T}}(k)$, where $\boldsymbol{x}(k)$ is a data vector drawn from the distribution. That is, the expression of the rules in online form can be approximated by slowly following $\boldsymbol{w}(k+1) = f(\boldsymbol{x}(k)\boldsymbol{x}^{\mathrm{T}}(k); \boldsymbol{w}(k))$ for subsequent observations of $\boldsymbol{x}$. Moller has pointed out [2] that this transition is infeasible if the equation contains $\boldsymbol{C}^{-1}$, because it is hard to replace the inverse matrix $\boldsymbol{C}^{-1}$ by an expression containing the input vector $\boldsymbol{x}$. However, this problem can be solved in another way [11, 12], in which $\boldsymbol{C}^{-1}$ is updated as

$$\widehat{\boldsymbol{C}}^{-1}(k+1) = \frac{k+1}{k}\left[\widehat{\boldsymbol{C}}^{-1}(k) - \frac{\widehat{\boldsymbol{C}}^{-1}(k)\,\boldsymbol{x}(k+1)\boldsymbol{x}^{\mathrm{T}}(k+1)\,\widehat{\boldsymbol{C}}^{-1}(k)}{k + \boldsymbol{x}^{\mathrm{T}}(k+1)\,\widehat{\boldsymbol{C}}^{-1}(k)\,\boldsymbol{x}^{\mathrm{T}}(k+1)}\right] \tag{8.57}$$

where $\widehat{\boldsymbol{C}}^{-1}(k)$ starts with $\widehat{\boldsymbol{C}}^{-1}(0) = \boldsymbol{I}$ and converges to $\boldsymbol{C}^{-1}$ as $k \to \infty$.Then, the CMCA system (8.55)–(8.56) has the online form as:

$$\boldsymbol{w}(k+1) = \boldsymbol{w}(k) + \gamma(k)\left\{\frac{[\boldsymbol{w}^{\mathrm{T}}(k)\boldsymbol{w}(k) + 1]\,\boldsymbol{Q}(k)\boldsymbol{w}(k)}{2\boldsymbol{w}^{\mathrm{T}}(k)\,\boldsymbol{Q}(k)\boldsymbol{w}(k)} - \boldsymbol{w}(k)\right\} \tag{8.58}$$

$$\lambda(k+1) = \lambda(k) + \gamma(k)\frac{\boldsymbol{w}^{\mathrm{T}}(k)\boldsymbol{w}(k) + 1}{2}\left[\frac{1}{\boldsymbol{w}^{\mathrm{T}}(k)\boldsymbol{Q}(k)\boldsymbol{w}(k)} - \lambda(k)\right] \tag{8.59}$$

$$\boldsymbol{Q}(k+1) = \frac{k+1}{\alpha k}\left[\boldsymbol{Q}(k) - \frac{\boldsymbol{Q}(k)\,\boldsymbol{x}(k+1)\,\boldsymbol{x}^{\mathrm{T}}(k+1)\,\boldsymbol{Q}(k)}{k + \boldsymbol{x}^{\mathrm{T}}(k+1)\,\boldsymbol{Q}(k)\,\boldsymbol{x}^{\mathrm{T}}(k+1)}\right] \tag{8.60}$$

where $0 < \alpha \leq 1$ denotes the forgetting factor and $\gamma(k)$ is the learning rate. If all training samples come from a stationary process, we choose $\alpha = 1$. $\boldsymbol{Q}(k) = \boldsymbol{C}^{-1}(k)$ starts with $\boldsymbol{Q}(0) = \boldsymbol{I}$. Here, we refer to the rule (8.55)–(8.56) and its online form (8.58)–(8.60) as "fMCA," where *f* means fast. In the rest of this section, the online form (which is used in the implementation) and the differential matrix form (which is used in the convergence analysis) of a rule have the same name, and we will not emphasize this again. If we further approximate $\boldsymbol{w}^{\mathrm{T}}\boldsymbol{w} \approx 1$ (which fulfills in the vicinity of the stationary points) in (8.55)–(8.56), we can obtain *q* simplified CMCA system

$$\dot{\boldsymbol{w}} = \frac{\boldsymbol{C}^{-1}\boldsymbol{w}}{2\boldsymbol{w}^{\mathrm{T}}\boldsymbol{C}^{-1}\boldsymbol{w}} - \boldsymbol{w} \tag{8.61}$$

$$\dot{\lambda} = \frac{1}{\boldsymbol{w}^{\mathrm{T}}\boldsymbol{C}^{-1}\boldsymbol{w}} - \lambda \tag{8.62}$$

and the online form is given as:

$$\boldsymbol{w}(k+1) = \boldsymbol{w}(k) + \gamma(k)\left\{\frac{\boldsymbol{Q}(k)\,\boldsymbol{w}(k)}{\boldsymbol{w}^{\mathrm{T}}(k)\,\boldsymbol{Q}(k)\,\boldsymbol{w}(k)} - \boldsymbol{w}(k)\right\} \tag{8.63}$$

$$\lambda(k+1) = \lambda(k) + \gamma(k)\left[\frac{1}{\boldsymbol{w}^{\mathrm{T}}(k)\,\boldsymbol{Q}(k)\,\boldsymbol{w}(k)} - \lambda(k)\right] \tag{8.64}$$

where $\mathbf{Q}(k)$ is updated by (8.60). In the following, we will refer to this algorithm as "aMCA," where *a* means adaptive.

8.3.2.2 Coupled PCA Algorithms

It is known that in unified rules, MCA rules can be derived from PCA rules by changing the sign or using the inverse of the covariance matrix, and vice versa. Here we propose to derive unified algorithms by deriving CPCA rules from CMCA rules. Suppose that the covariance matrix $\mathbf{C}$ has an eigen pair $(\boldsymbol{w}, \lambda)$; then it holds that [13] $\boldsymbol{C}\boldsymbol{w} = \lambda\boldsymbol{w}$ and $\boldsymbol{C}^{-1}\boldsymbol{w} = \lambda^{-1}\boldsymbol{w}$, which means that the minor eigen pair of $\boldsymbol{C}$ is also the principal eigen pair of the inverse matrix $\boldsymbol{C}^{-1}$, and vice versa. Therefore, by replacing $\boldsymbol{C}^{-1}$ with $\boldsymbol{C}$ in fMCA and aMCA rules, respectively, we obtain two modified rules to extract the principal eigen pair of $\boldsymbol{C}$, which is also the minor eigen pair of $\boldsymbol{C}^{-1}$. The modified rules are given as:

$$\dot{\boldsymbol{w}} = \frac{\boldsymbol{C}\boldsymbol{w}\,(\boldsymbol{w}^{\mathrm{T}}\boldsymbol{w}+1)}{2\boldsymbol{w}^{\mathrm{T}}\boldsymbol{C}\boldsymbol{w}} - \boldsymbol{w} \tag{8.65}$$

$$\dot{\lambda} = \frac{\boldsymbol{w}^{\mathrm{T}}\boldsymbol{w}+1}{2}\left(\boldsymbol{w}^{\mathrm{T}}\boldsymbol{C}\boldsymbol{w} - \lambda\right) \tag{8.66}$$

and

$$\dot{\boldsymbol{w}} = \frac{\boldsymbol{C}\boldsymbol{w}}{2\boldsymbol{w}^{\mathrm{T}}\boldsymbol{C}\boldsymbol{w}} - \boldsymbol{w} \tag{8.67}$$

$$\dot{\lambda} = \boldsymbol{w}^{\mathrm{T}}\boldsymbol{C}\boldsymbol{w} - \lambda. \tag{8.68}$$

Since the covariance matrix $\boldsymbol{C}$ is usually unknown in advance, we use its estimate at time k by $\widehat{\boldsymbol{C}}(k)$ suggested in [11], which is

$$\widehat{\boldsymbol{C}}(k+1) = \alpha\frac{k}{k+1}\widehat{\boldsymbol{C}}(k) + \frac{1}{k+1}\boldsymbol{x}(k+1)\boldsymbol{x}^{\mathrm{T}}(k+1) \tag{8.69}$$

where $\widehat{\boldsymbol{C}}(k)$ starts with $\widehat{\boldsymbol{C}}(0) = \boldsymbol{x}(0)\boldsymbol{x}^{\mathrm{T}}(0)$ (or $\boldsymbol{I}$). Actually, (8.57) is obtained from (8.69) by using the SM-formula. Then, the online form of (8.65)–(8.66) and (8.67)–(8.68) is given as:

$$\boldsymbol{w}(k+1) = \boldsymbol{w}(k) + \gamma(k)\left\{\frac{\left[\boldsymbol{w}^{\mathrm{T}}(k)\,\widehat{\boldsymbol{C}}(k)\boldsymbol{w}(k)+1\right]\widehat{\boldsymbol{C}}(k)\boldsymbol{w}(k)}{2\boldsymbol{w}^{\mathrm{T}}(k)\widehat{\boldsymbol{C}}(k)\boldsymbol{w}(k)} - \boldsymbol{w}(k)\right\} \tag{8.70}$$

$$\lambda(k+1) = \lambda(k) + \gamma(k)\frac{\boldsymbol{w}^{\mathrm{T}}(k)\boldsymbol{w}(k)+1}{2}\left[\boldsymbol{w}^{\mathrm{T}}(k)\widehat{\boldsymbol{C}}(k)\boldsymbol{w}(k) - \lambda(k)\right] \tag{8.71}$$

and

$$\boldsymbol{w}(k+1) = \boldsymbol{w}(k) + \gamma(k)\left\{\frac{\widehat{\boldsymbol{C}}(k)\,\boldsymbol{w}(k)}{\boldsymbol{w}^{\mathrm{T}}(k)\,\widehat{\boldsymbol{C}}(k)\,\boldsymbol{w}(k)} - \boldsymbol{w}(k)\right\} \tag{8.72}$$

$$\lambda(k+1) = \lambda(k) + \gamma(k)\left[\boldsymbol{w}^{\mathrm{T}}(k)\,\widehat{\boldsymbol{C}}(k)\,\boldsymbol{w}(k) - \lambda(k)\right] \tag{8.73}$$

respectively. Here we rename this algorithm deduced from fMCA and aMCA as "fPCA" and "aPCA," respectively. Finally, we obtain two unified and coupled algorithms. The first one is fMCA + fGPCA, and the second one is aMCA + aPCA. These two unified algorithms are capable of both PCA and MCA by using the original or inverse of covariance matrix.

8.3.2.3 Multiple Eigen Pairs Estimation

In some engineering practice, it is required to estimate the eigen-subspace or multiple eigen pairs. As introduced in [4], by using the nested orthogonal complement structure of the eigen-subspace, the problem of estimating the $p(\leq n)$-dimensional principal/minor subspace can be reduced to multiple principal/minor eigenvectors estimation. The following shows how to estimate there maining $p-1$ principal/minor eigen pairs.

For the CMCA case, consider the following equations:

$$\widehat{\boldsymbol{C}}_j = \widehat{\boldsymbol{C}}_{j-1} + \eta\lambda_{j-1}\boldsymbol{w}_{j-1}\boldsymbol{w}_{j-1}^{\mathrm{T}},\ j = 2,\ldots,p \tag{8.74}$$

where $\widehat{\boldsymbol{C}}_1 = \widehat{\boldsymbol{C}}$ and η is larger than the largest eigenvalue of $\widehat{\boldsymbol{C}}$, and $(\boldsymbol{w}_{j-1}, \lambda_{j-1})$ is the $(j-1)$th minor eigen pair of $\widehat{\boldsymbol{C}}$ that has been extracted. It is found that

$$\begin{aligned}\widehat{C}_j w_q &= (\widehat{C}_{j-1} + \eta\lambda_{j-1} w_{j-1}^{\mathrm{T}}) w_q \\ &= (\widehat{C}_1 + \eta \sum_{r=1}^{j-1} \lambda_r w_r w_r^{\mathrm{T}}) w_q \\ &= \widehat{C}_1 w_q + \eta \sum_{r=1}^{j-1} \lambda_r w_r w_r^{\mathrm{T}} w_q \\ &= \begin{cases} \widehat{C}_1 w_q + \eta\lambda_q w_q = (1+\eta)\lambda_q w_q & \text{for } q = 1,\ldots,j-1 \\ \widehat{C}_1 w_q = \lambda_q w_q & \text{for } q = j,\ldots,p \end{cases}.\end{aligned} \tag{8.75}$$

Suppose that matrix $\widehat{C}_1$ has eigenvectors $w_1, w_2, \ldots, w_n$ corresponding to eigenvalues $(0<)\ \sigma_1 < \sigma_2 < \cdots < \sigma_n$, and then matrix C_j has eigenvectors $w_j, \ldots, w_n$, $w_1, \ldots, w_{j-1}$ corresponding to eigenvalues $(0<)\sigma_j < \cdots < \sigma_n < (1+\eta)\sigma_1 < \cdots < (1+\eta)\sigma_{j-1}$. In this case, σ_j is the smallest eigenvalue of C_j. Based on the SM-formula, we have

$$\begin{aligned}Q_j = C_j^{-1} &= (C_{j-1} + \eta\lambda_{j-1} w_{j-1} w_{j-1}^{\mathrm{T}})^{-1} \\ &= C_{j-1}^{-1} - \frac{\eta\lambda_{j-1} C_{j-1}^{-1} w_{j-1} w_{j-1}^{\mathrm{T}} C_{j-1}^{-1}}{1 + \eta\lambda_{j-1} w_{j-1}^{\mathrm{T}} \widehat{C}_{j-1}^{-1} w_{j-1}} \\ &= Q_{j-1} - \frac{\eta\lambda_{j-1} Q_{j-1} w_{j-1} w_{j-1}^{\mathrm{T}} Q_{j-1}}{1 + \eta\lambda_{j-1} w_{j-1}^{\mathrm{T}} Q_{j-1} w_{j-1}},\ j = 2,\ldots,p.\end{aligned} \tag{8.76}$$

Thus, by replacing $\widehat{Q}$ with $\widehat{Q}_j$ in (8.58)–(8.59) or (8.63)–(8.64), they can be used to estimate the jth minor eigen pair (w_j, λ_j) of $\widehat{C}$.

For the CPCA case, consider the following equations

$$C_j = C_{j-1} - \lambda_{j-1} w_{j-1} w_{j-1}^{\mathrm{T}},\ j = 2,\ldots,p \tag{8.77}$$

where (w_{j-1}, λ_{j-1}) is the $(j-1)$th principal eigen pair that has been extracted. It is found that

$$\begin{aligned}\widehat{C}_j w_q &= (\widehat{C}_{j-1} - \lambda_{j-1} w_{j-1} w_{j-1}^{\mathrm{T}}) w_q \\ &= (\widehat{C}_1 - \sum_{r=1}^{j-1} \lambda_r w_r w_r^{\mathrm{T}}) w_q \\ &= \widehat{C}_1 w_q - \sum_{r=1}^{j-1} \lambda_r w_r w_r^{\mathrm{T}} w_q \\ &= \begin{cases} 0 & \text{for } q = 1,\ldots,j-1 \\ \widehat{C}_1 w_q = \lambda_q w_q & \text{for } q = j,\ldots,p \end{cases}.\end{aligned} \tag{8.78}$$

Suppose that the matrix $\widehat{\boldsymbol{C}}_1$ has eigenvectors $\boldsymbol{w}_1, \boldsymbol{w}_2, \ldots, \boldsymbol{w}_n$ corresponding to eigenvalues $\sigma_1 > \sigma_2 > \cdots > \sigma_n(>0)$, and then the matrix $\boldsymbol{C}_j$ has eigenvectors $\boldsymbol{w}_j, \ldots, \boldsymbol{w}_n, \boldsymbol{w}_1, \ldots, \boldsymbol{w}_{j-1}$ corresponding to eigenvalues $\sigma_j > \cdots > \sigma_n > \hat{\sigma}_1 = \cdots = \hat{\sigma}_{j-1}(=0)$. In this case, σ_j is the largest eigenvalue of $\boldsymbol{C}_j$. Thus, by replacing $\widehat{\boldsymbol{C}}$ with $\widehat{\boldsymbol{C}}_j$ in (8.70)–(8.71) or (8.72)–(8.73), they can be used to estimate the jth principal eigen pair $(\boldsymbol{w}_j, \lambda_j)$ of $\widehat{\boldsymbol{C}}$.

8.3.3 Analysis of Convergence and Self-stabilizing Property

The major work of convergence analysis of coupled rules is to find the eigenvalues of the Jacobian

$$\boldsymbol{J}(\boldsymbol{w}_1, \lambda_1) = \begin{pmatrix} \frac{\partial \dot{\boldsymbol{w}}}{\partial \boldsymbol{w}^{\mathrm{T}}} & \frac{\partial \dot{\boldsymbol{w}}}{\partial \lambda} \\ \frac{\partial \dot{\lambda}}{\partial \boldsymbol{w}^{\mathrm{T}}} & \frac{\partial \dot{\lambda}}{\partial \lambda} \end{pmatrix} \tag{8.79}$$

of the differential equations for a stationary point $(\boldsymbol{w}_1, \lambda_1)$. For fMCA rule, after some manipulations (see Appendix B in [6]), we get

$$\boldsymbol{J}_{fMCA}(\boldsymbol{w}_1, \lambda_1) = \begin{pmatrix} \boldsymbol{C}^{-1}\lambda_1 - \boldsymbol{I} - \boldsymbol{w}_1\boldsymbol{w}_1^{\mathrm{T}} & \boldsymbol{0} \\ -2\lambda_1\boldsymbol{w}_1^{\mathrm{T}} & -1 \end{pmatrix}. \tag{8.80}$$

The Jacobian can be simplified by an orthogonal transformation with

$$\boldsymbol{U} = \begin{pmatrix} \overline{\boldsymbol{W}} & \boldsymbol{0} \\ \boldsymbol{0}^{\mathrm{T}} & 1 \end{pmatrix}. \tag{8.81}$$

The transformed Jacobian $\boldsymbol{J}^* = \boldsymbol{U}^{\mathrm{T}}\boldsymbol{J}\boldsymbol{U}$ has the same eigenvalues as $\boldsymbol{J}$. In the vicinity of a stationary point $(\boldsymbol{w}_1, \lambda_1)$, we approximate $\overline{\boldsymbol{W}}^{\mathrm{T}}\boldsymbol{w} \approx \boldsymbol{e}_1$ and obtain

$$\boldsymbol{J}^*_{fMCA}(\boldsymbol{w}_1, \lambda_1) = \begin{pmatrix} \bar{\boldsymbol{\Lambda}}^{-1}\lambda_1 - \boldsymbol{I} - \boldsymbol{e}_1\boldsymbol{e}_1^{\mathrm{T}} & \boldsymbol{0} \\ -2\lambda_1\boldsymbol{e}_1^{\mathrm{T}} & -1 \end{pmatrix}. \tag{8.82}$$

The eigenvalues α of $\boldsymbol{J}^*$ are determined as $\det(\boldsymbol{J}^* - \alpha\boldsymbol{I}) = 0$, which are

$$\alpha_1 = \alpha_{n+1} = -1, \; \alpha_j = \frac{\lambda_1}{\lambda_j} - 1 \overset{\lambda_1 \ll \lambda_j}{\approx} -1, \; j = 2, \ldots, n. \tag{8.83}$$

Since stability requires $\alpha < 0$ and thus $\lambda_1 < \lambda_j$, $j = 2, \ldots, n$, we find that only minor eigen pairs are stable stationary points, while all others are saddles or repellers. What's more, if we further assume $\lambda_1 \ll \lambda_j$, all eigenvalues are $\alpha \approx -1$. Hence, the system converges with approximately equal speed in all its eigen directions, and this speed is widely independent of the eigenvalues λ_j of the covariance matrix [2]. That is to say, the speed stability problem does not exist in fMCA algorithm.

Similarly, for aMCA rule, we analyze the stability by finding the eigenvalues of

$$J^*_{aMCA}(\boldsymbol{w}_1, \lambda_1) = \begin{pmatrix} \bar{\boldsymbol{\Lambda}}^{-1}\lambda_1 - \boldsymbol{I} - 2\boldsymbol{e}_1\boldsymbol{e}_1^{\mathrm{T}} & \boldsymbol{0} \\ -2\lambda_1\boldsymbol{e}_1^{\mathrm{T}} & -1 \end{pmatrix} \tag{8.84}$$

which are

$$\alpha_1 = -2,\ \alpha_{n+1} = -1,\ \alpha_j = \frac{\lambda_1}{\lambda_j} - 1,\ j = 2, \ldots, n. \tag{8.85}$$

The situation of aMCA is similar to that of fMCA, and the only difference is that the first eigenvalue of Jacobian is $\alpha_1 = -1$ for fMCA and $\alpha_1 = -2$ for aMCA. Thus, the convergence speed of fMCA and aMCA is almost the same.

Similarly, the transformed Jacobian functions of fPCA and aPCA are given as:

$$J^*_{fPCA}(\boldsymbol{w}_1, \lambda_1) = \begin{pmatrix} \bar{\boldsymbol{\Lambda}}^{-1}\lambda_1 - \boldsymbol{I} - \boldsymbol{e}_1\boldsymbol{e}_1^{\mathrm{T}} & \boldsymbol{0} \\ 2\lambda_1\boldsymbol{e}_1^{\mathrm{T}} & -1 \end{pmatrix} \tag{8.86}$$

and

$$J^*_{aPCA}(\boldsymbol{w}_1, \lambda_1) = \begin{pmatrix} \bar{\boldsymbol{\Lambda}}^{-1}\lambda_1 - \boldsymbol{I} - 2\boldsymbol{e}_1\boldsymbol{e}_1^{\mathrm{T}} & \boldsymbol{0} \\ 2\lambda_1\boldsymbol{e}_1^{\mathrm{T}} & -1 \end{pmatrix} \tag{8.87}$$

respectively. And the eigenvalues of (8.86) and (8.87) are given as:

$$\alpha_1 = \alpha_{n+1} = -1,\ \alpha_j = \frac{\lambda_j}{\lambda_n} - 1 \overset{\lambda_n \gg \lambda_j}{\approx} -1,\ j = 1, \ldots, n-1 \tag{8.88}$$

$$\alpha_1 = -2, \alpha_{n+1} = -1,\ \alpha_j = \frac{\lambda_j}{\lambda_n} - 1 \overset{\lambda_n \ll \lambda_j}{\approx} -1,\ j = 1, \ldots, n-1 \tag{8.89}$$

respectively. We can see that only principal eigen pairs are stable stationary points, while all others are saddles or repellers. We can further assume $\lambda_1 \gg \lambda_j$ and thus $\alpha_j \approx -1\ (j \neq 1)$ for fPCA and aPCA.

The analysis of the self-stabilizing property of the proposed algorithms is omitted here. For details, see [6].

8.3.4 Simulation Experiments

In this section, we provide several experiments to illustrate the performance of the proposed algorithms in comparison with some well-known coupled algorithms and unified algorithms. Experiments 1 and 2 mainly show the stability of proposed CMCA and CPCA algorithms in comparison with existing CMCA and CPCA algorithms, respectively. In experiment 3, the self-stabilizing property of the proposed algorithm is shown. In experiment 4, we compare the performance of aMCA and aPCA with that of two unified algorithms. Experiments 5 and 6 illustrate some examples of practical applications.

In experiments 1–4, all algorithms are used to extract the minor or principal component from a high-dimensional input data sequence, which is generated from $\boldsymbol{x} = \boldsymbol{B} \cdot \boldsymbol{y}(t)$, where each column of $\boldsymbol{B} \in \Re^{30\times30}$ is Gaussian with variance 1/30, and $\boldsymbol{y}(t) \in \Re^{30\times1}$ is Gaussian and randomly generated.

In all experiments, to measure the estimation accuracy, we compute the norm of eigenvector estimation (weight vector) $\|\boldsymbol{w}(k)\|$ and the projection $[\psi(k)]$ of the weight vector onto the true eigenvector at each step:

$$\psi(k) = \frac{|\boldsymbol{w}^{\mathrm{T}}(k)\boldsymbol{w}_1|}{\|\boldsymbol{w}(k)\|}$$

where $\boldsymbol{w}_1$ is the true minor (for MCA) or principal (for PCA) eigenvector with unit length.

Unless otherwise stated, we set the initial conditions of experiments 1–4 as follows: (1) The weight vector is initialized with a random vector (unit length). (2) The learning rate $\gamma(k)$ starts at $\gamma(0) = 10^{-2}$ and decays exponentially toward zero with a final value $\gamma(k_{\max}) = 10^{-4}$. (3) We set $\alpha = 1$ (if used), and $\lambda(0) = 0.001$ for all cMCA and cPCA algorithms.

In experiments 1 and 2, $k_{\max}$ = 20,000 training steps are executed for all algorithms. In order to test the stability of the proposed algorithms, after 10,000 training steps, we drastically change the input signals; thus, the eigen information changed suddenly. All algorithms start to extract the new eigen pair since k = 10001. The learning rate for nMCA is 10 times smaller than that for the others. Then, 20 times of Monte Carlo simulation are executed for all experiments.

Figure 8.1 shows the time course of the projection of minor weight vector. We can see that in all rules except mMCA the projection converges toward unity; thus, these weight vectors align with the true eigenvector. The convergence speed of mMCA is lower than that of the others and the projection of mMCA cannot converge toward unity within 10,000 steps. We can also find that the convergence speed of fMCA and aMCA rules is similar, and higher than that of the others. We can also find that, at time step k = 10,001, where the input signals changed suddenly, all algorithms start to extract the new eigen pair. Figure 8.2 shows the time course of weight vector length. We can find that the vector length of nMCA converges to a nonunit length. The convergence speed and the stability of fMCA

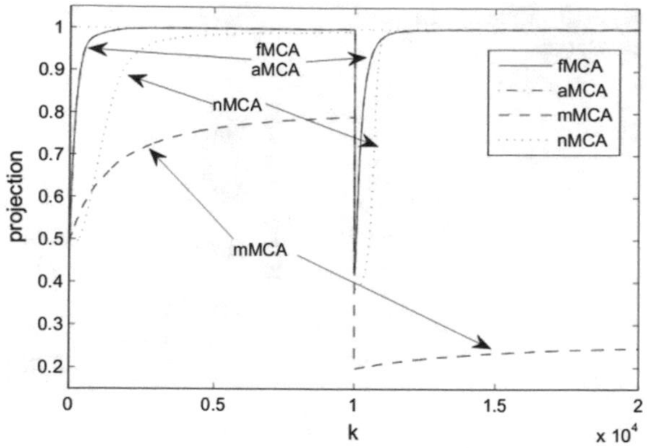

Fig. 8.1 Projection of weight vector onto the true minor eigenvector

and aMCA are higher and better than that of the others. It can be seen that the convergence speed of aMCA is a bit higher than that of fMCA.

Figure 8.3 shows the time course of the minor eigenvalue estimation. We can see that mMCA cannot extract the minor eigenvalue as effective as the other algorithms after the input signals changed. From Figs. 8.1 to 8.3, we can conclude that the performance of fMCA and aMCA is better than that of the other cMCA algorithms. Moreover, nMCA contains $\boldsymbol{C}$ and $\boldsymbol{C}^{-1}$ simultaneously in the equations, and we can prove that mMCA also has the speed stability problem though it is a coupled rule. These may be the reason why our algorithms perform better than nMCA and mCMA.

In experiment 2, we compare the performance of fPCA and aPCA with that of ALA and nPCA. The time course of the projection and the eigenvector length of principal weight vector are shown in Figs. 8.4 and 8.5, and the principal eigenvalue estimation is shown in Fig. 8.6, respectively. In Fig. 8.5, the curves for fPCA and aPCA are shown in a subfigure because of its small amplitude. We can see that the convergence speed of fPCA and aPCA is similar to that of nPCA and ALA, but fPCA and aPCA have less fluctuations over time compared with nPCA and ALA.

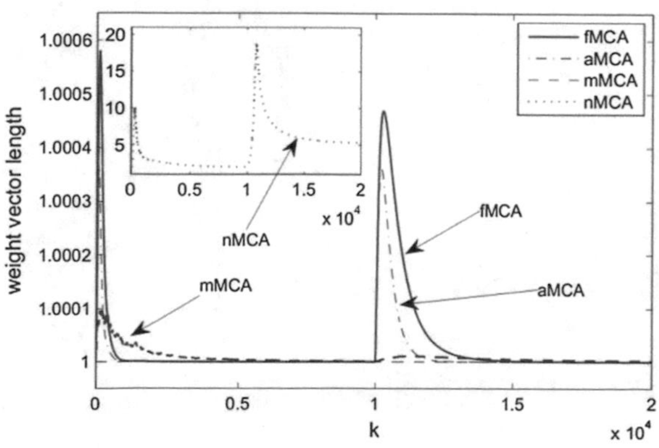

Fig. 8.2 Weight vector length

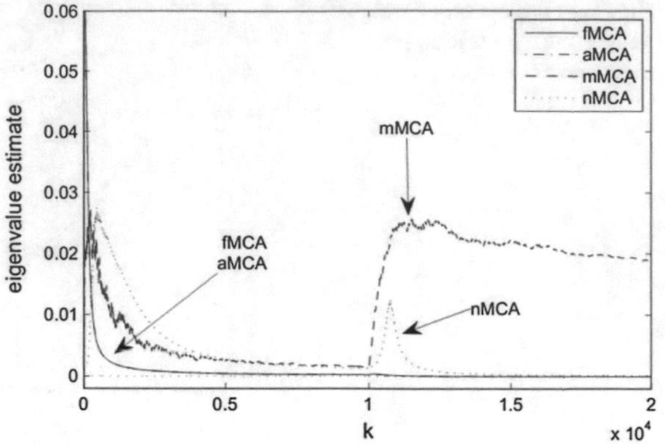

Fig. 8.3 Minor eigenvalue estimation

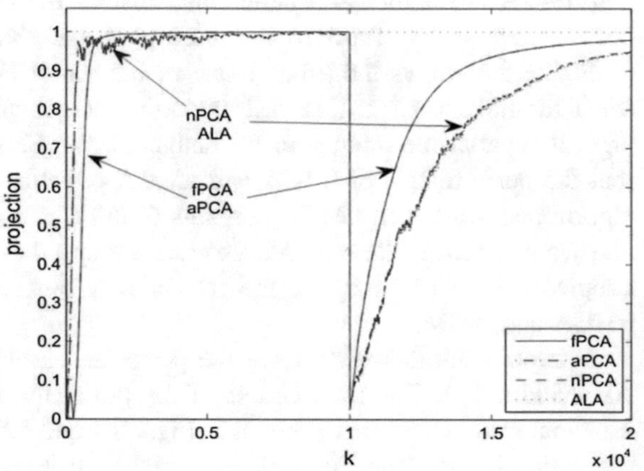

Fig. 8.4 Projection of weight vector onto the true principal eigenvector

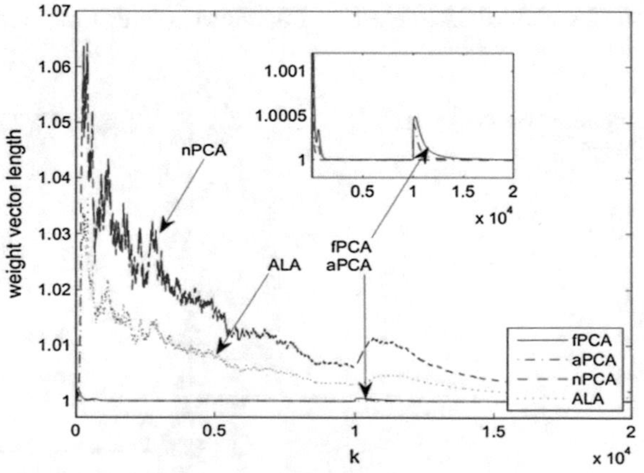

Fig. 8.5 Weight vector length

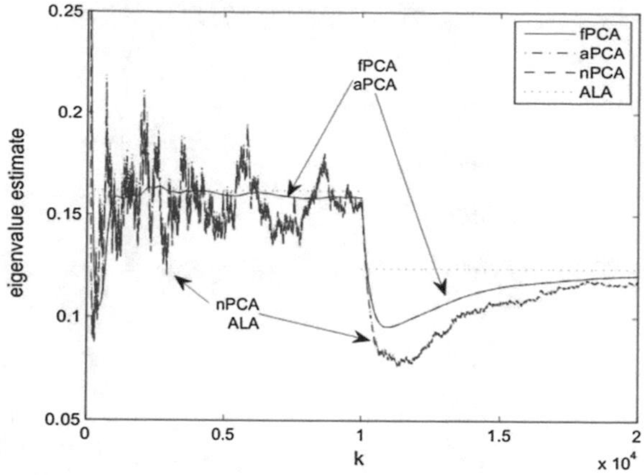

Fig. 8.6 Principal eigenvalue estimation

This is actually because that in fPCA and aPCA the covariance matrix $\boldsymbol{C}$ is updated by (8.69) while in nPCA and ALA$\boldsymbol{C}$ that is updated by $\boldsymbol{C}(k) = \boldsymbol{x}(k)\,\boldsymbol{x}^{\mathrm{T}}(k)$.

Experiment 3 is used to test the self-stabilizing property of the proposed algorithms. Figure 8.7 shows the time course of weight vector length estimation of fMCA, aMCA, fPCA, and aPCA which are initialized with nonunit length. We can find that all algorithms converge to unit length rapidly, which shows the self-stabilizing property of eigenvector estimates. The self-stabilizing property of eigenvalue estimates is shown in Figs. 8.3 and 8.6. From the results of experiments 1–3, we can see that the performance off MCA and fPCA is similar to that of aMCA and aPCA, respectively. Thus in experiment 4, we only compare the performance of aMCA and aPCA with that of two unified algorithms which were proposed in recent years, i.e., (1) kMCA + kPCA [14], where k means this algorithm was

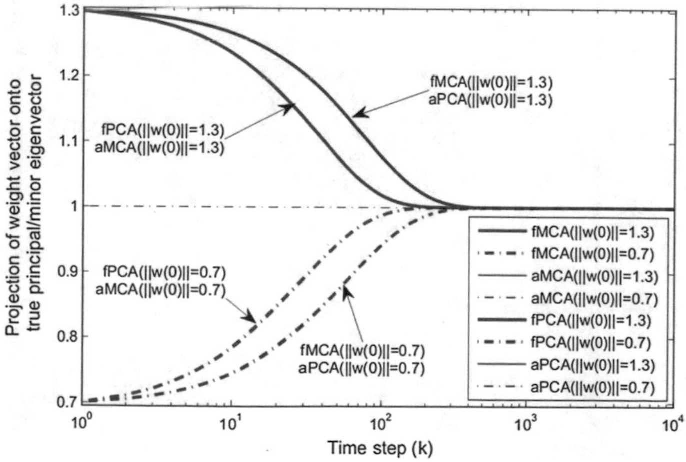

Fig. 8.7 Weight vector length

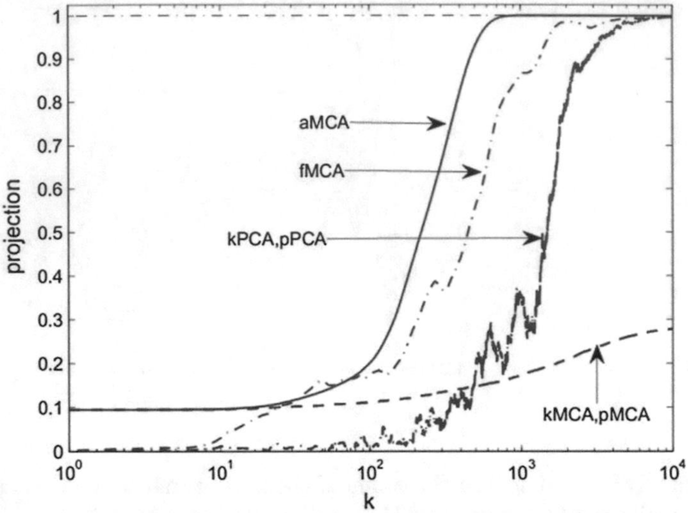

Fig. 8.8 Projection of weight vector onto the true principal/minor eigenvector

proposed by Kong;(2) *p*MCA + *p*PCA [15], where *p* means this algorithm was proposed by Peng. The time course of the projection of weight vector onto the true principal/minor eigenvector and the weight vector length is shown in Figs. 8.8 and 8.9, respectively. In Fig. 8.9, the first 1000 steps of aMCA and kMCA are shown in a subfigure. We can see that the proposed algorithms perform better the existing unified algorithms.

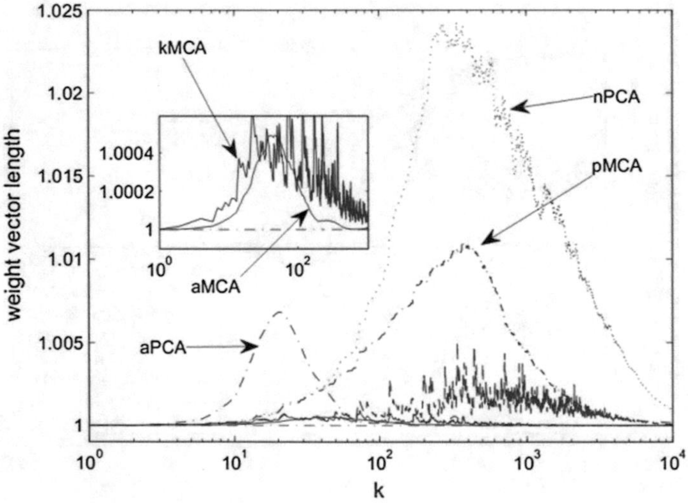

Fig. 8.9 Weight vector length

In summary, we propose a novel method to derive neural network algorithms based on a special information criterion. We firstly obtain two CMCA algorithms based on the modified Newton's method. Then, two CPCA rules are obtained from the CMCA rules. In this case, two unified and coupled algorithms are obtained, which are capable of both PCA and MCA and can also mitigate the speed-stability problem. The proposed algorithms converge faster and are more stable than existing algorithms. Moreover, all of the proposed algorithms are self-stabilized.

8.4 Adaptive Coupled Generalized Eigen Pairs Extraction Algorithms

In [4], based on Moller's work, Nguyen developed two well-performed quasi-Newton-type algorithms to extract generalized eigen pairs. Actually, Nguyen's algorithms are the generalization of Moller's coupled learning algorithms. But with DDT approach, Nguyen also reported the explicit convergence analysis for their learning rules, i.e., the region within which the initial estimate of the eigen pair must be chosen to guarantee the convergence to the desired eigen pair. However, as stated in [4], the GMCA algorithm proposed in [4] may lose robustness when the smallest eigenvalue of the matrix pencil is far less than 1.

Motivated by the efficacy of the coupled learning rules in [2] and [4] for the HEP and GHEP, we will introduce novel coupled algorithms proposed by us to estimate the generalized eigen pair information in this section. Based on a novel generalized information criterion, we have obtained an adaptive GMCA algorithm, as well as an adaptive GPCA algorithm by modifying the GMCA algorithm. It is worth noting that the procedure of obtaining the algorithms in this section is easier than the existing methods, for that it does not need to calculate the inverse of the Hessian matrix when deriving the new algorithms. It can be seen that our algorithms do not involve the reciprocal of the estimated eigenvalue in equations. Thus, they are numerically more robust than Nguyen's algorithms even when the smallest eigenvalue of the matrix pencil is far less than 1. Compared with Nguyen's algorithms, it is much easier to choose step size for online implementation of the algorithms.

8.4.1 A Coupled Generalized System for GMCA and GPCA

A. Generalized information criterion and coupled generalized system

Generally speaking, neural network model-based algorithms are often derived by optimizing some cost function or information criterion [2, 16]. As pointed out in [17], any criterion may be used if the maximum or minimum (possibly under a constraint) coincides with the desired principal or minor directions or subspace. In [2], Moller pointed out that the freedom of choosing an information criterion is greater if Newton's method is applied. In that case, it suffices to find a criterion of which the stationary points coincide with the desired solutions. Moller first proposed a special criterion which involves both eigenvector and eigenvalue estimates [2]. Based on Moller's work, Nguyen [4] first proposed to derive novel generalized eigen pair extraction algorithms by finding the stationary points of a generalized information criterion which is actually the generalization of Moller's information criterion.

In this section, for a given matrix pencil ($\boldsymbol{R}_y$, $\boldsymbol{R}_x$), we propose a generalized information criterion based on the criteria introduced in [2] and [4] as

$$p(\boldsymbol{w}, \lambda) = \boldsymbol{w}^H \boldsymbol{R}_y \boldsymbol{w} - \lambda \boldsymbol{w}^H \boldsymbol{R}_x \boldsymbol{w} + \lambda. \tag{8.90}$$

We can see that

$$\begin{pmatrix} \frac{\partial p}{\partial \boldsymbol{w}} \\ \frac{\partial p}{\partial \lambda} \end{pmatrix} = \begin{pmatrix} 2\boldsymbol{R}_y \boldsymbol{w} - 2\lambda \boldsymbol{R}_x \boldsymbol{w} \\ -\boldsymbol{w}^H \boldsymbol{R}_x \boldsymbol{w} + 1 \end{pmatrix}. \tag{8.91}$$

Thus, the stationary points $(\bar{\boldsymbol{w}}, \bar{\lambda})$ are defined by

$$\begin{cases} \boldsymbol{R}_y \bar{\boldsymbol{w}} = \bar{\lambda} \boldsymbol{R}_x \bar{\boldsymbol{w}} \\ \bar{\boldsymbol{w}}^H \boldsymbol{R}_x \bar{\boldsymbol{w}} = 1 \end{cases}, \tag{8.92}$$

from which we can conclude that $\bar{\boldsymbol{w}}^H \boldsymbol{R}_y \bar{\boldsymbol{w}} = \bar{\lambda} \bar{\boldsymbol{w}}^H \boldsymbol{R}_x \bar{\boldsymbol{w}} = \bar{\lambda}$. These imply that a stationary point $(\bar{\boldsymbol{w}}, \bar{\lambda})$ of (8.90) is a generalized eigen pair of the matrix pencil ($\boldsymbol{R}_y$, $\boldsymbol{R}_x$). The Hessian of the criterion is given as:

$$\boldsymbol{H}(\boldsymbol{w}, \lambda) = \begin{pmatrix} \frac{\partial^2 p}{\partial \boldsymbol{w}^2} & \frac{\partial^2 p}{\partial \boldsymbol{w} \partial \lambda} \\ \frac{\partial^2 p}{\partial \lambda \partial \boldsymbol{w}} & \frac{\partial^2 p}{\partial \lambda^2} \end{pmatrix} = 2 \begin{pmatrix} \boldsymbol{R}_y - \lambda \boldsymbol{R}_x & -\boldsymbol{R}_x \boldsymbol{w} \\ -\boldsymbol{w}^H \boldsymbol{R}_x & 0 \end{pmatrix}. \tag{8.93}$$

After applying the Newton's method, the equation used to obtain the system can be written as:

$$\begin{pmatrix} \dot{\boldsymbol{w}} \\ \dot{\lambda} \end{pmatrix} = -\boldsymbol{H}^{-1}(\boldsymbol{w}, \lambda) \begin{pmatrix} \frac{\partial p}{\partial \boldsymbol{w}} \\ \frac{\partial p}{\partial \lambda} \end{pmatrix}, \tag{8.94}$$

where $\dot{\boldsymbol{w}}$ and $\dot{\lambda}$ are the derivatives of $\boldsymbol{w}$ and λ with respect to time t, respectively. Based on the above equation, Nguyen [4] obtained their algorithms by finding the inverse matrix of the Hessian $\boldsymbol{H}^{-1}(\boldsymbol{w}, \lambda)$. Premultiplying both sides of the above equation by $\boldsymbol{H}(\boldsymbol{w}, \lambda)$, it yields

$$\boldsymbol{H}(\boldsymbol{w}, \lambda) \begin{pmatrix} \dot{\boldsymbol{w}} \\ \dot{\lambda} \end{pmatrix} = - \begin{pmatrix} \frac{\partial p}{\partial \boldsymbol{w}} \\ \frac{\partial p}{\partial \lambda} \end{pmatrix}. \tag{8.95}$$

In this section, all our later algorithms are built on this newly proposed Eq. (8.95). Substituting (8.91) and (8.93) into (8.95), we get

$$2\begin{pmatrix} \boldsymbol{R}_y - \lambda \boldsymbol{R}_x & -\boldsymbol{R}_x \boldsymbol{w} \\ -\boldsymbol{w}^H \boldsymbol{R}_x & 0 \end{pmatrix}\begin{pmatrix} \dot{\boldsymbol{w}} \\ \dot{\lambda} \end{pmatrix} = -\begin{pmatrix} 2\boldsymbol{R}_y \boldsymbol{w} - 2\lambda \boldsymbol{R}_x \boldsymbol{w} \\ -\boldsymbol{w}^H \boldsymbol{R}_x \boldsymbol{w} + 1 \end{pmatrix}. \tag{8.96}$$

From (8.96), we can get

$$(\boldsymbol{R}_y - \lambda \boldsymbol{R}_x)\dot{\boldsymbol{w}} - \boldsymbol{R}_x \boldsymbol{w}\dot{\lambda} = -(\boldsymbol{R}_y - \lambda \boldsymbol{R}_x)\boldsymbol{w} \tag{8.97}$$

$$-2\boldsymbol{w}^H \boldsymbol{R}_x \dot{\boldsymbol{w}} = \boldsymbol{w}^H \boldsymbol{R}_x \boldsymbol{w} - 1. \tag{8.98}$$

Premultiplying both sides of (8.97) by $(\boldsymbol{R}_y - \lambda \boldsymbol{R}_x)^{-1}$ gives the following:

$$\dot{\boldsymbol{w}} = (\boldsymbol{R}_y - \lambda \boldsymbol{R}_x)^{-1} \boldsymbol{R}_x \boldsymbol{w}\dot{\lambda} - \boldsymbol{w}. \tag{8.99}$$

Substituting (8.99) into (8.98), we have

$$-2\boldsymbol{w}^H \boldsymbol{R}_x \left((\boldsymbol{R}_y - \lambda \boldsymbol{R}_x)^{-1} \boldsymbol{R}_x \boldsymbol{w}\dot{\lambda} - \boldsymbol{w} \right) = \boldsymbol{w}^H \boldsymbol{R}_x \boldsymbol{w} - 1. \tag{8.100}$$

Thus,

$$\dot{\lambda} = \frac{\boldsymbol{w}^H \boldsymbol{R}_x \boldsymbol{w} + 1}{2\boldsymbol{w}^H \boldsymbol{R}_x (\boldsymbol{R}_y - \lambda \boldsymbol{R}_x)^{-1} \boldsymbol{R}_x \boldsymbol{w}}. \tag{8.101}$$

Substituting (8.101) into (8.99), we get

$$\dot{\boldsymbol{w}} = \frac{(\boldsymbol{R}_y - \lambda \boldsymbol{R}_x)^{-1} \boldsymbol{R}_x \boldsymbol{w} \left(\boldsymbol{w}^H \boldsymbol{R}_x \boldsymbol{w} + 1 \right)}{2\boldsymbol{w}^H \boldsymbol{R}_x (\boldsymbol{R}_y - \lambda \boldsymbol{R}_x)^{-1} \boldsymbol{R}_x \boldsymbol{w}} - \boldsymbol{w}. \tag{8.102}$$

By approximating $\boldsymbol{w}^H \boldsymbol{R}_x \boldsymbol{w} = 1$ in the vicinity of the stationary point $(\boldsymbol{w}_1, \lambda_1)$, we get a coupled generalized system as:

$$\dot{\boldsymbol{w}} = \frac{(\boldsymbol{R}_y - \lambda \boldsymbol{R}_x)^{-1} \boldsymbol{R}_x \boldsymbol{w}}{\boldsymbol{w}^H \boldsymbol{R}_x (\boldsymbol{R}_y - \lambda \boldsymbol{R}_x)^{-1} \boldsymbol{R}_x \boldsymbol{w}} - \boldsymbol{w}, \tag{8.103}$$

$$\dot{\lambda} = \frac{1}{\boldsymbol{w}^H \boldsymbol{R}_x (\boldsymbol{R}_y - \lambda \boldsymbol{R}_x)^{-1} \boldsymbol{R}_x \boldsymbol{w}} - \lambda. \tag{8.104}$$

B. Coupled generalized systems for GMCA and GPCA

Let $\boldsymbol{\Lambda}$ be a diagonal matrix containing all generalized eigenvalues of the matrix pencil $(\boldsymbol{R}_y, \boldsymbol{R}_x)$, i.e., $\boldsymbol{\Lambda} = \mathrm{diag}\{\lambda_1, \ldots, \lambda_N\}$. Let $\boldsymbol{V} = [\boldsymbol{v}_1, \ldots, \boldsymbol{v}_N]$, where $\boldsymbol{v}_1, \ldots, \boldsymbol{v}_N$

are the generalized eigenvectors associated with the generalized eigenvalues $\lambda_1, \ldots, \lambda_N$. It holds that $\boldsymbol{V}^H \boldsymbol{R}_x \boldsymbol{V} = \boldsymbol{I}$, $\boldsymbol{V}^H \boldsymbol{R}_y \boldsymbol{V} = \boldsymbol{\Lambda}$. Hence, $\boldsymbol{R}_x = (\boldsymbol{V}^H)^{-1} \boldsymbol{V}^{-1}$ and $\boldsymbol{R}_y = (\boldsymbol{V}^H)^{-1} \boldsymbol{\Lambda} \boldsymbol{V}^{-1}$, and

$$(\boldsymbol{R}_y - \lambda \boldsymbol{R}_x)^{-1} = \boldsymbol{V}(\boldsymbol{\Lambda} - \lambda \boldsymbol{I})^{-1} \boldsymbol{V}^H. \tag{8.105}$$

If we consider $\boldsymbol{w} \approx \boldsymbol{v}_1$ and $\lambda \approx \lambda_1 \ll \lambda_j (2 \leq j \leq N)$ in the vicinity of the stationary point $(\boldsymbol{w}_1, \lambda_1)$, then we have $\lambda_j - \lambda \approx \lambda_j$. In that case, $\boldsymbol{V}^H \boldsymbol{R}_x \boldsymbol{w} \approx e_1 = [1, 0, \ldots, 0]^H$ and

$$\begin{aligned} \boldsymbol{\Lambda} - \lambda \boldsymbol{I} &= \operatorname{diag}\{\lambda_1 - \lambda, \ldots, \lambda_N - \lambda\} \\ &\approx \operatorname{diag}\{\lambda_1 - \lambda, \lambda_2, \ldots, \lambda_N\} \\ &= \boldsymbol{\Lambda} - \lambda \boldsymbol{e}_1 \boldsymbol{e}_1^H, \end{aligned} \tag{8.106}$$

where diag{•} is the diagonal function. Substituting (8.106) into (8.105), we get the following:

$$\begin{aligned} (\boldsymbol{R}_y - \lambda \boldsymbol{R}_x)^{-1} &= \boldsymbol{V}(\boldsymbol{\Lambda} - \lambda \boldsymbol{I})^{-1} \boldsymbol{V}^H \\ &\approx [(\boldsymbol{V}^H)^{-1} (\boldsymbol{\Lambda} - \lambda \boldsymbol{e}_1 \boldsymbol{e}_1^H) \boldsymbol{V}^{-1}]^{-1} \\ &= [\boldsymbol{R}_y - \lambda (\boldsymbol{V}^H)^{-1} \boldsymbol{e}_1 \boldsymbol{e}_1^H \boldsymbol{V}^{-1}]^{-1} \\ &\approx \left[\boldsymbol{R}_y - \lambda (\boldsymbol{V}^H)^{-1} (\boldsymbol{V}^H \boldsymbol{R}_x \boldsymbol{w})(\boldsymbol{V}^H \boldsymbol{R}_x \boldsymbol{w})^H \boldsymbol{V}^{-1}\right]^{-1} \\ &= \left[\boldsymbol{R}_y - \lambda (\boldsymbol{R}_x \boldsymbol{w})(\boldsymbol{R}_x \boldsymbol{w})^H\right]^{-1}. \end{aligned} \tag{8.107}$$

It can be seen that

$$\begin{aligned} &\left[\boldsymbol{R}_y - \lambda_1 (\boldsymbol{R}_x \boldsymbol{v}_1)(\boldsymbol{R}_x \boldsymbol{v}_1)^H\right] \boldsymbol{v}_1 \\ &\quad = \boldsymbol{R}_y \boldsymbol{v}_1 - (\lambda_1 \boldsymbol{R}_x \boldsymbol{v}_1)(\boldsymbol{v}_1^H \boldsymbol{R}_x \boldsymbol{v}_1) = 0. \end{aligned} \tag{8.108}$$

Since $\boldsymbol{R}_y \boldsymbol{v}_1 = \lambda_1 \boldsymbol{R}_x \boldsymbol{v}_1$ and $\boldsymbol{v}_1^H \boldsymbol{R}_x \boldsymbol{v}_1 = 1$. This means that matrix $\boldsymbol{R}_y - \lambda (\boldsymbol{R}_x \boldsymbol{w})(\boldsymbol{R}_x \boldsymbol{w})^H$ has an eigenvalue 0 associated with eigenvector $\boldsymbol{v}_1$. This is to say, the matrix $\boldsymbol{R}_y - \lambda (\boldsymbol{R}_x \boldsymbol{w})(\boldsymbol{R}_x \boldsymbol{w})^H$ is rank-deficient and hence cannot be inverted if $(\boldsymbol{w}, \lambda) = (\boldsymbol{v}_1, \lambda_1)$. To address this issue, we add a penalty factor $\varepsilon \approx 1$ in (8.107), and then it yields the following:

$$\begin{aligned} (\boldsymbol{R}_y - \lambda \boldsymbol{R}_x)^{-1} &\approx \left[\boldsymbol{R}_y - \varepsilon \lambda (\boldsymbol{R}_x \boldsymbol{w})(\boldsymbol{R}_x \boldsymbol{w})^H\right]^{-1} \\ &= \boldsymbol{R}_y^{-1} + \frac{\varepsilon \lambda \boldsymbol{R}_y^{-1} \boldsymbol{R}_x \boldsymbol{w} \boldsymbol{w}^H \boldsymbol{R}_x \boldsymbol{R}_y^{-1}}{1 - \varepsilon \lambda \boldsymbol{w}^H \boldsymbol{R}_x \boldsymbol{R}_y^{-1} \boldsymbol{R}_x \boldsymbol{w}}, \end{aligned} \tag{8.109}$$

The last step of (8.109) is obtained by using the SM-formula (Sherman–Morrison formula) [13]. Substituting (8.107) into (8.103), we get the following:

$$\dot{\boldsymbol{w}} = \frac{\left(\boldsymbol{R}_y^{-1} + \frac{\varepsilon\lambda \boldsymbol{R}_y^{-1}\boldsymbol{R}_x\boldsymbol{w}\boldsymbol{w}^H\boldsymbol{R}_x\boldsymbol{R}_y^{-1}}{1-\varepsilon\lambda\boldsymbol{w}^H\boldsymbol{R}_x\boldsymbol{R}_y^{-1}\boldsymbol{R}_x\boldsymbol{w}}\right)\boldsymbol{R}_x\boldsymbol{w}}{\boldsymbol{w}^H\boldsymbol{R}_x\left(\boldsymbol{R}_y^{-1} + \frac{\varepsilon\lambda \boldsymbol{R}_y^{-1}\boldsymbol{R}_x\boldsymbol{w}\boldsymbol{w}^H\boldsymbol{R}_x\boldsymbol{R}_y^{-1}}{1-\varepsilon\lambda\boldsymbol{w}^H\boldsymbol{R}_x\boldsymbol{R}_y^{-1}\boldsymbol{R}_x\boldsymbol{w}}\right)\boldsymbol{R}_x\boldsymbol{w}} - \boldsymbol{w}. \tag{8.110}$$

Multiplying the numerator and denominator of (8.110) by $1 - \varepsilon\lambda\boldsymbol{w}^H\boldsymbol{R}_x\boldsymbol{R}_y^{-1}\boldsymbol{R}_x\boldsymbol{w}$ simultaneously, and after some manipulations, we get

$$\dot{\boldsymbol{w}} = \frac{\boldsymbol{R}_y^{-1}\boldsymbol{R}_x\boldsymbol{w}}{\boldsymbol{w}^H\boldsymbol{R}_x\boldsymbol{R}_y^{-1}\boldsymbol{R}_x\boldsymbol{w}} - \boldsymbol{w}. \tag{8.111}$$

Similarly, substituting (8.107) into (8.104), we can get

$$\dot{\lambda} = \frac{1}{\boldsymbol{w}^H\boldsymbol{R}_x\boldsymbol{R}_y^{-1}\boldsymbol{R}_x\boldsymbol{w}} - \varepsilon\lambda. \tag{8.112}$$

It can be seen that the penalty factor ε is not necessarily needed in the equations. Or in other words, we can approximate $\varepsilon = 1$ in future equations. Thus, we get the following:

$$\dot{\lambda} = \frac{1}{\boldsymbol{w}^H\boldsymbol{R}_x\boldsymbol{R}_y^{-1}\boldsymbol{R}_x\boldsymbol{w}} - \lambda. \tag{8.113}$$

Thus, (8.111) and (8.113) are the coupled systems for the GMCA case.

It is known that the ith principal generalized eigenvector $\boldsymbol{v}_i$ of the matrix pencil $(\boldsymbol{R}_y, \boldsymbol{R}_x)$ is also the ith minor generalized eigenvector of the matrix pencil $(\boldsymbol{R}_x, \boldsymbol{R}_y)$. Hence, the problem of extracting principal generalized subspace of the inversed matrix pencil $(\boldsymbol{R}_y, \boldsymbol{R}_x)$ is equivalent to that of extracting minor generalized subspace of the matrix pencil $(\boldsymbol{R}_x, \boldsymbol{R}_y)$, and vice versa [4]. Therefore, by swapping $\boldsymbol{R}_x$ and $\boldsymbol{R}_y$, $\boldsymbol{R}_x^{-1}$ and $\boldsymbol{R}_y^{-1}$ in (8.111) and (8.113), we obtain a modified system

$$\dot{\boldsymbol{w}} = \frac{\boldsymbol{R}_x^{-1}\boldsymbol{R}_y\boldsymbol{w}}{\boldsymbol{w}^H\boldsymbol{R}_y\boldsymbol{R}_x^{-1}\boldsymbol{R}_y\boldsymbol{w}} - \boldsymbol{w}, \tag{8.114}$$

$$\dot{\lambda} = \boldsymbol{w}^H\boldsymbol{R}_y\boldsymbol{R}_x^{-1}\boldsymbol{R}_y\boldsymbol{w} - \lambda, \tag{8.115}$$

to extract the minor eigen pair of matrix pencil $(\boldsymbol{R}_x, \boldsymbol{R}_y)$ as well as the principal eigen pair of matrix pencil $(\boldsymbol{R}_y, \boldsymbol{R}_x)$.

As was pointed out in [4], by using the nested orthogonal complement structure of the generalized eigen-subspace, the problem of estimating the $p\,(\leq N)$-dimensional minor/principal generalized subspace can be reduced to multiple GHEPs of estimating the generalized eigen pairs associated with the smallest/largest generalized eigenvalues of certain matrix pencils. In the following, we will show how to estimate the remaining $p - 1$ minor/principal eigen pairs. In the GMCA case, consider the following equations:

$$\boldsymbol{R}_j = \boldsymbol{R}_{j-1} + \rho \boldsymbol{R}_x \boldsymbol{w}_{j-1} \boldsymbol{w}_{j-1}^{\mathrm{T}} \boldsymbol{R}_y, \tag{8.116}$$

$$\boldsymbol{R}_j^{-1} = \boldsymbol{R}_{j-1}^{-1} - \frac{\rho \boldsymbol{R}_{j-1}^{-1} \boldsymbol{R}_x \boldsymbol{w}_{j-1} \boldsymbol{w}_{j-1}^{\mathrm{T}} \boldsymbol{R}_y \boldsymbol{R}_{j-1}^{-1}}{1 + \rho \boldsymbol{w}_{j-1}^{\mathrm{T}} \boldsymbol{R}_y \boldsymbol{R}_{j-1}^{-1} \boldsymbol{R}_x \boldsymbol{w}_{j-1}}, \tag{8.117}$$

where $j = 2, \ldots, p$, $\rho \geq \lambda_N / \lambda_1$, $\boldsymbol{R}_1 = \boldsymbol{R}_y$ and $\boldsymbol{w}_{j-1} = \boldsymbol{v}_{j-1}$ is the $(j - 1)$th minor generalized eigenvector extracted. It holds that

$$\begin{aligned} \boldsymbol{R}_j \boldsymbol{v}_q &= (\boldsymbol{R}_y + \rho \sum\nolimits_{i=1}^{j-1} \boldsymbol{R}_x \boldsymbol{v}_i \boldsymbol{v}_i^{\mathrm{T}} \boldsymbol{R}_y) \boldsymbol{v}_q \\ &= \boldsymbol{R}_y \boldsymbol{v}_q + \rho \sum\nolimits_{i=1}^{j-1} \boldsymbol{R}_x \boldsymbol{v}_i \boldsymbol{v}_i^{\mathrm{T}} \boldsymbol{R}_y \boldsymbol{v}_q \\ &= \lambda_q \boldsymbol{R}_x \boldsymbol{v}_q + \rho \lambda_q \sum\nolimits_{i=1}^{j-1} \boldsymbol{R}_x \boldsymbol{v}_i \boldsymbol{v}_i^{\mathrm{T}} \boldsymbol{R}_y \boldsymbol{v}_q \\ &= \begin{cases} (1+\rho)\lambda_q \boldsymbol{R}_x \boldsymbol{v}_q & \text{for } q = 1, \ldots, j-1 \\ \lambda_q \boldsymbol{R}_x \boldsymbol{v}_q & \text{for } q = j, \ldots, N \end{cases}. \end{aligned} \tag{8.118}$$

Thus, the matrix pencil ($\boldsymbol{R}_j$, $\boldsymbol{R}_x$) has eigenvalues $\lambda_j \leq \cdots \leq \lambda_N \leq (1+\rho)\lambda_1 \leq \cdots \leq (1+\rho)\lambda_{j-1}$ associated with eigenvectors $\boldsymbol{v}_j, \ldots, \boldsymbol{v}_N, \boldsymbol{v}_1 \ldots \boldsymbol{v}_{j-1}$. Equation (8.117) is obtained from (8.116) based on the SM-formula. That is to say, by replacing $\boldsymbol{R}_y$ with $\boldsymbol{R}_j$ and $\boldsymbol{R}_y^{-1}$ with $\boldsymbol{R}_j^{-1}$ in (8.111) and (8.113), we can estimate the jth minor generalized eigen pair ($\boldsymbol{v}_j$, λ_j).

In the GPCA case, consider the following equation

$$\boldsymbol{R}_j = \boldsymbol{R}_{j-1} - \boldsymbol{R}_x \boldsymbol{w}_{j-1} \boldsymbol{w}_{j-1}^{T} \boldsymbol{R}_y, \tag{8.119}$$

where $\boldsymbol{R}_1 = \boldsymbol{R}_y$, and $\boldsymbol{w}_{j-1} = \boldsymbol{v}_{N-j+1}$ is the $(j - 1)$th principal generalized eigenvector extracted. By replacing $\boldsymbol{R}_y$ with $\boldsymbol{R}_j$ in (8.114) and (8.115), we can estimate the jth principal generalized eigen pair $(\boldsymbol{v}_{N-j+1}, \lambda_{N-j+1})$.

8.4.2 Adaptive Implementation of Coupled Generalized Systems

In engineering practice, the matrices $\boldsymbol{R}_y$ and $\boldsymbol{R}_x$ are the covariance matrices of random input sequences $\{y(k)\}_{k \in z}$ and $\{x(k)\}_{k \in z}$, respectively. Thus, the matrix pencil ($\boldsymbol{R}_y$, $\boldsymbol{R}_x$) is usually unknown in advance, and even slowly changing over time if the signal is nonstationary. In that case, the matrices $\boldsymbol{R}_y$ and $\boldsymbol{R}_x$ are variables and thus need to be estimated with online approach. In this section, we propose to update $\boldsymbol{R}_y$ and $\boldsymbol{R}_x$ with:

$$\widehat{\boldsymbol{R}}_y(k+1) = \beta\widehat{\boldsymbol{R}}_y(k) + \boldsymbol{y}(k+1)\boldsymbol{y}^H(k+1), \tag{8.120}$$

$$\widehat{\boldsymbol{R}}_x(k+1) = \alpha\widehat{\boldsymbol{R}}_x(k) + \boldsymbol{x}(k+1)\boldsymbol{x}^H(k+1). \tag{8.121}$$

By using the MS-formula, $\boldsymbol{Q}_y(k) = \widehat{\boldsymbol{R}}_y^{-1}(k)$ and $\boldsymbol{Q}_x(k) = \widehat{\boldsymbol{R}}_x^{-1}(k)$ can be updated as:

$$\boldsymbol{Q}_y(k+1) = \frac{1}{\beta}\left(\boldsymbol{Q}_y(k) - \frac{\boldsymbol{Q}_y(k)\boldsymbol{y}(k+1)\boldsymbol{y}^H(k+1)\boldsymbol{Q}_y(k)}{\alpha + \boldsymbol{y}^H(k+1)\boldsymbol{Q}_y(k)\boldsymbol{y}(k+1)}\right), \tag{8.122}$$

$$\boldsymbol{Q}_x(k+1) = \frac{1}{\alpha}\left(\boldsymbol{Q}_x(k) - \frac{\boldsymbol{Q}_x(k)\boldsymbol{x}(k+1)\boldsymbol{x}^H(k+1)\boldsymbol{Q}_x(k)}{\alpha + \boldsymbol{x}^H(k+1)\boldsymbol{Q}_x(k)\boldsymbol{x}(k+1)}\right). \tag{8.123}$$

It is known that

$$\lim_{k\to\infty}\frac{1}{k}\widehat{\boldsymbol{R}}_y(k) = \boldsymbol{R}_y \tag{8.124}$$

$$\lim_{k\to\infty}\frac{1}{k}\widehat{\boldsymbol{R}}_x(k) = \boldsymbol{R}_x \tag{8.125}$$

when $\alpha = \beta = 1$. By replacing $\boldsymbol{R}_y, \boldsymbol{R}_x, \boldsymbol{R}_y^{-1}$ and $\boldsymbol{R}_x^{-1}$ in (8.111)–(8.115) with $\widehat{\boldsymbol{R}}_y(k)$, $\widehat{\boldsymbol{R}}_x(k)$, $\boldsymbol{Q}_y(k)$ and $\boldsymbol{Q}_x(k)$, respectively, we can easily obtain the online GMCA algorithm with normalized step as:

$$\tilde{\boldsymbol{w}}(k+1) = \eta_1\frac{\boldsymbol{Q}_y(k+1)\widehat{\boldsymbol{R}}_x(k+1)\boldsymbol{w}(k)}{\boldsymbol{w}^H(k)\widehat{\boldsymbol{R}}_x(k+1)\boldsymbol{Q}_y(k+1)\widehat{\boldsymbol{R}}_x(k+1)\boldsymbol{w}(k)} + (1-\eta_1)\boldsymbol{w}(k), \tag{8.126}$$

$$\boldsymbol{w}(k+1) = \frac{\tilde{\boldsymbol{w}}(k+1)}{\|\tilde{\boldsymbol{w}}(k+1)\|_{\widehat{\boldsymbol{R}}_x(k+1)}}, \tag{8.127}$$

$$\lambda(k+1) = \gamma_1\frac{1}{\boldsymbol{w}^H(k)\widehat{\boldsymbol{R}}_x(k+1)\boldsymbol{Q}_y(k+1)\widehat{\boldsymbol{R}}_x(k+1)\boldsymbol{w}(k)} + (1-\gamma_1)\lambda(k), \tag{8.128}$$

and the online GPCA algorithm with normalized step as:

$$\tilde{w}(k+1) = \eta_2 \frac{Q_x(k+1)\widehat{R}_y(k+1)w(k)}{w^H(k)\widehat{R}_y(k+1)Q_x(k+1)\widehat{R}_y(k+1)w(k)} + (1-\eta_2)w(k), \quad (8.129)$$

$$w(k+1) = \frac{\tilde{w}(k+1)}{\|\tilde{w}(k+1)\|_{\widehat{R}_y(k+1)}}, \quad (8.130)$$

$$\lambda(k+1) = \gamma_2 w^H(k)\,\widehat{R}_y(k+1)Q_x(k+1)\widehat{R}_y(k+1)w\,(k) + (1-\gamma_2)\lambda(k), \quad (8.131)$$

where η_1, η_2, γ_1, $\gamma_2 \in$ (0, 1] are the step sizes.

In the rest of this section, for convenience, we refer to the GPCA and GMCA algorithms proposed in [4] as nGPCA and nGMCA for short, respectively, where *n* means that these algorithms were proposed by Nguyen. Similarly, we refer to the algorithm in (8.126)–(8.128) as fGMCA and the algorithm in (8.129)–(8.131) as fGPCA for short.

At the end of this section, we discuss the computational complexity of our algorithms. Taking fGMCA as an example, the computation of $\widehat{R}_x(k)$ and $Q_y(k)$ requires $5N^2 + O(N)$ multiplications. Moreover, by using (8.121), we have the following:

$$\begin{aligned} &\widehat{R}_x(k+1)w(k) \\ &\quad = \left[\frac{k}{k+1}\widehat{R}_x(k) + \frac{1}{k+1}x(k+1)x^H(k+1)\right]w(k) \\ &\quad = \frac{k}{k+1}\widehat{R}_x(k)w(k) + \frac{1}{k+1}x(k+1)[x^H(k+1)w(k)], \end{aligned} \quad (8.132)$$

where

$$\widehat{R}_x(k)w(k) = \frac{\widehat{R}_x(k)\tilde{w}(k)}{\sqrt{\tilde{w}(k)^H\widehat{R}_x(k)\tilde{w}(k)}}. \quad (8.133)$$

Since $\widehat{R}_x(k)\tilde{w}(k)$ has been computed at the previous step when calculating the R_x-norm of w(k), the update of $\widehat{R}_x(k+1)w\,(k)$ requires only $O(N)$ multiplications. Thus, the updates of $w(k)$ and $\lambda(k)$ in fGMCA requires $2N^2 + O(N)$ multiplications. Hence, fGMCA requires a total of $7N^2 + O(N)$ multiplications at each iteration. In a similar way, we can see that fGPCA also requires a total of $7N^2 + O(N)$ multiplications at each iteration. Thus, the computational complexity of both fGMCA and fGPCA is less than that of nGMCA and nGPCA (i.e., $10N^2 + O(N)$).

8.4.3 Convergence Analysis

The convergence of neural network learning algorithms is a difficult topic for direct study and analysis, and as pointed out [18], from the application point of view. The DDT method is more reasonable for studying the convergence of algorithms than traditional method. Using the DDT approach, Nguyen first reported the explicit convergence analysis of coupled generalized eigen pair extraction algorithms [4]. In this section, we will also analyze the convergence of our algorithms with the DDT approach on the basis of [4].

The DDT system of fGMCA is given as:

$$\tilde{w}(k+1) = w(k) + \eta_1\left[\frac{Q_y\widehat{R}_x w(k)}{w^H(k)\widehat{R}_x Q_y \widehat{R}_x w(k)} - w(k)\right], \tag{8.134}$$

$$w(k+1) = \frac{\tilde{w}(k+1)}{\|\tilde{w}(k+1)\|_{R_x}}, \tag{8.135}$$

$$\lambda(k+1) = \lambda(k) + \gamma_1\left[\frac{1}{w^H(k)\widehat{R}_x Q_y \widehat{R}_x w(k)} - \lambda(k)\right]. \tag{8.136}$$

which is referred to as DDT System 1.

And the DDT system of fGPCA is given as:

$$\tilde{w}(k+1) = w(k) + \eta_2\left[\frac{Q_x\widehat{R}_y w(k)}{w^H(k)\widehat{R}_y Q_x \widehat{R}_y w(k)} - w(k)\right], \tag{8.137}$$

$$w(k+1) = \frac{\tilde{w}(k+1)}{\|\tilde{w}(k+1)\|_{R_y}}, \tag{8.138}$$

$$\lambda(k+1) = \lambda(k) + \gamma_2[w^H(k)\widehat{R}_y Q_x \widehat{R}_y w(k) - \lambda(k)]. \tag{8.139}$$

which is referred to as DDT System 2.

Similar to [4], we also denote by $\|u\|_R = \sqrt{u^H R u}$ the R-norm of a vector u, where $R \in C^{N\times N}$ and $u \in C^N$, $P_V^R(u) \in V$ is the R-orthogonal projection of u onto a subspace $V \in C^N$; i.e., $P_V^R(u)$ is the unique vector satisfying $\|u - P_V^R(u)\|_R = \min_{v\in V}\|u - v\|_R$, V_{λ_i} is the generalized eigen-subspace associated with the ith smallest generalized eigenvalue λ_i, i.e., $V_{\lambda_i} = \{v \in C^N | R_y v = \lambda_i R_x v\}$ $(i = 1, 2, \ldots, N)$. (Note that $V_{\lambda_i} = V_{\lambda_j}$ if $\lambda_i = \lambda_j$ for some $i \neq j$), $V_{<R>}^{\perp}$ is the R-orthogonal complement subspace of V for any subspace $V \subset C^N$, *i.e.*, $V_{<R>}^{\perp} = \{u \in C^N \mid <u, v>_R = v^H R u = 0, \forall v \in V\}$.

Next, we will present two theorems to show the convergence of our algorithms. In the following, two cases will be considered. In Case 1, $\lambda_1 = \lambda_2 = \cdots = \lambda_N$ and in Case 2, $\lambda_1 < \lambda_N$.

Theorem 8.1 (Convergence analysis of fGMCA) *Suppose that the sequence* $[\boldsymbol{w}(k), \lambda(k)]_{k=0}^{\infty}$ *is generated by DDT System 1 with any* $\eta_1, \gamma_1 \in (0, 1]$, *any initial* $\boldsymbol{R}_x$*-normalized vector* $\boldsymbol{w}(0) \notin (\boldsymbol{V}_{\lambda_1})^{\perp}_{<\boldsymbol{R}_x>}$, *and any* $\lambda(0) > 0$. *Then for Case 1, it holds that* $\boldsymbol{w}(k) = \boldsymbol{w}(0)$ *for all* $k \geq 0$, *which is also a generalized eigenvector associated with the generalized eigenvalue* λ_1 *of the matrix pencil* $(\boldsymbol{R}_y, \boldsymbol{R}_x)$, *and* $\lim_{k\to\infty} \lambda(k) = \lambda_1$. *For Case 2, it holds that*

$$\lim_{k\to\infty} \boldsymbol{w}(k) = \frac{\boldsymbol{P}_{\mathrm{V}_{\lambda_1}}^{\boldsymbol{R}_x}[\boldsymbol{w}(0)]}{\left\| \boldsymbol{P}_{\mathrm{V}_{\lambda_1}}^{\boldsymbol{R}_x}[\boldsymbol{w}(0)] \right\|_{\boldsymbol{R}_x}}, \tag{8.140}$$

$$\lim_{k\to\infty} \lambda(k) = \lambda_1. \tag{8.141}$$

Proof Case 1:

Since $\lambda_1 = \lambda_2 = \cdots = \lambda_N$ ensures $\boldsymbol{V}_{\lambda_1} = C^N$, we can verify that for all $k \geq 0$ that $\boldsymbol{w}(k) = \boldsymbol{w}(0) \neq \boldsymbol{0}$, which is also a generalized eigenvector associated with the generalized eigenvalue λ_1 of matrix pencil $(\boldsymbol{R}_y, \boldsymbol{R}_y)$. Moreover, from (8.128) we have $\lambda(k + 1) = (1 - \gamma_1)\lambda(k) + \gamma_1\lambda_1$ for all $k \geq 0$. Hence

$$\begin{aligned} \lambda(k+1) &= (1-\gamma_1)\lambda(k) + \gamma_1\lambda_1 = \cdots \\ &= (1-\gamma_1)^{k+1}\lambda(0) + \gamma_1\lambda_1[1 + (1-\gamma_1) + \cdots + (1-\gamma_1)^k] \\ &= (1-\gamma_1)^{k+1}\lambda(0) + \lambda_1[1 - (1-\gamma_1)^{k+1}] \\ &= \lambda_1 + (1-\gamma_1)^{k+1}[\lambda(0) - \lambda_1]. \end{aligned} \tag{8.142}$$

Since $\gamma_1 \in (0, 1]$, we can verify that $\lim_{k\to\infty} \lambda(k) = \lambda_1$.

Case 2: Suppose that the generalized eigenvalues of the matrix pencil $(\boldsymbol{R}_y, \boldsymbol{R}_x)$ have been ordered as $\lambda_1 = \cdots = \lambda_r < \lambda_{r+1} \leq \cdots \leq \lambda_N$ $(1 \leq r \leq N)$. Since $\{\boldsymbol{v}_1, \boldsymbol{v}_2, \ldots, \boldsymbol{v}_N\}$ is an $\boldsymbol{R}_x$-orthonormal basis of C^N, $\boldsymbol{w}(k)$ in DDT System 1 can be written uniquely as:

$$\boldsymbol{w}(k) = \sum_{i=1}^{N} z_i(k)\,\boldsymbol{v}_i, \; k = 0, 1, \ldots \tag{8.143}$$

where $z_i(k) = \langle \boldsymbol{w}(k), \boldsymbol{v}_i \rangle_{\boldsymbol{R}_x} = \boldsymbol{v}_i^H \boldsymbol{R}_x \boldsymbol{w}(k)$, $i = 1, 2, \ldots, N$.

First, we will prove by mathematical induction that for all $k > 0$, $\boldsymbol{w}(k)$ is well defined, $\boldsymbol{R}_x$-normalized, i.e.,

$$\boldsymbol{w}(k)^H \boldsymbol{R}_x \boldsymbol{w}(k) = \sum_{i=1}^{N} |z_i(k)|^2 = 1, \tag{8.144}$$

and $\boldsymbol{w}(k) \notin (\boldsymbol{V}_{\lambda_1})^{\perp}_{\langle \boldsymbol{R}_x \rangle}$, i.e., $[z_1(k), z_2(k),\ldots, z_r(k)] \neq 0$. Note that $\boldsymbol{w}(0) \notin (\boldsymbol{V}_{\lambda_1})^{\perp}_{\langle \boldsymbol{R}_x \rangle}$ is $\boldsymbol{R}_x$-normalized. Assume that $w(k)$ is well defined, $\boldsymbol{R}_x$-normalized, and $\boldsymbol{w}(k) \notin (\boldsymbol{V}_{\lambda_1})^{\perp}_{\langle \boldsymbol{R}_x \rangle}$ for some $k > 0$. By letting $\tilde{\boldsymbol{w}}(k+1) = \sum_{i=1}^{N} \tilde{z}_i(k+1)\mathbf{v}_i$, from (8.134) and (8.143), we have the following:

$$\tilde{z}_i(k+1) = z_i(k)\left\{1 + \eta_1 \left[\frac{1}{\lambda_i \boldsymbol{w}^H(k)\boldsymbol{R}_x \boldsymbol{R}_y^{-1} \boldsymbol{R}_x \boldsymbol{w}(k)} - 1\right]\right\}. \tag{8.145}$$

Since matrix pencil $\left(\boldsymbol{R}_x, \boldsymbol{R}_x \boldsymbol{R}_y^{-1} \boldsymbol{R}_x\right)$ has the same eigen pairs as $\left(\boldsymbol{R}_y, \boldsymbol{R}_x\right)$, and $\boldsymbol{w}(k)$ is $\boldsymbol{R}_x$-normalized, it follows that

$$\lambda_1 \leq \frac{\boldsymbol{w}^H(k)\boldsymbol{R}_x \boldsymbol{w}(k)}{\boldsymbol{w}^H(k)\boldsymbol{R}_x \boldsymbol{R}_y^{-1} \boldsymbol{R}_x \boldsymbol{w}(k)} = \frac{1}{\boldsymbol{w}^H(k)\boldsymbol{R}_x \boldsymbol{R}_y^{-1} \boldsymbol{R}_x \boldsymbol{w}(k)} \leq \lambda_N, \tag{8.146}$$

which is a generalization of the Rayleigh–Ritz ratio [19]. For $i = 1,\ldots, r$, (8.146) and (8.145) guarantee that

$$1 + \eta_1 \left[\frac{1}{\lambda_i \boldsymbol{w}^H(k)\boldsymbol{R}_x \boldsymbol{R}_y^{-1} \boldsymbol{R}_x \boldsymbol{w}(k)} - 1\right] = 1 + \eta_1 \left[\frac{1}{\lambda_1} \frac{1}{\boldsymbol{w}^H(k)\boldsymbol{R}_x \boldsymbol{R}_y^{-1} \boldsymbol{R}_x \boldsymbol{w}(k)} - 1\right] \geq 1, \tag{8.147}$$

and $[z_1(k + 1), z_2(k + 1), \ldots, z_r(k + 1)] \neq \mathbf{0}$. These imply that $\tilde{\boldsymbol{w}}(k+1) \neq 0$ and $\boldsymbol{w}(k+1) = \sum_{i=1}^{N} z_i(k+1)\, \boldsymbol{v}_i$ is well defined, $\boldsymbol{R}_x$-normalized, and $\boldsymbol{w}(k+1) \notin (\boldsymbol{V}_{\lambda_1})^{\perp}_{\langle \boldsymbol{R}_x \rangle}$, where

$$z_i(k+1) = \frac{\tilde{z}_i(k+1)}{\|\tilde{\boldsymbol{w}}(k+1)\|_{\boldsymbol{R}_x}}. \tag{8.148}$$

Therefore, $\boldsymbol{w}(k)$ is well defined, $\boldsymbol{R}_x$-normalized, and $\boldsymbol{w}(k) \notin (\boldsymbol{V}_{\lambda_1})^{\perp}_{\langle \boldsymbol{R}_x \rangle}$ for all $k \geq 0$.

Second, we will prove (8.125). Note that $\boldsymbol{w}(0) \notin (\boldsymbol{V}_{\lambda_1})^{\perp}_{\langle \boldsymbol{R}_x \rangle}$ implies the existence of some $m \in \{1,\ldots, r\}$ satisfying $z_m(0) \neq 0$, where $\lambda_1 = \cdots = \lambda_m = \cdots = \lambda_r$. From (8.145) and (8.148), we have $z_m(k + 1)/z_m(0) > 0$ for all $k \geq 0$. By using (8.145) and (8.148), we can see that for $i = 1,\ldots, r$, it holds that

$$
\begin{aligned}
\frac{z_i(k+1)}{z_m(k+1)} &= \frac{\tilde{z}_i(k+1)}{\|\tilde{\boldsymbol{w}}(k+1)\|_{\boldsymbol{R}_x}} \frac{\|\tilde{\boldsymbol{w}}(k+1)\|_{\boldsymbol{R}_x}}{\tilde{z}_m(k+1)} \\
&= \frac{z_i(k)}{z_m(k)} \cdot \frac{1+\eta_1\left[\frac{1}{\lambda_i \boldsymbol{w}^H(k)\boldsymbol{R}_x\boldsymbol{R}_y^{-1}\boldsymbol{R}_x\boldsymbol{w}(k)} - 1\right]}{1+\eta_1\left[\frac{1}{\lambda_m \boldsymbol{w}^H(k)\boldsymbol{R}_x\boldsymbol{R}_y^{-1}\boldsymbol{R}_x\boldsymbol{w}(k)} - 1\right]} \\
&= \frac{z_i(k)}{z_m(k)} = \cdots = \frac{z_i(0)}{z_m(0)}.
\end{aligned}
\tag{8.149}
$$

On the other hand, by using (8.145) and (8.148), we have for all $k \geq 0$ and $i = r + 1, \ldots, N$ that

$$
\begin{aligned}
\frac{|z_i(k+1)|^2}{|z_m(k+1)|^2} &= \frac{\tilde{z}_i(k+1)}{\|\tilde{\boldsymbol{w}}(k+1)\|_{\boldsymbol{R}_x}} \frac{\|\tilde{\boldsymbol{w}}(k+1)\|_{\boldsymbol{R}_x}}{\tilde{z}_m(k+1)} \\
&= \left[\frac{1+\eta_1\left(\frac{1}{\lambda_i \boldsymbol{w}^H(k)\boldsymbol{R}_x\boldsymbol{R}_y^{-1}\boldsymbol{R}_x\boldsymbol{w}(k)} - 1\right)}{1+\eta_1\left(\frac{1}{\lambda_m \boldsymbol{w}^H(k)\boldsymbol{R}_x\boldsymbol{R}_y^{-1}\boldsymbol{R}_x\boldsymbol{w}(k)} - 1\right)}\right]^2 \cdot \frac{|z_i(k)|^2}{|z_m(k)|^2} \\
&= \left[1 - \frac{\frac{1}{\lambda_1} - \frac{1}{\lambda_i}}{(\frac{1}{\eta_1} - 1)\boldsymbol{w}^H(k)\boldsymbol{R}_x\boldsymbol{R}_y^{-1}\boldsymbol{R}_x\boldsymbol{w}(k) + \frac{1}{\lambda_1}}\right]^2 \cdot \frac{|z_i(k)|^2}{|z_m(k)|^2} = \psi(k)\frac{|z_i(k)|^2}{|z_m(k)|^2},
\end{aligned}
\tag{8.150}
$$

where

$$
\psi(k) = \left[1 - \frac{\frac{1}{\lambda_1} - \frac{1}{\lambda_i}}{\left(\frac{1}{\eta_1} - 1\right)\boldsymbol{w}^H(k)\boldsymbol{R}_x\boldsymbol{R}_y^{-1}\boldsymbol{R}_x\boldsymbol{w}(k) + \frac{1}{\lambda_1}}\right]^2. \tag{8.151}
$$

For all $i = r + 1, \ldots, N$, together with $\eta_1 \in (0, 1]$ and $1/\lambda_1 - 1/\lambda_i > 0$, Eq. (8.146) guarantees that

$$
\begin{aligned}
1 - \frac{\frac{1}{\lambda_1} - \frac{1}{\lambda_i}}{\left(\frac{1}{\eta_1} - 1\right)\boldsymbol{w}^H(k)\boldsymbol{R}_x\boldsymbol{R}_y^{-1}\boldsymbol{R}_x\boldsymbol{w}(k) + \frac{1}{\lambda_1}} &\leq 1 - \frac{\frac{1}{\lambda_1} - \frac{1}{\lambda_{r+1}}}{\left(\frac{1}{\eta_1} - 1\right)\frac{1}{\lambda_1} + \frac{1}{\lambda_1}} \\
&= 1 - \eta_1\left(1 - \frac{\lambda_1}{\lambda_{r+1}}\right) < 1,
\end{aligned}
\tag{8.152}
$$

and

$$1-\frac{\frac{1}{\lambda_1}-\frac{1}{\lambda_i}}{\left(\frac{1}{\eta_1}-1\right)\boldsymbol{w}^H(k)\boldsymbol{R}_x\boldsymbol{R}_y^{-1}\boldsymbol{R}_x\boldsymbol{w}(k)+\frac{1}{\lambda_1}}\geq 1-\frac{\frac{1}{\lambda_1}-\frac{1}{\lambda_N}}{\left(\frac{1}{\eta_1}-1\right)\boldsymbol{w}^H(k)\boldsymbol{R}_x\boldsymbol{R}_y^{-1}\boldsymbol{R}_x\boldsymbol{w}(k)+\frac{1}{\lambda_1}}$$
$$=1-\frac{\frac{1}{\lambda_1}-\frac{1}{\lambda_N}}{\frac{1}{\eta_1}\frac{1}{\lambda_N}+\left(\frac{1}{\lambda_1}-\frac{1}{\lambda_N}\right)}>0. \tag{8.153}$$

From (8.152) and (8.153), we can verify that

$$0<\psi(k)<1,\ i=r+1,\ldots,N, \tag{8.154}$$

for all $k\geq 0$. Denote $\psi_{\max}=\max\{\psi(k)|k\geq 0\}$. Clearly $0<\psi_{\max}<1$. From (8.150), we have the following:

$$\frac{|z_i(k+1)|^2}{|z_m(k+1)|^2}\leq\psi_{\max}\frac{|z_i(k)|^2}{|z_m(k)|^2}\leq\cdots\leq\psi_{\max}^{k+1}\frac{|z_i(0)|^2}{|z_m(0)|^2}. \tag{8.155}$$

Since $\boldsymbol{w}(k)$ is $\boldsymbol{R}_x$-normalized, $|z_m(k)|^2\leq 1$ for all k $\geq$ 0, it follows from (8.155) that

$$\sum_{i=r+1}^{N}|z_i(k)|^2\leq\sum_{i=r+1}^{N}\frac{|z_i(k)|^2}{|z_m(k)|^2}\leq\cdots$$
$$\leq\psi_{\max}^{k}\sum_{i=r+1}^{N}\frac{|z_i(0)|^2}{|z_m(0)|^2}\to 0\text{ as }k\to\infty, \tag{8.156}$$

which along with (8.144) implies that

$$\lim_{k\to\infty}\sum_{i=1}^{r}|z_i(k)|^2=1. \tag{8.157}$$

Note that $z_m(k)/z_m(0)>0$ for all $k\geq 0$. Then, from (8.149) and (8.157) we have the following:

$$\lim_{k\to\infty}z_i(k)=\frac{z_i(0)}{\sqrt{\sum_{j=1}^{r}|z_j(0)|^2}},\quad i=1,2,\ldots,r. \tag{8.158}$$

Based on (8.156) and (8.158), (8.140) can be obtained as follows:

$$\lim_{k\to\infty}\boldsymbol{w}(k)=\sum_{i=1}^{r}\frac{z_i(0)}{\sqrt{\sum_{j=1}^{r}|z_j(0)|^2}}\boldsymbol{v}_i=\frac{\boldsymbol{P}_{\boldsymbol{V}_{\lambda_1}}^{\boldsymbol{R}_x}[\boldsymbol{w}(0)]}{\left\|\boldsymbol{P}_{\boldsymbol{V}_{\lambda_1}}^{\boldsymbol{R}_x}[\boldsymbol{w}(0)]\right\|_{\boldsymbol{R}_x}}. \tag{8.159}$$

Finally, we will prove (8.141). From (8.159), we can see that

$$\lim_{k\to\infty}\frac{1}{\boldsymbol{w}^H(k)\boldsymbol{R}_x\boldsymbol{R}_y^{-1}\boldsymbol{R}_x\boldsymbol{w}(k)}=\lambda_1. \tag{8.160}$$

That is, for any small positive δ, there exists a $K>0$ satisfying

$$\lambda_1-\delta<\frac{1}{\boldsymbol{w}^H(k)\boldsymbol{R}_x\boldsymbol{R}_y^{-1}\boldsymbol{R}_x\boldsymbol{w}(k)}<\lambda_1+\delta, \tag{8.161}$$

for all $k>K$. It follows from (8.128) that

$$\begin{aligned}\lambda(k)&>(1-\gamma_1)\lambda(k-1)+\gamma_1(\lambda_1-\delta)>\cdots>(1-\gamma_1)^{k-K}\lambda(K)+\gamma_1(\lambda_1-\delta)\\&\times\left[1+(1-\gamma_1)+\cdots+(1-\gamma_1)^{k-K}\right]=(1-\gamma_1)^{k-K}\lambda(K)+\gamma_1(\lambda_1-\delta)\\&\times\left[1-(1-\gamma_1)^{k-K}\right]=(\lambda_1-\delta)+(1-\gamma_1)^{k-K}[\lambda(K)-\lambda_1+\delta],\end{aligned} \tag{8.162}$$

and

$$\begin{aligned}\lambda(k)&<(1-\gamma_1)\lambda(k-1)+\gamma_1(\lambda_1+\delta)<\cdots<(1-\gamma_1)^{k-K}\lambda(K)+\gamma_1(\lambda_1+\delta)\\&\times\left[1+(1-\gamma_1)+\cdots+(1-\gamma_1)^{k-K-1}\right]=(1-\gamma_1)^{k-K}\lambda(K)+(\lambda_1+\delta)\\&\times\left[1-(1-\gamma_1)^{k-K}\right]=(\lambda_1+\delta)+(1-\gamma_1)^{k-K}[\lambda(K)-\lambda_1-\delta],\end{aligned} \tag{8.163}$$

for all $k>K$. Since $\gamma_1\in(0,\ 1]$, it is easy to verify from (8.162) and (8.163) that $\lim_{k\to\infty}\lambda(k)=\lambda_1$.

This completes the proof.

Theorem 8.2 (Convergence analysis of fGPCA) *Suppose that the sequence* $[\boldsymbol{w}(k),\lambda(k)]_{k=0}^{\infty}$ *is generated by DDT System 2 with any* $\eta_2,\gamma_2\in(0,1]$, *any initial* $\boldsymbol{R}_y$*-normalized vector* $\boldsymbol{w}(0)\notin(\boldsymbol{V}_{\lambda_N})^{\perp}_{<\boldsymbol{R}_x>}$, *and any* $\lambda(0)>0$. *Then for Case 1, it holds that* $\boldsymbol{w}(k)=\boldsymbol{w}(0)$ *for all* $k\geq 0$, *which is also a generalized eigenvector associated with the generalized eigenvalue* λ_N *of the matrix pencil* ($\boldsymbol{R}_y$, $\boldsymbol{R}_x$), *and* $\lim_{k\to\infty}\lambda(k)=\lambda_N$. *For Case 2, it holds that*

$$\lim_{k\to\infty}\boldsymbol{w}(k)=\sqrt{\frac{1}{\lambda_N}}\frac{\boldsymbol{P}_{\boldsymbol{V}_{\lambda_N}}^{\boldsymbol{R}_x}[\boldsymbol{w}(0)]}{\left\|\boldsymbol{P}_{\boldsymbol{V}_{\lambda_N}}^{\boldsymbol{R}_x}[\boldsymbol{w}(0)]\right\|_{\boldsymbol{R}_x}}, \tag{8.164}$$

$$\lim_{k\to\infty} \lambda(k) = \lambda_N. \tag{8.165}$$

The proof of Theorem 8.2 is similar to that of Theorem 8.1. A minor difference is that we need to calculate the $\boldsymbol{R}_y$-norm of $\boldsymbol{w}(k)$ at each step. Another minor difference is that in (8.146), it holds that matrix pencil $(\boldsymbol{R}_y\boldsymbol{R}_x^{-1}\boldsymbol{R}_y, \boldsymbol{R}_y)$ has the same eigen pairs as $(\boldsymbol{R}_y, \boldsymbol{R}_x)$ and $\boldsymbol{w}(k)$ is well defined, $\boldsymbol{R}_y$-normalized, and $\boldsymbol{w}\,(k) \notin (\boldsymbol{V}_{\lambda_N})^{\perp}_{\langle \boldsymbol{R}_x\rangle}$ for all $k \geq 0$. Therefore,

$$\lambda_1 \leq \frac{\boldsymbol{w}^H(k)\boldsymbol{R}_y\boldsymbol{R}_x^{-1}\boldsymbol{R}_y\boldsymbol{w}(k)}{\boldsymbol{w}^H(k)\boldsymbol{R}_y\boldsymbol{w}(k)} = \boldsymbol{w}^H(k)\boldsymbol{R}_y\boldsymbol{R}_x^{-1}\boldsymbol{R}_y\boldsymbol{w}(k) \leq \lambda_N. \tag{8.166}$$

Particularly, if λ_1 and λ_2 are distinct $(\lambda_1 < \lambda_2 \leq \cdots \leq \lambda_N)$, we have $\boldsymbol{V}_{\lambda_1} = \mathrm{span}\{\boldsymbol{V}_1\}$, $\boldsymbol{P}^{\boldsymbol{R}_x}_{\boldsymbol{V}_{\lambda_1}}[\boldsymbol{w}\,(0)] = \langle \boldsymbol{w}\,(0),\, \boldsymbol{V}_1\rangle_{\boldsymbol{R}_x}\boldsymbol{V}_1$, and $\left\|\boldsymbol{P}^{\boldsymbol{R}_x}_{\boldsymbol{V}_{\lambda_1}}[\boldsymbol{w}\,(0)]\right\|_{\boldsymbol{R}_x} = \left|\langle \boldsymbol{w}\,(0),\, \boldsymbol{V}_1\rangle_{\boldsymbol{R}_x}\right|$. Moreover, if λ_{N-1} and λ_N are distinct $(\lambda_1 \leq \cdots \leq \lambda_{N-1} < \lambda_N)$, we have $\boldsymbol{V}_{\lambda_N} = \mathrm{span}\{\boldsymbol{V}_N\}$, $\boldsymbol{P}^{\boldsymbol{R}_y}_{\boldsymbol{V}_{\lambda_N}}[\boldsymbol{w}(0)] = \langle \boldsymbol{w}(0), \boldsymbol{V}_N\rangle_{\boldsymbol{R}_y}\boldsymbol{V}_N$ and $\left\|\boldsymbol{P}^{\boldsymbol{R}_y}_{\boldsymbol{V}_{\lambda_N}}[\boldsymbol{w}\,(0)]\right\|_{\boldsymbol{R}_y} = \left|\langle \boldsymbol{w}\,(0),\, \boldsymbol{V}_N\rangle_{\boldsymbol{R}_y}\right|$. Hence, the following corollaries hold.

Corollary 8.1 *Suppose that* $\lambda_1 < \lambda_2 \leq \cdots \leq \lambda_N$. *Then the sequence* $[\boldsymbol{w}\,(k), \lambda(k)]_{k=0}^{\infty}$ *generated by DDT System 1 with any* $\eta_1, \gamma_1 \in (0,1]$, *any initial* $\boldsymbol{R}_x$*-normalized vector* $\boldsymbol{w}\,(0) \notin (\boldsymbol{V}_{\lambda_1})^{\perp}_{\langle \boldsymbol{R}_x\rangle}$, *and any* $\lambda(0) > 0$ *satisfies*

$$\lim_{k\to\infty} \boldsymbol{w}(k) = \frac{\langle \boldsymbol{w}(0), \boldsymbol{V}_1\rangle_{\boldsymbol{R}_x}\boldsymbol{V}_1}{\left|\langle \boldsymbol{w}(0), \boldsymbol{V}_1\rangle_{\mathbf{R}_x}\right|}, \tag{8.167}$$

$$\lim_{k\to\infty} \lambda(k) = \lambda_1. \tag{8.168}$$

Corollary 8.2 *Suppose that* $\lambda_1 \leq \cdots \leq \lambda_{N-1} < \lambda_N$. *Then the sequence* $[\boldsymbol{w}(k), \lambda(k)]_{k=0}^{\infty}$ *generated by DDT System 2 with any* $\eta_2, \gamma_2 \in (0,1]$, *any initial* $\boldsymbol{R}_y$*-normalized vector* $\boldsymbol{w}\,(0) \notin (\boldsymbol{V}_{\lambda_N})^{\perp}_{\langle \boldsymbol{R}_y\rangle}$, *and any* $\lambda(0) > 0$ *satisfies*

$$\lim_{k\to\infty} \boldsymbol{w}(k) = \sqrt{\frac{1}{\lambda_N}}\frac{\langle \boldsymbol{w}(0), \boldsymbol{V}_N\rangle_{\boldsymbol{R}_x}\boldsymbol{V}_N}{\left|\langle \boldsymbol{w}(0), \boldsymbol{V}_N\rangle_{\boldsymbol{R}_x}\right|}, \tag{8.169}$$

$$\lim_{k\to\infty} \lambda(k) = \lambda_N. \tag{8.170}$$

8.4.4 Numerical Examples

In this section, we present two numerical examples to evaluate the performance of our algorithms (fGMCA and fGPCA). The first estimates the principal and minor generalized eigenvectors from two random vector processes, which are generated by two sinusoids with additive noise. The second illustrates performance of our algorithms for the BSS problem. Besides nGMCA and nGPCA, we also compare with the following algorithms, which were proposed in the recent ten years:

(1) Gradient-based: adaptive version of ([4], Alg. 2) with negative (for GPCA) and positive (for GMCA) step sizes;
(2) Power-like: fast generalized eigenvector tracking [20] based on the power method;
(3) R-GEVE: reduced-rank generalized eigenvector extraction algorithm [21];
(4) Newton-type: adaptive version of Alg. I proposed in [22].

A. *Experiment 1*

In this experiment, the input samples are generated by:

$$y(n) = \sqrt{2}\,\sin(0.62\pi n + \theta_1) + \varsigma_1(n), \tag{8.171}$$

$$x(n) = \sqrt{2}\sin(0.46\pi n + \theta_2) + \sqrt{2}\sin(0.74\pi n + \theta_3) + \varsigma_2(n), \tag{8.172}$$

where θ_i (i = 1, 2, 3) are the initial phases, which follow uniform distributions within $[0, 2\pi]$, and $\zeta_1(n)$ and $\zeta_2(n)$ are zero-mean white noises with variance $\sigma_1^2 = \sigma_2^2 = 0.1$.

The input vectors $\{\boldsymbol{y}(k)\}$ and $\{\boldsymbol{x}(k)\}$ are arranged in blocks of size N = 8, i.e., $\boldsymbol{y}(k) = [y(k),\ldots, y(k-N+1)]^{\mathrm{T}}$ and $\boldsymbol{x}(k) = [x(k),\ldots, x(k-N+1)]^{\mathrm{T}}$, $k \geq N$. Define the $N \times N$ matrix pencil $(\overline{\boldsymbol{R}}_y, \overline{\boldsymbol{R}}_x)$ with the (p, q) entry (p, q = 1,2,…,N) of $\overline{\boldsymbol{R}}_y$ and $\overline{\boldsymbol{R}}_x$ given by

$$\left[\overline{\boldsymbol{R}}_y\right]_{pq} = \cos[0.62\pi\,(p-q)] + \delta_{pq}\sigma_1^2, \tag{8.173}$$

$$\left[\overline{\boldsymbol{R}}_x\right]_{pq} = \cos[0.46\pi\,(p-q)] + \cos[0.74\pi\,(p-q)] + \delta_{pq}\sigma_2^2. \tag{8.174}$$

For comparison, the direction cosine DC(k) is used to measure the accuracy of direction estimate. We also measure the numerical stability of all algorithms by the sample standard deviation of the direction cosine:

$$SSD(k) = \sqrt{\frac{1}{L-1}\sum_{j=1}^{L}\left[\mathrm{DC}_j(k) - \overline{\mathrm{DC}}(k)\right]^2}, \tag{8.175}$$

where $\mathrm{DC}_j(k)$ is the direction cosine of the jth independent run (j = 1, 2,…, L) and $\overline{\mathrm{DC}}(k)$ is the average over L = 100 independent runs.

In this example, we conduct two simulations. In the first simulation, we use fGMCA, nGMCA, and the other aforementioned algorithms to extract the minor

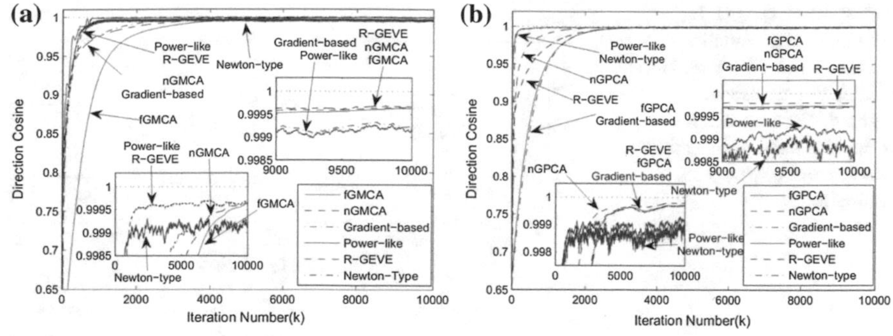

Fig. 8.10 Example 1: Direction cosine of the principal/minor generalized eigenvector. **a** First simulation. **b** Second simulation

generalized eigenvector of matrix pencil ($\boldsymbol{R}_y$, $\boldsymbol{R}_x$). Note that in gradient-based algorithm a positive step size is used, and the other algorithms are applied to estimate the principal generalized eigenvector of matrix pencil ($\boldsymbol{R}_y$, $\boldsymbol{R}_x$) which is also the minor generalized eigenvector of ($\boldsymbol{R}_y$, $\boldsymbol{R}_x$). In the second simulation, we use fGPCA, nGPCA, and the other algorithms to extract the principal generalized eigenvector of matrix pencil ($\boldsymbol{R}_y$, $\boldsymbol{R}_x$). Note that in gradient-based algorithm a negative step size is used. The sets of parameters used in simulations refer to [4], [22]. All algorithms have been initialized with $\widehat{\boldsymbol{R}}_x(0) = \widehat{\boldsymbol{R}}_y(0) = \boldsymbol{Q}_x(0) = \boldsymbol{Q}_y(0) = \boldsymbol{I}_N$ (if used) and $\boldsymbol{w}(0) = \boldsymbol{e}_1$, where $\boldsymbol{e}_1$ stands for the first columns of $\boldsymbol{I}_N$.

The experimental results are shown in Figs. 8.10 to 8.12 and Table 8.1.

Figures 8.10 and 8.11 depict the time course of direction cosine for generalized eigenvector estimation and sample standard deviation of the direction cosine. The results of minor and principal generalized eigenvalues estimation of all generalized eigen-pair extraction algorithms are shown in Fig. 8.12. We find that fGM(P)CA converge faster than nGMCA and nGPCA at the beginning steps, respectively, and

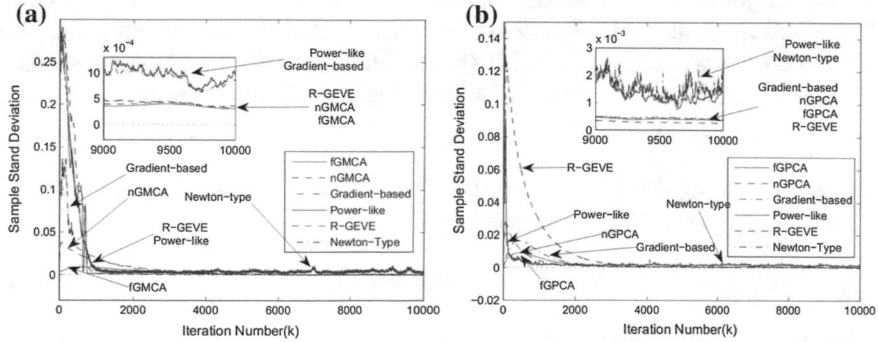

Fig. 8.11 Example 1: Sample standard deviation of the direction cosine. **a** First simulation. **b** Second simulation

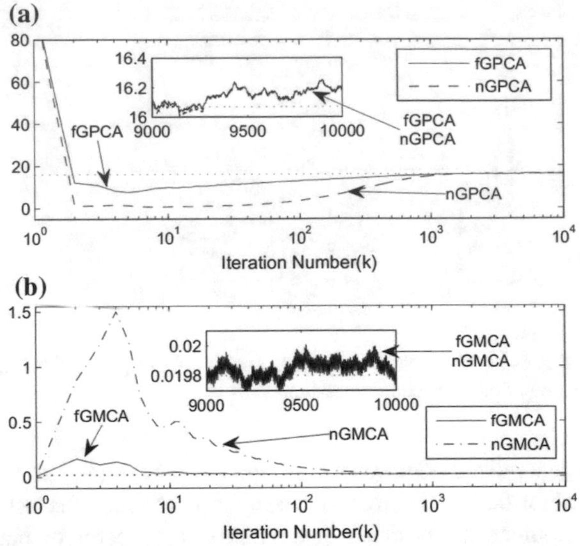

Fig. 8.12 Example 1: Generalized eigenvalues estimation. **a** First simulation: principal generalized eigenvalues estimation. **b** Second simulation: minor generalized eigenvalues estimation

Table 8.1 Computational complexity of all algorithms

Algorithm	fGM(P)CA	nGM(P)CA	Gradient-based
Complexity	$7N^2 + O(N)$	$10N^2 + O(N)$	$10N^2 + O(N)$
Algorithm	Power-like	R-GEVE	Newton-type
Complexity	$13N^2 + O(N)$	$6N^2 + O(N)$	$4N^2 + O(N)$

fGMCA and fGPCA have similar estimation accuracy as nGMCA and nGPCA, respectively. Figure 8.12 shows that all generalized eigen-pair extraction algorithms can extract the principal or minor generalized eigenvalue efficiently.

The computational complexities of all aforementioned algorithms are shown in Table 8.1. We find that Newton-type has the lowest computational complexity but the worst estimation accuracy and standard deviation. The Power-like has the highest computational complexity compared with the other algorithms. The nGM (P)CA and gradient-based algorithms have same computational complexity. The computational complexities of R-GEVE and the proposed algorithms are similar, which are lower than that of nGM(P)CA and gradient-based algorithms.

B. *Experiment 2*

We perform this experiment to show the performance of our algorithm for the BSS problem. Consider a linear BSS model [23]:

$$\boldsymbol{x}(n) = \boldsymbol{A}\boldsymbol{s}(n) + \boldsymbol{e}(n), \tag{8.176}$$

where $\boldsymbol{x}(n)$ is a r-dimensional vector of the observed signals at time k, $\boldsymbol{s}(n)$ is a l-dimensional vector of the unknown source signals, $\boldsymbol{A} \in \boldsymbol{R}^{l \times r}$ denotes the unknown mixing matrix, and $\boldsymbol{e}(n)$ is an unknown noise vector. In general, BSS problem is that of finding a separating matrix $\boldsymbol{W}$ such that the r-dimensional output signal vector $\boldsymbol{y} = \boldsymbol{W}^{\mathrm{T}}\boldsymbol{x}$ contains components that are as independent as possible. In this experiment, we compare the proposed algorithms with nGMCA and nGPCA algorithms, as well as batch-processing generalized eigenvalue decomposition method (EVD method in MATLAB software). We use the method given in [20, 22] to formulate the matrix pencil by applying FIR filtering. $z(n)$, the output of FIR filter, is given as

$$z(n) = \sum_{t=0}^{m} \tau(t)\boldsymbol{x}(n-t), \tag{8.177}$$

where $\tau(t)$ are the coefficients of the FIR filter. Let $\boldsymbol{R}_x = E[\boldsymbol{x}(k)\boldsymbol{x}^{\mathrm{T}}(k)]$ and $\boldsymbol{R}_z = E[z(k)z^{\mathrm{T}}(k)]$. It was shown in [20] that the separating matrix $\boldsymbol{W}$ can be found by extracting the generalized eigenvectors of matrix pencil $(\boldsymbol{R}_z, \boldsymbol{R}_x)$. Hence, the BSS problem can be formulated as finding the generalized eigenvectors associated with the two sample sequences $\boldsymbol{x}(k)$ and $z(k)$. Therefore, we can directly apply our algorithm to solve the BSS problem.

In the simulation, four benchmark signals are extracted from the file ABio7.mat provided by ICALAB [23], as shown in Fig. 8.13. We use the mixing matrix

$$\mathbf{A} = \begin{bmatrix} 2.7914 & -0.1780 & -0.4945 & 0.3013 \\ 1.3225 & -1.7841 & -0.3669 & 0.4460 \\ 0.0714 & -1.9163 & 0.4802 & -0.3701 \\ -1.7396 & 0.1302 & 0.9249 & -0.4007 \end{bmatrix}, \tag{8.178}$$

which was randomly generated. $\boldsymbol{e}[n]$ is a zero-mean white noise vector with covariance $10^{-5}\boldsymbol{I}$. Figure 8.14 shows the mixed signals. We use a simple FIR filter with coefficients $\tau = [1, -1]^{\mathrm{T}}$.

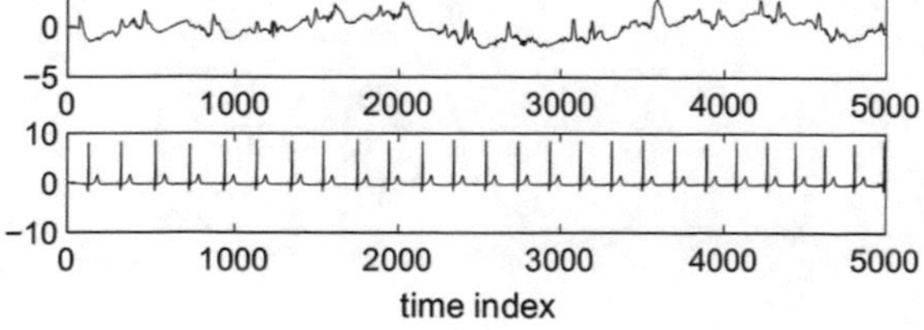

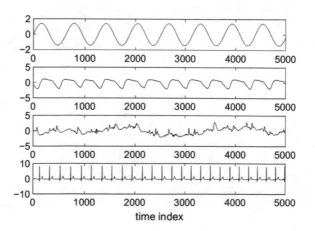

Fig. 8.13 Four original signals

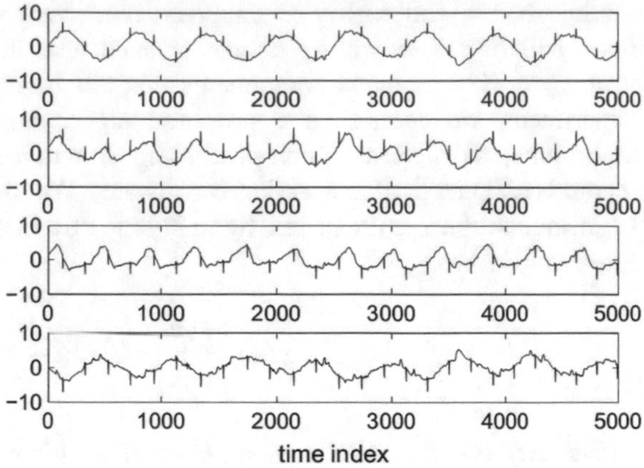

Fig. 8.14 Mixed signals

Suppose that the matrix pencil ($\boldsymbol{R}_z$, $\boldsymbol{R}_x$) has four eigenvectors $\boldsymbol{w}_1$, $\boldsymbol{w}_2$, $\boldsymbol{w}_3$, $\boldsymbol{w}_4$ associated with four eigenvalues $\sigma_1 < \sigma_2 < \sigma_3 < \sigma_4$. Thus, $\boldsymbol{B} = [\boldsymbol{w}_1, \boldsymbol{w}_2, \boldsymbol{w}_3, \boldsymbol{w}_4]$. We use fGPCA, nGPCA, and all other algorithms to extract the two principal generalized eigenvectors ($\boldsymbol{w}_3$ and $\boldsymbol{w}_4$). To extract the two minor generalized eigenvectors ($\boldsymbol{w}_1$ and $\boldsymbol{w}_2$), we use fGMCA, nGMCA, and gradient-based algorithms to extract the minor generalized eigenvectors of matrix pencil ($\boldsymbol{R}_z$, $\boldsymbol{R}_x$) and other algorithms to extract the principal generalized eigenvectors of matrix pencil ($\boldsymbol{R}_x$, $\boldsymbol{R}_z$). All parameters and initial values are the same as in Example 1.

Similar to Example 1, a total of $L = 100$ independent runs are evaluated in this example. The separating matrix $\boldsymbol{B}$ is calculated as $\boldsymbol{B} = (1/L)\sum_{j=1}^{L} \boldsymbol{B}_j$, where $\boldsymbol{B}_j$ is the separating matrix extracted from the jth independent run (j = 1, 2,…, L).

Figures 8.15 to 8.16 show the recovered signals by EVD and our method, respectively. Signals separated by other algorithms are similar to Figs. 8.15 and 8.16, which are not shown in these two figures. Table 8.2 shows the absolute values

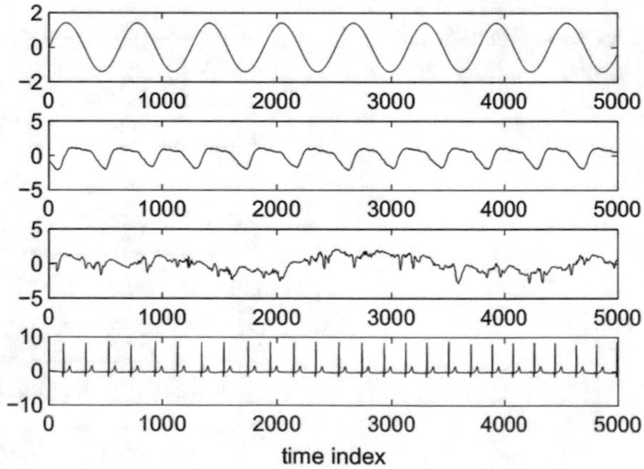

Fig. 8.15 Signals separated by EVD method

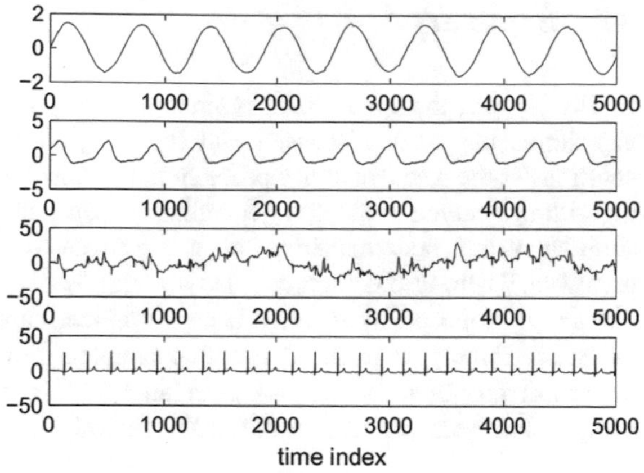

Fig. 8.16 Signals separated by proposed method

Table 8.2 Absolute values of correlation coefficients between sources and recovered signals

Method	Source 1	Source 2	Source 3	Source 4
EVD	1.0000	0.9998	0.9997	0.9989
fGM(P)CA	1.0000	0.9997	0.9992	0.9987
nGM(P)CA	1.0000	0.9996	0.9994	0.9987
Gradient-based	0.9983	0.9811	0.9989	0.9983
Power method	0.9998	0.9995	0.9991	0.9980
R-GEVE	0.9999	0.9995	0.9993	0.9988

of correlation coefficients between the sources and the recovered signals. The simulation results demonstrate that all methods can solve the BSS problem effectively, and our algorithms and the algorithms proposed in [4] can separate the signals more accurately than other algorithms. Moreover, the advantage of neural network model-based algorithms over EVD method for the BSS problem is that they are recursive algorithms and therefore can be implemented online, whereas EVD is a batch-processing method and therefore needs intensive computation.

In this section, we have derived a coupled dynamic system for GHEP based on a novel generalized information criterion. Compared with the existing work, the proposed approach is easier to obtain for that it does not need to calculate the inverse of the Hessian. Based on the dynamic system, a coupled GMCA algorithm (fGMCA) and a coupled GPCA algorithm (fGPCA) have been obtained. The convergence speed of fGMCA and fGPCA is similar to that of Nguyen's well-performed algorithms (nGMCA and nGPCA), but the computational complexity is less than that of Nguyen. Experiment results show that our algorithms have better numerical stability and can extract the generalized eigenvectors more accurately than the other algorithms.

8.5 Summary

In this chapter, the speed stability problem that plagues most noncoupled learning algorithms has been discussed and the coupled learning algorithms that are a solution for the speed stability problem have been analyzed. Moller's coupled PCA algorithm, Nguyen's coupled generalized eigen pair extraction algorithm, coupled singular value decomposition of a cross-covariance matrix, etc., have been reviewed. Then, unified and coupled algorithms for minor and principal eigen pair extraction proposed by us have been introduced, and their convergence has been analyzed. Finally, a fast and adaptive coupled generalized eigen pair extraction algorithm proposed by us has been analyzed in detail, and their convergence analysis has been proved via the DDT method.

References

1. Chen, L. H., & Chang, S. (1995). An adaptive learning algorithm for principal component analysis. *IEEE Transactions on Neural Networks, 6*(5), 1255–1263.
2. Moller, R., & Konies, A. (2004). Coupled principal component analysis. *IEEE Transactions on Neural Networks, 15*(1), 214–222.
3. Sanger, T. D. (1989). Optimal unsupervised learning in a single-layer linear feedforward neural network. *Neural Networks, 2*(6), 459–473.
4. Nguyen, T. D., & Yamada, I. (2013). Adaptive normalized quasi-Newton algorithms for extraction of generalized eigen-pairs and their convergence analysis. *IEEE Transactions on Signal Processing, 61*(6), 1404–1418.
5. Hou, L., & Chen, T. P. (2006). Online algorithm of coupled principal (minor) component analysis. *Journal of Fudan University, 45*(2), 158–169.
6. Feng, X. W., Kong, X. Y., Ma, H. G., & Liu, H. M. (2016). Unified and coupled self-stabilizing algorithm for minor and principal eigen-pair extraction. *Neural Processing Letter.* doi:10.1007/s11063-016-9520-3
7. Diamantaras, K. I., & Kung, S. Y. (1996). *Principal component neural networks. Theory and applications*. New York: Wiley.
8. Nguyen, T. D., Takahashi, N., & Yamada, I. (2013). An adaptive extraction of generalized eigensubspace by using exact nested orthogonal complement structure. *Multidimensional Systems and Signal Processing, 24*(3), 457–483.
9. Kaiser, A., Schenck, M., & Moller, R. (2010). Coupled singular value decomposition of a cross-covariance matrix. *International Journal of Neural Systems, 20*(4), 293–318.
10. Hyvarinen, A., Karhunen, J., & Oja, E. (2004). *Independent component analysis*. New Jersey: Wiley.
11. Miao, Y., & Hua, Y. (1998). Fast subspace tracking and neural network learning by a novel information criterion. *IEEE Transactions on Signal Processing, 46*(7), 1967–1979.
12. Ouyang, S., Ching, P., & Lee, T. (2002). Quasi-Newton algorithm for adaptive minor component extraction. *Electronics Letters, 38*(19), 1142–1144.
13. Golub, G. H., & Van Loan, C. F. (2012). *Matrix computations*. Baltimore: JHU Press.
14. Kong, X., Hu, C., & Han, C. (2012). A dual purpose principal and minor subspace gradient flow. *IEEE Transactions on Signal Processing, 60*(1), 197–210.
15. Peng, D., Yi, Z., & Xiang, Y. (2009). A unified learning algorithm to extract principal and minor components. *Digital Signal Processing, 19*(4), 640–649.

16. Ouyang, S., Ching, P., & Lee, T. (2003). Robust adaptive quasi-Newton algorithms for eigensubspace estimation. *IEE Proceedings-Vision, Image and Signal Processing, 150*(5), 321–330.
17. Higham, D. J., & Higham, N. J. (2005). *MATLAB Guide*. SIAM.
18. Kong, X. Y., Hu, C. H., & Han, C. Z. (2010). On the discrete-time dynamics of a class of self-stabilizing MCA extraction algorithms. *IEEE Transactions on Neural Networks, 21*(1), 175–181.
19. Horn, R. A., & Johnson, C. R. (2012). *Matrix analysis*. Cambridge: Cambridge University Press.
20. Tom, A. M. (2006). The generalized eigendecomposition approach to the blind source separation problem. *Digital Signal Processing, 16*(3), 288–302.
21. Attallah, S., & Abed-Meraim, K. (2008). A fast adaptive algorithm for the generalized symmetric eigenvalue problem. *IEEE Signal Processing Letters, 15*, 797–800.
22. Yang, J., Chen, X., & Xi, H. (2013). Fast adaptive extraction algorithm for multiple principal generalized eigenvectors. *International Journal of Intelligent Systems, 28*(3), 289–306.
23. Icalab. Available: http://www.bsp.brain.riken.go.jp/ICALAB/
24. Tanaka, T. (2009). Fast generalized eigenvector tracking based on the power method. *IEEE Signal Processing Letters, 16*(11), 969–972.
25. Feng, X. W., Kong, X. Y., Duan, Z. S., & Ma, H. G. (2016). Adaptive generalized eigen-pairs extraction algorithms and their convergence analysis. *IEEE Transactions on Signal Processing, 64*(11), 2976–2989.

Chapter 9
Singular Feature Extraction and Its Neural Networks

9.1 Introduction

From the preceding chapters, we have seen that in the wake of the important initiative work by Oja and Sanger, many neural network learning algorithms for PCA have been developed. However, the related field of neural networks that perform SVD, in contrast, has received relatively little attention. This is somewhat surprising since SVD is a crucial ingredient of regression and approximation methods, data compression, and other signal processing applications [1].

In this chapter, our goal is to discuss and analyze the SVD of a rectangular matrix or cross-correlation matrix and the neural network-based algorithms for SVD. It is well known that many signal processing tasks can be efficiently tackled by using SVD of a rectangular matrix or cross-correlation matrix [2]. Several iterative methods for SVD have been proposed by the use of purely matrix algebra [3–6], and these algorithms of updating SVD for tracking subspace can obtain the exact or approximate SVD of a cross-correlation matrix [2]. Recently, in order to get online algorithms, some sample-based rules have been proposed which can avoid the computation of the cross-covariance matrix and instead directly work on the data samples [2, 7–13]. This is advantageous especially for high-dimensional data where the cross-covariance matrices would consume a large amount of memory and their updates are computationally expensive in general [1]. A detailed discussion regarding the model and rationale can be found in [1, 12].

In [12], Diamantaras et al. proposed the cross-correlation neural network (CNN) models that can be directly used for extracting the cross-correlation features between two high-dimensional data streams. However, the CNN models are sometimes divergent for some initial states [14]. In [15], Sanger proposed double generalized Hebbian algorithm (DGHA) for SVD, which was derived from a twofold optimization problem. It adapts the left singular vector estimate by the

© Science Press, Beijing and Springer Nature Singapore Pte Ltd. 2017

X. Kong et al., *Principal Component Analysis Networks and Algorithms*,
DOI 10.1007/978-981-10-2915-8_9

generalized Hebbian algorithm, i.e., a PCA neural network, whereas it adapts the right singular vector estimate by the Widrow–Hoff learning rule. In [16], the cross-correlation asymmetric PCA (APCA) network was proposed and it consists of two sets of neurons that are laterally hierarchically connected. The APCA networks can be used to extract the singular values of the cross-correlation matrix of two stochastic signal vectors, or to implement the SVD of a general matrix. In [17–19], the so-called trace algorithm, "Riccati" algorithm, and their online algorithms were proposed. It should be noted that for one-unit case, the "trace" and "Riccati" algorithms coincide with the cross-coupled Hebbian rule [1]. The algorithm proposed by Helmke–Moore [20] resembles the "trace" algorithm, and if the weights are chosen mutually different, the system converges to the exact SVD of cross-correlation matrix up to permutations of the principal singular vectors [1]. In order to improve the convergence speed of the CNN models, Feng et al. proposed a novel CNN model [2, 13] in which the learning rate is independent of the singular value distribution of the cross-correlation matrix, and its state matrix maintains orthonormality if the initial state matrix is orthonormal. In order to resolve the speed stability problem that plagues most noncoupled learning algorithms, Kaiser et al. proposed a coupled online learning algorithms for the SVD of a cross-covariance matrix [1], which is called coupled SVD algorithms. In the coupled SVD rules, the singular value is estimated alongside the singular vectors, and the effective learning rates for the singular vector rules are influenced by the singular value estimates. In [21], we proposed a novel information criterion for principal singular subspace (PSS) tracking and derived a corresponding PSS gradient flow based on the information criterion. The proposed gradient flow has fast convergence speed, good suitability for data matrix close to singular, and excellent self-stabilizing property. Moreover, in [22], based on Kaiser's work, we proposed a novel information criterion and derive a fast and coupled algorithm from this criterion and using Newton's method, which can extract the principal singular triplet (PST) of a cross-correlation matrix between two high-dimensional data streams and can solve the speed stability problem that plagues most noncoupled learning rules.

In this chapter, we will review and discuss the existing singular feature extraction neural networks and their corresponding learning algorithms. Two singular feature extraction and corresponding neural-based algorithms proposed by us will be analyzed in detail. The remainder of this chapter is organized as follows. An overview of the singular feature extraction neural network-based algorithms is presented in Sect. 9.2. A novel information criterion for PSS tracking, its corresponding PSS gradient flow, convergence, and self-stabilizing property are discussed in Sect. 9.3. A novel coupled neural-based algorithm to extract the PST of a cross-correlation matrix between two high-dimensional data streams is presented in Sect. 9.4, followed by summary in Sect. 9.5.

9.2 Review of Cross-Correlation Feature Method

In this section, we will review SVD learning rules in the literature. In the following, let $\boldsymbol{A}$ denote an $m \times n$ real matrix with its SVD given by Kaiser et al. [1]

$$\boldsymbol{A} = \overline{\boldsymbol{U}}\,\overline{\boldsymbol{S}}\,\overline{\boldsymbol{V}}^{\mathrm{T}} + \overline{\boldsymbol{U}}_2\overline{\boldsymbol{S}}_2\overline{\boldsymbol{V}}_2^{\mathrm{T}}, \tag{9.1}$$

where $\overline{\boldsymbol{U}} = [\bar{\boldsymbol{u}}_1, \bar{\boldsymbol{u}}_2, \ldots, \bar{\boldsymbol{u}}_M] \in \Re^{m \times M}$ denotes the matrix composed of the left principal singular vectors, $\overline{\boldsymbol{S}} = \mathrm{diag}(\bar{\sigma}_1, \bar{\sigma}_2, \ldots, \bar{\sigma}_M) \in \Re^{M \times M}$ denotes the matrix with the principal singular values on its diagonal, and $\overline{\boldsymbol{V}} = [\bar{\boldsymbol{v}}_1, \bar{\boldsymbol{v}}_2, \ldots, \bar{\boldsymbol{v}}_M] \in \Re^{n \times M}$ denotes the matrix composed of the right principal singular vectors. These matrices are referred to as the principal portion of the SVD. Furthermore, $\overline{\boldsymbol{U}}_2 = [\bar{\boldsymbol{u}}_{M+1}, \bar{\boldsymbol{u}}_{M+2}, \ldots, \bar{\boldsymbol{u}}_p] \in \Re^{m \times (p-M)}$, $\overline{\boldsymbol{S}}_2 = diag(\bar{\sigma}_{M+1}, \bar{\sigma}_{M+2}, \ldots, \bar{\sigma}_p) \in \Re^{(p-M) \times (p-M)}$, and $\overline{\boldsymbol{V}}_2 = [\bar{\boldsymbol{v}}_{M+1}, \bar{\boldsymbol{v}}_{M+2}, \ldots, \bar{\boldsymbol{v}}_p] \in \Re^{n \times (p-M)}$ correspond to the minor portion of the SVD. Thus, $\widehat{\boldsymbol{A}} = \overline{\boldsymbol{U}}\,\overline{\boldsymbol{S}}\,\overline{\boldsymbol{V}}^{\mathrm{T}}$ is the best rank-M approximation of $\boldsymbol{A}$, where $M \leq p = \min\{m, n\}$. Left and right singular vectors are normal, i.e., $\|\bar{\boldsymbol{u}}_j\| = \|\bar{\boldsymbol{v}}_j\| = 1, \forall j$ and mutually orthogonal. Thus, $\overline{\boldsymbol{U}}^{\mathrm{T}}\overline{\boldsymbol{U}} = \overline{\boldsymbol{V}}^{\mathrm{T}}\overline{\boldsymbol{V}} = \boldsymbol{I}_M$ and $\overline{\boldsymbol{U}}_2^{\mathrm{T}}\overline{\boldsymbol{U}}_2 = \overline{\boldsymbol{V}}_2^{\mathrm{T}}\overline{\boldsymbol{V}}_2 = \boldsymbol{I}_{p-M}$. Moreover, we assume that the singular values are ordered and mutually different with respect to their absolute values such that $|\bar{\sigma}_1| > \cdots |\bar{\sigma}_M| > |\bar{\sigma}_{M+1}| \geq \cdots \geq |\bar{\sigma}_p|$. In the following, all considerations (e.g., concerning fixed points) depend on the principal portion of the SVD only.

In the following, the input samples are denoted as $\boldsymbol{x}_k \in \Re^n$, and the output samples are denoted as $\boldsymbol{y}_k \in \Re^m$. The data are assumed to be centered, so their covariance matrices become $\boldsymbol{C}_x = E[\boldsymbol{x}\boldsymbol{x}^{\mathrm{T}}]$ and $\boldsymbol{C}_y = E[\boldsymbol{y}\boldsymbol{y}^{\mathrm{T}}]$, respectively. Moreover, the cross-covariance matrix is defined as $\boldsymbol{A} = E[\boldsymbol{y}\boldsymbol{x}^{\mathrm{T}}]$. The vectors $\boldsymbol{u}$ and $\boldsymbol{v}$ denote the state vectors of the ODEs.

9.2.1 *Cross-Correlation Neural Networks Model and Deflation Method*

In [12], the cross-coupled Hebbian rule and the APCA networks were proposed, which were used to extract the singular values of the cross-correlation matrix of two stochastic signal vectors or to implement the SVD of a general matrix. The cross-correlation APCA network consists of two sets of neurons that are laterally hierarchically connected, whose topology is shown in Fig. 9.1.

The vectors $\boldsymbol{x}$ and $\boldsymbol{y}$ are, respectively, the n_1-dimensional and n_2-dimensional input signals. The $n_1 \times m$ matrix $\underline{\boldsymbol{W}} = [\underline{\boldsymbol{w}}_1, \ldots, \underline{\boldsymbol{w}}_m]$ and the $n_2 \times m$ matrix $\overline{\boldsymbol{W}} = [\bar{\boldsymbol{w}}_1, \ldots, \bar{\boldsymbol{w}}_m]$ are the feedforward weights, and the $n_2 \times m$ matrices $\underline{\boldsymbol{U}} = [\underline{\boldsymbol{u}}_1, \ldots, \underline{\boldsymbol{u}}_m]$ and $\overline{\boldsymbol{U}} = [\bar{\boldsymbol{u}}_1, \ldots, \bar{\boldsymbol{u}}_m]$ are the lateral connection weights, where $\underline{\boldsymbol{u}}_{\mathrm{i}} =$

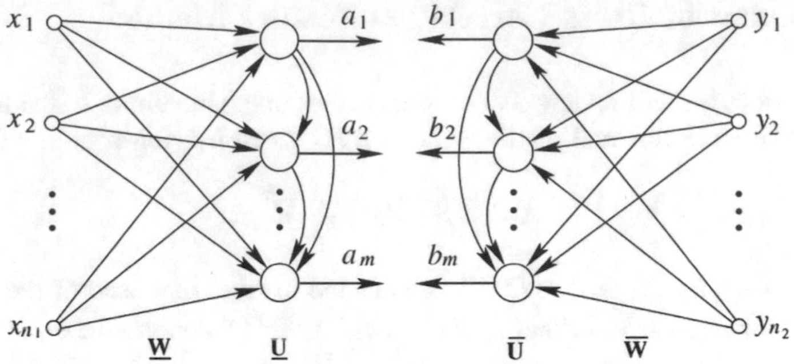

Fig. 9.1 Architecture of the cross-correlation APCA network

$(\underline{u}_{1i}, \ldots, \underline{u}_{mi})^{\mathrm{T}}$, $\bar{\boldsymbol{u}}_i = (\bar{u}_{1i}, \ldots, \bar{u}_{mi})^{\mathrm{T}}$, and $m \leq \min\{n_1, n_2\}$. This model is used to perform the SVD of $\boldsymbol{C}_{xy} = E[\boldsymbol{x}_k \boldsymbol{y}_k^{\mathrm{T}}]$.

The network has the following relations: $\boldsymbol{a} = \underline{\boldsymbol{W}}^{\mathrm{T}}\boldsymbol{x}$ and $\boldsymbol{b} = \overline{\boldsymbol{W}}^{\mathrm{T}}\boldsymbol{y}$, where $\boldsymbol{a} = (a_1, \ldots, a_m)^{\mathrm{T}}$ and $\boldsymbol{b} = (b_1, \ldots, b_m)^{\mathrm{T}}$. In [12], the following cross-correlation cost function was used

$$E_{\mathrm{APCA}}(\underline{\boldsymbol{w}}, \bar{\boldsymbol{w}}) = \frac{E[a_1(k) b_1(k)]}{\|\underline{\boldsymbol{w}}\| \|\bar{\boldsymbol{w}}\|} = \frac{\underline{\boldsymbol{w}}^{\mathrm{T}} \boldsymbol{C}_{xy} \bar{\boldsymbol{w}}}{\|\underline{\boldsymbol{w}}\| \|\bar{\boldsymbol{w}}\|}. \tag{9.2}$$

Maximizing the correlation, the solution is known to be $(\bar{\boldsymbol{w}}, \underline{\boldsymbol{w}}) = (\rho \boldsymbol{u}_1, \mu \boldsymbol{v}_1)$, where ρ and μ are any nonzero scalars; namely, the optimal weights are the principal left and right singular vectors of $\boldsymbol{C}_{xy}$. It is obvious that $\|\bar{\boldsymbol{w}}\|^2 = \rho^2$, $\|\underline{\boldsymbol{w}}\|^2 = \mu^2$, and $\max J_{\mathrm{APCA}} = J_{\max} = \sigma_1$. In [12], the following cross-coupled Hebbian rule

$$\underline{\boldsymbol{w}}_1(k+1) = \underline{\boldsymbol{w}}_1(k) + \beta[\boldsymbol{x}(k) - \underline{\boldsymbol{w}}_1(k) a_1(k)] b_1(k), \tag{9.3}$$

$$\bar{\boldsymbol{w}}_1(k+1) = \bar{\boldsymbol{w}}_1(k) + \beta[\boldsymbol{y}(k) - \bar{\boldsymbol{w}}_1(k) b_1(k)] a_1(k) \tag{9.4}$$

was proposed. By using the stochastic approximation theory and the Lyapunov method, it has been proved that for Algorithms (9.3) and (9.4), if $\sigma_1 > \sigma_2 \geq \sigma_3 \geq \cdots \geq \sigma_q \geq 0$, $q = \min\{m, n\}$, then with probability 1, $\bar{\boldsymbol{w}}(k) \to \pm \boldsymbol{u}_1$ and $\underline{\boldsymbol{w}}(k) \to \pm \boldsymbol{v}_1$ as $k \to \infty$.

After the principal singular component (PSC) has been extracted, a deflation transformation is introduced to nullify the principal singular value so as to make the next singular value principal. Thus, $\boldsymbol{C}_{xy}$ in the criterion (9.2) can be replaced by one of the following three transformed forms so as to extract the $(i + 1)$th PSC:

$$\boldsymbol{C}_{xy}^{(i+1)} = \boldsymbol{C}_{xy}^{(i)} (\boldsymbol{I} - \boldsymbol{v}_i \boldsymbol{v}_i^{\mathrm{T}}), \tag{9.5}$$

$$\boldsymbol{C}_{xy}^{(i+1)} = (\boldsymbol{I} - \boldsymbol{u}_i \boldsymbol{u}_i^{\mathrm{T}}) \boldsymbol{C}_{xy}^{(i)}, \tag{9.6}$$

$$\boldsymbol{C}_{\boldsymbol{xy}}^{(i+1)} = \left(\boldsymbol{I} - \boldsymbol{u}_i\boldsymbol{u}_i^{\mathrm{T}}\right)\boldsymbol{C}_{\boldsymbol{xy}}^{(i)}\left(\boldsymbol{I} - \boldsymbol{v}_i\boldsymbol{v}_i^{\mathrm{T}}\right), \tag{9.7}$$

for i = 1,2, … ,m − 1, where $\boldsymbol{C}_{\boldsymbol{xy}}^{(1)} = \boldsymbol{C}_{\boldsymbol{xy}}$. The deflation can be achieved by the transformation on the data:

$$\boldsymbol{x} \leftarrow \boldsymbol{x}, \boldsymbol{y} \leftarrow \boldsymbol{y} - \boldsymbol{v}_i\boldsymbol{v}_i^{\mathrm{T}}\boldsymbol{y}, \tag{9.8}$$

$$\boldsymbol{x} \leftarrow \boldsymbol{x} - \boldsymbol{u}_i\boldsymbol{u}_i^{\mathrm{T}}\boldsymbol{x}, \boldsymbol{y} \leftarrow \boldsymbol{y}, \tag{9.9}$$

$$\boldsymbol{x} \leftarrow \boldsymbol{x} - \boldsymbol{u}_i\boldsymbol{u}_i^{\mathrm{T}}\boldsymbol{x}, \boldsymbol{y} \leftarrow \boldsymbol{y} - \boldsymbol{v}_i\boldsymbol{v}_i^{\mathrm{T}}\boldsymbol{y}. \tag{9.10}$$

Assuming that the previous j − 1 have already been extracted, and using a deflation transformation, the two sets of neurons are trained with the cross-coupled Hebbian learning rules, which are given by Diamantaras and Kung [12]

$$\underline{\boldsymbol{w}}_j(k+1) = \underline{\boldsymbol{w}}_j(k) + \beta[\boldsymbol{x}(k) - \underline{\boldsymbol{w}}_j(k)a_j(k)]b_j'(k), \tag{9.11}$$

$$\bar{\boldsymbol{w}}_j(k+1) = \bar{\boldsymbol{w}}_j(k) + \beta[\boldsymbol{y}(k) - \bar{\boldsymbol{w}}_j(k)b_j(k)]a_j'(k), \tag{9.12}$$

for j = 1, … , m, where β is the learning rate selected as a small constant or according to the Robbins–Monro conditions, where

$$a_j' = a_j - \sum_{i=1}^{j-1} \underline{u}_{ij}a_i, \quad a_i = \underline{\boldsymbol{w}}_i^{\mathrm{T}}\boldsymbol{x}, \ i = 1,\ldots,j, \tag{9.13}$$

$$b_j' = b_j - \sum_{i=1}^{j-1} \bar{u}_{ij}b_i, \quad b_i = \bar{\boldsymbol{w}}_i^{\mathrm{T}}\boldsymbol{y}, \ i = 1,\ldots,j, \tag{9.14}$$

and the lateral weights should be equal to

$$\underline{u}_{ij} = \underline{\boldsymbol{w}}_i^{\mathrm{T}}\underline{\boldsymbol{w}}_j, \quad \bar{u}_{ij} = \bar{\boldsymbol{w}}_i^{\mathrm{T}}\bar{\boldsymbol{w}}_j, \ i = 1,\ldots,j-1. \tag{9.15}$$

The set of lateral connections among the units is called the lateral othogonaliztion network, and $\overline{\boldsymbol{U}}$ and $\underline{\boldsymbol{U}}$ are upper triangular matrices. By premultiplying (9.11) by $\underline{\boldsymbol{w}}_i^{\mathrm{T}}$ and (9.12) by $\bar{\boldsymbol{w}}_i^{\mathrm{T}}$, a local algorithm, called the lateral orthogonalization rule, for calculating $\underline{u}_{ij}$ and $\bar{u}_{ij}$, has been derived as follows:

$$\underline{u}_{ij}(k+1) = \underline{u}_{ij}(k) + \beta[a_i(k) - \underline{u}_{ij}(k)a_j(k)]b_j'(k), \tag{9.16}$$

$$\bar{u}_{ij}(k+1) = \bar{u}_{ij}(k) + \beta[b_i(k) - \bar{u}_{ij}(k)b_j(k)]a_j'(k). \tag{9.17}$$

The initial values can be selected as $\underline{u}_{ij}(0) = \underline{\boldsymbol{w}}_i^{\mathrm{T}}(0)\underline{\boldsymbol{w}}_j(0)$ and $\bar{u}_{ij}(0) = \bar{\boldsymbol{w}}_i^{\mathrm{T}}(0)\bar{\boldsymbol{w}}_j(0)$. However, this initial condition is not critical to the convergence of the algorithm [12]. It has been proved that $\underline{\boldsymbol{w}}_i$ and $\bar{\boldsymbol{w}}_i$ converge to the ith left and right principal singular vectors of $\boldsymbol{C}_{xy}$, respectively, and σ_i converges to its corresponding criterion E_{APCA}, as $t \to \infty$. That is, the algorithm extracts the first m principal singular values in the descending order and their corresponding left and right singular vectors. Like the APEX, the APCA algorithm can incrementally add nodes without retraining the learned nodes.

9.2.2 Parallel SVD Learning Algorithms on Double Stiefel Manifold

In this section, we will introduce several parallel SVD learning algorithms, which allow to simultaneously compute the SVD vectors. The considered neural algorithms were developed in [18, 20] and have been analyzed in [17]. These algorithms are utilized to train in an unsupervised way a three-layer neural network with the classical "buttery" topology [17–19]. The first layer has connection matrix $\boldsymbol{A}$, the second one has connection matrix $\boldsymbol{B}$, and the middle (hidden) layer provides network's output. It has been shown that when proper initial conditions are chosen, the associated learning trajectories lie on the double Stiefel manifold [17].

In this section, the following matrix set is useful for analysis purpose: $\mathrm{St}(m,n,\boldsymbol{K}) \stackrel{\mathrm{def}}{=} \{\boldsymbol{X} \in \boldsymbol{K}^{m \times n} | \boldsymbol{X}^*\boldsymbol{X} = \boldsymbol{I}_n\}$ with $m-1, n-1 \in N$; $K \in \Re$ or C. When $m = n$, the manifold coincides with the orthogonal group $O(m,\boldsymbol{K}) \stackrel{\mathrm{def}}{=} \{\boldsymbol{X} \in \boldsymbol{K}^{m \times m} | \boldsymbol{X}^*\boldsymbol{X} = \boldsymbol{I}_m\}$. Here, the product $O(m,\boldsymbol{K}) \times O(n,\boldsymbol{K})$ is referred to as double orthogonal group and the product $\mathrm{St}(m,p,\boldsymbol{K}) \times \mathrm{St}(n,p,\boldsymbol{K})$ as double Stiefel manifold.

Denote as $\boldsymbol{Z} \in C^{m \times n}$ the matrix whose SVD is to be computed and as $r \leq \min\{m,n\}$ the rank of $\boldsymbol{Z}$, the singular value decomposition can be written as $\boldsymbol{Z} = \boldsymbol{U}\boldsymbol{D}\boldsymbol{V}^*$, where $\boldsymbol{U} \in \Re^{m \times m}$ and $\boldsymbol{V} \in \Re^{n \times n}$ are orthogonal matrices and $\boldsymbol{D}$ is a pseudo-diagonal matrix whose elements are all zeros except for the first r diagonal entries, termed as singular values.

Denote as $\boldsymbol{A}(t) \in \Re^{m \times p}$ the network connection matrix-stream that should learn p left singular vectors and as $\boldsymbol{B}(t) \in \Re^{n \times p}$ the estimator for p right singular vectors of the SVD of matrix $\boldsymbol{Z} \in \Re^{m \times n}$, with $p \leq r \leq \min\{m,n\}$. For simplicity, parallel SVD learning algorithms in [17–19] are described in terms of their ordinary differential equations.

The algorithm WH2 [17] can be written as (also called as the "trace" algorithm):

$$\dot{\boldsymbol{A}} = \boldsymbol{Z}\boldsymbol{B} - \boldsymbol{A}\boldsymbol{B}^{\mathrm{T}}\boldsymbol{Z}^{\mathrm{T}}\boldsymbol{A}, \quad \boldsymbol{A}(0) = \boldsymbol{A}_0, \tag{9.18}$$

$$\dot{\boldsymbol{B}} = \boldsymbol{Z}^{\mathrm{T}}\boldsymbol{A} - \boldsymbol{B}\boldsymbol{A}^{\mathrm{T}}\boldsymbol{Z}\boldsymbol{B}, \quad \boldsymbol{B}(0) = \boldsymbol{B}_0. \tag{9.19}$$

Equations (9.18) and (9.19) have been derived by extending Brockett's work on isospectral flow systems [23] from single to double orthogonal group. The initial state $\boldsymbol{A}_0$, $\boldsymbol{B}_0$ can be freely chosen. In particular, one can consider the following choice: $\boldsymbol{A}_0 \in \mathrm{St}(m,p,\Re)$ and $\boldsymbol{B}_0 \in \mathrm{St}(n,p,\Re)$, for instance $\boldsymbol{A}_0 = \boldsymbol{I}_{m,p}$ and $\boldsymbol{B}_0 = \boldsymbol{I}_{n,p}$.

Then, for the algorithm WH2, it has been proved that: (1) If the initial states of the WH2 system belong to the Stiefel manifold, then the whole dynamics is double Stiefel manifold. (2) The steady states of WH2 learning system can be written as $\mathbf{A} = \mathbf{U}_p\mathbf{K}$ and $\boldsymbol{B} = \boldsymbol{V}_p\boldsymbol{K}$, where $\boldsymbol{K}$ is arbitrary in $O(p,\Re)$ and $\boldsymbol{U}_p$ and $\boldsymbol{V}_p$ denote the submatrices whose columns are p right and left singular vectors of matrix $\boldsymbol{Z}$, respectively [17].

Obviously, the WH2 algorithm does not actually compute the true SVD, but a SVD subspace of dimension p.

The WH3 learning system was derived as an extension of the well-known Oja's subspace rule. The algorithm WH3 can be written as [17]:

$$\dot{\boldsymbol{A}} = \boldsymbol{Z}\boldsymbol{B} - \boldsymbol{A}(\boldsymbol{A}^{\mathrm{T}}\boldsymbol{Z}\boldsymbol{B} + \boldsymbol{B}^{\mathrm{T}}\boldsymbol{Z}^{\mathrm{T}}\boldsymbol{A}), \quad \boldsymbol{A}(0) = \boldsymbol{A}_0, \tag{9.20}$$

$$\dot{\boldsymbol{B}} = \boldsymbol{Z}^{\mathrm{T}}\boldsymbol{A} - \boldsymbol{B}(\boldsymbol{A}^{\mathrm{T}}\boldsymbol{Z}\boldsymbol{B} + \boldsymbol{B}^{\mathrm{T}}\boldsymbol{Z}^{\mathrm{T}}\boldsymbol{A}), \quad \boldsymbol{B}(0) = \boldsymbol{B}_0. \tag{9.21}$$

For the algorithm WH3, it has been proved that under the hypotheses $A_0/\sqrt{2} \in \mathrm{St}(m,p,\Re)$ and $\boldsymbol{B}_0/\sqrt{2} \in \mathrm{St}(n,p,\Re)$, the learning equation WH3 can keep $\boldsymbol{A}(t)/\sqrt{2}$ and $\boldsymbol{B}(t)/\sqrt{2}$ within the Stiefel manifold.

The WH4 learning system can be written as (also called as the "Riccati" algorithm)

$$\dot{\boldsymbol{A}} = \boldsymbol{Z}\boldsymbol{B} - \frac{1}{2}\boldsymbol{A}(\boldsymbol{A}^{\mathrm{T}}\boldsymbol{Z}\boldsymbol{B} + \boldsymbol{B}^{\mathrm{T}}\boldsymbol{Z}^{\mathrm{T}}\boldsymbol{A}), \quad \boldsymbol{A}(0) = \boldsymbol{A}_0, \tag{9.22}$$

$$\dot{\boldsymbol{B}} = \boldsymbol{Z}^{\mathrm{T}}\boldsymbol{A} - \frac{1}{2}\boldsymbol{B}(\boldsymbol{A}^{\mathrm{T}}\boldsymbol{Z}\boldsymbol{B} + \boldsymbol{B}^{\mathrm{T}}\boldsymbol{Z}^{\mathrm{T}}\boldsymbol{A}), \quad \boldsymbol{B}(0) = \boldsymbol{B}_0. \tag{9.23}$$

For the algorithm WH4, it has been proved that under the hypotheses $\boldsymbol{A}_0 \in \mathrm{St}(m,p,\Re)$ and $\boldsymbol{B}_0 \in \mathrm{St}(n,p,\Re)$, the learning equations WH4 can keep $\boldsymbol{A}(t)$ and $\boldsymbol{B}(t)$ within the Stiefel manifold.

The structure of the stationary points of the WH3–WH4 algorithms is similar to the structure of the equilibria of the WH2 system. It has been proved that the steady states of WH3 and WH4 learning systems can be written as $\boldsymbol{A} = \boldsymbol{U}_p\boldsymbol{K}$ and $\boldsymbol{B} = \boldsymbol{V}_p\boldsymbol{K}$, where $\boldsymbol{K}$ is arbitrary in $O(p,\Re)$ and $\boldsymbol{U}_p$ and $\boldsymbol{V}_p$ denote the submatrices whose columns are p right and left singular vectors of the matrix $\boldsymbol{Z}$, respectively [17].

The HM dynamics [20] arising from the maximization of a specific metric (criterion) $\Phi_W: O(m, C) \times O(n, C) \rightarrow \Re$ is defined as:

$$\Phi_{\boldsymbol{W}}(\boldsymbol{A}, \boldsymbol{B}) \stackrel{\text{def}}{=} 2\text{Retr}(\boldsymbol{W}\boldsymbol{A}^*\boldsymbol{Z}\boldsymbol{B}), \tag{9.24}$$

where $\boldsymbol{W} \in \Re^{n \times m}$ is a weighting matrix, $\boldsymbol{Z} \in C^{m \times n}$, and $m \geq n$. The dynamic system, derived as a Riemannian gradient flow on $O(m, C) \times O(n, C)$, can be written as [17]

$$\dot{\boldsymbol{A}} = \boldsymbol{A}(\boldsymbol{W}^*\boldsymbol{B}^*\boldsymbol{Z}^*\boldsymbol{A} - \boldsymbol{A}^*\boldsymbol{Z}\boldsymbol{B}\boldsymbol{W}), \quad \boldsymbol{A}(0) = \boldsymbol{A}_0, \tag{9.25}$$

$$\dot{\boldsymbol{B}} = \boldsymbol{B}(\boldsymbol{W}\boldsymbol{A}^*\boldsymbol{Z}\boldsymbol{B} - \boldsymbol{B}^*\boldsymbol{Z}^*\boldsymbol{A}\boldsymbol{W}^*), \quad \boldsymbol{B}(0) = \boldsymbol{B}_0. \tag{9.26}$$

It has been proved that $\boldsymbol{A}(t) \in O(m, C)$ and $\boldsymbol{B}(t) \in O(n, C)$. In the particular case that $\boldsymbol{W} = -\boldsymbol{I}_{n,m}$ and the involved quantities are real-valued, the above system is equivalent to WH2 when $p = n$. Obviously, the Weingessel–Hornik SVD learning equations can be regarded as special cases of the Helmke–Moore system.

9.2.3 Double Generalized Hebbian Algorithm (DGHA) for SVD

In [15], Sanger presented two iterative algorithms for finding the SVD of a matrix P given only samples of the inputs $\boldsymbol{u}$ and outputs $\boldsymbol{y}$. The first algorithm is the double generalized Hebbian algorithm (DGHA), and it is described by the following two coupled difference equations:

$$\Delta \boldsymbol{G} = \gamma(\boldsymbol{z}\boldsymbol{y}^{\mathrm{T}} - \mathrm{LT}[\boldsymbol{z}\boldsymbol{z}^{\mathrm{T}}]\boldsymbol{G}), \tag{9.27}$$

$$\Delta \boldsymbol{N}^{\mathrm{T}} = \gamma(\boldsymbol{z}\boldsymbol{u}^{\mathrm{T}} - \mathrm{LT}[\boldsymbol{z}\boldsymbol{z}^{\mathrm{T}}]\boldsymbol{N}^{\mathrm{T}}), \tag{9.28}$$

where LT[·] is an operator that makes the diagonal elements of its matrix argument zeros, $\boldsymbol{y} = \boldsymbol{P}\boldsymbol{u}$, $z = \boldsymbol{G}\boldsymbol{y}$, and γ is a learning rate constant. Obviously, in single-component case, Eq. (9.27) is derived from the objective function $J_{\max}^{\boldsymbol{g}} = (\boldsymbol{g}^{\mathrm{T}}\boldsymbol{C}_y\boldsymbol{g})/\|\boldsymbol{g}\|^2$, and its fixed points are $\boldsymbol{g}^* = \bar{\boldsymbol{g}}_1[1]$. Equation (9.28) is derived from the objective function $J_{\min}^{\boldsymbol{n}} = \frac{1}{2}E\left[\|\boldsymbol{x} - \boldsymbol{n}\boldsymbol{g}^{\mathrm{T}}\boldsymbol{y}\|^2\right]$, and its fixed points is $\boldsymbol{n}^* = \bar{\sigma}_1^{-1}\bar{\boldsymbol{n}}_1[1]$, where $\|\boldsymbol{g}^*\| = 1$ and $\|\boldsymbol{n}^*\| = \bar{\sigma}_1^{-1}$.

Equation (9.27) is the Generalized Hebbian Algorithm which finds the eigenvectors of the autocorrelation matrix of its inputs $\boldsymbol{y}$. For random uncorrelated inputs $\boldsymbol{u}$, the autocorrelation of $\boldsymbol{y}$ is $\mathrm{E}[\boldsymbol{y}\boldsymbol{y}^T] = \boldsymbol{L}^T\boldsymbol{S}^2\boldsymbol{L}$. So Eq. (9.27) will cause $\boldsymbol{G}$ to converge to the matrix $\boldsymbol{L}$ composed of left singular vectors. Equation (9.28) is related to the Widrow–Hoff LMS rule for approximating $\boldsymbol{u}^T$ from z, but it enforces orthogonality of the columns of $\boldsymbol{N}$. Equations (9.27) and (9.28) together cause $\boldsymbol{N}$ to converge to $\boldsymbol{R}^{\mathrm{T}}\boldsymbol{S}^{-1}$, so that the combination $\mathbf{NG} = \boldsymbol{R}^{\mathrm{T}}\boldsymbol{S}^{-1}L$ is an approximation to the plant inverse.

It has been proved that if $\boldsymbol{y} = \boldsymbol{Pu}$, $z = \boldsymbol{Gy}$, and $E[\boldsymbol{uu}^{\mathrm{T}}] = \boldsymbol{I}$, then Eqs. (9.27) and (9.28) converge to the left and right singular vectors of P [15].

The second algorithm is the Orthogonal Asymmetric Encoder (OAE) which is described by

$$\Delta \boldsymbol{G} = \gamma(\hat{z}y^{\mathrm{T}} - \mathrm{LT}[\hat{z}\hat{z}^{\mathrm{T}}]\boldsymbol{G}), \tag{9.29}$$

$$\Delta \boldsymbol{N}^{\mathrm{T}} = \gamma(\boldsymbol{Gy} - \mathrm{LT}[\boldsymbol{GG}^{\mathrm{T}}]\hat{z}u^{\mathrm{T}}), \tag{9.30}$$

where $\hat{z} = \boldsymbol{N}^{\mathrm{T}}\boldsymbol{u}$. Obviously, for single-component case, Eqs. (9.29) and (9.30) are derived from the objective function $J_{\min} = \frac{1}{2}E\left[\|\boldsymbol{y} - \boldsymbol{gn}^{\mathrm{T}}\boldsymbol{x}\|^2\right]$, and their fixed points are $\boldsymbol{g}^* = \rho\bar{\boldsymbol{g}}_1$ and $\boldsymbol{n}^* = \rho^{-1}\bar{\sigma}_1\bar{\boldsymbol{n}}_1$, where $\rho \neq 0$ is an arbitrary constant [1].

It has been proved that Eqs. (9.29) and (9.30) converge to the left and right singular vectors of $\boldsymbol{P}$.

9.2.4 Cross-Associative Neural Network for SVD(CANN)

In [2], Feng et al. proposed a novel CNN model to improve the convergence speed of the CNN models, in which the learning rate is independent of the singular value distribution of the cross-correlation matrix, and its state matrix maintains orthonormality if the initial state matrix is orthonormal. Later, based on [2], Feng et al. also proposed a novel CNN model for finding the PSS of the cross-correlation matrix between two high-dimensional data streams and introduced a novel non-quadratic criterion (NQC) for searching the optimum weights of two linear neural network (LNN). An adaptive algorithm based on the NQC for tracking the PSS of the cross-correlation matrix between two high-dimensional vector sequences was developed, and the NQC algorithm provides fast online learning of the optimum weights for two LNNs.

In order to improve the cost surface for the PSS and the convergence of gradient searching, a novel NQC for the PSS was presented in [2]. Given $\boldsymbol{U} \in \Re^{M\times r}$ and $\boldsymbol{V} \in \Re^{N\times r}$ in the domain $\{(\boldsymbol{U}, \boldsymbol{V})|\boldsymbol{U}^{\mathrm{T}}\boldsymbol{CV} > 0\}$, the following framework for PSS was proposed:

$$\begin{aligned} &\min_{\boldsymbol{U},\boldsymbol{V}} J_{\mathrm{NQC}}(\boldsymbol{U}, \boldsymbol{V}) \\ &J_{\mathrm{NQC}}(\boldsymbol{U}, \boldsymbol{V}) = -\mathrm{tr}[\ln(\boldsymbol{U}^{\mathrm{T}}\boldsymbol{CV})] + \frac{1}{2}\mathrm{tr}[\boldsymbol{U}^{\mathrm{T}}\boldsymbol{U} + \boldsymbol{V}^{\mathrm{T}}\boldsymbol{V}]. \end{aligned} \tag{9.31}$$

The landscape of NQC is depicted by the following two theorems.

Theorem 9.1 $(\boldsymbol{U}, \boldsymbol{V})$ *is a stationary point of* $J_{\text{NQC}}(\boldsymbol{U}, \boldsymbol{V})$ *in the domain* $\{(\boldsymbol{U}, \boldsymbol{V}) | \boldsymbol{U}^{\text{T}} \boldsymbol{C} \boldsymbol{V} > 0\}$ *if and only if* $\boldsymbol{U} \ddot{=} \boldsymbol{L}_r$ *and* $\boldsymbol{V} \ddot{=} \boldsymbol{R}_r$*, where* $\boldsymbol{L}_r \in \Re^{M \times r}$ *and* $\boldsymbol{R}_r \in \Re^{N \times r}$ *consist of the left and right singular vectors of C, respectively. Note that* $(\boldsymbol{U}, \boldsymbol{V}) \ddot{=} (\boldsymbol{L}_r, \boldsymbol{R}_r)$ *shows a stationary set of* $J_{\text{NQC}}(\boldsymbol{U}, \boldsymbol{V})$. And for the definition of "$\ddot{=}$," see [2].

Theorem 9.2 *In the domain* $\{(\boldsymbol{U}, \boldsymbol{V}) | \boldsymbol{U}^{\text{T}} \boldsymbol{C} \boldsymbol{V} > 0\}$, $J_{\text{NQC}}(\boldsymbol{U}, \boldsymbol{V})$ *has a global minimum that is achieved if and only if* $\boldsymbol{U} \ddot{=} \boldsymbol{L}_s$ *and* $\boldsymbol{V} \ddot{=} \boldsymbol{R}_s$*. And the global minimum is* $J_{\text{NQC}}(\boldsymbol{U}, \boldsymbol{V}) = \sum_{i=1}^{r} \ln \sigma_i - r$*. All the other stationary points* $(\boldsymbol{U}, \boldsymbol{V})(\ddot{\neq}(\boldsymbol{L}_s, \boldsymbol{R}_s))$ *are saddle (unstable) points of* $J_{\text{NQC}}(\boldsymbol{U}, \boldsymbol{V})$*. In fact* $(\boldsymbol{U}, \boldsymbol{V}) \ddot{=} (\boldsymbol{L}_s, \boldsymbol{R}_s)$ *shows a global minimizer set of* $J_{\text{NQC}}(\boldsymbol{U}, \boldsymbol{V})$*, where* $\boldsymbol{L}_s$ *and* $\boldsymbol{R}_s$ *are the left and right singular vector matrix associated with signal, respectively.*

One can apply the gradient descent searching to the unconstrained minimization of $J_{\text{NQC}}(\boldsymbol{U}, \boldsymbol{V})$. In [2], a batch algorithm and a recursive algorithm were derived.

Given the gradient of $J_{\text{NQC}}(\boldsymbol{U}, \boldsymbol{V})$ with respect to **U** and **V**, the following gradient descent rule can be used for updating $\boldsymbol{U}(k)$ and $\boldsymbol{V}(k)$:

$$\boldsymbol{U}(k) = (1 - \eta)\boldsymbol{U}(k-1) + \eta \boldsymbol{C}(k)\boldsymbol{V}(k) \times (\boldsymbol{U}^{\text{T}}(k)\boldsymbol{C}(k)\boldsymbol{V}(k))^{-1}, \tag{9.32}$$

$$\boldsymbol{V}(k) = (1 - \eta)\boldsymbol{V}(k-1) + \eta \boldsymbol{C}^{\text{T}}(k)\boldsymbol{U}(k) \times (\boldsymbol{V}^{\text{T}}(k)\boldsymbol{C}^{\text{T}}(k)\boldsymbol{U}(k))^{-1}, \tag{9.33}$$

$$\boldsymbol{C}(k) = \alpha \boldsymbol{C}(k-1) + \boldsymbol{x}(k)\boldsymbol{y}^{\text{T}}(k), \tag{9.34}$$

where $0 < \eta < 1$ denotes the learning rate, and $0 < \alpha \leq 1$ is the forgetting factor. This batch implementation, however, is mainly suitable for the adaptive singular subspace estimation and tracking, where the cross-correlation matrix $\boldsymbol{C}(k)$ is explicitly involved in computations [2]. For online learning of neural networks, it is expected that the network should learn the PSS directly from the input data sequences $\boldsymbol{x}(k)$ and $\boldsymbol{y}(k)$ [2].

Following the projection approximation method of [24], the recursive implementation of the NQC algorithm was derived in [2], and the algorithm can be summarized as follows:

Initializations:

$\boldsymbol{P}(0) = \varepsilon \boldsymbol{I}_r$($\varepsilon$ is a very large positive number)
$\widehat{\boldsymbol{U}}(0) = \boldsymbol{0}$, and $\widehat{\boldsymbol{V}}(0) = \boldsymbol{0}$
$\boldsymbol{U}(0)$ = a random $M \times r$ matrix with very small Frobenius norm
$\boldsymbol{V}(0)$ = a random $N \times r$ matrix with very small Frobenius norm

Update:

$$\begin{cases} \boldsymbol{g}(k) = \boldsymbol{U}^{\mathrm{T}}(k-1)\boldsymbol{x}(k) \\ \boldsymbol{h}(k) = \boldsymbol{V}^{\mathrm{T}}(k-1)\boldsymbol{y}(k) \\ \hat{\boldsymbol{g}}(k) = \boldsymbol{P}(k-1)\boldsymbol{g}(k) \\ \hat{\boldsymbol{h}}(k) = \boldsymbol{P}^{\mathrm{T}}(k-1)\boldsymbol{h}(k) \\ \gamma(k) = \frac{1}{(\alpha + \boldsymbol{h}^{\mathrm{T}}(k)\hat{\boldsymbol{g}}(k))} \\ \boldsymbol{P}(k) = \alpha^{-1}\left[\boldsymbol{P}(k-1) - \gamma(k)\hat{\boldsymbol{g}}(k)\boldsymbol{h}^{\mathrm{T}}(k)\right] \\ \widehat{\boldsymbol{U}}(k) = \widehat{\boldsymbol{U}}(k-1) + \gamma(k)\left[\boldsymbol{x}(k) - \widehat{\boldsymbol{U}}(k)\boldsymbol{g}(k)\right]\hat{\boldsymbol{h}}^{\mathrm{T}}(k) \\ \widehat{\boldsymbol{V}}(k) = \widehat{\boldsymbol{V}}(k-1) + \gamma(k)\left[\boldsymbol{y}(k) - \widehat{\boldsymbol{V}}(k)\boldsymbol{h}(k)\right]\hat{\boldsymbol{g}}^{\mathrm{T}}(k) \\ \boldsymbol{U}(k) = (1-\eta)\boldsymbol{U}(k-1) + \eta\widehat{\boldsymbol{U}}(k) \\ \boldsymbol{V}(k) = (1-\eta)\boldsymbol{V}(k-1) + \eta\widehat{\boldsymbol{V}}(k). \end{cases} \tag{9.35}$$

The above update equation yields an online learning algorithm for two linear neural networks. $\boldsymbol{U}^{\mathrm{T}}(k)$ and $\boldsymbol{V}^{\mathrm{T}}(k)$ denote the weight matrices of these linear neural networks. In [2], the convergence of batch algorithm was proved via the Lyapunov theory.

In [2], it has been proved that the iterative algorithm for computing $\boldsymbol{U}^{\mathrm{T}}(k)\boldsymbol{U}(k)$ and $\boldsymbol{V}^{\mathrm{T}}(k)\boldsymbol{V}(k)$ can be written as

$$\begin{cases} \boldsymbol{U}^{\mathrm{T}}(k)\boldsymbol{U}(k) = (1-2\eta)\boldsymbol{U}^{\mathrm{T}}(k-1)\boldsymbol{U}(k-1) + 2\eta\boldsymbol{I}_r \\ \boldsymbol{V}^{\mathrm{T}}(k)\boldsymbol{V}(k) = (1-2\eta)\boldsymbol{V}^{\mathrm{T}}(k-1)\boldsymbol{V}(k-1) + 2\eta\boldsymbol{I}_r \end{cases}, \tag{9.36}$$

where an appropriate learning rate is $0 < \eta < 0.5$. Since (9.36) is linear, the learning rate can be selected as a fixed constant close to 0.5. For example, $\eta = 0.49$. That is, the learning rate can be selected as a constant independent of the singular value distribution of the underlying matrix, which evidently increases the convergence speed of the CNN.

9.2.5 Coupled SVD of a Cross-Covariance Matrix

It is known that there exists the speed stability problem in most noncoupled learning algorithms. In order to resolve the speed stability problem that plagues most non-coupled learning algorithms, Kaiser et al. proposed a coupled online coupled learning algorithms for the SVD of a cross-covariance matrix [1]. In this algorithm, the singular value is estimated alongside the singular vectors, and the effective learning rates for the singular vector rules are influenced by the singular value estimates.

9.2.5.1 Single-Component Learning Rules

In [1], Kaiser et al. proposed an information criterion, which is designed in a way that its stationary points coincide with the singular triplets of the general rectangular matrix $\boldsymbol{A}$. However, since the stationary points correspond to the saddle points, Newton's method was applied to derive a learning rule to turn even saddle points into attractors which guarantees the stability of the resulting learning rules.

The information criterion is

$$\rho = \sigma^{-1}\boldsymbol{u}^{\mathrm{T}}\boldsymbol{A}\boldsymbol{v} - \frac{1}{2}\boldsymbol{u}^{\mathrm{T}}\boldsymbol{u} - \frac{1}{2}\boldsymbol{v}^{\mathrm{T}}\boldsymbol{v} + \ln\sigma. \tag{9.37}$$

From the gradient of (9.37), it is clear that the stationary points of (9.37) are the singular triplets of A, and $\|\boldsymbol{u}\| = \|\boldsymbol{v}\| = 1$ for these points. Since the Hessian of (9.37) is indefinite for all stationary points, the first principal singular triplet is only a saddle point of (9.37), and hence, gradient-based methods would not converge to this solution. Here, Newton's method has been applied to derive a learning rule which converges toward the principal singular triplet. In [1], the derived individual component equations are as follows:

$$\begin{cases} \dot{\boldsymbol{u}} = \sigma^{-1}(\boldsymbol{A}\boldsymbol{v} - \boldsymbol{u}\boldsymbol{u}^{\mathrm{T}}\boldsymbol{A}\boldsymbol{v}) + \frac{1}{2}(\boldsymbol{u}^{\mathrm{T}}\boldsymbol{u} - 1)\boldsymbol{u} \\ \dot{\boldsymbol{v}} = \sigma^{-1}(\boldsymbol{A}^{\mathrm{T}}\boldsymbol{u} - \boldsymbol{v}\boldsymbol{v}^{\mathrm{T}}\boldsymbol{A}^{\mathrm{T}}\boldsymbol{u}) + \frac{1}{2}(\boldsymbol{v}^{\mathrm{T}}\boldsymbol{v} - 1)\boldsymbol{v} \\ \dot{\sigma} = \boldsymbol{u}^{\mathrm{T}}\boldsymbol{A}\boldsymbol{v} - \frac{1}{2}\sigma(\boldsymbol{u}^{\mathrm{T}}\boldsymbol{u} + \boldsymbol{v}^{\mathrm{T}}\boldsymbol{v}). \end{cases} \tag{9.38}$$

By using a linear stability analysis which is based on the ED of the stability matrix, i.e., the system's Jacobian, evaluated at an equilibrium point, i.e., the kth principal singular triplet $(\bar{\boldsymbol{u}}_k, \bar{\boldsymbol{v}}_k, \bar{\sigma}_k)$, it has been proved that only the first principal singular triplet is stable [1]. Moreover, it has been proved that the system converges with approximately equal speed in all its eigen directions and is widely independent of σ.

If one further approximates $\boldsymbol{u}^{\mathrm{T}}\boldsymbol{u} \approx 1$ and $\boldsymbol{v}^{\mathrm{T}}\boldsymbol{v} \approx 1$ in (9.38), then the approximated system can be obtained

$$\begin{cases} \dot{\boldsymbol{u}} = \sigma^{-1}(\boldsymbol{A}\boldsymbol{v} - \boldsymbol{u}\boldsymbol{u}^{\mathrm{T}}\boldsymbol{A}\boldsymbol{v}) \\ \dot{\boldsymbol{v}} = \sigma^{-1}(\boldsymbol{A}^{\mathrm{T}}\boldsymbol{u} - \boldsymbol{v}\boldsymbol{v}^{\mathrm{T}}\boldsymbol{A}^{\mathrm{T}}\boldsymbol{u}) \\ \dot{\sigma} = \boldsymbol{u}^{\mathrm{T}}\boldsymbol{A}\boldsymbol{v} - \sigma. \end{cases} \tag{9.39}$$

It can be easily shown that the approximated system has the same equilibria as the original system. The convergence properties are widely unchanged; i.e., the first principal singular triplet is the only stable equilibrium, all eigenvalues are widely independent of the singular values, and the convergence speed is approximately the same in all its eigen directions [1].

Ignoring the factor σ^{-1} in (9.39) yields the cross-coupled Hebbian rule in its averaged form

$$\begin{cases} \dot{\boldsymbol{u}} = (\boldsymbol{A}\boldsymbol{v} - \boldsymbol{u}\boldsymbol{u}^{\mathrm{T}}\boldsymbol{A}\boldsymbol{v}) \\ \dot{\boldsymbol{v}} = (\boldsymbol{A}^{\mathrm{T}}\boldsymbol{u} - \boldsymbol{v}\boldsymbol{v}^{\mathrm{T}}\boldsymbol{A}^{\mathrm{T}}\boldsymbol{u}). \end{cases} \tag{9.40}$$

By analyzing the stability of these ODEs, it has been shown that their convergence speed strongly depends on the singular values of $\boldsymbol{A}$.

It is worth noting that for the online learning rule, the learning rates have to be selected. In the cross-coupled Hebbian rule, the convergence speed depends on the principal singular value of $\boldsymbol{A}$, which leads to a problem in the selection of the learning rate. Thus, there exists a typical speed stability dilemma. Obviously, the influence of the singular value estimate (9.39) on the equations of the singular vector estimates potentially improves the convergence properties.

A stochastic online algorithm can be derived from the ODEs (5.39) by formally replacing $\boldsymbol{A}$ by $\boldsymbol{y}_t\boldsymbol{x}_t^{\mathrm{T}}$ and introducing a small learning rate γ_t that decrease to zero as $t \to \infty$. The stochastic online algorithm is as follows:

$$\begin{cases} \boldsymbol{u}_{t+1} = \boldsymbol{u}_t + \gamma_t\sigma_t^{\ 1}(\boldsymbol{y}_t - \eta_t\boldsymbol{u}_t)\xi_t \\ \boldsymbol{v}_{t+1} = \boldsymbol{v}_t + \gamma_t\sigma_t^{-1}(\boldsymbol{x}_t - \eta_t\boldsymbol{v}_t)\eta_t \\ \sigma_{t+1} = \sigma_t + \gamma_t(\eta_t\xi_t - \sigma_t). \end{cases} \tag{9.41}$$

where the auxiliary variables $\xi_t = \boldsymbol{v}_t^{\mathrm{T}}\boldsymbol{x}_t$ and $\eta_t = \boldsymbol{u}_t^{\mathrm{T}}\boldsymbol{y}_t$ have been introduced. If the estimate of σ_t and the factor σ_t^{-1} in (9.41) are omitted, the learning rule (9.41) coincides with the cross-coupled Hebbian rule:

$$\begin{cases} \boldsymbol{u}_{t+1} = \boldsymbol{u}_t + \gamma_t(\boldsymbol{y}_t - \eta_t\boldsymbol{u}_t)\xi_t \\ \boldsymbol{v}_{t+1} = \boldsymbol{v}_t + \gamma_t\sigma_t^{-1}\boldsymbol{x}_t - \eta_t\boldsymbol{v}_t)\eta_t, \end{cases} \tag{9.42}$$

where the singular value estimate is not required in these learning rules.

9.2.5.2 Multiple Component Learning Rules

For the estimation of multiple principal singular triplets, the above single singular component analyzer has to be combined using some decorrelation method. These methods can be interpreted as descendents of the Gram–Schmidt orthonormalization method. In [1], four methods were introduced, i.e., full Gram–Schmidt orthonormalization method, first-order Gram–Schmidt approximation, the deflation, and double deflation. Applying the Gram–Schmidt method would lead to perfectly orthonormal left and right singular vectors. However, the complexity of this method is in the order of mM^2 for the left and nM^2 for the right singular vectors. First-order approximation of Gram–Schmidt orthonormalization reduces the computational effort to the order of mM and nM. Deflation methods have the same order of complexity but approximately halving the number of computation steps.

Here, we take multiple component learning of the coupled SVD rule using first-order approximation of Gram–Schmidt orthonormalization as an example. The algorithms can be summarized as follow:

$$\hat{\boldsymbol{u}}_k = \boldsymbol{u}_k + \gamma\left[\sigma_k^{-1}\left(\boldsymbol{y} - \sum_{j=1}^{k}\eta_j\boldsymbol{u}_j\right)\xi_k - \eta_k\sum_{j=1}^{k-1}\sigma_j^{-1}\xi_j\boldsymbol{u}_j\right], \tag{9.43}$$

$$\hat{\boldsymbol{v}}_k = \boldsymbol{v}_k + \gamma\left[\sigma_k^{-1}\left(\boldsymbol{x} - \sum_{j=1}^{k}\eta_j\boldsymbol{v}_j\right)\xi_k - \eta_k\sum_{j=1}^{k-1}\sigma_j^{-1}\xi_j\boldsymbol{v}_j\right], \tag{9.44}$$

$$\sigma_k'' = \sigma_k + \gamma(\eta_k\xi_k - \sigma_k), \tag{9.45}$$

where $\xi_k = \boldsymbol{v}_k^{\mathrm{T}}\boldsymbol{x}$, $\eta_k = \boldsymbol{u}_k^{\mathrm{T}}\boldsymbol{y}$. The weight vector of the next time step is either the same as this intermediate vector, i.e.,

$$\boldsymbol{u}_k'' = \hat{\boldsymbol{u}}_k, \quad \boldsymbol{v}_k'' = \hat{\boldsymbol{v}}_k, \tag{9.46}$$

or a normalized version:

$$\boldsymbol{u}_k'' = \frac{\hat{\boldsymbol{u}}_k}{\|\hat{\boldsymbol{u}}_k\|}, \quad \boldsymbol{v}_k'' = \frac{\hat{\boldsymbol{v}}_k}{\|\hat{\boldsymbol{v}}_k\|}. \tag{9.47}$$

For $k = 1$, these rules coincide with the coupled SVD rule. For other estimation rules of multiple principal singular triplets, see [1].

9.3 An Effective Neural Learning Algorithm for Extracting Cross-Correlation Feature

In this section, a novel information criterion for PSS tracking will be proposed and a corresponding PSS gradient flow will be derived based on the information criterion. The global asymptotic stability and self-stabilizing property of the PSS gradient flow will be analyzed.

In particular, for a matrix close to singular, it does not converge and the columns of state matrix do not orthogonalize. For a matrix close to singular, the algorithm in [2] can converge. However, it has slower convergence speed, and there appears to be residual deviations from the orthogonality for its columns of state matrix.

The objective of this section is to obtain more effective learning algorithm for extracting cross-correlation feature between two high-dimensional data streams. Firstly, we propose a novel information criterion (NIC) formulation. Secondly, based on it, we derive a PSS tracking gradient flow. Then, the landscape of the information criterion formulation, self-stabilizing property, and the global asymptotical convergence of the PSS gradient flow will be analyzed in detail. Finally, simulation experiments are carried to testify the effectiveness of algorithms.

9.3.1 Preliminaries

9.3.1.1 Definitions and Properties

Definition 9.1 Given an $r \times r$ matrix $\boldsymbol{B}$, then its EVD is represented as $\boldsymbol{B} = \boldsymbol{\Phi\Psi\Phi}^{-1}$, where $\boldsymbol{\Phi}$ denotes an $r \times r$ matrix formed by all its eigenvectors, and $\boldsymbol{\Psi} = \mathrm{diag}(\lambda_1, \ldots, \lambda_r) > 0$ is a diagonal matrix formed by all its eigenvalues.

Property 9.1 *The trace of a matrix* $\boldsymbol{B}$ *can be computed by* $\mathrm{tr}(\boldsymbol{B}) = \sum_{i=1}^{r} \lambda_i$.

Property 9.2 *If* $\boldsymbol{A}$ *is an* $m \times n$ *matrix and* $\boldsymbol{B}$ *is an* $n \times m$ *matrix, then it holds that* $\mathrm{tr}(\boldsymbol{AB}) = \mathrm{tr}(\boldsymbol{BA})$.

Property 9.3 *Let* $\widetilde{\boldsymbol{U}}$ *and* $\widetilde{\boldsymbol{V}} \in \Re^{N \times r}$ *be two different matrices satisfying* span $(\widetilde{\boldsymbol{U}})$ = span $(\widetilde{\boldsymbol{V}})$ *for* $r \leq N$, *then there always exists an* $r \times r$ *rank-full matrix* $\boldsymbol{B}$ *such that* $\widetilde{\boldsymbol{U}} = \widetilde{\boldsymbol{V}\boldsymbol{B}}$.

Property 9.4 *Given two different matrices* $\boldsymbol{U} = [\tilde{\boldsymbol{u}}_1, \ldots, \tilde{\boldsymbol{u}}_r]$ *and* $\boldsymbol{V} = [\tilde{\boldsymbol{v}}_1, \ldots, \tilde{\boldsymbol{v}}_r]$ $\in \Re^{N \times r}$, *if* $N > r$ *and rank* $(\widetilde{\boldsymbol{U}}^{\mathrm{T}}\widetilde{\boldsymbol{V}}) = r$, *then* span$(\widetilde{\boldsymbol{U}})$ = *span* $(\widetilde{\boldsymbol{V}})$.

Property 9.5 *Known* $\boldsymbol{P}_r, \boldsymbol{\Lambda}$ *and* $\hat{j}_i$, *then there is* $\boldsymbol{\Lambda P}_r = \boldsymbol{P}_r\widehat{\boldsymbol{\Lambda}}$, *where* $\boldsymbol{P}_r$ *is an* $N \times r$ *permutation matrix in which each column has exactly one nonzero element equal to 1 and each row has, at most, one nonzero element* ($N \geq r$), $\boldsymbol{\Lambda}$ *is a* $N \times N$ *diagonal matrix given by* $\mathrm{diag}(\lambda_1, \ldots, \lambda_N)$, $\hat{j}_i$ *is an integer such that the permutation matrix* $\boldsymbol{P}_r$ *has exactly the nonzero entry equal to 1 in row* $\hat{j}_i$ *and column i, and* $\widehat{\boldsymbol{\Lambda}} = \mathrm{diag}(\lambda_{\hat{j}_1}, \ldots, \lambda_{\hat{j}_r})$.

About the details of these property and definition mentioned above, we can see [2, 21, 25].

9.3.1.2 Some Formulations Relative to PSS

Consider an M-dimensional sequence $\boldsymbol{x}(t)$ and an N-dimensional sequence $\boldsymbol{y}(t)$ with the sample size k large enough. Without loss of generality, let $M \geq N$. If $\boldsymbol{x}(t)$ and $\boldsymbol{y}(t)$ are jointly stationary, their cross-correlation matrix [2, 26] can be estimated by

$$\boldsymbol{C}(k) = \frac{1}{k}\sum_{j=1}^{k} \boldsymbol{x}(j)\boldsymbol{y}^{\mathrm{T}}(j) \in \Re^{M\times N}, \tag{9.48}$$

And if $\boldsymbol{x}(t)$ and $\boldsymbol{y}(t)$ are jointly nonstationary and slowly time-varying, then their cross-correlation matrix can be estimated by

$$\boldsymbol{C}(k) = \sum_{j=1}^{k} \alpha^{k-j}\boldsymbol{x}(j)\boldsymbol{y}^{\mathrm{T}}(j) \in \Re^{M\times N}, \tag{9.49}$$

where $0<\alpha<1$ denotes the forgetting factor which makes the past data samples less weighted than the most recent ones. The exact value for α depends on specific applications. Generally speaking, for slowly time-varying system, α is chosen close to one, whereas for fast time-varying system, α should be chosen close to zero [2, 26].

Let $\sigma_i, \boldsymbol{l}_i$ and $r_i, i = 1,\ldots,N$ denote the singular values, the corresponding left and right singular vectors of $\boldsymbol{C}$, respectively. We shall arrange the orthonormal singular vectors $\boldsymbol{l}_1, \boldsymbol{l}_2, \ldots, \boldsymbol{l}_M$ and $\boldsymbol{r}_1, \boldsymbol{r}_2, \ldots, \boldsymbol{r}_N$ such that the associated singular values are in nonincreasing order, i.e., $\sigma_1 \geq \sigma_2 \geq \cdots \geq \sigma_N \geq 0$. Note that since these left singular vectors $\boldsymbol{l}_{N+1}, \boldsymbol{l}_{N+2}, \ldots, \boldsymbol{l}_M$ are associated with the null subspace of $\mathbf{C}$, we shall not consider them. Let $\boldsymbol{\Lambda} = \mathrm{diag}(\sigma_1, \ldots, \sigma_N)$, $\boldsymbol{L} = [\boldsymbol{l}_1, \ldots, \boldsymbol{l}_N]$, and $\boldsymbol{R} = [\boldsymbol{r}_1, \ldots, \boldsymbol{r}_N]$. Then, the SVD of $\mathbf{C}$ is described by $\boldsymbol{C} = \sum_{i=1}^{N} \sigma_i \boldsymbol{l}_i \boldsymbol{r}_i^{\mathrm{T}} = \boldsymbol{L}\boldsymbol{\Lambda}\boldsymbol{R}^{\mathrm{T}}$. Usually, all $\boldsymbol{l}_1, \boldsymbol{l}_2, \ldots, \boldsymbol{l}_r$ and $\boldsymbol{r}_1, \boldsymbol{r}_2, \ldots, \boldsymbol{r}_r$ are called the PSC, and $\boldsymbol{L}_{ps} = [\boldsymbol{l}_1, \ldots, \boldsymbol{l}_r]$ and $\boldsymbol{R}_{ps} = [\boldsymbol{r}_1, \ldots, \boldsymbol{r}_r]$ are called the left and right singular vector matrices associated with the signal, respectively. The associated principal singular values can construct a diagonal matrix $\overline{\boldsymbol{\Lambda}} = \mathrm{diag}(\sigma_1, \ldots, \sigma_r)$, where r denotes the number of the PSCs. An efficient estimation can be achieved by Akaike information criterion [27] based on the distribution of the singular values. In some applications [28, 29], we are required only to find a PSS spanned by $\boldsymbol{l}_1, \ldots, \boldsymbol{l}_r$ or $\boldsymbol{r}_1, \ldots, \boldsymbol{r}_r$, given by $\boldsymbol{L}_{ps}\boldsymbol{Q}_1$ or $\boldsymbol{R}_{ps}\boldsymbol{Q}_2$, where $\boldsymbol{Q}_1, \boldsymbol{Q}_2$ is $r \times r$ orthogonal matrices.

Consider the following two linear transformations:

$$\boldsymbol{u}(k) = \boldsymbol{U}^{\mathrm{T}}\boldsymbol{x}(k) \in \Re^{r\times 1}, \tag{9.50}$$

$$\boldsymbol{v}(k) = \boldsymbol{V}^{\mathrm{T}}\boldsymbol{y}(k) \in \Re^{r\times 1}, \tag{9.51}$$

where $\boldsymbol{U} \in \Re^{M\times r}$ and $\boldsymbol{V} \in \Re^{N\times r}$ denote the optimal weight matrices whose columns span the same space as $\boldsymbol{L}_{ps}$ and $\boldsymbol{R}_{ps}$, respectively; $\boldsymbol{u}(k)$ and $\boldsymbol{v}(k)$ are the low-dimensional representation of $\boldsymbol{x}(t)$ and $\boldsymbol{y}(k)$, respectively. If $\boldsymbol{U} = \boldsymbol{L}_{ps}$ and

$V = R_{ps}$, then $u(k)$ and $v(k)$ are PCs of $x(t)$ and $y(k)$. In [2], the relation between the PSC and PSS is given. For the convenience of analysis, one definition of the matrix equivalency $\widetilde{U} = \widetilde{VP}$ was provided in [2], where $\widetilde{U}$ and $\widetilde{V} \in \Re^{N\times r}$ are matrices, and P is a $r \times r$ permutation matrix. Another definition of the matrix equivalency $\widetilde{U} = \widetilde{VQ}$ was also provided in [2], where $\widetilde{U}$ and $\widetilde{V} \in \Re^{N\times r}$ are two column-orthonormal matrices, and Q is a $r \times r$ orthonormal matrix. For detail, see [2]. In this part, we only consider those points $(U, V)(\doteqdot(L_{ps}, R_{ps}))$ satisfying $\|U\| \neq 0$ and $\|V\| \neq 0$.

9.3.2 *Novel Information Criterion Formulation for PSS*

In this part, we will propose a NIC, based on which we can derive a PSS tracking algorithm.

9.3.2.1 Novel Information Criterion Formulation for PSS

Given $U \in \Re^{M\times r}$ and $V \in \Re^{N\times r}$ in the domain $\{(U, V)| \|U\|_F \neq 0, \|V\|_F \neq 0\}$, we present a nonquadratic criterion (NQC) for PSS as follows:

$$\min_{U,V} J_{\mathrm{NQC}}(U, V)$$

$$J_{\mathrm{NQC}}(U, V) = -\mathrm{tr}\Big\{(U^{\mathrm{T}}CV)(\|U\|_F\|V\|_F)^{-1}\Big\} + \frac{1}{2}\mathrm{tr}\{[I - U^{\mathrm{T}}U]^2\} + \frac{1}{2}\mathrm{tr}\{[I - V^{\mathrm{T}}V]^2\}. \tag{9.52}$$

It is worth noting that this criterion is referred as novel because it is different from all existing PSS criteria [2, 9, 10, 12, 13], etc., or all existing PSA criteria [24, 30], etc.

From (5), we can see that $J_{\mathrm{NQC}}(U, V)$ has a lower bound and approaches infinity from the above as $\mathrm{tr}(U^{\mathrm{T}}U) \to \infty$ or (and) $\mathrm{tr}(V^{\mathrm{T}}V) \to \infty$. Obviously, the gradient searching algorithm can be derived based on the above NQC, which will be discussed in the latter section.

If U and V are expanded by the left and right singular vector bases into

$$U = L^{\mathrm{T}}\widetilde{U} \quad V = R^{\mathrm{T}}\widetilde{V}, \tag{9.53}$$

respectively, then we can find the NQC for the expanded coefficient matrices

$$\min_{\widetilde{U},\widetilde{V}} \tilde{J}_{\mathrm{NQC}}(\widetilde{U},\widetilde{V})$$
$$\tilde{J}_{\mathrm{NQC}}(\widetilde{U},\widetilde{V}) = -\mathrm{tr}\left\{\left(\widetilde{U}^{\mathrm{T}}\Lambda\widetilde{V}\right)\left(\left\|\widetilde{U}\right\|_F\left\|\widetilde{V}\right\|_F\right)^{-1}\right\} + \frac{1}{2}\mathrm{tr}\{[I - \widetilde{U}^{\mathrm{T}}\widetilde{U}]^2\} + \frac{1}{2}\mathrm{tr}\{[I - \widetilde{V}^{\mathrm{T}}\widetilde{V}]^2\}, \tag{9.54}$$

where $\widetilde{U}$ and $\widetilde{V} \in \Re^{N\times r}$ are two expanded coefficient matrices. Obviously, (9.54) represents an equivalent form of (9.52). The landscape of this novel criterion is depicted by the following two theorems. Since the matrix differential will be used extensively, interested readers can refer to [31] for more detail.

9.3.2.2 Landscape of Nonquadratic Criterion

The landscape of (9.52) is depicted by the following two theorems.

Theorem 9.3 *(U, V) is a stationary point of $J_{\mathrm{NQC}}(U, V)$ in the domain $\{(U, V)|\|U\|_F \neq 0, \|V\|_F \neq 0\}$ if and only if $U \doteq L_r$ and $V \doteq R_r$, where $L_r \in \Re^{M\times r}$ and $\mathbf{R}_r \in \Re^{N\times r}$ consist of the r left and right singular vectors of C, respectively.*

Note that $(U, V) \doteq (L_r, R_r)$ shows a stationary set of $J_{\mathrm{NQC}}(U, V)$.

It can be seen that Theorem 9.3 is equivalent to the following Corollary 9.1. So we will only provide the proof of Corollary 9.1.

Corollary 9.1 *$(\widetilde{U}, \widetilde{V})$ is a stationary point of $\tilde{J}_{\mathrm{NQC}}(\widetilde{U}, \widetilde{V})$ in the domain $\{(\widetilde{U}, \widetilde{V})|\left\|\widetilde{U}\right\|_F \neq 0, \left\|\widetilde{V}\right\|_F \neq 0\}$ if and only if $\widetilde{U} \doteq P_r$ and $\widetilde{V} \doteq P_r$, where $P_r \in \Re^{N\times r}$ is a permutation matrix and consists of r eigenvectors of $\mathbf{\Lambda}$, respectively.*

Proof The gradient of $\tilde{J}_{\mathrm{NQC}}(\widetilde{U}, \widetilde{V})$ for PSS tracking with respect to $\widetilde{U}$ and $\widetilde{V}$ exists and is given by

$$\nabla_{\widetilde{U}}\tilde{J}_{\mathrm{NQC}}(\widetilde{U},\widetilde{V}) = -\left\{\Lambda\widetilde{V}\left(\left\|\widetilde{U}\right\|\left\|\widetilde{V}\right\|\right)^{-1} - \widetilde{U}(\widetilde{U}^{\mathrm{T}}\Lambda\widetilde{V})\left(\left\|\widetilde{U}\right\|^3\left\|\widetilde{V}\right\|\right)^{-1}\right\} - \widetilde{U}[I - \widetilde{U}^{\mathrm{T}}\widetilde{U}], \tag{9.55}$$

$$\nabla_{\widetilde{V}}\tilde{J}_{\mathrm{NQC}}(\widetilde{U},\widetilde{V}) = -\left\{\Lambda^{\mathrm{T}}V(\|U\|\|V\|)^{-1} - V(U^{\mathrm{T}}\Lambda V)(\|U\|\|V\|)^3\right\}^{-1} V[I - V^{\mathrm{T}}V], \tag{9.56}$$

where $\nabla_{\widetilde{U}}$ and $\nabla_{\widetilde{U}}$ denote $\partial\tilde{J}_{\mathrm{NQC}}/\partial\widetilde{U}$ and $\partial\tilde{J}_{\mathrm{NQC}}/\partial\widetilde{V}$, respectively.

Given a point in $(\boldsymbol{P}_r\boldsymbol{Q}_1, \boldsymbol{P}_r\boldsymbol{Q}_2)$ in $(\tilde{\boldsymbol{U}}, \tilde{\boldsymbol{V}}) \ddot{=} (\boldsymbol{P}_r, \boldsymbol{P}_r)$, then we have

$$\begin{aligned}\nabla_{\tilde{U}} \tilde{J}_{\text{NQC}}(\boldsymbol{P}_r\boldsymbol{Q}_1, \boldsymbol{P}_r\boldsymbol{Q}_2) &= -\{\boldsymbol{\Lambda}\boldsymbol{P}_r\boldsymbol{Q}_2(\|\boldsymbol{P}_r\boldsymbol{Q}_1\|\|\boldsymbol{P}_r\boldsymbol{Q}_2\|)^{-1} \\ &\quad - \boldsymbol{P}_r\boldsymbol{Q}_1(\boldsymbol{Q}_1^{\mathrm{T}}\boldsymbol{P}_r^{\mathrm{T}}\boldsymbol{\Lambda}\boldsymbol{P}_r\boldsymbol{Q}_2)(\|\boldsymbol{P}_r\boldsymbol{Q}_1\|^3\|\boldsymbol{P}_r\boldsymbol{Q}_2\|)^{-1}\} \\ &\quad - \boldsymbol{P}_r\boldsymbol{Q}_1[\boldsymbol{I} - \boldsymbol{Q}_1^{\mathrm{T}}\boldsymbol{P}_r^{\mathrm{T}}\boldsymbol{P}_r\boldsymbol{Q}_1] = -\{\boldsymbol{\Lambda}\boldsymbol{P}_r\boldsymbol{Q}_2 - \boldsymbol{\Lambda}\boldsymbol{P}_r\boldsymbol{Q}_2\} = 0.\end{aligned} \tag{9.57}$$

Similarly, we can get the following equation:

$$\nabla_{\tilde{V}} \tilde{J}_{\text{NQC}}(\boldsymbol{P}_r\boldsymbol{Q}_1, \boldsymbol{P}_r\boldsymbol{Q}_2) = 0. \tag{9.58}$$

Conversely, $\tilde{J}_{\text{NQC}}(\tilde{\boldsymbol{U}}, \tilde{\boldsymbol{V}})$ for PSS tracking at a stationary point should satisfy $\nabla_{\tilde{U}} \tilde{J}_{\text{NQC}}(\tilde{\boldsymbol{U}}, \tilde{\boldsymbol{V}}) = 0$ and $\nabla_{\tilde{V}} \tilde{J}_{\text{NQC}}(\tilde{\boldsymbol{U}}, \tilde{\boldsymbol{V}}) = 0$, which yields

$$\tilde{\boldsymbol{U}}(\tilde{\boldsymbol{U}}^{\mathrm{T}}\boldsymbol{\Lambda}\tilde{\boldsymbol{V}})\left(\left\|\tilde{\boldsymbol{U}}\right\|^3\left\|\tilde{\boldsymbol{V}}\right\|\right)^{-1} = \boldsymbol{\Lambda}\tilde{\boldsymbol{V}}\left(\left\|\tilde{\boldsymbol{U}}\right\|\left\|\tilde{\boldsymbol{V}}\right\|\right)^{-1} + \tilde{\boldsymbol{U}}[\boldsymbol{I} - \tilde{\boldsymbol{U}}^{\mathrm{T}}\tilde{\boldsymbol{U}}], \tag{9.59}$$

$$\tilde{\boldsymbol{V}}(\tilde{\boldsymbol{U}}^{\mathrm{T}}\boldsymbol{\Lambda}\tilde{\boldsymbol{V}})\left(\left\|\tilde{\boldsymbol{U}}\right\|\left\|\tilde{\boldsymbol{V}}\right\|^3\right)^{-1} = \boldsymbol{\Lambda}^{\mathrm{T}}\tilde{\boldsymbol{V}}\left(\left\|\tilde{\boldsymbol{U}}\right\|\left\|\tilde{\boldsymbol{V}}\right\|\right)^{-1} + \tilde{\boldsymbol{V}}[\boldsymbol{I} - \tilde{\boldsymbol{V}}^{\mathrm{T}}\tilde{\boldsymbol{V}}]. \tag{9.60}$$

Premultiplying both sides of (9.59) and (9.60) by $\tilde{\boldsymbol{U}}^{\mathrm{T}}$ and $\tilde{\boldsymbol{V}}^{\mathrm{T}}$, respectively, we have

$$\tilde{\boldsymbol{U}}^{\mathrm{T}}\tilde{\boldsymbol{U}} = \boldsymbol{I}_r, \tag{9.61}$$

$$\tilde{\boldsymbol{V}}^{\mathrm{T}}\tilde{\boldsymbol{V}} = \boldsymbol{I}_r, \tag{9.62}$$

which implies that the columns of $\tilde{\boldsymbol{U}}$ and $\tilde{\boldsymbol{V}} \in R^{N \times r}$ are column-orthonormal at a stationary point of $\tilde{J}_{\text{NQC}}(\tilde{\boldsymbol{U}}, \tilde{\boldsymbol{V}})$ for PSS tracking.

From (9.61), we have

$$\text{rank}(\tilde{\boldsymbol{U}}^{\mathrm{T}}\boldsymbol{\Lambda}\tilde{\boldsymbol{U}}) = r. \tag{9.63}$$

Moreover, premultiplying both sizes of (9.60) by $\tilde{\boldsymbol{U}}^{\mathrm{T}}$, we have

$$\text{rank}\left(\tilde{\boldsymbol{U}}^{\mathrm{T}}\tilde{\boldsymbol{V}}(\tilde{\boldsymbol{U}}^{\mathrm{T}}\boldsymbol{\Lambda}\tilde{\boldsymbol{V}})\left(\left\|\tilde{\boldsymbol{U}}\right\|\left\|\tilde{\boldsymbol{V}}\right\|^3\right)^{-1}\right) = \text{rank}\left\{\tilde{\boldsymbol{U}}^{\mathrm{T}}\boldsymbol{\Lambda}^{\mathrm{T}}\tilde{\boldsymbol{V}}\left(\left\|\tilde{\boldsymbol{U}}\right\|\left\|\tilde{\boldsymbol{V}}\right\|\right)^{-1} + \tilde{\boldsymbol{U}}^{\mathrm{T}}\tilde{\boldsymbol{V}}[\boldsymbol{I} - \tilde{\boldsymbol{V}}^{\mathrm{T}}\tilde{\boldsymbol{V}}]\right\} = r. \tag{9.64}$$

From (9.64), it follows that

$$\operatorname{rank}(\widetilde{\boldsymbol{U}}^{\mathrm{T}}\widetilde{\boldsymbol{V}}(\widetilde{\boldsymbol{U}}^{\mathrm{T}}\boldsymbol{\Lambda}\widetilde{\boldsymbol{V}})) = \operatorname{rank}(\widetilde{\boldsymbol{U}}^{\mathrm{T}}\boldsymbol{\Lambda}^{\mathrm{T}}\widetilde{\boldsymbol{V}}) = \operatorname{rank}(\widetilde{\boldsymbol{U}}^{\mathrm{T}}\widetilde{\boldsymbol{V}}) = r. \tag{9.65}$$

From Property (2.4) and (9.65), we have

$$\operatorname{span}(\widetilde{\boldsymbol{U}}) = \operatorname{span}(\widetilde{\boldsymbol{V}}) = \operatorname{span}(\boldsymbol{\Lambda}\widetilde{\boldsymbol{U}}) = \operatorname{span}(\boldsymbol{\Lambda}\widetilde{\boldsymbol{V}}). \tag{9.66}$$

From Property (2.3), it follows that

$$\widetilde{\boldsymbol{U}} = \widetilde{\boldsymbol{V}}\boldsymbol{Q}, \text{ i.e. } \widetilde{\boldsymbol{U}} \doteq \widetilde{\boldsymbol{V}}. \tag{9.67}$$

Substituting (9.67) into (9.59) and from $\tilde{\boldsymbol{U}}\tilde{\boldsymbol{U}}^{\mathrm{T}} = \tilde{\boldsymbol{V}}\tilde{\boldsymbol{V}}^{\mathrm{T}}$, we have

$$\boldsymbol{\Lambda}\tilde{\boldsymbol{V}} = \tilde{\boldsymbol{V}}\tilde{\boldsymbol{V}}^{\mathrm{T}}\boldsymbol{\Lambda}\tilde{\boldsymbol{V}}. \tag{9.68}$$

Let $\widetilde{\boldsymbol{V}} = [\breve{\boldsymbol{v}}_1^{\mathrm{T}}, \ldots, \breve{\boldsymbol{v}}_N^{\mathrm{T}}]^{\mathrm{T}}$, where $\breve{\boldsymbol{v}}_i (i = 1, \ldots, N)$ is a row vector, and take $\boldsymbol{B} = \widetilde{\boldsymbol{V}}^{\mathrm{T}}\boldsymbol{\Lambda}\tilde{\boldsymbol{V}}$ that is a $r \times r$ symmetric positive definite matrix. Then, an alternative form of (9.68) is:

$$\sigma_i \breve{\boldsymbol{v}}_i = \breve{\boldsymbol{v}}_i \boldsymbol{B} (i = 1, 2, \ldots, N). \tag{9.69}$$

Obviously, (9.69) is the EVD of $\boldsymbol{B}$. Since $\boldsymbol{B}$ is a $r \times r$ symmetric positive definite matrix, it has only r nonzero orthonormal left eigenvectors, i.e., $\breve{\boldsymbol{V}}$ has only r nonzero orthonormal row vectors. Moreover, all the r nonzero row vectors in $\breve{\boldsymbol{V}}$ form an orthonormal matrix, which shows that $\widetilde{\boldsymbol{V}}$ can always be represented as

$$\widetilde{\boldsymbol{V}} = \breve{\boldsymbol{P}}_r \boldsymbol{Q}_2, \tag{9.70}$$

i.e.,

$$\widetilde{\boldsymbol{V}} \doteq \breve{\boldsymbol{P}}_r. \tag{9.71}$$

Similarly, we can get

$$\widetilde{\boldsymbol{U}} \doteq \breve{\boldsymbol{P}}_r. \tag{9.72}$$

Since in the domain $\{(\widetilde{\boldsymbol{U}}, \widetilde{\boldsymbol{V}}) | \widetilde{\boldsymbol{U}}^{\mathrm{T}}\boldsymbol{\Lambda}\widetilde{\boldsymbol{V}} > 0\}$,

$$\operatorname{rank}(\widetilde{\boldsymbol{U}}^{\mathrm{T}}\boldsymbol{\Lambda}\widetilde{\boldsymbol{V}}) = \operatorname{rank}(\boldsymbol{Q}_1^{\mathrm{T}}\widehat{\boldsymbol{P}}_r^{\mathrm{T}}\boldsymbol{\Lambda}\breve{\boldsymbol{P}}_r\boldsymbol{Q}_2) = r. \tag{9.73}$$

From Property (2.5), we have

$$\operatorname{rank}(\widehat{\boldsymbol{P}}_r^{\mathrm{T}}\boldsymbol{\Lambda}\breve{\boldsymbol{P}}_r) = \operatorname{rank}(\widehat{\boldsymbol{P}}_r^{\mathrm{T}}\breve{\boldsymbol{P}}_r\breve{\boldsymbol{\Lambda}}) = r, \tag{9.74}$$

where the $r \times r$ diagonal matrix $\breve{\boldsymbol{\Lambda}}$ is similar to $\widehat{\boldsymbol{\Lambda}}$ in Property (9.5).

This means that

$$\operatorname{rank}(\widehat{\boldsymbol{P}}_r^{\mathrm{T}}\breve{\boldsymbol{P}}_r) = r, \tag{9.75}$$

or equivalently

$$\widehat{\boldsymbol{P}}_r \dot{=} \breve{\boldsymbol{P}}_r. \tag{9.76}$$

Thus, (9.71) and (9.72) can always be rewritten as

$$(\widetilde{\boldsymbol{U}}, \widetilde{\boldsymbol{V}}) \ddot{=} (\boldsymbol{P}_r, \boldsymbol{P}_r). \tag{9.77}$$

This completes the proof.

Clearly, $(\widetilde{\boldsymbol{U}}, \widetilde{\boldsymbol{V}}) \ddot{=} (\boldsymbol{P}_r, \boldsymbol{P}_r)$ shows a stationary set of $\tilde{J}_{\mathrm{NQC}}(\widetilde{\boldsymbol{U}}, \widetilde{\boldsymbol{V}})$.

Theorem (9.3) establishes a property for all stationary points of $J_{\mathrm{NQC}}(\boldsymbol{U}, \boldsymbol{V})$. The next theorem further distinguishes the global minimizer set obtained by spanning the PSS from the other stationary points that are saddle (unstable) points.

Theorem 9.4 *In the domain* $\{(\boldsymbol{U}, \boldsymbol{V}) | \|\boldsymbol{U}\|_F \neq 0, \|\boldsymbol{V}\|_F \neq 0\}$, $J_{\mathrm{NQC}}(\boldsymbol{U}, \boldsymbol{V})$ *has a global minimum that is achieved if and only if* $\boldsymbol{U} \ddot{=} \boldsymbol{L}_{ps}$ *and* $\boldsymbol{V} \ddot{=} \boldsymbol{R}_{ps}$. *And the global minimum is* $J_{\mathrm{NQC}}(\boldsymbol{U}, \boldsymbol{V}) = -\sum_{i=1}^{r} \sigma_i$. *All the other stationary points* $(\boldsymbol{U}, \boldsymbol{V}) \neq (\boldsymbol{L}_{ps}, \boldsymbol{R}_{ps})$ *are saddle (unstable) points of* $J_{\mathrm{NQC}}(\boldsymbol{U}, \boldsymbol{V})$.

In fact, $(\boldsymbol{U}, \boldsymbol{V}) \ddot{=} (\boldsymbol{L}_{ps}, \boldsymbol{R}_{ps})$ shows a global minimizer set of $J_{\mathrm{NQC}}(\boldsymbol{U}, \boldsymbol{V})$.

It can also be seen that Theorem 9.4 is equivalent to the following Corollary 9.2. So we will only provide the proof of Corollary 9.2.

Corollary 9.2 *In the domain* $\{(\widetilde{\boldsymbol{U}}, \widetilde{\boldsymbol{V}}) | \left\|\widetilde{\boldsymbol{U}}\right\|_F \neq 0, \left\|\widetilde{\boldsymbol{V}}\right\|_F \neq 0\}$, $\tilde{J}_{\mathrm{NQC}}(\widetilde{\boldsymbol{U}}, \widetilde{\boldsymbol{V}})$ *has a global minimum that is achieved if and only if* $\widetilde{\boldsymbol{U}} \ddot{=} \overline{\boldsymbol{P}}$ *and* $\widetilde{\boldsymbol{V}} \ddot{=} \overline{\boldsymbol{P}}$, *where* $\overline{\boldsymbol{P}} \dot{=} \begin{pmatrix} \boldsymbol{I}_r \\ \boldsymbol{0} \end{pmatrix}$, $\boldsymbol{I}_r$ *is a* $r \times r$ *identity matrix. And the global minimum is* $\tilde{J}_{\mathrm{NQC}}(\widetilde{\boldsymbol{U}}, \widetilde{\boldsymbol{V}}) = -\sum_{i=1}^{r} \sigma_i$. *All the other stationary points* $(\widetilde{\boldsymbol{U}}, \widetilde{\boldsymbol{V}})(\neq (\overline{\boldsymbol{P}}, \overline{\boldsymbol{P}}))$ *are saddle (unstable) points of* $\tilde{J}_{\mathrm{NQC}}(\widetilde{\boldsymbol{U}}, \widetilde{\boldsymbol{V}})$.

Proof Since $\tilde{J}_{\mathrm{NQC}}(\widetilde{\boldsymbol{U}}, \widetilde{\boldsymbol{V}})$ for PSS tracking is bounded from below and unbounded from above as $\operatorname{tr}(\widetilde{\boldsymbol{U}}^{\mathrm{T}}\widetilde{\boldsymbol{U}}) \to \infty$ and/or $\operatorname{tr}(\widetilde{\boldsymbol{V}}^{\mathrm{T}}\widetilde{\boldsymbol{V}}) \to \infty$, the global minimum can only be achieved by a stationary point of $\tilde{J}_{\mathrm{NQC}}(\widetilde{\boldsymbol{U}}, \widetilde{\boldsymbol{V}})$. By computing $\tilde{J}_{\mathrm{NQC}}(\widetilde{\boldsymbol{U}}, \widetilde{\boldsymbol{V}})$ for

PSS tracking in the stationary point set for the domain $\{(\widetilde{\boldsymbol{U}},\widetilde{\boldsymbol{V}})|\widetilde{\boldsymbol{U}}^{\mathrm{T}}\widetilde{\boldsymbol{U}}\neq 0, \widetilde{\boldsymbol{V}}^{\mathrm{T}}\widetilde{\boldsymbol{V}}\neq 0\}$, we can directly verify that a global minimum of $\tilde{J}_{\mathrm{NQC}}(\widetilde{\boldsymbol{U}},\widetilde{\boldsymbol{V}})$ for PSS tracking is achieved if and only if

$$(\widetilde{\boldsymbol{U}},\widetilde{\boldsymbol{V}})\in(\overline{\boldsymbol{P}}\boldsymbol{Q}_1,\overline{\boldsymbol{P}}\boldsymbol{Q}_2). \tag{9.78}$$

By substituting (9.78) into (9.54) for PSS and performing some algebraic manipulations, we can get the global minimum of $\tilde{J}_{\mathrm{NQC}}(\widetilde{\boldsymbol{U}},\widetilde{\boldsymbol{V}})$ for PSS tracking as

$$\tilde{J}_{\mathrm{NQC}}(\widetilde{\boldsymbol{U}},\widetilde{\boldsymbol{V}})=-\sum\nolimits_{i=1}^{r}\sigma_i. \tag{9.79}$$

Moreover, we can determine a stationary point of $\tilde{J}_{\mathrm{NQC}}(\widetilde{\boldsymbol{U}},\widetilde{\boldsymbol{V}})$ for PSS tracking as saddle (unstable) in such a way that within an infinitesimal neighborhood near the stationary point, there is a point $(\boldsymbol{U}',\boldsymbol{V}')$ such that its value $\tilde{J}_{\mathrm{NQC}}(\boldsymbol{U}',\boldsymbol{V}')$ for PSS tracking is less than $\tilde{J}_{\mathrm{NQC}}(\widetilde{\boldsymbol{U}},\widetilde{\boldsymbol{V}})$ for PSS tracking.

Let $\boldsymbol{P}_r\neq\overline{\boldsymbol{P}}$. Then, there, at least, exists a nonzero element in the row vectors from $r+1$ to N for $\boldsymbol{P}_r$. Since $\overline{\boldsymbol{P}}$ and $\boldsymbol{P}_r$ are two permutation matrices, from Property 9.5, there exist certainly two diagonal matrices $\overline{\boldsymbol{\Lambda}}$ and $\widehat{\boldsymbol{\Lambda}}$ such that $\overline{\boldsymbol{P}}^{\mathrm{T}}\boldsymbol{\Lambda}\overline{\boldsymbol{P}}=\overline{\boldsymbol{P}}^{\mathrm{T}}\overline{\boldsymbol{P}}\,\overline{\boldsymbol{\Lambda}}$ and $\boldsymbol{P}_r^{\mathrm{T}}\boldsymbol{\Lambda}\boldsymbol{P}_r=\mathbf{P}_r^{\mathrm{T}}\mathbf{P}_r\widehat{\boldsymbol{\Lambda}}$. This yields

$$\overline{\boldsymbol{P}}^{\mathrm{T}}\boldsymbol{\Lambda}\overline{\boldsymbol{P}}=\overline{\boldsymbol{\Lambda}}, \tag{9.80}$$

$$\boldsymbol{P}_r^{\mathrm{T}}\boldsymbol{\Lambda}\boldsymbol{P}_r=\widehat{\boldsymbol{\Lambda}}. \tag{9.81}$$

Thus, it holds that

$$\mathrm{tr}(\overline{\boldsymbol{P}}^{\mathrm{T}}\boldsymbol{\Lambda}\overline{\boldsymbol{P}})=\sum_{i=1}^{r}\sigma_i, \tag{9.82}$$

$$\mathrm{tr}(\boldsymbol{P}_r^{\mathrm{T}}\boldsymbol{\Lambda}\boldsymbol{P}_r)=\sum_{i=1}^{r}\sigma_{\hat{j}_i}. \tag{9.83}$$

If $\sigma_{\hat{j}_i}(i=1,\ldots,r)$ are rearranged in a nonincreasing order, i.e., $\tilde{\sigma}_1\geq\tilde{\sigma}_2\geq\cdots\geq\tilde{\sigma}_r$, then there exist $\sigma_i\geq\tilde{\sigma}_i(i=2,\ldots,r)$ and $\sigma_1>\tilde{\sigma}_1$ for $\boldsymbol{P}_r\neq\overline{\boldsymbol{P}}$. This means that

$$\mathrm{tr}(\overline{\boldsymbol{P}}^{\mathrm{T}}\boldsymbol{\Lambda}\overline{\boldsymbol{P}})>\mathrm{tr}(\boldsymbol{P}_r^{\mathrm{T}}\boldsymbol{\Lambda}\boldsymbol{P}_r). \tag{9.84}$$

Since

$$\begin{aligned}\tilde{J}_{\text{NQC}}(\boldsymbol{P}_r\boldsymbol{Q}_1,\boldsymbol{P}_r\boldsymbol{Q}_2) &= -\text{tr}\Big\{(\boldsymbol{Q}_1^{\text{T}}\boldsymbol{P}_r^{\text{T}}\boldsymbol{\Lambda}\boldsymbol{P}_r\boldsymbol{Q}_2)(\|\boldsymbol{P}_r\boldsymbol{Q}_1\|\|\boldsymbol{P}_r\boldsymbol{Q}_2\|)^{-1}\Big\} \\ &\quad + \frac{1}{2}\text{tr}\{\boldsymbol{I}-(\boldsymbol{Q}_1^{\text{T}}\boldsymbol{P}_r^{\text{T}}\boldsymbol{P}_r\boldsymbol{Q}_1)\} + \frac{1}{2}\text{tr}\{\boldsymbol{I}-(\boldsymbol{Q}_2^{\text{T}}\boldsymbol{P}_r^{\text{T}}\boldsymbol{P}_r\boldsymbol{Q}_2)\} \\ &= -\text{tr}\{(\boldsymbol{P}_r^{\text{T}}\boldsymbol{\Lambda}\boldsymbol{P}_r)\},\end{aligned} \tag{9.85}$$

$$\begin{aligned}\tilde{J}_{\text{NQC}}(\overline{\boldsymbol{P}}\boldsymbol{Q}_1,\overline{\boldsymbol{P}}\boldsymbol{Q}_2) &= -\text{tr}\Big\{(\boldsymbol{Q}_1^{\text{T}}\overline{\boldsymbol{P}}^{\text{T}}\boldsymbol{\Lambda}\overline{\boldsymbol{P}}Q_2)(\|\boldsymbol{P}_r\boldsymbol{Q}_1\|\|\boldsymbol{P}_r\boldsymbol{Q}_2\|)^{-1}\Big\} \\ &\quad + \frac{1}{2}\text{tr}\{\boldsymbol{I}-(\boldsymbol{Q}_1^{\text{T}}\overline{\boldsymbol{P}}^{\text{T}}\overline{\boldsymbol{P}}\boldsymbol{Q}_1)\} + \frac{1}{2}\text{tr}\{\boldsymbol{I}-(\boldsymbol{Q}_2^{\text{T}}\overline{\boldsymbol{P}}^{\text{T}}\overline{\boldsymbol{P}}\boldsymbol{Q}_2)\} \\ &= -\text{tr}\{(\overline{\boldsymbol{P}}^{\text{T}}\boldsymbol{\Lambda}\overline{\boldsymbol{P}})\}.\end{aligned} \tag{9.86}$$

Thus, we have

$$\tilde{J}_{\text{NQC}}(\boldsymbol{P}_r\boldsymbol{Q}) > \tilde{J}_{\text{NQC}}(\overline{\boldsymbol{P}}\boldsymbol{Q}), \tag{9.87}$$

which means that the set $\{(\boldsymbol{P}_r\boldsymbol{Q}_1,\boldsymbol{P}_r\boldsymbol{Q}_2)|\boldsymbol{P}_r \neq \overline{\boldsymbol{P}}\}$ is not a global minimizer set.

Since $\boldsymbol{P}_r \neq \overline{\boldsymbol{P}}$, we can always select a column $\overline{\boldsymbol{P}}_i(1 \leq i \leq r)$ from $\overline{\boldsymbol{P}} = [\overline{\boldsymbol{P}}_1,\ldots,\overline{\boldsymbol{P}}_r]$ such that

$$\overline{\boldsymbol{P}}_i^{\text{T}}\boldsymbol{P}_r = 0, \tag{9.88}$$

otherwise, $\boldsymbol{P}_r \doteq \overline{\boldsymbol{P}}$. Moreover, we can always select a column $\boldsymbol{P}_{r,j}(1 \leq j \leq r)$ from $\boldsymbol{P}_r = [\boldsymbol{P}_{r,1},\ldots,\boldsymbol{P}_{r,r}]$ such that

$$\boldsymbol{P}_{r,j}^{\text{T}}\overline{\boldsymbol{P}} = 0. \tag{9.89}$$

Otherwise $\boldsymbol{P}_r \doteq \overline{\boldsymbol{P}}$. Let $\overline{\boldsymbol{P}}_i$ have nonzero element only in row $\bar{j}_i$ and $\boldsymbol{P}_{r,j}$ have nonzero entry only in row $\hat{j}_j$. Obviously, $\bar{j}_i < \hat{j}_j$ and $\sigma_{\hat{j}_j} > \sigma_{\bar{j}_i}$; otherwise, $\boldsymbol{P}_r \doteq \overline{\boldsymbol{P}}$. Define an orthonormal matrix as $\boldsymbol{B} = [\boldsymbol{P}_{r,1},\ldots,(\boldsymbol{P}_{r,i}+\varepsilon\overline{\boldsymbol{P}}_i)/\sqrt{1+\varepsilon^2},\ldots,\boldsymbol{P}_{r,r}]$, where ε is a positive infinitesimal. Since $\boldsymbol{P}_{r,j}$ and $\bar{\text{P}}_{\text{i}}$ have one nonzero entry, it follows that

$$\boldsymbol{\Lambda}\boldsymbol{B} = [\sigma_{\hat{j}_1}\boldsymbol{P}_{r,1},\ldots,(\sigma_{\hat{j}_j}\boldsymbol{P}_{r,i}+\sigma_{\bar{j}_i}\varepsilon\overline{\boldsymbol{P}}_i)/\sqrt{1+\varepsilon^2},\ldots,\sigma_{\hat{j}_r}\boldsymbol{P}_{r,r}]. \tag{9.90}$$

Combining (9.88), (9.89), and (9.90), we have

$$\boldsymbol{B}^{\text{T}}\boldsymbol{\Lambda}\boldsymbol{B} = \text{diag}[\sigma_{\hat{j}_1},\ldots,(\sigma_{\hat{j}_j}+\varepsilon^2\sigma_{\bar{j}_i})/(1+\varepsilon^2),\ldots,\sigma_{\hat{j}_r}]. \tag{9.91}$$

Since $\boldsymbol{P}_r$ is an $N \times r$ permutation matrix, we have

$$\boldsymbol{P}_r^{\mathrm{T}} \boldsymbol{\Lambda} \boldsymbol{P}_r = \mathrm{diag}[\sigma_{\hat{j}_1}, \ldots, \sigma_{\hat{j}_j}, \ldots, \sigma_{\hat{j}_r}]. \tag{9.92}$$

Then, it follows that

$$\begin{aligned} \boldsymbol{B}^{\mathrm{T}} \boldsymbol{\Lambda} \boldsymbol{B} - \boldsymbol{P}_r^{\mathrm{T}} \boldsymbol{\Lambda} \boldsymbol{P}_r &= \mathrm{diag}[\sigma_{\hat{j}_1}, \ldots, (\sigma_{\hat{j}_j} + \varepsilon \sigma_{\bar{j}_i})/(1+\varepsilon^2), \ldots, \sigma_{\hat{j}_r}] - \mathrm{diag}[\sigma_{\hat{j}_1}, \ldots, \sigma_{\hat{j}_j}, \ldots, \sigma_{\hat{j}_r}] \\ &= \mathrm{diag}[0, \ldots, 0, (-\sigma_{\hat{j}_j} + \sigma_{\bar{j}_i})\varepsilon^2/(1+\varepsilon^2), 0, \ldots, 0]. \end{aligned} \tag{9.93}$$

Since $\sigma_{\hat{j}_j} < \sigma_{\bar{j}_i}$, $\boldsymbol{B}^{\mathrm{T}} \boldsymbol{\Lambda} \boldsymbol{B} - \boldsymbol{P}_r^{\mathrm{T}} \boldsymbol{\Lambda} \boldsymbol{P}_r$ is a nonzero positive semi-definite matrix. Thus, we have

$$\begin{aligned} \tilde{J}_{\mathrm{NQC}}(\boldsymbol{B}\boldsymbol{Q}_1, \boldsymbol{B}\boldsymbol{Q}_2) &= -\mathrm{tr}\left\{ (\boldsymbol{Q}_1^T \boldsymbol{B}^T \boldsymbol{\Lambda} \boldsymbol{B} \boldsymbol{Q}_2)(\|\boldsymbol{B}_r \boldsymbol{Q}_1\|_F \|\boldsymbol{B}_r \boldsymbol{Q}_2\|_F)^{-1} \right\} + \frac{1}{2}\mathrm{tr}\{\boldsymbol{I} - (\boldsymbol{Q}_1^T \boldsymbol{B}^T \boldsymbol{B} \boldsymbol{Q}_1)\} + \frac{1}{2}\mathrm{tr}\{\boldsymbol{I} - (\boldsymbol{Q}_2^T \boldsymbol{B}^T \boldsymbol{B} \boldsymbol{Q}_2)\} \\ &= -\mathrm{tr}\{\boldsymbol{B}^T \boldsymbol{\Lambda} \boldsymbol{B}\} < \tilde{J}_{\mathrm{NQC}}(\boldsymbol{P}_r \boldsymbol{Q}_1, \boldsymbol{P}_r \boldsymbol{Q}_2) = -\mathrm{tr}\{\boldsymbol{P}_r^T \boldsymbol{\Lambda} \boldsymbol{P}_r\}. \end{aligned}$$

This means that $\{(\boldsymbol{P}_r \boldsymbol{Q}_1, \boldsymbol{P}_r \boldsymbol{Q}_2) | \boldsymbol{P}_r \neq \overline{\boldsymbol{P}}\}$ is a saddle (unstable) point set.

This completes the proof.

It can be easily shown that $(\widetilde{\boldsymbol{U}}, \widetilde{\boldsymbol{V}}) \doteq (\overline{\boldsymbol{P}}, \overline{\boldsymbol{P}})$ denotes a global minimizer set of $\tilde{J}_{\mathrm{NQC}}(\widetilde{\boldsymbol{U}}, \widetilde{\boldsymbol{V}})$.

9.3.2.3 Remarks and Comparisons

In the following, we make some remarks on the novel NQC (9.52).

Remark 9.1 From Theorems 9.3 and 9.4, it is obvious that the minimum of $J_{\mathrm{NQC}}(\boldsymbol{U}, \boldsymbol{V})$ for PSS automatically orthonormalizes the columns of $\boldsymbol{U}$ and $\boldsymbol{V}$, and at the minimizer of $J_{\mathrm{NQC}}(\boldsymbol{U}, \boldsymbol{V})$ for PSS, $\boldsymbol{U}$ and $\boldsymbol{V}$ only produce an arbitrary orthonormal basis of the PSS but not the multiple PSCs. However, $\boldsymbol{U}\boldsymbol{U}^{\mathrm{T}} = \boldsymbol{L}_{ps}\boldsymbol{L}_{ps}^{\mathrm{T}}$ and $\boldsymbol{V}\boldsymbol{V}^{\mathrm{T}} = \boldsymbol{R}_{ps}\boldsymbol{R}_{ps}^{\mathrm{T}}$ are two orthogonal projection on the PSS and can be uniquely determined.

Remark 9.2 $J_{\mathrm{NQC}}(\boldsymbol{U}, \boldsymbol{V})$ for PSS has a global minimizer set and no local ones. Thus, the iterative algorithms, like the gradient descent search algorithm for finding the global minimum, are guaranteed to globally converge to the desired PSS for a proper initialization of $\boldsymbol{U}$ and $\boldsymbol{V}$ in the domain $\boldsymbol{\Omega}$. The presence of the saddle points does not cause any problem of convergence because they are avoided through random perturbations of $\boldsymbol{U}$ and $\boldsymbol{V}$ in practice.

9.3.3 Adaptive Learning Algorithm and Performance Analysis

9.3.3.1 Adaptive Learning Algorithm

At time instant k, we have $\boldsymbol{x}(1), \boldsymbol{x}(2), \ldots, \boldsymbol{x}(k)$ and $\boldsymbol{y}(1), \boldsymbol{y}(2), \ldots, \boldsymbol{y}(k)$ available and are interested in estimating $\boldsymbol{U}(k)$ and $\boldsymbol{V}(k)$. Our objective is to establish a fast adaptive algorithm for calculating recursively an estimate of the PSS at time instant k from the known estimate at k-1 and the newly observed samples $\boldsymbol{x}(k)$ and $\boldsymbol{y}(k)$. We will apply the gradient descent searching to minimize $J_{\mathrm{NQC}}(\boldsymbol{U}, \boldsymbol{V})$.

Given the gradient of $J_{\mathrm{NQC}}(\boldsymbol{U}, \boldsymbol{V})$ with respect to $\boldsymbol{U}$ and $\boldsymbol{V}$, we have the following gradient descent rule for updating $\boldsymbol{U}(k)$ and $\boldsymbol{V}(k)$:

$$\begin{aligned} \boldsymbol{U}(k+1) = \boldsymbol{U}(k) + \mu\{ & \boldsymbol{C}(k)\boldsymbol{V}(k)(\|\boldsymbol{U}(k)\|_F\|\boldsymbol{V}(k)\|_F)^{-1} \\ & - \boldsymbol{U}(k)(\boldsymbol{U}(k)^{\mathrm{T}}\boldsymbol{C}(k)\boldsymbol{V}(k))(\|\boldsymbol{U}(k)\|_F^3\|\boldsymbol{V}(k)\|_F)^{-1}\} + \mu\boldsymbol{U}(k)[\boldsymbol{I} - \boldsymbol{U}(k)^{\mathrm{T}}\boldsymbol{U}(k)], \end{aligned} \tag{9.94}$$

$$\begin{aligned} \boldsymbol{V}(k+1) = \boldsymbol{V}(k) + \mu\{ & \boldsymbol{C}(k)^{\mathrm{T}}\boldsymbol{U}(k)(\|\boldsymbol{U}(k)\|_F\|\boldsymbol{V}(k)\|_F)^{-1} \\ & - \boldsymbol{V}(k)(\boldsymbol{U}(k)^{\mathrm{T}}\boldsymbol{C}(k)\boldsymbol{V}(k))(\|\boldsymbol{U}(k)\|_F\|\boldsymbol{V}(k)\|_F^3)^{-1}\} + \mu\boldsymbol{V}(k)[\boldsymbol{I} - \boldsymbol{V}(k)^{\mathrm{T}}\boldsymbol{V}(k)], \end{aligned} \tag{9.95}$$

where $0 < \mu < 1$ denotes the learning step size. In Eqs. (9.94) and (9.95), if we replace $\boldsymbol{C}$ by $\boldsymbol{x}(k)\boldsymbol{y}^{\mathrm{T}}(k)$, we can obtain a nonlinear stochastic learning rule:

$$\begin{aligned} \boldsymbol{U}(k+1) = \boldsymbol{U}(k) + \mu\{ & \boldsymbol{x}(k)\boldsymbol{y}^{\mathrm{T}}(k)\boldsymbol{V}(k)(\|\boldsymbol{U}(k)\|_F\|\boldsymbol{V}(k)\|_F)^{-1} \\ & - \boldsymbol{U}(k)(\boldsymbol{U}^{\mathrm{T}}(k)\boldsymbol{x}(k)\boldsymbol{y}^{\mathrm{T}}(k)\boldsymbol{V}(k))(\|\boldsymbol{U}(k)\|_F^3\|\boldsymbol{V}(k)\|_F)^{-1}\} + \mu\boldsymbol{U}(k)[\boldsymbol{I} - \boldsymbol{U}^{\mathrm{T}}(k)\boldsymbol{U}(k)], \end{aligned} \tag{9.96}$$

$$\begin{aligned} \boldsymbol{V}(k+1) = \boldsymbol{V}(k) + \mu\{ & (\boldsymbol{x}(k)\boldsymbol{y}^{\mathrm{T}}(k))^{\mathrm{T}}\boldsymbol{U}(k)(\|\boldsymbol{U}(k)\|_F\|\boldsymbol{V}(k)\|_F)^{-1} \\ & - \boldsymbol{V}(k)(\boldsymbol{U}^{\mathrm{T}}(k)\boldsymbol{x}(k)\boldsymbol{y}^{\mathrm{T}}(k)\boldsymbol{V}(k))(\|\boldsymbol{U}(k)\|_F\|\boldsymbol{V}(k)\|_F^3)^{-1}\} + \mu\boldsymbol{V}(k)[\boldsymbol{I} - \boldsymbol{V}^{\mathrm{T}}(k)\boldsymbol{V}(k)]. \end{aligned} \tag{9.97}$$

Equations (9.96) and (9.97) constitute our PSS tracking algorithm for extracting cross-correlation features between two high-dimensional data streams. Our algorithm has a computational complexity of $(1+3r)MN + 3(M+N)r^2$ flops per update, which is at the same order as that of algorithms in [2]. The operations involved in (9.96) and (9.97) are simple matrix addition, multiplication, and inverse, which are easy for systolic array implementation [26].

9.3.3.2 Convergence Analysis

We now study the convergence property of the gradient rule (9.96) and (9.97). If $\boldsymbol{x}(t)$ and $\boldsymbol{y}(t)$ are two jointly stationary processes and the learning rate μ is small enough, then the discrete-time difference in Eqs. (9.96) and (9.97) approximates the following continuous-time ODE [32, 33]

$$\begin{aligned}\frac{\mathrm{d}\boldsymbol{U}(t)}{\mathrm{d}t} = &+\{\boldsymbol{C}(t)\boldsymbol{V}(t)\left(\|\boldsymbol{U}(t)\|_F\|\boldsymbol{V}(t)\|_F\right)^{-1}\\ &-\boldsymbol{U}(t)(\boldsymbol{U}(t)^{\mathrm{T}}\boldsymbol{C}(t)\boldsymbol{V}(t))(\|\boldsymbol{U}(t)\|_F^3\|\boldsymbol{V}(t)\|_F)^{-1}\}+\mu\boldsymbol{U}(t)[\boldsymbol{I}-\boldsymbol{U}(t)^{\mathrm{T}}\boldsymbol{U}(t)],\end{aligned} \tag{9.98}$$

$$\begin{aligned}\frac{\mathrm{d}\boldsymbol{V}(t)}{\mathrm{d}t} = &+\{\boldsymbol{C}(t)^{\mathrm{T}}\boldsymbol{U}(t)\left(\|\boldsymbol{U}(t)\|_F\|\boldsymbol{V}(t)\|_F\right)^{-1}\\ &-\boldsymbol{V}(t)(\boldsymbol{U}(t)^{\mathrm{T}}\boldsymbol{C}(t)\boldsymbol{V}(t))(\|\boldsymbol{U}(t)\|_F\|\boldsymbol{V}(t)\|_F^3)^{-1}\}+\mu\boldsymbol{V}(t)[\boldsymbol{I}-\boldsymbol{V}(t)^{\mathrm{T}}\boldsymbol{V}(t)].\end{aligned} \tag{9.99}$$

By analyzing the global convergence of (9.98) and (9.99), we can establish the condition for the global convergence of (9.98) and (9.99). Next, we study the convergence of (9.98) and (9.99) via the Lyapunov function theory.

Obviously, we can see that $J_{\mathrm{NQC}}(\boldsymbol{U},\boldsymbol{V})$ is a Lyapunov function for the ODE (9.98) and (9.99). To show this, let us define a region $\Omega=\{(\boldsymbol{U},\boldsymbol{V})|J_{\mathrm{NQC}}(\boldsymbol{U},\boldsymbol{V})<\infty\}=\{(\boldsymbol{U},\boldsymbol{V})|0<\|\boldsymbol{U}\|_F<\infty \text{ and } 0<\|\boldsymbol{V}\|_F<\infty\}$. Within this region, $J_{\mathrm{NQC}}(\boldsymbol{U},\boldsymbol{V})$ is continuous and has a continuous first-order derivative.

Theorem 9.5 *Given the ODE (9.98), (9.99) and* $(\boldsymbol{U}(0),\boldsymbol{V}(0))\in\Omega$, *then* $(\boldsymbol{U}(t),\boldsymbol{V}(t))$ *for tracking PSS converges to a point in the set* $(\boldsymbol{U},\boldsymbol{V})\ddot{=}(\boldsymbol{L}_{ps},\boldsymbol{R}_{ps})$ *with probability 1 as* $t\to\infty$.

Proof For PSS tracking algorithm, the energy function associated with (9.98) and (9.99) can be given as follows:

$$\begin{aligned}E(\boldsymbol{U},\boldsymbol{V}) = &-\mathrm{tr}\left\{(\boldsymbol{U}^{\mathrm{T}}\boldsymbol{C}\boldsymbol{V})\left[\|\boldsymbol{U}\|_F\|\boldsymbol{V}\|_F\right]^{-1}\right\}+\frac{1}{2}\mathrm{tr}\{[\boldsymbol{I}-\boldsymbol{U}^{\mathrm{T}}\boldsymbol{U}]^2\}\\ &+\frac{1}{2}\mathrm{tr}\{[\boldsymbol{I}-\boldsymbol{V}^{\mathrm{T}}\boldsymbol{V}]^2\}.\end{aligned} \tag{9.100}$$

Clearly, we can see that when $\|\boldsymbol{U}\|_F\to\infty$ and/or $\|\boldsymbol{V}\|_F\to\infty, E(\boldsymbol{U},\boldsymbol{V})\to\infty$. Thus, it can be seen that $G_c=\{\boldsymbol{U},\boldsymbol{V};E(\boldsymbol{U},\boldsymbol{V})<c\}$ is bounded for each c. The gradient of $E(\boldsymbol{U},\boldsymbol{V})$ with respect to $\boldsymbol{U},\boldsymbol{V}$ is given by

$$\begin{aligned}\nabla_U E(\boldsymbol{U},\boldsymbol{V}) = &-\{\boldsymbol{C}(k)\boldsymbol{V}(k)(\|\boldsymbol{U}(k)\|_F\|\boldsymbol{V}(k)\|_F)^{-1}\\ &-\boldsymbol{U}(k)(\boldsymbol{U}(k)^{\mathrm{T}}\boldsymbol{C}(k)\boldsymbol{V}(k))(\|\boldsymbol{U}(k)\|_F^3\|\boldsymbol{V}(k)\|_F)^{-1}\}-\boldsymbol{U}(k)[\boldsymbol{I}-\boldsymbol{U}(k)^{\mathrm{T}}\boldsymbol{U}(k)],\end{aligned} \tag{9.101}$$

$$\nabla_V E(\boldsymbol{U},\boldsymbol{V}) = -\{\boldsymbol{C}(k)^{\mathrm{T}}\boldsymbol{U}(k)(\|\boldsymbol{U}(k)\|_F\|\boldsymbol{V}(k)\|_F)^{-1} \\ - \boldsymbol{V}(k)(\boldsymbol{U}(k)^{\mathrm{T}}\boldsymbol{C}(k)\boldsymbol{V}(k))(\|\boldsymbol{U}(k)\|_F\|\boldsymbol{V}(k)\|_F^3)^{-1}\} - \boldsymbol{V}(k)[\boldsymbol{I} - \boldsymbol{V}(k)^{\mathrm{T}}\boldsymbol{V}(k)]. \tag{9.102}$$

Clearly, Eqs. (9.98) and (9.99) for PSS tracking are equivalent to the following equation:

$$\frac{\mathrm{d}\boldsymbol{U}}{\mathrm{d}t} = -\nabla_U E(\boldsymbol{U},\boldsymbol{V}) \text{ and } \frac{\mathrm{d}\boldsymbol{V}}{\mathrm{d}t} = -\nabla_V E(\boldsymbol{U},\boldsymbol{V}). \tag{9.103}$$

Differentiating $E(\boldsymbol{U},\boldsymbol{V})$ along the solution of (9.98) and (9.99) for PSS tracking algorithm yields

$$\frac{\mathrm{d}E(\boldsymbol{U},\boldsymbol{V})}{\mathrm{d}t} = \mathrm{tr}\left(\frac{\mathrm{d}\boldsymbol{U}^{\mathrm{T}}}{\mathrm{d}t}\cdot\nabla_{\mathrm{U}}E(\boldsymbol{U},\boldsymbol{V}) + \frac{\mathrm{d}\boldsymbol{V}^{\mathrm{T}}}{\mathrm{d}t}\cdot\nabla_V E(\boldsymbol{U},\boldsymbol{V})\right) \\ = -\mathrm{tr}\left(\frac{\mathrm{d}\boldsymbol{U}^{\mathrm{T}}}{\mathrm{d}t}\cdot\frac{\mathrm{d}\boldsymbol{U}}{\mathrm{d}t} + \frac{\mathrm{d}\boldsymbol{V}^{\mathrm{T}}}{\mathrm{d}t}\cdot\frac{\mathrm{d}\boldsymbol{V}}{\mathrm{d}t}\right). \tag{9.104}$$

This means that $E(\boldsymbol{U},\boldsymbol{V})$ is a Lyapunov function, and $(\mathrm{d}\boldsymbol{U}(t)/\mathrm{d}t) \to 0$ and $(\mathrm{d}\boldsymbol{V}(t)/\mathrm{d}t) \to 0$, i.e., $\nabla_{\mathrm{U}}E(\boldsymbol{U}(t),\boldsymbol{V}(t)) \to 0$ and $\nabla_V E(\boldsymbol{U}(t),\boldsymbol{V}(t)) \to 0$ as $t \to \infty$. Thus, $(\boldsymbol{U}(t),\boldsymbol{V}(t))$ globally converges to a stationary point of $E(\boldsymbol{U},\boldsymbol{V})$. Moreover, $E(\boldsymbol{U},\boldsymbol{V})$ achieves its unique global minimum in set $\{(\boldsymbol{L}_{ps}\boldsymbol{Q}_1, \boldsymbol{R}_{ps}\boldsymbol{Q}_2)\}$. Since all the other stationary points of $E(\boldsymbol{U},\boldsymbol{V})$ are saddle and also unstable, we can see that $(\boldsymbol{U}(t),\boldsymbol{V}(t))$ converges to a point in set $\{(\boldsymbol{L}_{ps}\boldsymbol{Q}_1, \boldsymbol{R}_{ps}\boldsymbol{Q}_2)\}$ with probability 1, as $t \longrightarrow \infty$.

This completes the proof.

9.3.3.3 Self-Stability Property Analysis

Next, we will study the self-stability property of (9.96) and (9.97).

Theorem 9.6 *If the learning factor μ is small enough and the input vector $\boldsymbol{x}(k)$ and $\boldsymbol{y}(k)$ are bounded, then $\|\boldsymbol{U}(k)\|_F$ and $\|\boldsymbol{V}(k)\|_F$ in learning algorithm* (9.96) *and* (9.97) *for tracking the PSS approach to $\sqrt{r}$, respectively.*

Proof Since the learning factor μ is small enough and the input vector is bounded, we have

$$\|\boldsymbol{U}(k+1)\|_F^2 = \mathrm{tr}[\boldsymbol{U}^{\mathrm{T}}(k+1)\boldsymbol{U}(k+1)] \approx \mathrm{tr}\{\boldsymbol{U}^{\mathrm{T}}(k)\boldsymbol{U}(k) + 2\mu\boldsymbol{U}^{\mathrm{T}}(k)\boldsymbol{U}(k)[\boldsymbol{I} - \boldsymbol{U}^{\mathrm{T}}(k)\boldsymbol{U}(k)]\} \\ = \mathrm{tr}[\boldsymbol{U}^{\mathrm{T}}(k)\boldsymbol{U}(k)] - 2\mu\cdot\mathrm{tr}[(\boldsymbol{U}^{\mathrm{T}}(k)\boldsymbol{U}(k))^2 - \boldsymbol{U}^{\mathrm{T}}(k)\boldsymbol{U}(k)]. \tag{9.105}$$

Note that in the previous equation, the second-order terms associated with the learning factor have been neglected.

It holds that

$$
\begin{aligned}
\|U(k+1)\|_F^2/\|U(k)\|_F^2 &= \frac{\mathrm{tr}[U^{\mathrm{T}}(k)U(k)] - 2\mu\cdot\mathrm{tr}[(U^{\mathrm{T}}(k)U(k))^2 - U^{\mathrm{T}}(k)U(k)]}{\mathrm{tr}[U^{\mathrm{T}}(k)U(k)]} \\
&= 1-2\mu\left[\frac{\mathrm{tr}[(U^{\mathrm{T}}(k)U(k))^2}{\mathrm{tr}[U^{\mathrm{T}}(k)U(k)]}-1\right] = 1-2\mu\left[\frac{\mathrm{tr}\{M^2\}}{\mathrm{tr}\{M\}}-1\right] = 1-2\mu\cdot\left(\frac{\mathrm{tr}\{\Phi\Psi\Phi^{-1}\Phi\Psi\Phi^{-1}\}}{\mathrm{tr}\{\Phi\Psi\Phi^{-1}\}}-1\right) \\
&= 1-2\mu\cdot\left(\frac{\mathrm{tr}\{\Phi\Psi^2\Phi^{-1}\}}{\mathrm{tr}\{\Phi\Psi\Phi^{-1}\}}-1\right) = 1-2\mu\cdot\left(\frac{\mathrm{tr}\{\Phi^{-1}\Phi\Psi^2\}}{\mathrm{tr}\{\Phi^{-1}\Phi\Psi\}}-1\right) = 1-2\mu\cdot\left(\frac{\mathrm{tr}\{\Psi^2\}}{\mathrm{tr}\{\Psi\}}-1\right) \\
&= 1-2\mu\cdot\left(\frac{\sum_{i=1}^r \zeta_i^2}{\sum_{i=1}^r \zeta_i}-1\right) = \begin{cases} >1 & \text{for } \sum_{i=1}^r \zeta_i^2 < \sum_{i=1}^r \zeta_i \\ =1 & \text{for } \zeta_i = 1, (i=1,2,\ldots,r) \\ <1 & \text{for } \sum_{i=1}^r \zeta_i^2 > \sum_{i=1}^r \zeta_i \end{cases} = \begin{cases} >1 & \text{for } \|\mathrm{U}(k)\|_F < \sqrt{r} \\ =1 & \text{for } \|\mathrm{U}(k)\|_F = \sqrt{r} \\ <1 & \text{for } \|\mathrm{U}(k)\|_F > \sqrt{r}, \end{cases}
\end{aligned}
\tag{9.106}
$$

where $M = U(k)^{\mathrm{T}}U(k)$ is a $r \times r$ matrix, and its EVD is represented as $M = \Phi\Psi\Phi^{-1}$.

The self-stabilizing property of $\|V(k)\|_F$ in Algorithms (9.96) and (9.97) for tracking the PSS can be proved similarly.

This completes the proof.

Remark 9.3 In (9.98) and (9.99), if $U^{\mathrm{T}}(k)U(k) = I$ and $V^{\mathrm{T}}(k)V(k) = I$ hold, we can obtain Eqs. (9.19) and (9.20) in [12], respectively. So we can see that the algorithm in [12] is a special case of our algorithm for PSS tracking, or we can regard our algorithm for PSS tracking as an extension of the algorithm in [12].

9.3.4 Computer Simulations

In this section, we will provide several interesting experiments to illustrate the performance of our PSS algorithms. Generate randomly a 9-dimensional Gaussian white sequence $y(k)$, and $x(k) = Ay(k)$, where A is an ill-conditioned matrix:

$$
A = [u_0, \ldots, u_8]\mathrm{diag}(10, 10, 10, 10^{-3}, 10^{-3}, 10^{-3}, 10^{-7}, 10^{-7}, 10^{-7})[v_0, \ldots, v_8]^{\mathrm{T}},
\tag{9.107}
$$

and u_i and v_i (i = 0,…,8) are the *i*th components of 11-dimensional and 9-dimensional orthogonal discrete cosine basis functions, respectively. In order to measure the convergence speed and precision of learning algorithm, we compute the norm of a state matrix at the *k*th update:

$$
\rho(U(k)) = \|U(k)\|_F \quad \text{and } \rho(V(k)) = \|V(k)\|_F,
\tag{9.108}
$$

and the direction cosine at the *k*th update:

$$\text{Direction Cosine(k)} = \begin{cases} \left|\boldsymbol{U}^{\mathrm{T}}(k)\cdot\boldsymbol{L}_{ps}\right|/\|\boldsymbol{U}(k)\|\cdot\left\|\boldsymbol{L}_{ps}\right\| & LPSS \\ \left|\boldsymbol{V}^{\mathrm{T}}(k)\cdot\boldsymbol{R}_{ps}\right|/\|\boldsymbol{V}(k)\|\cdot\left\|\boldsymbol{R}_{ps}\right\| & RPSS. \end{cases} \tag{9.109}$$

Also, we use the following index parameter:

$$\text{dist}(\boldsymbol{U}(k)) = \left\|\boldsymbol{U}^{\mathrm{T}}(k)\boldsymbol{U}(k) - \boldsymbol{I}_r\right\|_F \quad \text{and} \quad \text{dist}(\boldsymbol{V}(k)) = \left\|\boldsymbol{V}^{\mathrm{T}}(k)\boldsymbol{V}(k) - \boldsymbol{I}_r\right\|_F, \tag{9.110}$$

which measures the deviation of a state matrix from the orthogonality. Clearly, if direction cosine(k) converges to 1, then state matrices $\boldsymbol{U}(k)$ (or$\boldsymbol{V}(k)$) must approach the direction of the left PSS (or right PSS). If dist($\boldsymbol{U}(k)$) (or dist($\boldsymbol{V}(k)$) converges to zero, then it means that $\boldsymbol{U}(k)$ (or$\boldsymbol{V}(k)$) is an orthonormal basis of the left PSS (or right PSS).

In theory, the cross-correlation matrix $\boldsymbol{A}$ given by (9.107) has nine nonzero singular values among which the totally distinct three singular values 10, 10^{-3}, and 10^{-7} have multiplicity 3. The data matrix is ill conditioned, since its condition number is 10^8 [2]. The PSS spanned by the first 5 singular components will be tracked. The sample size is 1000, i.e., $\boldsymbol{x}(1),\ldots,\boldsymbol{x}(1000)$, and $\boldsymbol{y}(1),\ldots,\boldsymbol{y}(1000)$.

Here, we will compare performance of our algorithm for PSS tracking with that of the recursive implementation algorithm in [2] and the algorithm in [12]. Figure 9.2a, b are the simulation result for our algorithm and the algorithm in [2]. Figure 9.3a,b are the simulation result for our algorithm and the algorithm in [12].

In order to further compare the performance of our PSS tracking algorithm and the algorithm in [12], another experiment is conducted. Here, the vector data sequence are generated by $\boldsymbol{x}(k) = \boldsymbol{B}\cdot\mathbf{y}(k)$, where $\boldsymbol{B}$ is given by:

$$\boldsymbol{B} = [\boldsymbol{u}_0,\ldots,\boldsymbol{u}_8]\cdot\text{rand}n(9,9)\cdot[\mathbf{v}_0,\ldots,\boldsymbol{v}_8]^{\mathrm{T}}, \tag{9.111}$$

where $\boldsymbol{u}_i$ and $\boldsymbol{v}_i$(i = 0,…,8) are the *i*th components of 11-dimensional and 9-dimensional orthogonal discrete cosine basis functions, respectively. In this simulation, we let $\boldsymbol{y}_t \in \Re^{9\times 1}$ be Gaussian, spatially and temporally white and randomly generated. Here, a PSS with dimension 5 is tracked. Figure 9.4a, b shows the simulation results for our algorithm and the algorithm in [12] with the initial weight modulus value normalized to $\|\boldsymbol{U}_0\| = \|\boldsymbol{V}_0\| = \sqrt{5} = 2.236$, where the learning factors of our PSS tracking algorithm and the algorithm in [12] are 0.03 and 0.01, respectively.

From Fig. 9.2, we can see that our algorithm for PSS tracking and the recursive implementation algorithm in [2] both can converge to an orthonormal basis of the PSS of the cross-correlation matrix between two high-dimensional data streams. However, it is obvious that our algorithm for PSS tracking has faster convergence speed and higher solution precision. In Fig. 9.3, the norm of the state matrix and the parameter dist($\boldsymbol{U}(k)$) (or dist($\boldsymbol{V}(k)$)) for algorithm in [12] do not converge. In

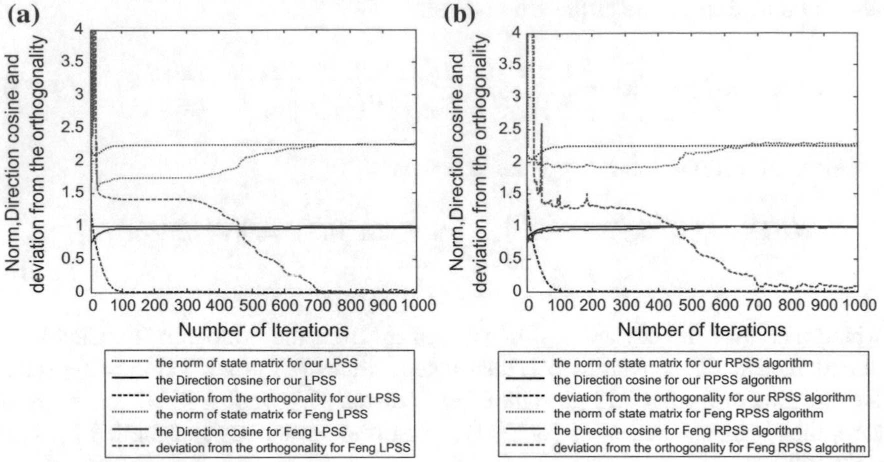

Fig. 9.2 **a** The experiment on the *left* PSS tracking.**b** The experiment on the *right* PSS tracking with $\|\boldsymbol{U}_0\| = \|\boldsymbol{V}_0\| = \sqrt{5} = 2.236$, $\mu_{\text{our}} = 0.03, \mu_{\text{feng}} = 0.49$ with $\|\boldsymbol{U}_0\| = \|\boldsymbol{V}_0\| = \sqrt{5} = 2.236$, $\mu_{\text{our}} = 0.03, \mu_{\text{feng}} = 0.49$

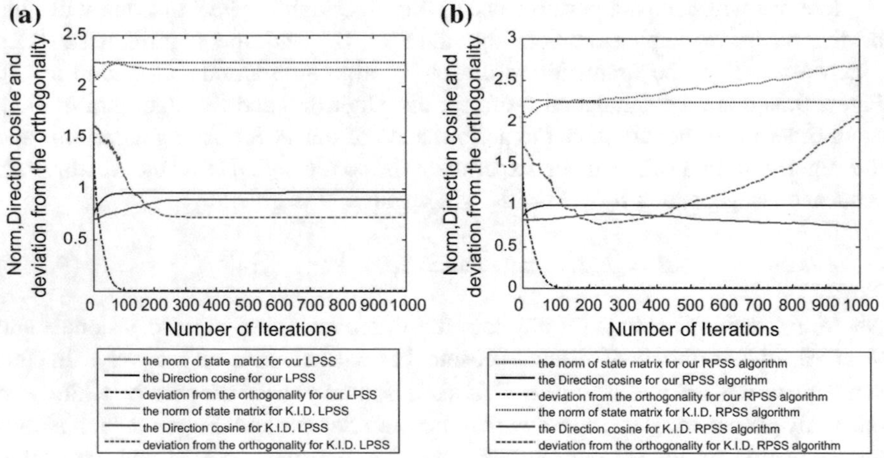

Fig. 9.3 **a** The experiment on the *left* PSS tracking. **b** The experiment on the *right* PSS tracking with $\|\boldsymbol{U}_0\| = \|\boldsymbol{V}_0\| = \sqrt{5} = 2.236, \mu_{\text{our}} = 0.03, \mu_{\text{K.D.}} = 0.01$ with $\|\boldsymbol{U}_0\| = \|\boldsymbol{V}_0\| = \sqrt{5} = 2.236$, $\mu_{\text{our}} = 0.03, \mu_{\text{K.D.}} = 0.01$

Fig. 9.4, although the three parameters of the algorithm in [12] can converge, their convergence speed is much slower than that of our PSS tracking algorithm. From Figs. 9.2, 9.3, and 9.4, it is obvious that our algorithm for PSS tracking outperforms other congener algorithms.

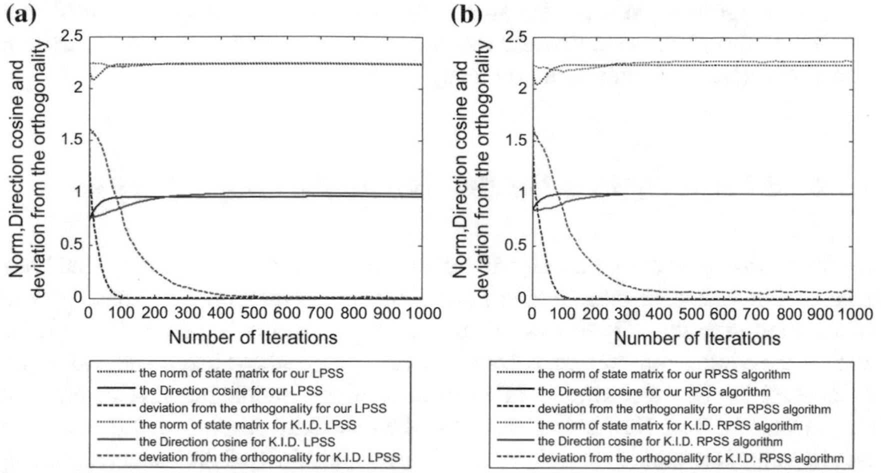

Fig. 9.4 **a** The experiment on the *left* PSS tracking. **b** The *right* PSS tracking with $\|\boldsymbol{U}_0\| = \|\boldsymbol{V}_0\| = 1.0, \mu_{\text{our}} = 0.03, \mu_{\text{K.D.}} = 0.01$ with $\|\boldsymbol{U}_0\| = \|\boldsymbol{V}_0\| = \sqrt{5} = 2.236, \mu_{\text{our}} = 0.03, \mu_{\text{K.D.}} = 0.01$

In this section, a novel information criterion for PSS tracking has been introduced, and based on it, a PSS gradient flow has been derived. The averaging equation of the algorithm for PSS tracking exhibits a single global minimum that is achieved if and only if its state matrix spans the PSS of the cross-correlation matrix between two high-dimensional vector sequences. Simulations have shown that the PSS tracking gradient flow can make the corresponding state matrix tend to column-orthonormal basis of the PSS and have also shown that the PSS tracking flow has fast convergence speed and can work satisfactorily.

9.4 Coupled Cross-Correlation Neural Network Algorithm for Principal Singular Triplet Extraction of a Cross-Covariance Matrix

In this section, we propose a novel coupled neural network learning algorithm to extract the principal singular triplet (PST) of a cross-correlation matrix between two high-dimensional data streams. We firstly introduce a novel information criterion (NIC), in which the stationary points are singular triplet of the cross-correlation matrix. Then, based on Newton's method, we obtain a coupled system of ordinary differential equations (ODEs) from the NIC. The ODEs have the same equilibria as the gradient of NIC; however, only the first PST of the system is stable (which is also the desired solution), and all others are (unstable) saddle points. Based on the system, we finally obtain a fast and stable algorithm for PST extraction. The proposed algorithm can solve the speed stability problem that plagues most non-coupled learning rules. Moreover, the proposed algorithm can also be used to extract multiple PSTs effectively by using sequential method.

9.4.1 A Novel Information Criterion and a Coupled System

We know that gradient-based algorithms can be derived by maximizing the variance of the projected data or minimizing the reconstruction error based on an information criterion. Thus, it is required that the stationary points of the information criterion must be attractors. However, the gradient-based method is not suitable for the NIC since the first PST of the NIC is a saddle point. Different from the gradient method, Newton's method has the beneficial property that it turns even saddle points into attractors, which guarantees the stability of the resulting learning rules [34]. In this case, the learning rule can be derived using an information criterion which is subject to neither minimization nor maximization [1]. Moreover, Newton's method has higher convergence speed than gradient method. In this section, a coupled system is derived from an NIC based on Newton's method. The NIC is defined as

$$p = \boldsymbol{u}^{\mathrm{T}}\boldsymbol{A}\boldsymbol{v} - \frac{1}{2}\sigma\boldsymbol{u}^{\mathrm{T}}\boldsymbol{u} - \frac{1}{2}\sigma\boldsymbol{v}^{\mathrm{T}}\boldsymbol{v} + \sigma \tag{9.112}$$

The gradient of (9.112) is determined through

$$\nabla p = \left[\left(\frac{\partial p}{\partial \boldsymbol{u}}\right)^{\mathrm{T}}, \left(\frac{\partial p}{\partial \boldsymbol{v}}\right)^{\mathrm{T}}, \frac{\partial p}{\partial \sigma}\right]^{\mathrm{T}} \tag{9.113}$$

which has the components

$$\frac{\partial p}{\partial \boldsymbol{u}} = A\boldsymbol{v} - \sigma\boldsymbol{u} \tag{9.114}$$

$$\frac{\partial p}{\partial \boldsymbol{v}} = \mathrm{A}^{\mathrm{T}}\boldsymbol{u} - \sigma\boldsymbol{v} \tag{9.115}$$

$$\frac{\partial p}{\partial \sigma} = -\frac{1}{2}\boldsymbol{u}^{\mathrm{T}}\boldsymbol{u} - \frac{1}{2}\boldsymbol{v}^{\mathrm{T}}\boldsymbol{v} + 1 \tag{9.116}$$

It is clear that the stationary points of (9.112) are also the singular triplets of $\boldsymbol{A}$, and we can also conclude that $\boldsymbol{u}^{\mathrm{T}}\boldsymbol{A}\boldsymbol{v} = \sigma$ and $\boldsymbol{u}^{\mathrm{T}}\boldsymbol{u} = \boldsymbol{v}^{\mathrm{T}}\boldsymbol{v} = 1$.

In Newton descent, the gradient is premultiplied by the inverse Hessian. The Hessian $\boldsymbol{H} = \nabla\nabla^{\mathrm{T}}p$ of (9.112) is

$$\boldsymbol{H} = \begin{pmatrix} -\sigma\boldsymbol{I}_m & \boldsymbol{A} & -\boldsymbol{u} \\ \boldsymbol{A}^{\mathrm{T}} & -\sigma\boldsymbol{I}_n & -\boldsymbol{v} \\ -\boldsymbol{u}^{\mathrm{T}} & -\boldsymbol{v}^{\mathrm{T}} & 0 \end{pmatrix}, \tag{9.117}$$

where $\boldsymbol{I}_m$ and $\boldsymbol{I}_n$ are identity matrices of dimension m and n, respectively. In the following, we determine an approximation for the inverse Hessian in the vicinity of the PST.

Newton's method requires the inverse of the Hessian (9.117) in the vicinity of the stationary point, here the PST. The inversion of the Hessian is simplified by a similarity transformation, using the orthogonal matrix

$$\boldsymbol{T} = \begin{pmatrix} \overline{\boldsymbol{U}} & \boldsymbol{0}_{mn} & \boldsymbol{0}_m \\ \boldsymbol{0}_{nm} & \overline{\boldsymbol{V}} & \boldsymbol{0}_n \\ \boldsymbol{0}_m^{\mathrm{T}} & \boldsymbol{0}_n^{\mathrm{T}} & 1 \end{pmatrix}, \tag{9.118}$$

Here, $\overline{\boldsymbol{U}} = [\bar{u}_1, \ldots, \bar{u}_m]$ and $\overline{\boldsymbol{V}} = [\bar{v}_1, \ldots, \bar{v}_n]$ are matrices which contain all left and right singular vectors in their columns, respectively. For the computation of the transformed matrix $\boldsymbol{H}^* = \boldsymbol{T}^{\mathrm{T}}\boldsymbol{H}\boldsymbol{T}$, we exploit $\boldsymbol{A}^{\mathrm{T}}\overline{\boldsymbol{U}} = \overline{\boldsymbol{V}}\,\overline{\boldsymbol{S}}^{\mathrm{T}}$ and $\boldsymbol{A}\overline{\boldsymbol{V}} = \overline{\boldsymbol{U}}\,\overline{\boldsymbol{S}}$, where $\overline{\boldsymbol{S}}$ is a $n \times n$ matrix whose first N diagonal elements are the singular values arranged in decreasing order, and all remaining elements of $\overline{\boldsymbol{S}}$ are zero. Moreover, in the vicinity of the stationary point $(\bar{\boldsymbol{u}}_1, \bar{\boldsymbol{v}}_1, \bar{\sigma}_1)$, we can approximate $\bar{\boldsymbol{u}}^{\mathrm{T}}\boldsymbol{u} \approx e_m$ and $\bar{\boldsymbol{v}}^{\mathrm{T}}\boldsymbol{v} \approx e_n$, where e_m and e_n are unit vectors of the specified dimensions with a 1 as the first element. We obtain

$$\boldsymbol{H}_0^* = \begin{pmatrix} -\sigma\boldsymbol{I}_m & \overline{\boldsymbol{S}} & -\boldsymbol{e}_m \\ \overline{\boldsymbol{S}}^{\mathrm{T}} & -\sigma\boldsymbol{I}_n & -\boldsymbol{e}_n \\ -\boldsymbol{e}_m^{\mathrm{T}} & -\boldsymbol{e}_n^{\mathrm{T}} & 0 \end{pmatrix}, \tag{9.119}$$

Next, by approximating $\sigma \approx \bar{\sigma}_1$ and by assuming $|\bar{\sigma}_1| \geq |\bar{\sigma}_j|, \forall j = 2, \ldots, M$, the expression $\overline{\boldsymbol{S}}$ can be approximated as $\overline{\boldsymbol{S}} = \sigma\sigma^{-1}\overline{\boldsymbol{S}} \approx \sigma\boldsymbol{e}_m\boldsymbol{e}_n^{\mathrm{T}}$. In this case, (9.119) yields

$$\boldsymbol{H}^* = \begin{pmatrix} -\sigma\boldsymbol{I}_m & \sigma\boldsymbol{e}_m\boldsymbol{e}_n^{\mathrm{T}} & -\boldsymbol{e}_m \\ \sigma\boldsymbol{e}_n\boldsymbol{e}_m^{\mathrm{T}} & -\sigma\boldsymbol{I}_n & -\boldsymbol{e}_n \\ -\boldsymbol{e}_m^{\mathrm{T}} & -\boldsymbol{e}_n^{\mathrm{T}} & 0 \end{pmatrix}. \tag{9.120}$$

As introduced in [35], suppose that an invertible matrix of size $(j+1) \times (j+1)$ is of the form

$$\boldsymbol{R}_{j+1} = \begin{pmatrix} \boldsymbol{R}_j & \boldsymbol{r}_j \\ \boldsymbol{r}_j^{\mathrm{T}} & \rho_j \end{pmatrix} \tag{9.121}$$

where $\boldsymbol{r}_j$ is a vector of size j and ρ_j is a real number. If $\boldsymbol{R}_j$ is invertible and $\boldsymbol{R}_j^{-1}$ is known, then $\boldsymbol{R}_{j+1}^{-1}$ can be determined from

$$\boldsymbol{R}_{j+1}^{-1} = \begin{pmatrix} \boldsymbol{R}_j^{-1} & \boldsymbol{0}_j \\ \boldsymbol{0}_j^{\mathrm{T}} & 0 \end{pmatrix} + \frac{1}{\beta_j}\begin{pmatrix} \boldsymbol{b}_j\boldsymbol{b}_j^{\mathrm{T}} & \boldsymbol{b}_j \\ \boldsymbol{b}_j^{\mathrm{T}} & 1 \end{pmatrix}, \tag{9.122}$$

where $\boldsymbol{b}_j = -\boldsymbol{R}_j^{-1}\boldsymbol{r}_j$ and $\beta_j = \rho_j + \boldsymbol{r}_j^{\mathrm{T}}\boldsymbol{b}_j$. Here, it is obvious that $r_j = \left[\boldsymbol{e}_m^{\mathrm{T}}, \boldsymbol{e}_n^{\mathrm{T}}\right]^{\mathrm{T}}, \rho = 0$ and

$$\boldsymbol{R}_j = \begin{pmatrix} -\sigma\boldsymbol{I}_m & \sigma\boldsymbol{e}_m\boldsymbol{e}_n^{\mathrm{T}} \\ \sigma\boldsymbol{e}_n\boldsymbol{e}_m^{\mathrm{T}} & -\sigma\boldsymbol{I}_n \end{pmatrix}. \tag{9.123}$$

Based on the method introduced in [36, 37], we obtain

$$\boldsymbol{R}_j^{-1} = \begin{pmatrix} -\sigma\boldsymbol{I}_m & \sigma\boldsymbol{e}_m\boldsymbol{e}_n^{\mathrm{T}} \\ \sigma\boldsymbol{e}_n\boldsymbol{e}_m^{\mathrm{T}} & -\sigma\boldsymbol{I}_n \end{pmatrix}^{-1} = -\sigma^{-1}\begin{pmatrix} (\boldsymbol{I}_m - \boldsymbol{e}_m\boldsymbol{e}_m^{\mathrm{T}})^{-1} & \boldsymbol{e}_m\boldsymbol{e}_n^{\mathrm{T}}(\boldsymbol{I}_n - \boldsymbol{e}_n\boldsymbol{e}_n^{\mathrm{T}})^{-1} \\ \boldsymbol{e}_n\boldsymbol{e}_m^{\mathrm{T}}(\boldsymbol{I}_m - \boldsymbol{e}_m\boldsymbol{e}_m^{\mathrm{T}})^{-1} & (\boldsymbol{I}_n - \boldsymbol{e}_n\boldsymbol{e}_n^{\mathrm{T}})^{-1} \end{pmatrix}. \tag{9.124}$$

It was found that $\boldsymbol{I}_m - \boldsymbol{e}_m\boldsymbol{e}_m^{\mathrm{T}}$ and $\boldsymbol{I}_n - \boldsymbol{e}_n\boldsymbol{e}_n^{\mathrm{T}}$ are singular, and thus, we add a penalty term $\varepsilon \approx 1$ to (9.124), and then, it yields

$$\begin{aligned} \boldsymbol{R}_j^{-1} &\approx -\sigma^{-1}\begin{pmatrix} (\boldsymbol{I}_m - \varepsilon\boldsymbol{e}_m\boldsymbol{e}_m^{\mathrm{T}})^{-1} & \boldsymbol{e}_m\boldsymbol{e}_n^{\mathrm{T}}(\boldsymbol{I}_n - \varepsilon\boldsymbol{e}_n\boldsymbol{e}_n^{\mathrm{T}})^{-1} \\ \boldsymbol{e}_n\boldsymbol{e}_m^{\mathrm{T}}(\boldsymbol{I}_m - \varepsilon\boldsymbol{e}_m\boldsymbol{e}_m^{\mathrm{T}})^{-1} & (\boldsymbol{I}_n - \varepsilon\boldsymbol{e}_n\boldsymbol{e}_n^{\mathrm{T}})^{-1} \end{pmatrix} \\ &\approx -\sigma^{-1}\begin{pmatrix} \boldsymbol{I}_m + \eta\boldsymbol{e}_m\boldsymbol{e}_m^{\mathrm{T}} & (1+\eta)\boldsymbol{e}_m\boldsymbol{e}_n^{\mathrm{T}} \\ (1+\eta)\boldsymbol{e}_n\boldsymbol{e}_m^{\mathrm{T}} & \boldsymbol{I}_n + \eta\boldsymbol{e}_n\boldsymbol{e}_n^{\mathrm{T}} \end{pmatrix}, \end{aligned} \tag{9.125}$$

where $\eta = \varepsilon/(1-\varepsilon)$. Thus,

$$\boldsymbol{b}_j = -\boldsymbol{R}_j^{-1}\boldsymbol{r}_j = -2(1+\eta)\sigma^{-1}[\boldsymbol{e}_m^{\mathrm{T}}, \boldsymbol{e}_n^{\mathrm{T}}]^{\mathrm{T}}, \tag{9.126}$$

$$\beta_j = \rho_j + \boldsymbol{r}_j^{\mathrm{T}}\boldsymbol{b}_j = 4(1+\eta)\sigma^{-1}. \tag{9.127}$$

Substituting (9.125) to (9.127) into (9.122), it yields

$$(\boldsymbol{H})^{*-1} = \boldsymbol{R}_{j+1}^{-1} \tag{9.128}$$

From the inverse transformation $\boldsymbol{H}^{-1} = \boldsymbol{T}\boldsymbol{H}^{*-1}\boldsymbol{T}^{\mathrm{T}}$ and by approximating $\frac{\sigma}{4(1+\sigma)} = \frac{1}{4}\sigma(1-\varepsilon) \approx 0$, we get

$$H^{-1} \approx \begin{pmatrix} -\sigma^{-1}(I_m - uu_m^{\mathrm{T}}) & 0_{mn} & -\frac{1}{2}u \\ 0_{mn} & -\sigma^{-1}(I_n - vv_n^{\mathrm{T}}) & -\frac{1}{2}v \\ -\frac{1}{2}u^{\mathrm{T}} & -\frac{1}{2}v^{\mathrm{T}} & 0 \end{pmatrix} \tag{9.129}$$

where we have approximated the unknown singular vectors by $\bar{u}_1 \approx u$ and $\bar{v}_1 \approx v$, and 0_{mn} is a zero matrix of size $m \times n$, and 0_{nm} denotes its transpose.

The Newton's method for SVD is defined as

$$\begin{bmatrix} \dot{u} \\ \dot{v} \\ \dot{\sigma} \end{bmatrix} = -H^{-1}\nabla p = - - H^{-1}\begin{bmatrix} \frac{\partial p}{\partial u} \\ \frac{\partial p}{\partial v} \\ \frac{\partial p}{\partial \sigma} \end{bmatrix} \tag{9.130}$$

By applying the gradient of p in (9.114)–(9.115) and the inverse Hessian (9.129) to (9.130), we have a system of ODEs

$$\dot{u} = \sigma^{-1}Av - \sigma^{-1}uu^{\mathrm{T}}Av - \frac{1}{2}u + \frac{3}{4}uu^{\mathrm{T}}u - \frac{1}{4}uv^{\mathrm{T}}v, \tag{9.131}$$

$$\dot{v} = \sigma^{-1}A^{\mathrm{T}}u - \sigma^{-1}vu^{\mathrm{T}}Av - \frac{1}{2}v + \frac{3}{4}vv^{\mathrm{T}}v - \frac{1}{4}vu^{\mathrm{T}}u, \tag{9.132}$$

$$\dot{\sigma} = u^{\mathrm{T}}Av - \frac{1}{2}\sigma(u^{\mathrm{T}}u + v^{\mathrm{T}}v). \tag{9.133}$$

It is straightforward to show that this system of ODEs has the same equilibria as the gradient of p. However, only the first PST is stable (See Sect. 9.4.2).

9.4.2 Online Implementation and Stability Analysis

A stochastic online algorithm can be derived from the ODEs in (9.131)–(9.133) by formally replacing A with $x(k)y(k)^{\mathrm{T}}$ and by introducing a small learning rate γ, where k denotes the discrete-time step. Under certain conditions, the online algorithm has the same convergence goal as the ODEs. We introduce the auxiliary variable $\xi(k) = u^{\mathrm{T}}(k)x(k)$ and $\varsigma(k) = v^{\mathrm{T}}(k)y(k)$ and then obtain the update equations from (9.131) to (9.133):

$$\begin{aligned} u(k+1) = u(k) + \gamma\{\sigma(k)^{-1}\varsigma(k)[x(k) - \xi(k)u(k)] - \frac{1}{2}u(k) + \frac{3}{4}u(k)u^{\mathrm{T}}(k)u(k) \\ - \frac{1}{4}u(k)v^{\mathrm{T}}(k)v(k)\}, \end{aligned} \tag{9.134}$$

$$\boldsymbol{v}(k+1)=\boldsymbol{v}(k)+\gamma\left\{\sigma(k)^{-1}\xi(k)[\boldsymbol{y}(k)-\varsigma(k)\boldsymbol{v}(k)]-\frac{1}{2}\boldsymbol{v}(k)+\frac{3}{4}\boldsymbol{v}(k)\boldsymbol{v}^{\mathrm{T}}(k)\boldsymbol{v}(k)-\frac{1}{4}\boldsymbol{v}(k)\boldsymbol{u}^{\mathrm{T}}(k)\boldsymbol{u}(k)\right\}, \tag{9.135}$$

$$\sigma(k+1)=\sigma(k)+\gamma\left\{\xi(k)\varsigma(k)-\frac{1}{2}\sigma(k)[\boldsymbol{u}^{\mathrm{T}}(k)\boldsymbol{u}(k)+\boldsymbol{v}^{\mathrm{T}}(k)\boldsymbol{v}(k)]\right\}. \tag{9.136}$$

Stability is a crucial property of the learning rules, as it guarantees convergence. The stability of proposed algorithm can be proven by analyzing the Jacobian of the averaged learning rule in (9.131)–(9.133), evaluated at the qth singular triplet, i.e., $(\bar{\boldsymbol{u}}_q,\bar{\boldsymbol{v}}_q,\bar{\sigma}_q)$. A learning rule is stable if its Jacobian is negative definite. Jacobian of the original ODE system (9.131)–(9.133) is

$$J(\bar{\boldsymbol{u}}_q,\bar{\boldsymbol{v}}_q,\bar{\sigma}_q)=\begin{pmatrix}-\boldsymbol{I}_m+\frac{1}{2}\bar{\boldsymbol{u}}_q\bar{\boldsymbol{u}}_q^{\mathrm{T}} & \bar{\sigma}_q^{-1}\boldsymbol{A}-\frac{3}{2}\bar{\boldsymbol{u}}_q\bar{\boldsymbol{v}}_q^{\mathrm{T}} & \boldsymbol{0}_m\\ \bar{\sigma}_q^{-1}\boldsymbol{A}^{\mathrm{T}}-\frac{3}{2}\bar{\boldsymbol{v}}_q\bar{\boldsymbol{u}}_q^{\mathrm{T}} & -\boldsymbol{I}_n+\frac{1}{2}\bar{\boldsymbol{v}}_q\bar{\boldsymbol{v}}_q^{\mathrm{T}} & \boldsymbol{0}_n\\ \boldsymbol{0}_m^{\mathrm{T}} & \boldsymbol{0}_n^{\mathrm{T}} & 1\end{pmatrix} \tag{9.137}$$

The Jacobian (9.137) has $M-1$ eigenvalue pairs $\alpha_i=\bar{\sigma}_i/\bar{\sigma}_q-1,\alpha_{M+i}=-\bar{\sigma}_i/\bar{\sigma}_q-1,\forall i\neq q$ and $i\neq 1$. A double eigenvalues $\alpha_q=\alpha_{M+q}=-0.5$, and all other eigenvalues $\alpha_r=-1$. Since a stable equilibrium requires $|\bar{\sigma}_i|/|\bar{\sigma}_q|<1,\forall i\neq q$, and consequently $|\bar{\sigma}_q|>|\bar{\sigma}_i|,\forall i\neq q$, which is only provided by choosing $q=1$, i.e., the first PST $(\bar{\boldsymbol{u}}_1,\bar{\boldsymbol{v}}_1,\bar{\sigma}_1)$. Moreover, if $|\bar{\sigma}_1|\gg|\bar{\sigma}_j|,\forall j\neq 1$, all eigenvalues (except $\alpha_1=\alpha_{M+1}=-0.5$) are $\alpha_i\approx-1$, so the system converges with approximately equal speed in almost all of its eigen directions and is widely independent of the singular values.

9.4.3 Simulation Experiments

9.4.3.1 Experiment 1

In this experiment, we will conduct a simulation to test the ability of PST extraction of proposed algorithm and also compare its performance with that of some other algorithms, i.e., the coupled algorithm [1] and two noncoupled algorithms [2, 21]. Same as in Sect. 9.3.4, here, the vector data sequence is generated by $\boldsymbol{x}(k)=\boldsymbol{A}\cdot\mathbf{y}(k)$, where $\boldsymbol{A}$ is given by $\boldsymbol{A}=[\boldsymbol{u}_0,\ldots,\boldsymbol{u}_8]\cdot\mathrm{rand}n(9,9)\cdot[\boldsymbol{v}_0,\ldots,\boldsymbol{v}_8]^{\mathrm{T}}$, where $\boldsymbol{u}_i$ and $\boldsymbol{v}_i$(i = 0,…,8) are the ith components of 11-dimensiaonl and 9-dimensional orthogonal discrete cosine basis functions, respectively. In this simulation, we let $\mathbf{y}_t\in\Re^{9\times 1}$ be Gaussian, spatially and temporally white and randomly generated. In order to measure the convergence speed and precision of learning algorithms, we compute the direction cosine between the state vectors, i.e., $\boldsymbol{u}(k)$ and $\boldsymbol{v}(k)$, and the true principal singular vector, i.e., $\bar{\boldsymbol{u}}_1$ and $\bar{\boldsymbol{v}}_1$, at the kth update:

$$\mathrm{DC}(\boldsymbol{u}(k)) = \frac{|\boldsymbol{u}^{\mathrm{T}}(k) \cdot \bar{\boldsymbol{u}}_1|}{\|\boldsymbol{u}(k)\| \cdot \|\bar{\boldsymbol{u}}_1\|}, \tag{9.138}$$

$$\mathrm{DC}(\boldsymbol{v}(k)) = \frac{|\boldsymbol{v}^{\mathrm{T}}(k) \cdot \bar{\boldsymbol{v}}_1|}{\|\boldsymbol{v}(k)\| \cdot \|\bar{\boldsymbol{v}}_1\|}. \tag{9.139}$$

Clearly, if direction cosines (9.138) and (9.139) converge to 1, then state vectors $\boldsymbol{u}(k)$ and $\boldsymbol{v}(k)$ must approach the direction of the true left and right singular vectors, respectively. For coupled algorithms, we define the left and right singular error at the kth update:

$$\epsilon_L(k) = \left\|\sigma(k)^{-1}\boldsymbol{A}^{\mathrm{T}}\boldsymbol{u}(k) - \boldsymbol{v}(k)\right\| \tag{9.140}$$

$$\epsilon_R(k) = \left\|\sigma(k)^{-1}\boldsymbol{A}\boldsymbol{v}(k) - \boldsymbol{u}(k)\right\| \tag{9.141}$$

If these two singular errors converge to 0, then the singular value estimate $\sigma(k)$ must approach the true singular value as $k \to \infty$. In this experiment, the learning rate is chosen as $\gamma = 0.02$ for all rules. The initial values $\boldsymbol{u}(0)$ and $\boldsymbol{v}(0)$ are set to be orthogonal to $\bar{\boldsymbol{u}}_1$ and $\bar{\boldsymbol{v}}_1$. Experiment results are shown in Figs. 9.5, 9.6, and 9.7.

From Fig. 9.5a, b, it is observed that all algorithms can effectively extract both of the left and right principal singular vectors of a cross-correlation matrix between two data streams. The coupled algorithms have higher convergence speed than Feng's algorithm. Compared with Kong's algorithm, the coupled algorithms have similar convergence speed in the whole process but higher convergence speed in the beginning steps. In Fig. 9.6a, b, we find that all left and right state vectors of all algorithms converge to a unit length, and coupled algorithms have higher convergence speed than noncoupled algorithms. What is more, the principal singular value of the cross-correlation matrix can also be estimated in coupled algorithms, which is actually an advantage of coupled algorithms over noncoupled algorithms. This is very helpful in some engineering applications when singular value estimation is required. Figures 9.7a, b verified the efficiency of principal singular value estimation of coupled algorithms.

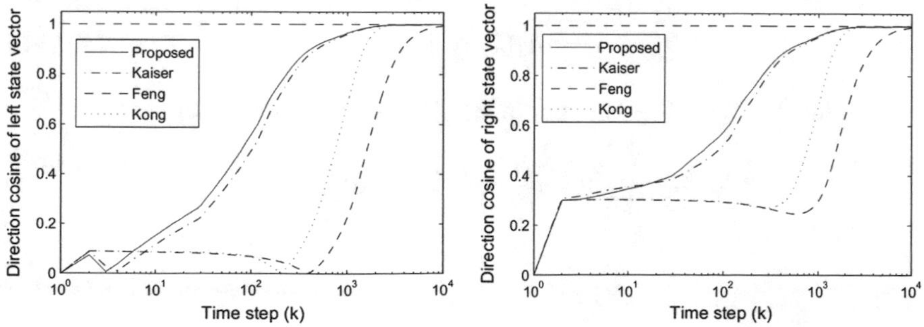

(a) Direction cosine between $\boldsymbol{u}(k)$ and $\bar{\boldsymbol{u}}_1$ (b) Direction cosine between $\boldsymbol{v}(k)$ and $\bar{\boldsymbol{v}}_1$

Fig. 9.5 Direction cosine of *left* and *right* singular vector estimation

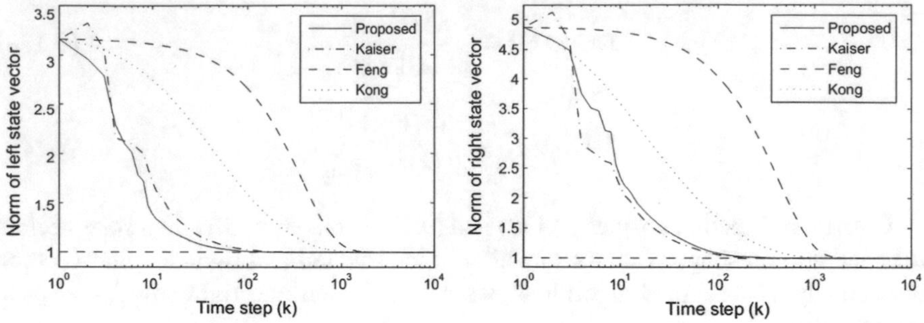

(a) Norm of left state vector (b) Norm of right state vector

Fig. 9.6 Norm of *left* and *right* state vectors

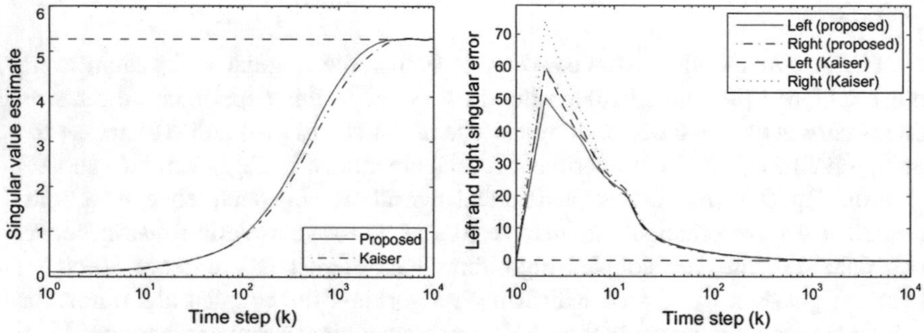

(a) Principal singular value estimation (b) Estimation errors

Fig. 9.7 Principal singular value estimation and their errors

9.4.3.2 Experiment 2

Next, we will use the proposed algorithm to extract multiple principal singular components, i.e., the first 3 PSTs of a cross-covariance matrix, in this experiment. The initial conditions are set to be the same as that in Experiment 1. The method of multiple component extractions is a sequential method, which was introduced in [13]:

$$\mathbf{A}_1(k) = \mathbf{A}(k) \tag{9.142}$$

$$\begin{aligned} \mathbf{A}_i(k) &= \mathbf{A}_{i-1}(k-1) - \mathbf{u}_{i-1}(k)\mathbf{u}_{i-1}^{\mathrm{T}}(k)\mathbf{A}_1(k)\mathbf{v}_{i-1}(k)\mathbf{v}_{i-1}^{\mathrm{T}}(k) \\ &= \mathbf{A}_1(k) - \sum_{j=1}^{i-1}\mathbf{u}_j(k)\mathbf{u}_j^{\mathrm{T}}(k)\mathbf{A}_1(k)\mathbf{v}_j(k)\mathbf{v}_j^{\mathrm{T}}(k) \end{aligned} \tag{9.143}$$

where $\boldsymbol{A}(k)$ can be updated as

$$\mathbf{A}(k) = \frac{k-1}{k}\mathbf{A}(k-1) + \frac{1}{k}x(k)y^{\mathrm{T}}(k) \tag{9.144}$$

By replacing $\boldsymbol{A}$ with $\boldsymbol{A}_i(k)$ instead of $\xi(k)\varsigma(k)$ in (9.131)–(9.133) at each step, then the ith triplet $(\boldsymbol{u}_i, \boldsymbol{v}_i, \sigma_i)$ can be estimated. In this experiment, we set $\alpha = 1$.

Figure 9.8 shows the direction cosine between the first 3 (left and right) principal singular vectors and the true (left and right) singular vectors, and their norms of first 3 (left and right) principal singular vectors. Figure 9.9 shows the first 3 principal singular value estimation. Figure 9.10 shows that the left and right principal singular estimation errors. Figures 9.8, 9.9, and 9.10 verify the ability of multiple component analysis of proposed algorithm.

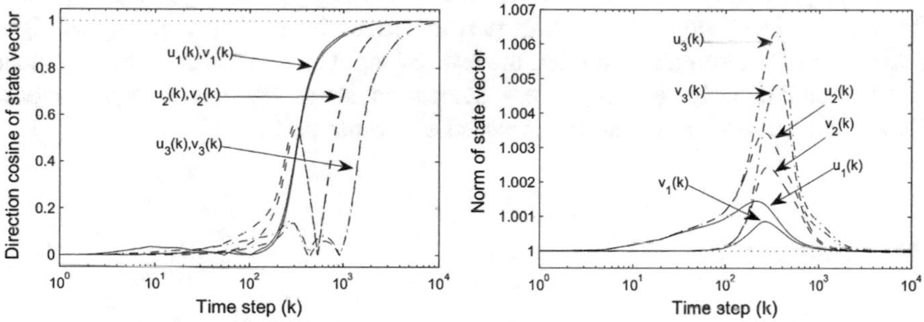

(a) Direction cosine (b) Vectors norm

Fig. 9.8 Direction cosine and norm of state vectors

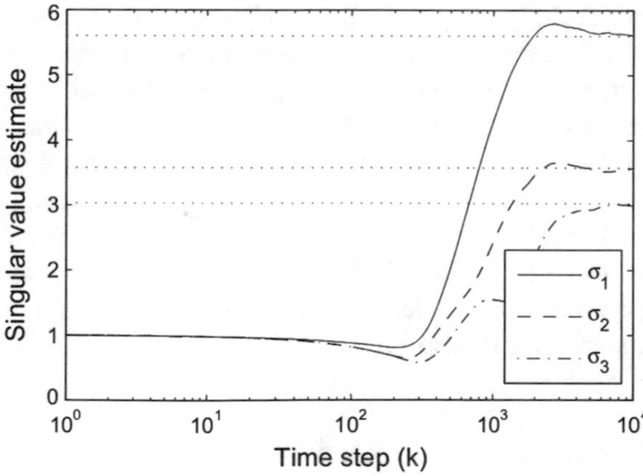

Fig. 9.9 The first 3 principal singular value estimation

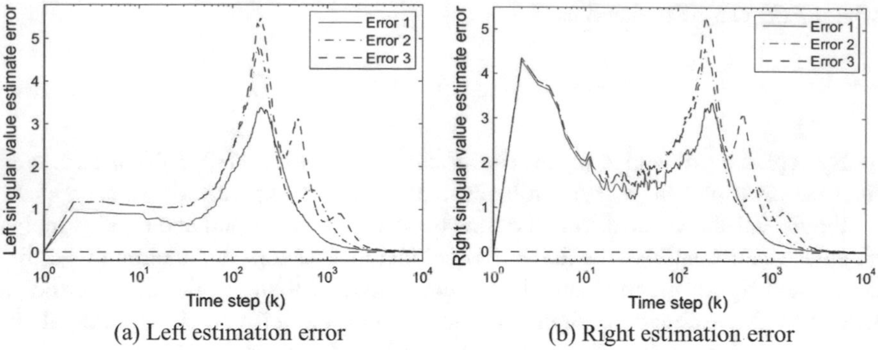

Fig. 9.10 *Left* and *right* principal singular estimation errors

In this section, a novel CSVD algorithm is presented by finding the stable stationary point of an NIC via Newton's method. The proposed algorithm can solve the speed stability problem and thus perform much better than noncoupled algorithms. CSVD algorithms can track the left and right singular vectors and singular value simultaneously, which is very helpful in some engineering applications. Experiment results show that the proposed algorithm performs well.

9.5 Summary

In this chapter, we have reviewed SVD learning algorithms in the literature. These algorithms include single-component learning rules and symmetrical learning rules. The former extract the first principal singular left and right singular vectors and can be extended to multiple component rules using sequential decorrelation methods such as Gram–Schmidt or deflation. And the latter update multiple vectors simultaneously or extract the principal singular subspace. Several well-known neural network-based SVD algorithms, e.g., cross-coupled Hebbian rule, DGHA rule, linear approximation APCA rule, "trace" algorithm, "Riccati" algorithm, CANN rule, and coupled SVD algorithm, have been analyzed briefly. Then, an information criterion for singular feature extraction and corresponding neural-based algorithms proposed by us have been analyzed in details. Finally, a novel coupled neural-based algorithm is introduced to extract the principal singular triplet (PST) of a cross-correlation matrix between two high-dimensional data streams. The algorithm can solve the speed stability problem that plagues most noncoupled learning rules. Moreover, the algorithm can also be used to extract multiple PSTs effectively by using sequential method.

References

1. Kaiser, A., Schenck, M., & Moller, R. (2010). Coupled singular value decomposition of a cross-covariance matrix. *International Journal of Neural Systems, 20*(4), 293–318.
2. Feng, D. Z., Zhang, X. D., & Bao, Z. (2004). A neural network learning for adaptively extracting cross-correlation features between two high-dimensional data streams. *IEEE Transactions on Neural Networks, 15*(6), 1541–1554.
3. Bunch, J. R., & Nielsen, C. P. (1978). Updating the singular value decomposition. *Numerical Mathematics, 31*(31), 111–129.
4. Comon, P., & Golub, G. H. (1990). Tracking a few extreme singular values and vectors in signal processing. *Processing of the IEEE, 78*(8), 1318–1343.
5. Ferali, W., & Proakis, J. G. (1990). Adaptive SVD algorithm for covariance matrix eigenstructure computation. In *Proceedings of 1990 IEEE International Conference on Acoustic, Speech and Signal Processing* (pp. 176–179).
6. Mooden, M., Dooren, P. V., & Vandewalle, J. (1992). A singular value decomposition updating algorithm for subspace tracking. *SIAM Journal of Matrix Analysis and Applications, 13*(4), 1015–1038.
7. Yuile, A. L., Kammen, D. M., & Cohen, D. S. (1985). Quadrature and the development of orientation selective cortical cells by Hebb rules. *Biology Cybernetics, 61*(3), 183–194.
8. Samardzija, N., & Waterland, R. L. (1989). A neural network for computing eigenvectors and eigenvalues. *Biology Cybernetics, 65*(4), 211–214.
9. Cichocki, A., & Unbehauen, R. (1992). Neural network for computing eigenvalues and eigenvectors. *Biology Cybernetics, 68*(2), 155–159.
10. Cichocki, A. (1992). Neural network for singular value decomposition. *Electronic Letter, 28* (8), 784–786.
11. Cichocki, A., & Unbehauen, R. (1994). *Neural networks for optimization and signal processing*. New York: Wiley.
12. Diamantaras, K. I., & Kung, S. Y. (1994). Cross-correlation neural network models. *IEEE Transactions on Signal Processing, 42*(11), 3218–3223.
13. Feng, D. Z., Bao, Z., & Zhang, X. D. (2001). A cross-associative neural network for SVD of nonsquared data matrix in signal processing. *IEEE Transactions on Neural Networks, 12*(9), 1215–1221.
14. Feng, D. Z., & Bao, Z. (1997). Modified cross-correlation neural networks. *Chinese Journal of Electronic, 6*(3), 63–70.
15. Sanger, T. D. (1994).Two iterative algorithms for computing the singular value decomposition from input/output samples. In J. D. Cowan, G. Tesauro, & J. Alspector (Eds.), *Advances in neural information processing systems*, Vol. 6 (pp. 144–151). Morgan Kaufmann.
16. Diamantaras, K. I., & Kung, S. Y. (1994). Multilayer neural networks for reduced-rank approximation. *IEEE Transactions on Neural Networks, 5*(5), 684–697.
17. Fiori, S. (2003). Singular value decomposition learning on double Stiefel manifolds. *International Journal of Neural Systems, 13*(2), 1–16.
18. Weingessel, A., & Hornik, K. (1997). SVD algorithms: APEX-like versus subspace methods. *Neural Processing Letter, 5*(3), 177–184.
19. Weingessel, A. (1999). *An analysis of learning algorithms in PCA and SVD Neural Networks.* PhD thesis, Technische University at Wien.
20. Helmke, U., & Moore, J. B. (1992). Singular-value decomposition via gradient and self-equivalent flows. *Linear Algebra and Application, 169*(92), 223–248.
21. Kong, X. Y., Ma, H. G., An, Q. S., & Zhang, Q. (2015). An effective neural learning algorithm for extracting cross-correlation feature between two high-dimensional data streams. *Neural Processing Letter, 42*(2), 459–477.
22. Feng, X. W., Kong, X. Y., & Ma, H. G. (2016). Coupled cross-correlation neural network algorithm for principal singular triplet extraction of a cross-covariance matrix. *IEEE/CAA Journal of Automatica Sinica, 3*(2), 141–148.

23. Brockett, R. W. (1991). Dynamical systems that sort lists, diagonalize matrices and solve linear programming problems. *Linear Algebra and its Applications, 146*, 79–91.
24. Yang, B. (1995). Projection approximation subspace tracking. *IEEE Transactions on Signal Processing, 43*(1), 95–107.
25. Golub, G. H., & Van Loan, C. F. (1989). *Matrix computations*. Baltimore, MD: The Johns Hopkins University Press.
26. Haykin, S. (1991). *Adaptive filter theory* (2nd ed.). Englewood Cliffs, NJ: Prentice-Hall.
27. Wax, M., & Kailath, T. (1985). Detection of signals by information theoretic criteria. *IEEE Transactions on Acoustic, Speech and Signal Processing, 33*(2), 387–392.
28. Schmidt R. (1986). Multiple emitter location and signal parameter estimation. *IEEE Transactions on Antennas Propagation, AP-34*(3), 276–281.
29. Roy, R., Paulraj, A., & Kailath, T. (1986). ESPRIT-A subspace rotation approach to estimation of parameters of cissoids in noise. *IEEE Transactions on Acoustic, Speech and Signal Processing, 34*(10), 1340–1342.
30. Kong, X. Y., Hu, C. H., & Han, C. Z. (2012). A dual purpose principal and minor subspace gradient flow. *IEEE Transactions on Signal Processing, 60*(1), 197–210.
31. Magnus, J. R., & Neudecker, H. (1991). *Matrix differential calculus with applications in statistics and econometrics* (2nd ed.). New York: Wiley.
32. Kushner, H. J., & Clark, D. S. (1976). *Stochastic approximation methods for constrained and unconstrained systems*. New York: Springer-Verlag.
33. Ljung, L. (1977). Analysis of recursive stochastic algorithms. *IEEE Transactions on Automatic Control, 22*(4), 551–575.
34. Moller, R., & Konies, A. (2004). Coupled principal component analysis. *IEEE Transactions on Neural Networks, 15*(1), 214–222.
35. Noble, B., & Daniel, J. W. (1988). *Applied linear algebra* (3rd ed.). Englewood Cliffs, NJ: Prentice Hall.
36. Hotelling, H. (1943). Some new methods in matrix calculation. *The Annals of Mathematical Statistics, 14*(1), 1–34.
37. Hotelling, H. (1943). Further points on matrix calculation and simultaneous equations. *The Annals of Mathematical Statistics, 14*(4), 440–441.
38. Moller, R., & Homann, H. (2004). An extension of neural gas to local PCA. *Neurocomputing, 62(1)*, 305–326.
39. Golub, G. H., & Reinsch, C. (1970). Singular value decomposition and least squares solutions. *Numerical Mathematics, 14*(5), 403–420.
40. Moller, R. (2006). First-order approximation of Gram-Schmidt orthonormalization beats deflation in coupled PCA learning rules. *Neurocomputing, 69*(13–15), 1582–1590.
41. Baldi, P. F., & Hornik, K. (1995). Learning in linear neural networks: A survey. *IEEE Transactions on Neural Networks, 6*(4), 837–858.
42. Hasan, M. (2008). A logarithmic cost function for principal singular component analysis. In *IEEE International Conference on Acoustic Speech and Signal Processing* (pp. 1933–1936).
43. Haykin, S. (1999). *Neural networks. A comprehensive foundation* (2nd ed.). Englewood Cliffs, NJ: Prentice Hall.
44. Rumelhart, D. E., Hinton, G. E., & Williams, R. J. (1986). Learning representations by back-propagating errors. *Nature, 323*(1), 533–536.
45. Baldi, P. F., & Hornik, K. (1989). Neural networks and principal component analysis: learning from examples without local minima. *Neural Networks, 2*(1), 53–58.
46. Smith, S. T. (1991). Dynamical systems that perform the singular value decomposition. *System Control Letter, 16*(5), 319–327.
47. Hori, G. (2003). A general framework for SVD flows and joint SVD flows. In *Proceeding of IEEE International Conference on Acoustic Speech and Signal Processing*, Vol. 2. (pp. 693–696).
48. Hasan, M. A. (2008). Low-rank approximations with applications to principal singular component learning systems. In *Proceedings of 2008 IEEE Conference on Decision and Control* (pp. 3293–3298).

49. Luo, F. L., Unbehauen, R., & Li, Y. D. (1997). Real-time computation of singular vectors. *Application Mathematics and Computation, 86*(2–3), 197–214.
50. Chong, E. K. P., & Zak, S. H. (2001). *An introduction to optimization* (2nd ed.). New York: Wiley.
51. Martinetz, T. M., Berkovich, S. G., & Schulten, K. J. (1993). Neural-gas' network for vector quantization and its application to time-series prediction. *IEEE Transactions on Neural Networks, 4*(4), 558–569.
52. Brockett, R. W. (1991). Dynamical systems that sort lists, diagonalize matrices, and solve linear programming problems. *Linear Algebra and its Applications, 146*, 79–91.

Printed in the United States
By Bookmasters